Fortschritte der Chemie organischer Naturstoffe

Progress in the Chemistry of Organic Natural Products

38

Founded by L. Zechmeister
Edited by W. Herz, H. Grisebach, G. W. Kirby

Authors:
H. D. Fischer, N. H. Fischer,
R. W. Franck, E. J. Olivier

Springer-Verlag
Wien New York 1979

Dr. W. Herz, Professor of Chemistry, Department of Chemistry,
The Florida State University, Tallahassee, Florida, U.S.A.

Prof. Dr. H. Grisebach, Biologisches Institut II, Lehrstuhl für Biochemie der Pflanzen,
Albert-Ludwigs-Universität, Freiburg i. Br., Federal Republic of Germany

G. W. Kirby, Sc. D., Regius Professor of Chemistry, Chemistry Department,
The University, Glasgow, Scotland

With 5 Figures

© 1979 by Springer-Verlag/Wien

Softcover reprint of the hardcover 1st editon 1979

Library of Congress Catalog Card Number AC 39-1015

ISSN 0071-7886

ISBN-13: 978-3-7091-8550-6   e-ISBN-13: 978-3-7091-8548-3
DOI: 10.1007/978-3-7091-8548-3

Contents

List of Contributors

Fischer, Helga D., Associate, Department of Chemistry, Louisiana State University, Baton Rouge, LA 70803, U.S.A.

Fischer, Prof. Dr. N. H., Department of Chemistry, Louisiana State University, Baton Rouge, LA 70803, U.S.A.

Franck, Prof. R. W., Ph. D., Department of Chemistry, Fordham University, Bronx, NY 10458, U.S.A.

Olivier, E. J., Doctoral Candidate, Department of Chemistry, Louisiana State University, Baton Rouge, LA 70803, U.S.A.

List of Contributors

The Mitomycin Antibiotics

By R. W. FRANCK, Department of Chemistry, Fordham University,
Bronx, N.Y., U.S.A.

With 1 Figure

Contents

I. Isolation and Structure

The mitomycins were first obtained in 1956 from *Streptomyces caespitosus* by HATA and coworkers in Japan (*1, 2*). The isolation involved alumina chromatography of the chloroform extracts of the concentrated aqueous filtrates of the fermentation broth. Although there were many active components in the eluted fractions, only two antibiotics, mitomycins A and B were obtained in crystalline form. Later fractionations of broths of *S. caespitosus* yielded mitomycin C (*3*). By 1958, the outstanding antitumor activity of mitomycin C was a subject of great interest in spite of the fact that there existed no firm evidence as to its structure (4). In 1960, N-methylmitomycin C, called porfiromycin, was isolated from cultures of *Streptoverticillatium ardus* (5). In 1962 the isolation of the same four antibiotics from *Streptomyces verticillatus* was reported in the U.S. (6). Additionally, a fifth, but inactive member of mitomycin family, mitiromycin was obtained. The fractionation of the *S. verticillatus* broths to afford the crystalline antibiotics was accomplished by partition chromatography on a diatomaceous earth support.

(1a) R=H Mitomycin A   (2) R=H Mitomycin C

(b)   R = SO$_2$—⟨benzene⟩—Br   (3) R=CH$_3$ Porfiromycin

(c)   R = CH$_3$

The antibiotics are all purple pigments and were characterized as aminoquinones by uv analysis. Mitomycins A und B (both C$_{16}$H$_{19}$N$_3$O$_6$) and mitiromycin (C$_{16}$H$_{17}$N$_3$O$_5$) gave λ max (CH$_3$OH) 218 nm (ε 17,400), 320 nm (ε 10,400) and 520 nm (ε 1,400). These data compare favorably to a simple model compound, 2-dimethylamino-5-methoxybenzoquinone, λ max (CH$_3$OH) 218 nm (ε 18,500), 305 nm (ε 13,900) and 490 nm (ε 3,900). Mitomycin C, (C$_{15}$H$_{18}$N$_4$O$_5$), and porfiromycin, (Cl$_6$H$_{20}$N$_4$O$_5$), gave λ max (CH$_3$OH) 217 nm (ε 24,600), 360 nm (ε 23,000) and 555 nm (ε 209) which compares well with 2,5-bisdimethylaminobenzequinone, λ max (CH$_3$OH) 222 nm (ε 24,000), 365 nm (ε 21,400) and 513 nm (ε 407). Since the structures of these antibiotics were *sui generis,* extensive chemical degradation studies were undertaken by groups at Kyowa Hakko (7, 8), Lederle Laboratories (9) and Wayne State University (10) in order to obtain known or recognizable structural fragments. Hydrolysis studies, the structural results of which will be discussed below in a section on *Transformation Products,* suggested the presence of an aziridine and an incipient indole. The aminohydrin which resulted from ring-opening of the aziridine was subjected to a semipinacolic deamination which resulted in the formation of a carbonyl conjugated to the indole. This carbonyl derivative yielded β-alanine, the only known degradation product ever obtained in these studies, upon KMnO$_4$ oxidation. A carbamate function was located on a hydroxymethyl group attached to the incipient indole. The data available from degradation and spectroscopy was sufficient for the Lederle workers to propose the correct structural formulae for the mitomycins. However, a simultaneously completed X-ray crystal structure was finally required to serve as a basis for interpreting all the chemistry and stereochemistry of the antibiotics.

N-Brosylmitomycin A was shown by TULINSKY and VAN DEN HENDE (*11, 12*) to have the relative and absolute configuration shown in (**1**). Thus, mitomycin C which can be prepared from mitomycin A by displacement of the 7-methoxyl with ammonia, was assigned structure (**2**). Both mitomycin A and C could be converted to porfiromycin, (**3**). N-Alkylation using CH$_3$I and K$_2$CO$_3$ serves to convert (**1a**) to N-methyl-mitomycin A (**1c**) which can be ammonolyzed to (**3**) while N-methylation of (**2**) yields (**3**) directly.

Mitomycin B was isolated from the same cultures as the other members of the series, but has a different relative stereochemistry as shown in (**4**). The X-ray determination of structure of a heavy atom derivative of mitomycin B served as the basis for assigning the relative stereochemistry (*13*). However, the crystallographers inadvertently derived the absolute configuration shown in (**5**). Thus, the initial X-ray disclosure, now revised, contradicted the report that N-methylmitomycin A and mitomycin B have both been degraded to the identical indoloquinone (**6**) where the different relative stereochemistries at carbons (**9**) and

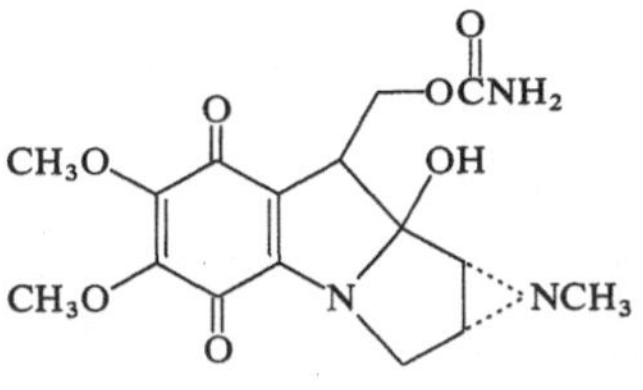

(4) Mitomycin B 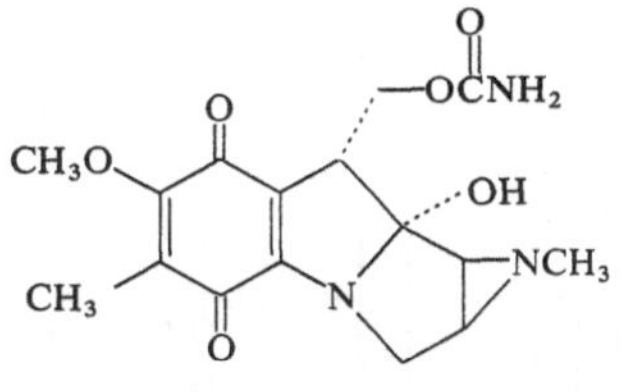   (5)

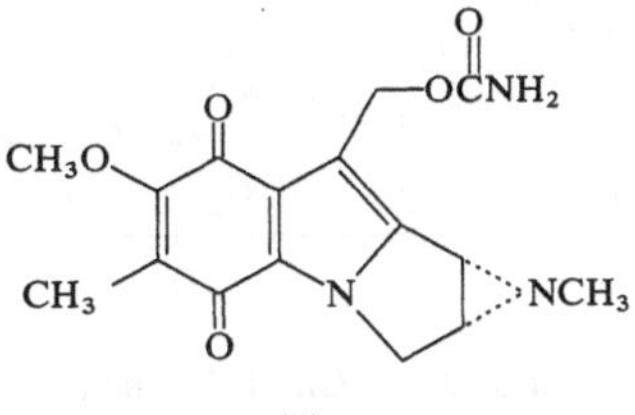

(6)

1*

(**1a**) have been eliminated (*14*). Thus, the degradation experiment demonstrated the identity of aziridine absolute configuration in the A and
B series and required the assignment of structure (**4**) to mitomycin B.

The 100 Mhz pmr spectrum of mitomycin C is reproduced in Fig. 1
(*15*). Worthy of comment is the rather small coupling between hydrogens
on C-1 and C-2 (J = 4.5 Hz) even though the dihedral angle between
them is 0°. This small value is a consequence of the aziridine fusion
changing the hybridizations and bond angles so as to diminish coupling.
Furthermore, H_2 is coupled to its neighbor $H_{3'}$ and not to H_3. This result
can be interpreted in terms of the H_3-H_2 dihedral angle being 90°. Also
noteworthy is the great chemical shift difference, 1 ppm between H_3 and
$H_{3'}$. It can be seen that the upfield $H_{3'}$ is *syn* to the N_4 lone pair and
reasonably far removed from any anisotropic effects of the quinone
carbonyl. Complementary to this analysis, it can be seen that H_3 is *anti*
to the N_4 lone pair and is really quite close to the quinone carbonyl. This
difference in shift is useful in assigning stereochemistry to mitiromycin
(*vide infra*). The C_{13} spectral data is reproduced in Table 1 (*15*). Note
that all 15 carbons are clearly distinguishable and assignable. This
property of the spectrum should become important if a biosynthetic
study using C_{13} labeled precursors is carried out.

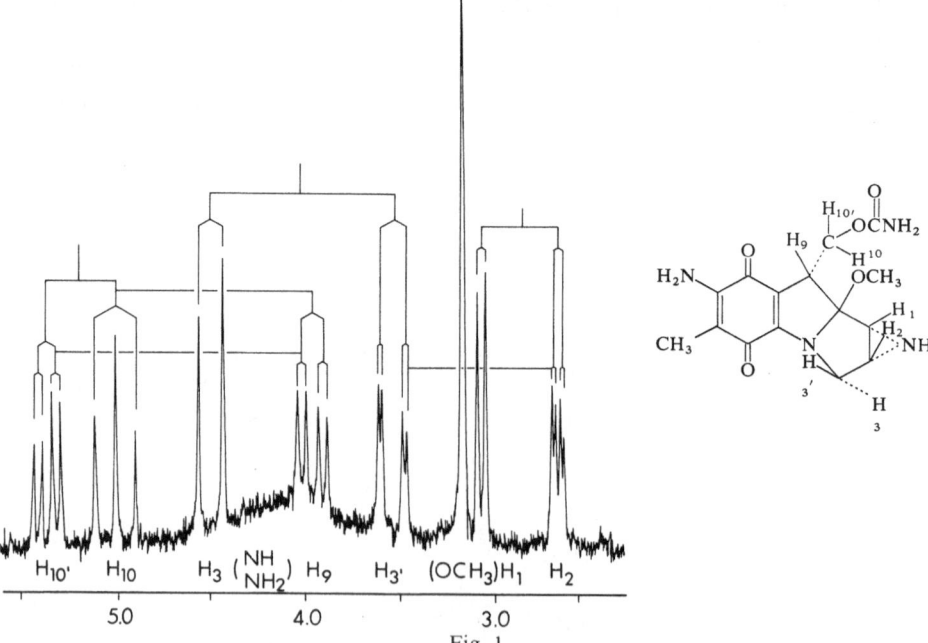

Fig. 1

(From J. W. Lown and A. Begleiter, Cand. J. Chem. **52**, 2331—2336, 1974. Reproduced
by permission of the National Research Council of Canada)

References, pp. 41—45

Table 1. ^{13}C *Chemical Shifts in p.p.m. From TMS of Antibiotics*
*as 0.224 M Solutions in Pyridine-d_5 at 25.15 MHz**

(From J. W. LOWN and A. BEGLEITER, Cand. J. Chem. **52,** 2331—2336, 1974.
Reproduced by permission of the National Research Council of Canada)

Mitomycin C

Chemical Shift	Assignment
178.6 s	8
176.8 s	5
158.0 s	10a
156.0 s	5a
150.0 s	7
111.0 s	9a
107.0 s	6
104.5 s	8a
62.7 t	10
50.7 t	3
49.7 (q)	9a-OCH$_3$
44.5 d	9
36.8 d	1
32.8 d	2
8.7 q	6-CH$_3$

* Lock signal pyridine-d_5. Chemical shift data using a 4 K data set are accurate to
±0.05 p.p.m.

The mass spectra for all the mitomycins except mitiromycin are
quite similar (*16*). The molecular ions are weak; a major fragment ion
is due to loss of methanol (mitomycins A, C, porfiromycin) or water
(mitomycin B) to form the mitosene (**7**). An intense ion also observed
is due to loss of carbamate (M-61), which yields ion (**8**) as shown in
Scheme I. Ion (**8**) further fragments to yield ion (**9**) or (**10**). An important
fragment for biosynthetic studies (*vide infra*) corresponds to C_4H_8N,
(m/e 70) observed in N-methylaziridine derivatives, mitomycin B,
N-methylmitomycin A and porfiromycin. The origin of the peak,
attributed to breakdown of ion (**10**), was confirmed by synthesizing
N-CD$_3$ mitomycin A which gave a fragment peak shifted by 3 mass
units (to m/e 73).

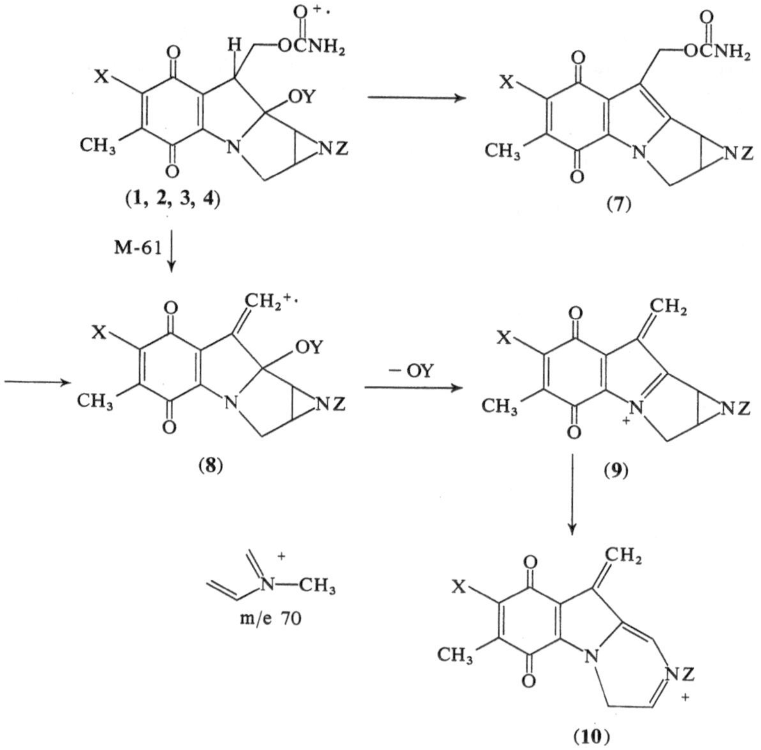

Scheme I

A biologically inactive member of the antibiotic family called mitiro-mycin has the pentacyclic oxazine structure (11) where the carbamate nitrogen has replaced the angular oxygen function present in the other members of the series (17). The location of the carbamate function within ring E prevents easy elimination; thus the mass spectrum of mitiromycin does not reveal the intense M-61 peak. Instead, it shows a strong molecular ion and a prominent peak due to loss of the entire oxazine function as shown. Unfortunately, the scheme is not as simple as depicted because an additional hydrogen is lost in the fragmentation. The m/e 70 peak due to the C_4H_8N fragment is observed as in other N-methyl mitomycins.

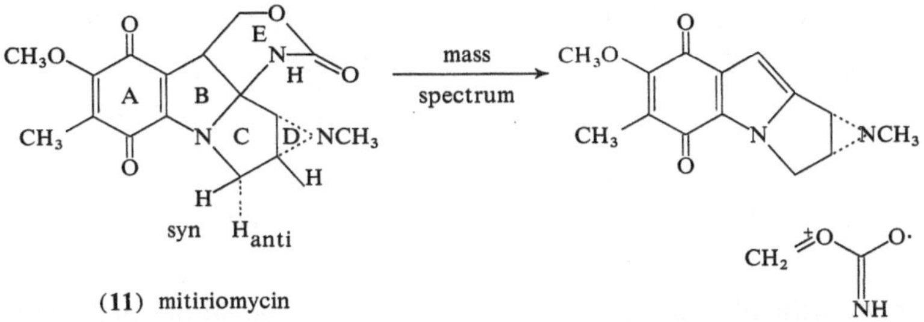

(11) mitiriomycin

The relative stereochemistry assigned by the present author to mitiromycin is the same as mitomycin B, based on an examination of Dreiding models and on published nmr data. Models reveal that the oxazine ring E must be *cis*-fused to the pyrroline. Further the proton at C_2 in mitiromycin is coupled to only the upfield proton on C_3 ($J = 1.5$ Hz) which is the proton labeled H_{syn} in the figure because it is *syn* to the N_4 lone pair. This pattern is typical of the mitomycins. The opposite fusion of the aziridine ring would require coupling of H-2 to that proton on C_3 which is labeled H_{anti} and has a different chemical shift. Also in the opposite configuration of the aziridine ring, the angle between H_{anti} and H-2 would have resulted in a coupling larger than that observed.

In general then, the spectral data for the mitomycins is quite comprehensive and well-understood. The one caveat is that the stereochemical differentiation between the A and B series has not yet been made through an analysis of spectral data.

There is a report of three other mitomycin-like molecules which have been isolated from *Streptomyces reticulu var shimofusaensis* and identified as antibiotics G-253 B 1, B 2 and C 1. No structural or spectral information has been published on these compounds (*18*).

II. Transformation Products

The most extensive and careful studies of mitomycin chemistry are those resulting from acid hydrolysis. Interestingly, the major product of dilute HCl (0.1—0.05 N) hydrolysis for short time periods of all the mitomycins is a mixture of *cis* and *trans*-1-hydroxy-2-amino derivatives (12) and (13), with the *cis* product predominating. The stereochemical assignment of (12) is based upon its ability to form a cyclic carbamate upon

(12)   (14)

treatment with phosgene. Such a carbamate can be stable only when it is *cis*-fused to the five-membered ring C. Furthermore, when (12) is subjected to semipinaolic deamination (X = OCH_3), ketone (14) was produced. Ketone formation requires the migration of a vicinal hydrogen *trans* to the diazotized amine leaving group; hence the vicinal hydroxyl must be *cis* (9). Conversely, *trans* aminohydrin (13) (X − OCH_3) upon deamination yields diol (15) of undetermined stereochemistry.

(13)   (15)

In contrast to the dominance of *cis* aminohydrin products in aqueous acid hydrolysis, the major product of methanol-acetic acid solvolysis of mitomycin C is the *trans*-methoxyamine (16) along with small amounts of the *cis* product (17) (10, 19). The early experimenters who detected only a single product, (16), assumed that it had the *trans* stereochemistry since the latter could be accounted for by a simple stereospecific migration of methoxyl, with inversion at C_1, which was concerted with aziridine ring opening. Later workers converted epimer (17) to an N-acetate which was identical with an acetylated, methylated product of (12), thus confirming the *cis* assignment of (17) directly and showing that (16) must be the *trans*-isomer because it is isomeric with (17). The unusual dominance of retention of configuration at the carbon subject to nucleophilic attack upon opening of the mitomycin aziridine ring by external nucleophiles has no satisfactory rationalization. Any mechanism requiring a double inversion involving angular methoxy migration (or participation) followed by external nucleophilic attack does not explain the observation that the indolic aziridine (18) is also converted to an aminohydrin with predominant *cis* stereochemistry (20).

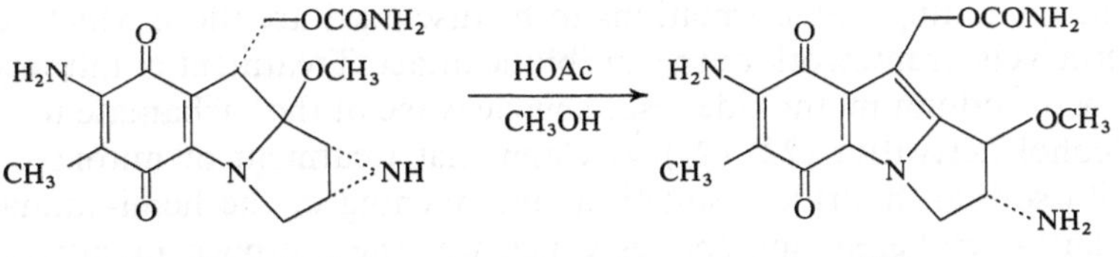

(2) mitomycin C   (16)

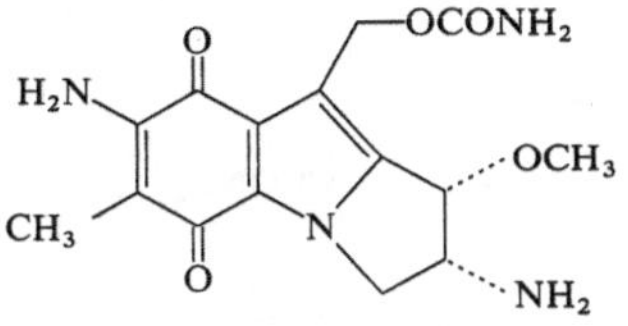

(17) 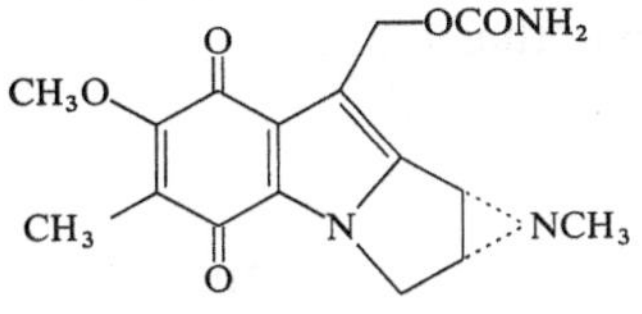   (18)

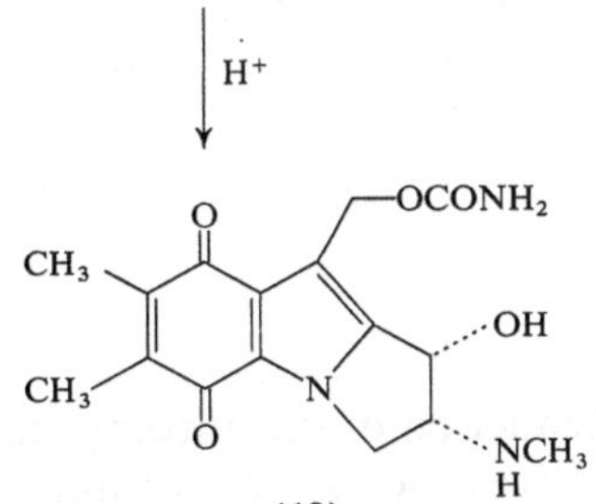

(19)

Extended dilute acid hydrolysis of all the mitomycins also causes replacement of the labile substituent on the quinone ring and yields (20) as the principal product. Stronger acid (6 N HCl) cleaves the carbamate in addition to the other labile functions and (21) becomes the principal product (9, 10).

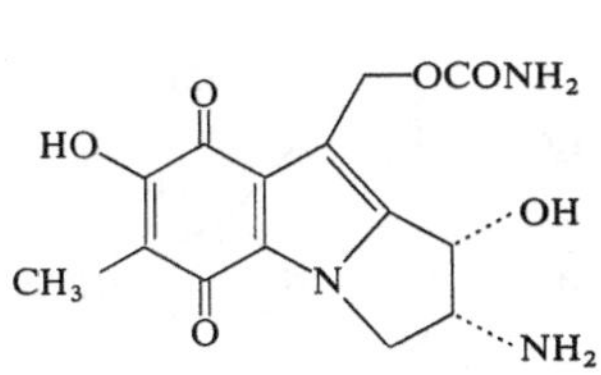

(20) 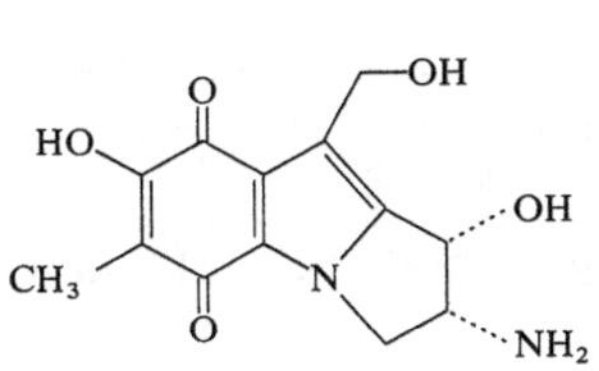   (21)

The remaining transformations to be discussed are those which leave the tetracyclic framework of the antibiotic intact. Treatment of mitomycin C (2) with sodium methoxide results in cleavage of the carbamate to yield an alcohol derivative (22) (21). A claim that treatment of mitomycin B (4) with sodium hydride resulted in ring-opening of the hemi-aminal to yield an 8-membered ring species which was then trapped by alkylating agents, e. g. methyl iodide, to yield (23) is incorrect (22). In fact, simple etherification of the angular hydroxyl occurred to form mitomycin B methyl ether (23, 24).

(22)   (23)

Reduction of the mitomycins with sodium borohydride afforded 9 a-H derivatives such as (24) of unknown stereochemistry (Scheme II). Reductions with lithium aluminum hydride gave two different results depending on whether the aziridine nitrogen was free (mitomycin C and A) or methylated (mitomycin B and porfiromycin). In the former case, the initial product is (25) which can be reoxidized to decarbamoylmitomycin (26). There was also obtained some 9a-H compound (27). The 9a-H compounds (28) and (29) become the sole products when the N-methyl compounds are reduced with LAH (25). Of course, mitomycin B should yield C-9 epimers of the products obtained from porfiromycin such as (30). However, the article describing this work contains no characterization of the products.

(4) mitomycin B   (30)

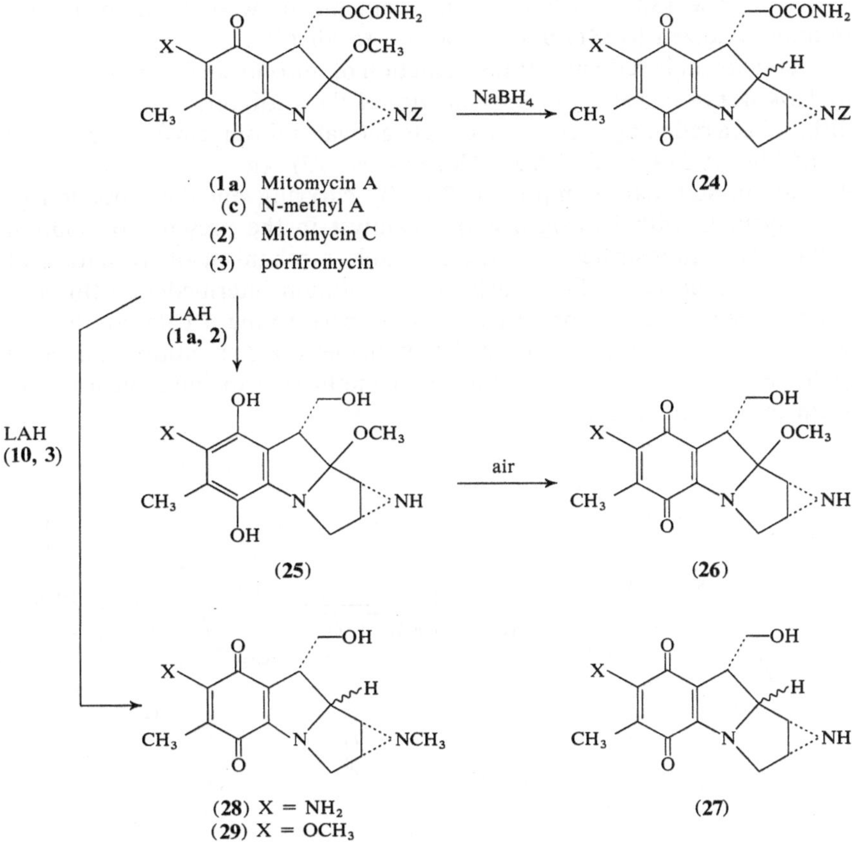

Scheme II

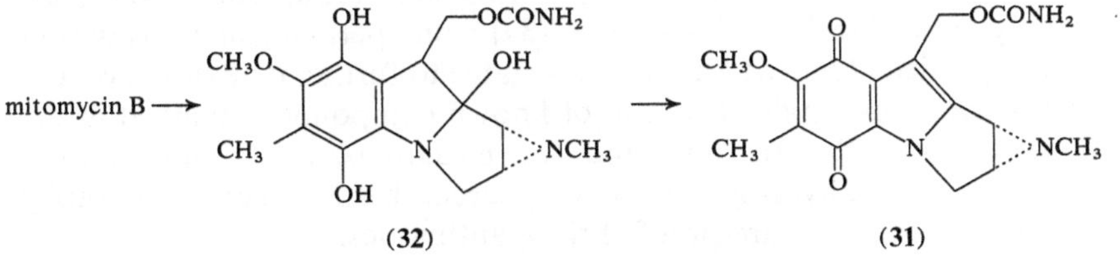

An unexpected reductive transformation of the mitomycins resulted upon catalytic hydrogenation of mitomycin B (**4** and other N-methyl-mitomycins). The product isolated upon workup was (**31**), the result of the elimination of water (or methanol). This product, which is not in a lower oxidation state than (**4**) must arise from initial reduction to form hydroquinone (32) which loses the elements of water and then must become oxidized to (31) upon exposure to air (26).

A remarkable reductive transformation of mitomycin C was uncovered by HORNEMAN in a study of the behavior of the antibiotic towards sodium dithionite, a reducing agent often used to activate mitomycin for mechanism of action studies (vide infra). HORNEMAN (77) was able to isolate the 9a-sodium sulfonate compound (2a). By carrying out the reduction of mitomycin C with hydrogen and a catalyst in the presence of sodium sulfite, he demonstrated that the probable mechanism of formation of (2a) proceeded via sulfite attack on an iminium intermediate (2b). The reverse reaction could be carried out by performing a catalytic hydrogenation on (2a) in the presence of sodium methoxide to afford (2) in good yield. Both the addition of sulfite and methoxide to iminium ion (2b) seem to be stereospecific.

1. H_2/Pt Na_2SO_3, 2. O_2

1. H_2/Pt $NaOCH_3$, 2. O_2

(2)   (2a)

(2b)

The only oxidations of mitomycins that yielded characterizable products were performed on hydrolysis products (33a) and (33b) (10). Ozonolysis of (33a) afforded pyrrole (34) while permanganate oxidation of (33b) yielded aldehyde (35) which is stable to further oxidation because it is a vinylogous amide. The lack of known compounds obtained in the above-described degradations and the meagre structural information derivable from the hydrolyses make it apparent that X-ray crystallography was required for structure-proof of these antibiotics.

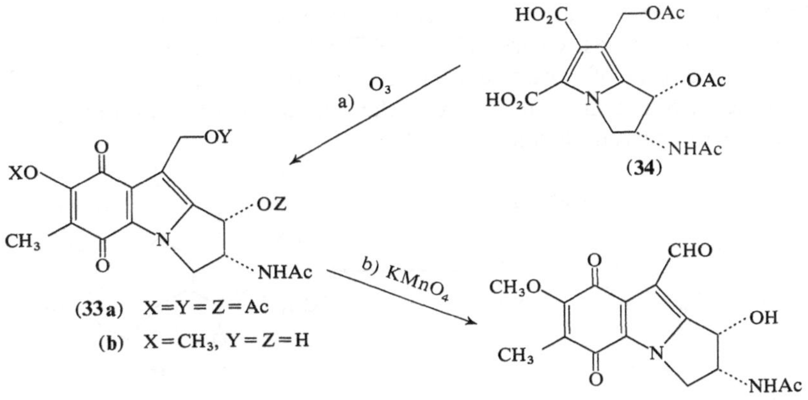

(34)

(33a) X=Y=Z=Ac

(b) X=CH₃, Y=Z=H

(35)

Of great interest to pharmaceutical chemists are modifications of the aziridine substituent group. The aziridine nitrogen has been alkylated, acylated and sulfonylated to produce a group of mitomycin analogs substituted at N-1 (25). A further series of analogs was produced by replacement of the C-7 methoxyl of the quinone by a variety of primary and secondary amines (25). Hundreds of derivatives of the general formula (36) have been produced for bioassay by exploiting these two procedures.

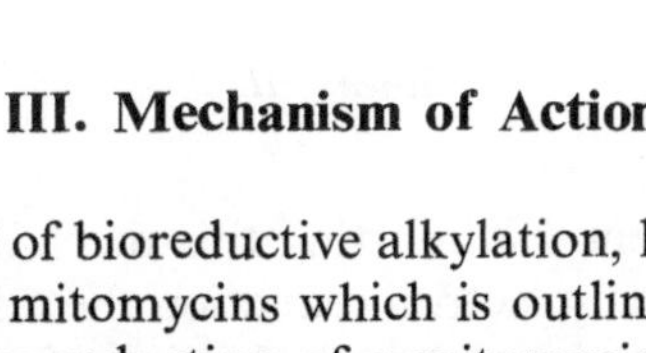

(36)

III. Mechanism of Action

MOORE, in a review of bioreductive alkylation, has proposed a mechanism of action for the mitomycins which is outlined in Scheme III (27). The initial step requires reduction of a mitomycin to the hydroquinone (37) which loses methanol (or water in mitomycin B) to yield indole (38). Then in two consecutive eliminations promoted by the phenolic oxygens the aziridine and the carbamate are cleaved to form an indolic analog of an *ortho* quinone methide (40). Now the methide termini of (40) are subject to nucleophilic attack by purines of complementary strands of DNA. This would result in a cross-linked structure such as (41) where

DNA_1 and DNA_2 can no longer separate in order to participate in replication processes. Cross-linking of DNA by an activated mitomycin was also the basis of a much earlier proposal by IYER and SZYBALSKI (28). The older mechanism differs only in the postulated oxidation level of the activated mitomycin species.

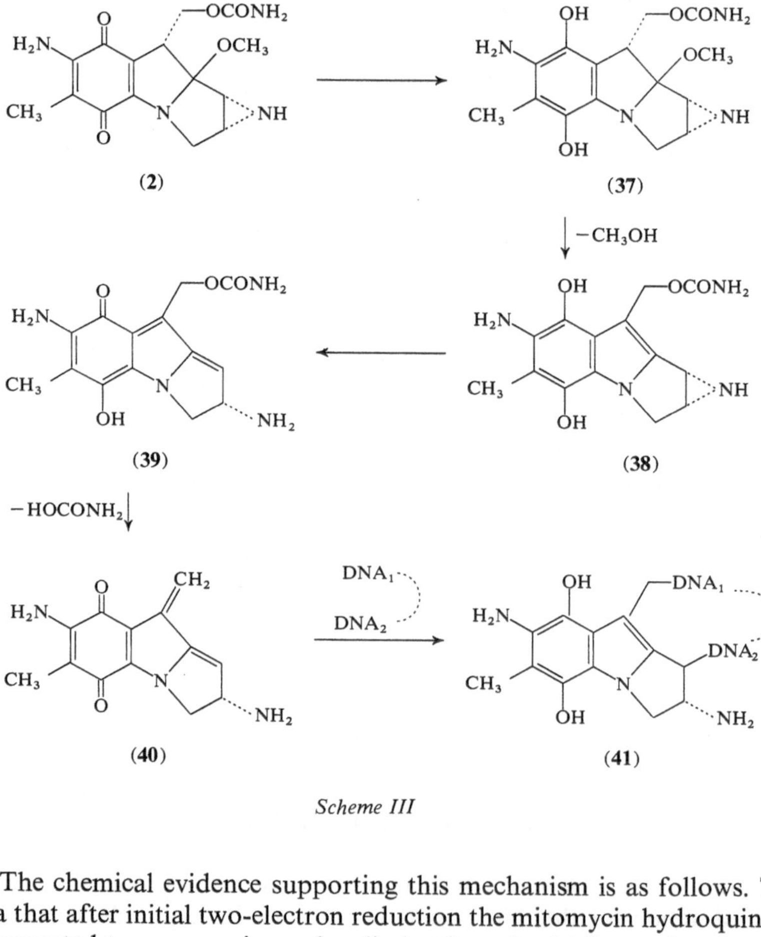

Scheme III

The chemical evidence supporting this mechanism is as follows. The idea that after initial two-electron reduction the mitomycin hydroquinone is converted to a new quinone by elimination of aziridine and carbamate group received remarkable confirmation in the electrochemical studies of LOWN et al. (29, 30). When the mitomycins are reduced, there takes place an initial 2-electron, 2 H^+ reduction which corresponds in potential to a quinone-hydroquinone conversion. A second 2-electron, 2 H^+ reduction then occurs, the potential of which corresponds to that of an authentic indoloquinone, indolohydroquinone reduction, this potential was readily determined by studying the reduction of a known mitomycin degradation product (42). Although the MOORE pathway does not have a free indoloquinone among its several intermediates, it is clear that the addition of a nucleophile to quinone methide (40) could lead to a species such as (43a) which if not attacked by a second nucleophile, can be protonated to yield indoloquinone (43b). This species (43b) structurally similar to (42), would account for the addition of the second pair of electrons and the reduction potential observed.

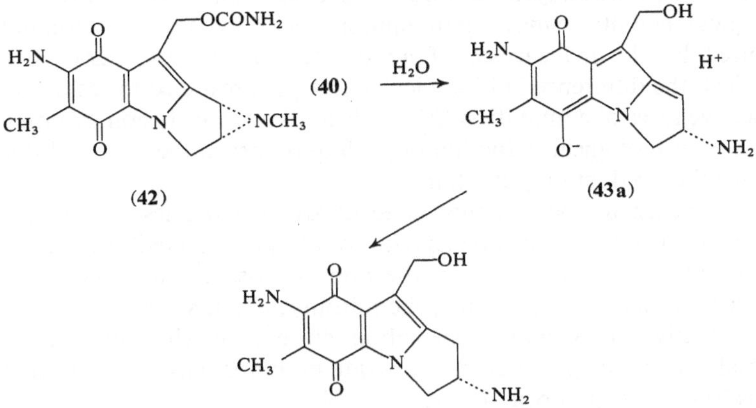

(42)   (40) $\xrightarrow{H_2O}$ (43a)

(43b)

Additional evidence for the hypothesis that a second reducible species other than the original quinone is the biologically active state of the mitomycins has been obtained by TOMASZ (31). Initiation of *in vitro* activity of the mitomycins requires a reduction, with sodium thiosulfate being used in most experiments. Furthermore, controlled addition of $Na_2S_2O_4$ results in greater binding of antibiotic to nucleotide than when excess $Na_2S_2O_4$ is added in one portion. This observation is consistent with the involvement of a quinone methide such as (40). Clearly, if excess reducing agent is present while (37) is converted to (40), then the reducing agent could convert the quinone methide (40) to hydroquinone (44) thus competing with nucleophilic attack of DNA on the quinone methide.

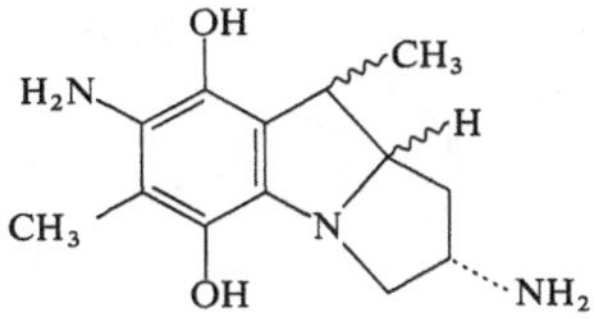

(44)

KINOSHITA et al. (25) have examined the antibacterial activity of
mitomycins with A-C stereochemistry which differ only in the nature of
their C-7 substituent and, concomitantly, in reduction potential. A
reasonable correlation was found in that the more easily reduced derivatives (with a less negative potential) were more potent antibiotics and
were more rapidly activated by a reducing liver homogenate in antibiotic
assays. One puzzling datum in this analysis is the observation that mitomycin B derivatives having reduction potentials essentially identical with
those of their mitomycin A and C relatives exhibited greatly reduced
biological activity. Since upon elimination of water or methanol, as
required by the mechanism of action, the B and A-C series become
identical, the difference in biological activity is presumed to reside in their
respective rates of elimination (25). This tentative conclusion assumes that
the rates of biological reduction of both quinones in the A-C and B series
parallel their reduction potentials.

In contrast to the evidence cited above that defines a reduction requirement, LOWN (32) has observed cross-linking of DNA by mitomycin
C without reduction. However, the rate of cross-linking is considerably
slower than that observed when the mitomycin is first reduced. Also, this
non-reductive cross-linking was observed at somewhat lower pH than
normal. Thus, this pathway may operate in tumor cells where the average
pH is lower than in normal tissue.

The requirement for the complete conjugation in an indole that is
fundamental to the proposed mechanism of action is confirmed by the
observation that the transformation product (24) which cannot easily
become an indole has no significant biological activity. The requirement
for attachment of good leaving groups to C_{10} and C_1, the carbamate and
aziridine, so that a quinone methide can form, is demonstrated by the
diminished biological activity of derivatives (22) and (12) in which the
leaving group abilities at C_{10} and C_1 are decreased from those of the
natural product (19, 22, 25).

(24)   (22)

(12)

An attempt has been made to use a multicategory pattern classification technique to develop a structure-activity relationship for 16 mitomycin derivatives of general formula (45). The descriptors used for the X substituent were F, a field or inductive effect constant for aromatic substituents and Vw, the van der Waals volume of the group. For the Y substituent, arbitrary constants for each of the 3 possible groups, H, OCH_3, and OH were used. The Z group was described by E_s, the standard Taft steric constant. The function shown below correctly classified 15 of the mitomycins by their activity against solid sarcoma 180 with 5 categories of effectiveness, $3+$, $2+$, $+$, $\pm$, $-$.

$$L \text{ (activity)} = -4.53 \text{ F-X} \text{ ——— } 2.64 \text{ Vw-X} + 0.77 \text{ E}_s\text{-Z} +$$
$$2.48 \text{ Y}_{OMe} + 2.28 \text{ Y}_{OH} + 1.42$$

Thus increasing size and electron-withdrawing power of substituents decrease activity of the mitomycins as agents against one specific tumor. Experimentation with a greater variety of substituted mitomycins is probably required to give further credence to this novel analysis of data (33).

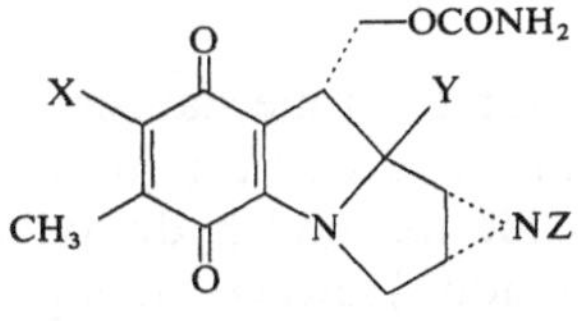

(45)

There has been a great deal of *in vitro* experimentation designed to elucidate the details of the interaction of the mitomycins with natural and synthetic polynucleotides. Two assays have been used to demonstrate cross-linking caused by the antibiotic. A density gradient centrifugation method demonstrated that some mitomycin-treated DNA molecules are spontaneously renaturable under conditions where untreated DNA is irreversibly denatured (28). An ethidium fluorescence method that detects double-stranded DNA also demonstrated the existence of intact DNA which had been protected by mitomycin treatment prior to being subjected to conditions of irreversible denaturation (32). The locus of alkylation of the polynucleotide seems to be the guanine nucleus. Evidence for attack on guanine is the fact that DNA's with higher G-C/A-T ratios incorporate more mitomycin than those with higher A-T/G-C ratios. Furthermore, only poly-G of all the synthetic polynucleotides tested incorporates mitomycins (34, 35). Unfortunately, mitomycin C does not alkylate guanine, guanosine or 5″ GMP. Furthermore, when the DNA-mitomycin complex is degraded, no guanine-mitomycin linked product can be detected (*31*). These failures to isolate a covalently linked antibiotic-nucleotide fragment might suggest non-covalent cross-linking in the mitomycin-DNA complex. However, after extensive probing of the complex using every dissociation method known to cleave non-covalent complexes, Tomasz concluded that the cross-linking must be covalent (*31*). A hypothesis that requires intercalation of the mitomycin in the DNA strands prior to covalent cross-linking served to rationalize the failure of mitomycin to alkylate small molecules. The nucleophilic atom of the guanine in the polynucleotide which is responsible for the putative link to mitomycin has not been identified. An experiment which normally results in the release of tritium from T-labeled C-8 of guanine when the heterocycle is alkylated at N-7 indicates that mitomycin does not attack N-7 (*36*). In addition, the ethidium fluorescence experiments described above have been used to contrast the action of dimethyl sulfate (a known G-7 alkylator) with mitomycin C (*32*).

In conclusion, the hypothetical intermediacy of a cross-linking quinone-methide is reasonably convincing, but conclusive evidence that a mitomycin fragment is covalently linked to a fragment from a cell is still not available.

The only other suggestion for a mechanism of action is due to Lown (*32*). He has demonstrated that reduced mitomycin C in the presence of oxygen generates superoxide ion and hydroxyl radicals which degrade DNA by single-strand scission. However, this proposal does not require cross-linking or any sort of covalent attachment of mitomycin to DNA. Since most antitumor agents are now recognized to exert multiple mechanisms of action on the cell target, the identification of both a free radical pathway and the bioreductive alkylation sequence is not to be viewed as inconsistent or contradictory.

Clinical aspects of mitomycin therapy have been reviewed recently by Crooke and Bradner (*37*). The following paragraph is based on their article. Mitomycin C is currently used in the treatment of end-stage patients with adenocarcinoma of the stomach, pancreas or colon when surgery or radiotherapy is not longer indicated.

Results of therapy by tumor types are summarized in Table 2 where an objective response is defined as a 50% decrease in the multiple of the two longest dimensions of a measureable mass. Thus, this table eliminates most studies performed in countries other than the U.S. Crooke and Bradner suggest that the omitted results would skew the data to the positive. The mean remission times are about 3 months.

Although mitomycin C is orally active, the usual dosage is 0.05 mg/Kg/day I.V. for six days, then every other day till evidence of toxicity. The most significant toxicity in man is myelo suppression. The toxicity incidence for the dosage schedule cited above is approximately 60% of the patients with leukopenia (< 4000 white blood cell count) and about 70% with thrombocytopenia (< 400,000 platelets). Often toxic effects of mitomycin C therapy include anorexia, nausea, vomiting and diarrhea which are usually mild.

Table 2. *Objective Responses to Mitomycin C in Solid Tumors*

Disease site and type	Totals objective responses/trials	%
Breast	26/75	34.7
Stomach	25/93	26.9
Colo-rectal	46/248	18.5
Pancreas	11/53	20.8
Biliary tree	7/15	46.7
Lung	13/77	16.9
Head and neck	18/86	20.9
Cervix	4/20	20.0
Ovary	6/27	22.2
Bladder	5/24	20.8
Sarcoma	3/39	7.7
Melanoma	9/65	13.0
Other	10/59	16.9

IV. Biosynthesis of the Mitomycins

The essentials of the biosynthesis of the mitomycins are known, but some important details remain to be worked out. It is clear that the contiguous chain of 6 carbons from C_{10} to C_3 is derived from D-glucosamine (*38, 39, 40*). The incorporation of the aminosugar was demonstrated by the finding that $^{14}C_6$ of glucosamine is located at C_{10} of mitomycin while a $1\text{-}^{14}C$, 6-^3H doubly-labeled substrate is incorporated intact into the antibiotic. The location of activity at C_{10} was proven by periodate cleavage of a mitomycin hydrolysis product which liberates C_{10} as formaldehyde.

Carbons 1, 2 and 3 and the aziridine N can be detected mass spectrometrically since the m/e 70 peak results from cleavage of that grouping. Thus it can be demonstrated by ^{15}N, ^{13}C double-labeling experiments that the nitrogen of the aziridine is also derived from an intact glucosamine (*40*). The evidence for this conclusion is simply that the C_{13}/N_{15} ratio in the antibiotic is identical with the C_{13}/N_{15} ratio in the precursor.

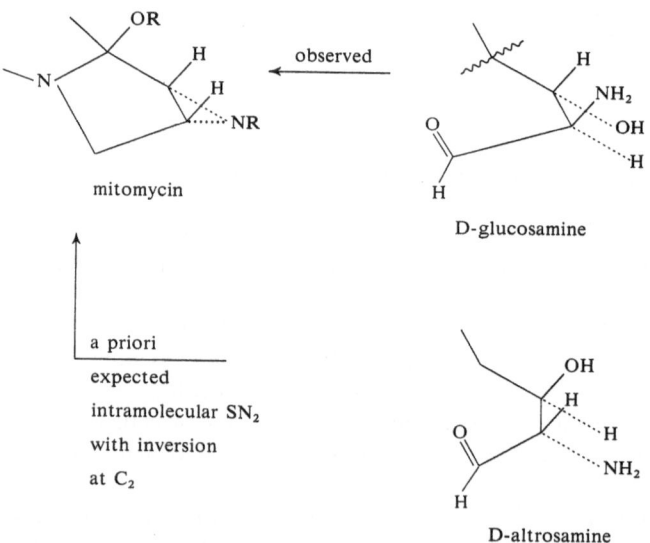

(2)

D-glucosamine

From this it follows (Scheme IV) that the absolute stereochemistry at C_2 in mitomycin, which originates in C_2 of glucosamine, is the inverse of that of its precursor; while the configuration at C_1 of mitomycin which originates in C_3 of glucosamine is retained. This result is opposite to the simplest prediction one would make, namely one of C-2 retention and C-3 inversion, if aziridine synthesis were in fact an intramolecular S_N2 displacement by the amino group of an adjacent OH activated to become a leaving group.

mitomycin

observed

D-glucosamine

a priori

expected

intramolecular SN_2

with inversion

at C_2

D-altrosamine

Scheme IV

D-Mannosamine, whose C-2 saccharide stereochemistry is inverted from that of glucosamine and thus corresponds to the absolute configuration of mitomycin is not incorporated as well as glucosamine.

D-Galactosamine, with the same configuration as glucosamine at the carbons which become the aziridine in mitomycin is incorporated, though less efficiently than glucosamine (*39*). Other 2-aminohexose epimers, D-altrosamine with both saccharide C-2 and C-3 opposite to that of glucosamine, and D-allosamine, with only C-3 opposite that of glucosamine, have not been assayed as biosynthetic precursors.

The carbon of the carbamate function of the mitomycins is derived from the urea carbonyl of citrulline (**46**) (*40*). Incorporation of label from ureido-^{14}C-citrulline was more efficient than incorporation of the guanido carbon of arginine (**47**).

NHCONH$_2$

NHC(NH)NH$_2$

—NH$_3^+$

—NH$_3^+$

CO$_2^-$

CO$_2^-$

(**46**) citrulline

(**47**) arginine

The O-methyls of the ethers derive from methionine, as do the N-methyls of porfiromycin and mitomycin B (*38, 41, 42*). The quinone portion is subject to a less informative assay because only two carbons of the seven that constitute the toluquinone are released by Kuhn-Roth oxidation which yields acetic acid derived from carbons 6 and 6a (*42, 43*).

Incorporation of glucose and pyruvic acid is modest, < 1% taken up in mitomycin C, compared to that of glucosamine in the non-quinoid section of mitomycin C, 5%. With 1-^{14}C pyruvic acid as a precursor, the labeled carbon becomes the C-6a methyl group of the quinone, not C-6. But less than 1/3 of the label is accounted for by Kuhn-Roth oxidation. With 2-^{14}C pyruvic acid, 40% of the label is found in the acetic acid derived from C-6 and C-6a and most of the label is found at C-6. Glucose labeled at C-1 or C-6 and pyruvic acid labeled at C-3 do not furnish labeled acetic acid from Kuhn-Roth oxidation while glucose labeled at C-2, -3, or -4 yielded labeled acetic acid (*43*). The explanation for this observation is that glucose serves as the biosynthetic precursor for pyruvic acid. Thus HORNEMANN tentatively suggests that pyruvic acid derived saccharide affords dehydroquinic acid (**48**) with the pyruvate incorporated as shown in Scheme V, which combines with glucosamine form the mitomycin skeleton. The source of nitrogen for N-4 remains obscure at this time (*40*).

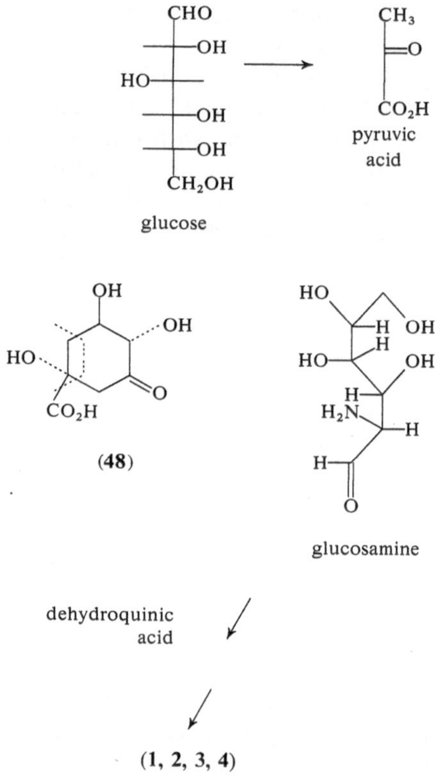

CHO
—OH
HO—
—OH
—OH
CH₂OH

glucose

CH₃
=O
CO₂H
pyruvic
acid

(48)

glucosamine

dehydroquinic
acid

(1, 2, 3, 4)

Scheme V

V. Synthetic Studies

The discussion of synthetic efforts in the mitomycin area will, for convenience, be sub-divided into two classifications. The larger number of studies deals with derivatives related to the indolic series (trivially named mitosenes) where the aziridine and carbamate are conjugated with the quinone. The rationale for synthesizing mitosenes is that the difficult functionality and stereochemical problems of introducing the labile angular oxygen found in the natural series are avoided and that the mitosenes are biologically active degradation products readily obtainable from the mitomycins. A smaller number of investigations aims at the mitosanes in which the aziridine and carbamate functions are insulated from the quinone by a dihydroindole.

mitosane

mitosene

pyrrolo(1,2-a)indole

This section will review only those reports which develop systems that are closely related to the natural product with more features than the parent tricyclic framework. An exhaustive survey of every synthesis of a tricyclic pyrrole(1,2a)indole need not be presented because a recent review by KAMETANI encompasses all the relevant work in the field (44). Thus, to be included in the following discussion, the work described will usually have some relevance to elaboration of the quinone, aziridine or carbamate.

A major success in mitosene synthesis can be found in the early work of the Lederle Laboratories group led by M. J. WEISS. Their approach is shown in Scheme VI. Key carbon-carbon bond-forming steps involved acylation of a nitro-activated methyl group with oxalate (49→50), Dieckman condensation of diester (52→53) and Vilsmaier-Haack formylation of an indole (54→55). Further critical manipulations included elaboration of the quinone functionality by oxidation with Fremy's salt followed by Thiele acetoxylation, air oxidation and diazomethane methylation (56→57→58) (45). In addition their technique for carbamate introduction (59→60) namely, carboxylation with phenylchloroformate and subsequent amination with ammonia, is the only known successful method for either mitosenes or mitosanes.

Using ketone (54) as a starting material, REMERS has been able to synthesize a group of 1- and 1,2-substituted mitosenes exemplified by (61) and (62). The key to this work was reduction and blocking of the carbonyl group of (54) prior to Vilsmaier formylation of the indole. Some modifications of methods for quinone elaboration were developed by REMERS as well (46, 47). Note that (62) is the regioisomer of acid hydrolysis product (12) followed by acetylation of mitomycin A. Compound (62) possesses quite good antibiotic activity although no anticancer activity. Quinone (58) has also been obtained (Scheme VII) in several steps

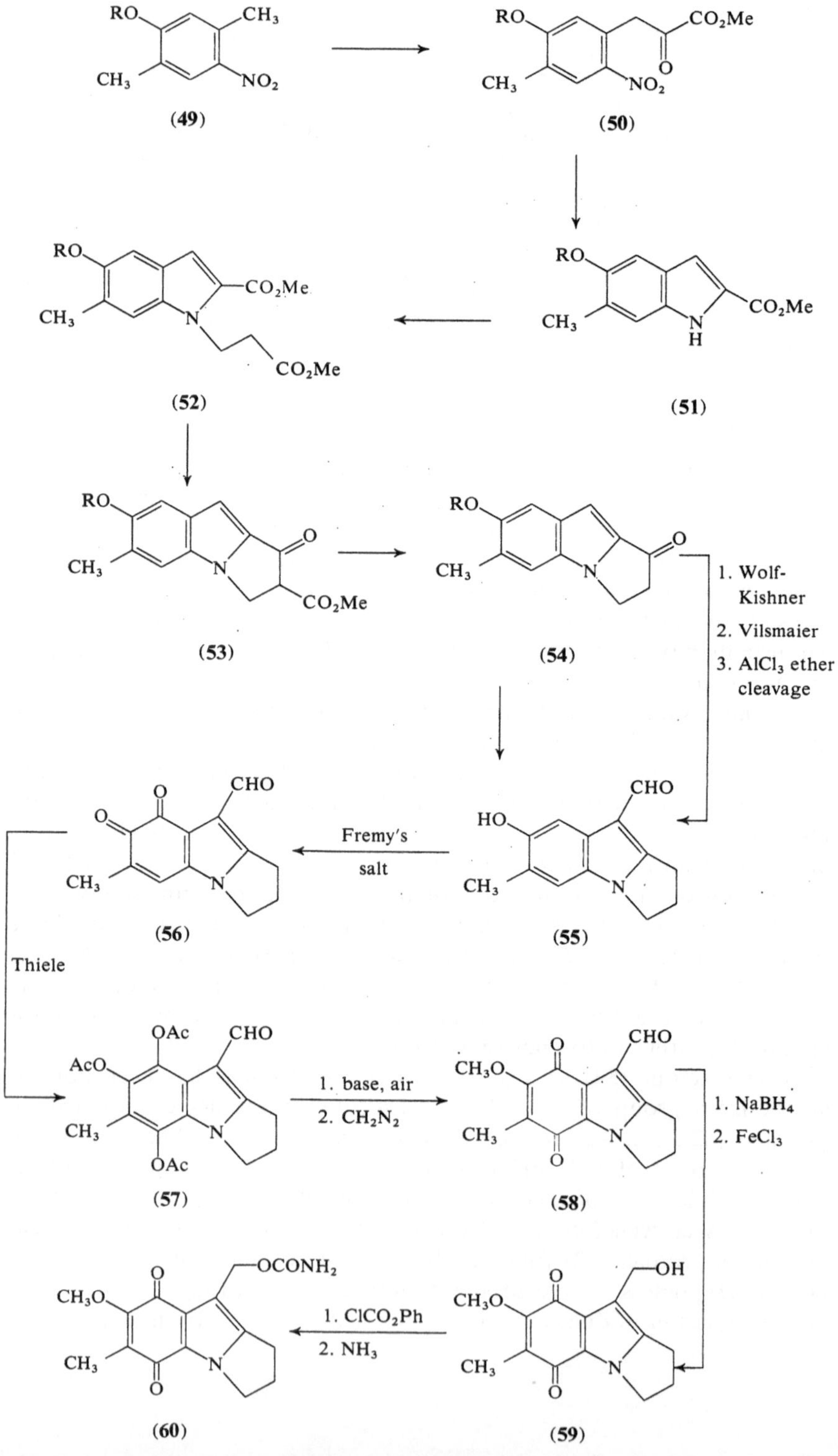

R. W. Franck:

(49)

RO, CH₃, CH₃, NO₂

(50)

RO, CO₂Me, CH₃, NO₂, O

(52)

RO, CH₃, CO₂Me, N, CO₂Me

(51)

RO, CH₃, CO₂Me, N, H

(53)

RO, CH₃, N, O, CO₂Me

(54)

RO, CH₃, N, O

1. Wolf-Kishner
2. Vilsmaier
3. AlCl₃ ether cleavage

(56)

O, O, CHO, CH₃, N

Fremy's salt

(55)

CHO, HO, CH₃, N

Thiele

(57)

OAc, CHO, AcO, CH₃, N, OAc

1. base, air
2. CH₂N₂

(58)

CHO, CH₃O, CH₃, N, O

1. NaBH₄
2. FeCl₃

(60)

OCONH₂, CH₃O, CH₃, N, O

1. ClCO₂Ph
2. NH₃

(59)

OH, CH₃O, CH₃, N, O

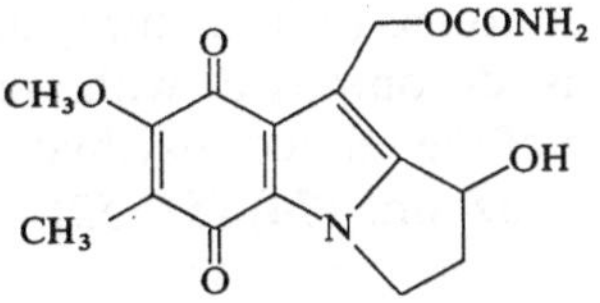

(61)

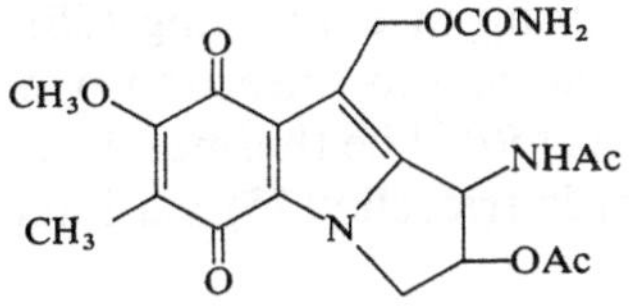

(62)

from nitrile (**63**), which in turn was prepared via an intramolecular S_NAr reaction of nitrile (**64**) (*48*).

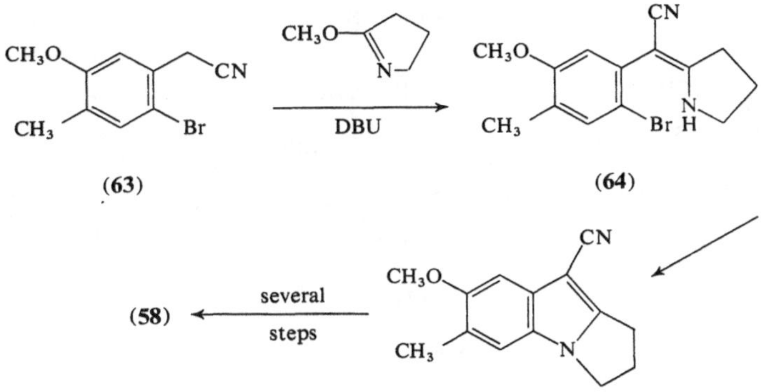

(63) (64) (65)

Scheme VII

A one-step synthesis of (**68**) based on the Nenitzescue indole synthesis, is outlined below (*49*). Although this method is appealing because of its simplicity, the route is marred because isomer (**69**) is actually

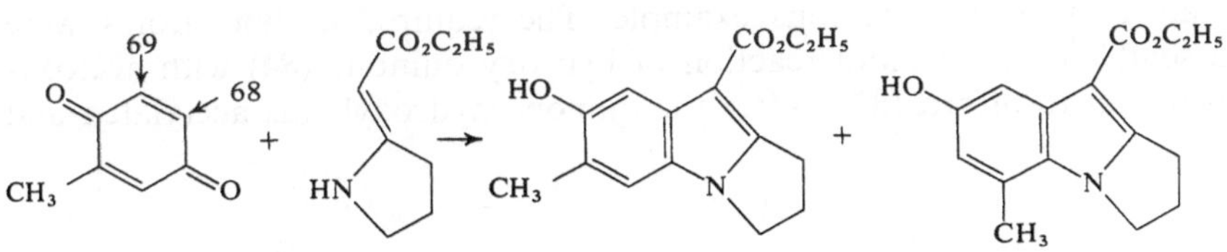

(66) (67) (68) (69)

formed as the predominant product. Quinone (**70**) has been obtained from nitroindoline (**71a**) by a series of oxidations of the aniline (**71b**) (Scheme VIII). The methodology was identical with that used by the

Lederle group in converting (55) to (58). Indoline (71), in turn, derives from a carbenoid insertion when (72) is decomposed with sodium methoxide (50). The thermal decomposition of the related tosylhydrazone (73) affords tricyclic (75) via intermediate ozazoline (74) (51, 52).

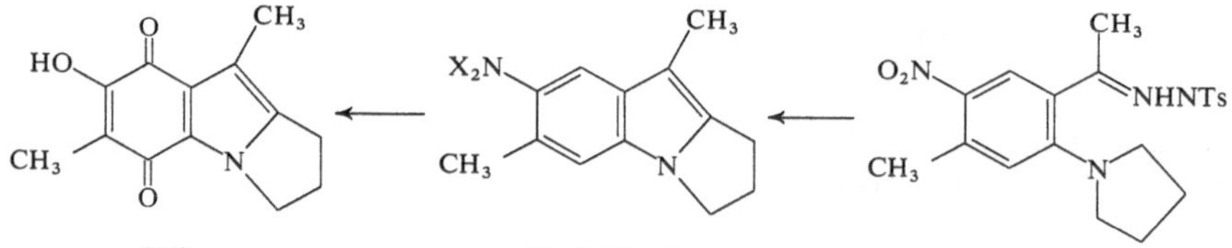

(70)   (71 a) X=O   (72)
                             (b) X=H

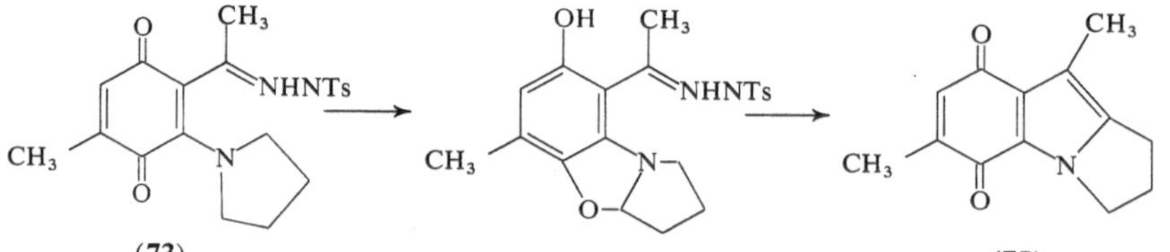

(73)   (74)   (75)

Scheme VIII

An interesting functionalized mitosene (76) is derived from hydroxy-proline in an elegant use of dipolar addition to munchnone (79) as shown in Scheme IX (53). The use of munchnone chemistry has been exploited to synthesize the other pyrrole ring of potential mitosenes as shown below in the synthesis of pyrrolo(1,2-a)indole (83) from (82) (54).

A novel route to a benzomitosene illustrated in Scheme X uses an intact quinone throughout, thus requiring no adjustment of oxidation state as in every previous example. The required carbon atoms were assembled via an aldol reaction of hydroxy quinone (84) with acetoxy-pentanal to afford (85). After the quinone hydroxyl was acetylated and

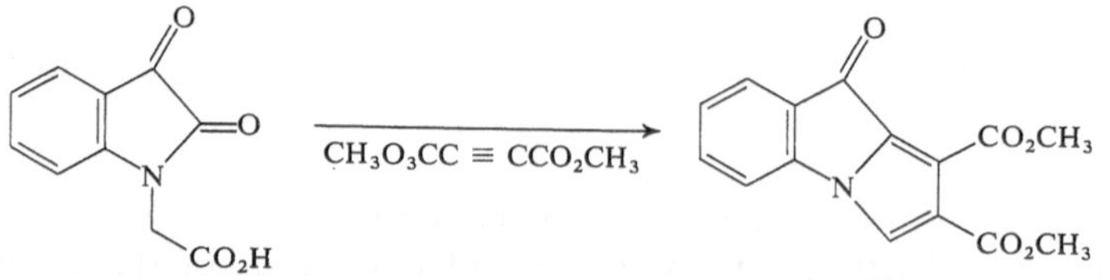

$$CH_3O_3CC \equiv CCO_2CH_3$$

(82)   (83)

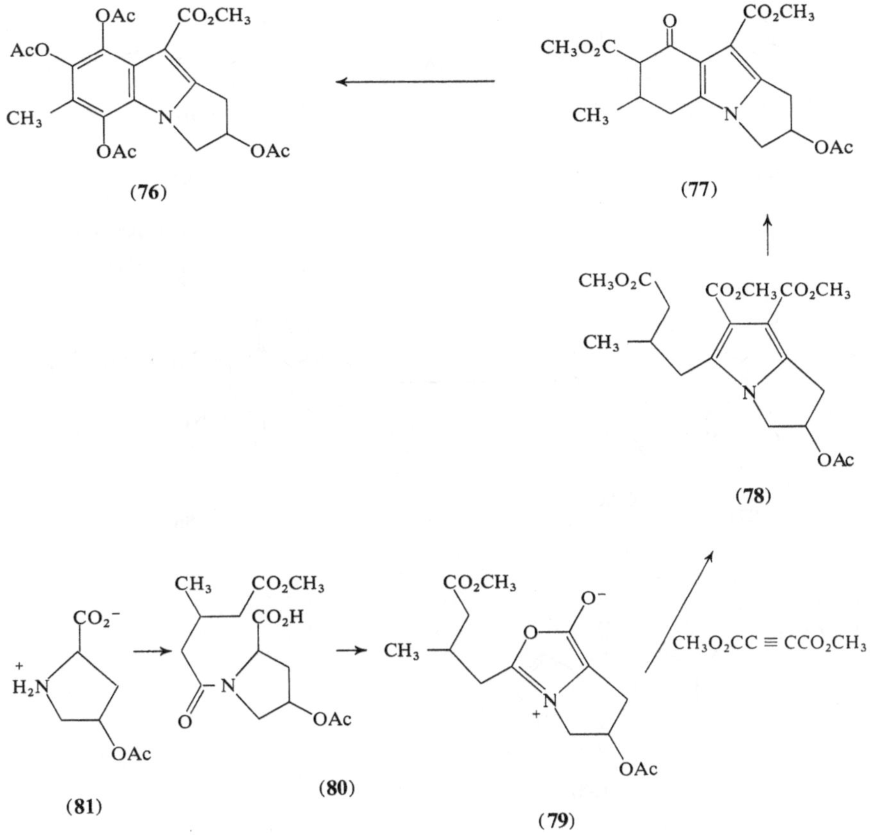

Scheme IX

then displaced by an azide, the key step, the thermal decomposition of azidoquinone (86) to yield indoloquinone (87) was carried out. It is interesting to note that in the alkylation of the ambient ion derived from indoloquinone (87) the major product (88) is the result of N-alkylation rather than C-alkylation (55).

A transannular cyclization of an azacyclo-octanone derivative to form the N-4, C-9a bond has been carried out by two groups as shown in Scheme XI (56, 57, 58). The intermediate in both cyclizations must have been a 9a-oxygenated species such as (96) which resembles the natural functionality. Unfortunately, aromatization to the indole is not impeded, as it must be in order for the angular oxygen to remain intact (vide infra).

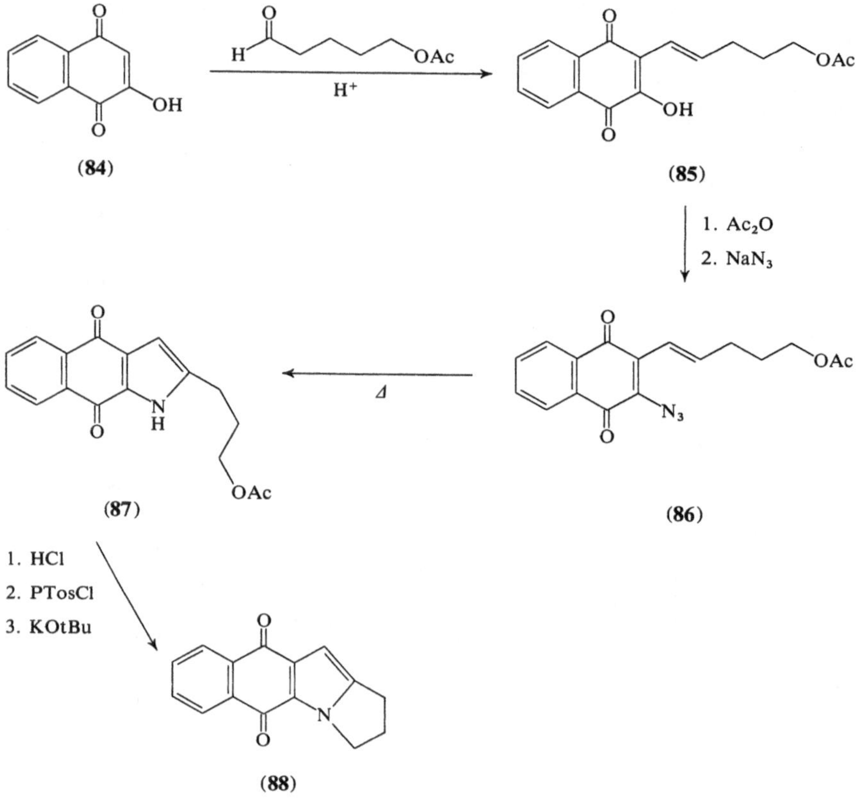

(84)   (85)   (86)   (87)   (88)

1. Ac₂O
2. NaN₃

Δ

1. HCl
2. PTosCl
3. KOtBu

Scheme X

There are fewer studies directed toward the introduction of an azi-ridine function into the mitosenes. Early reports from the Lederle group detail many unsuccessful attempts to utilize 1,2-disubstituted mitosenes such as (97) as aziridine precursors (59).

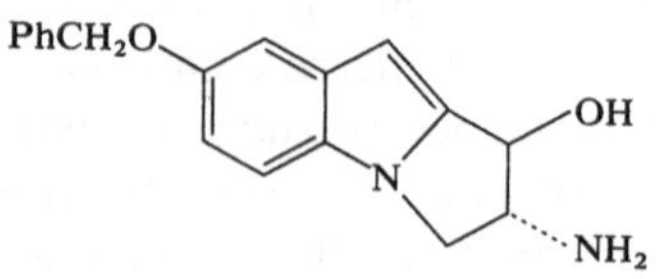

(97)

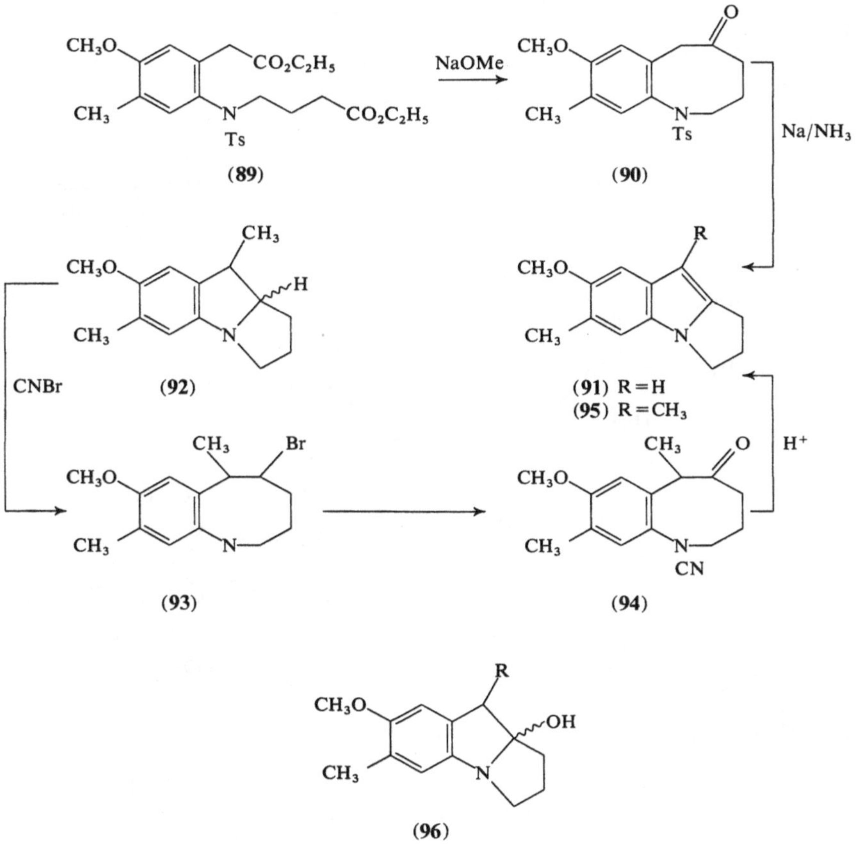

Scheme XI

While the group at Fordham University (*60*) was unable to functionalize alkene (**99**) which is available from acylation of anion (**98**), a Kyowa Fermentation team was able to prepare (**100**) by regiospecific addition of IN_3 followed by selective reduction of the azide to the amine (Scheme XII). First, the amine was blocked as a methyl carbamate which, upon treatment with methoxide, yielded the oxazoline (**101**). Base treatment of the unblocked amine (**100**) followed by carbomethoxylation afforded (**102**) in unstated yield (*61, 62*). Vinylindole (**99b**) was later obtained by pyrolysis of acetoxyester (**103**) which in turn derived from nitrile (**65**) *via* $Pb(OAc)_4$ acetoxylation and nitrile hydrolysis (*63*).

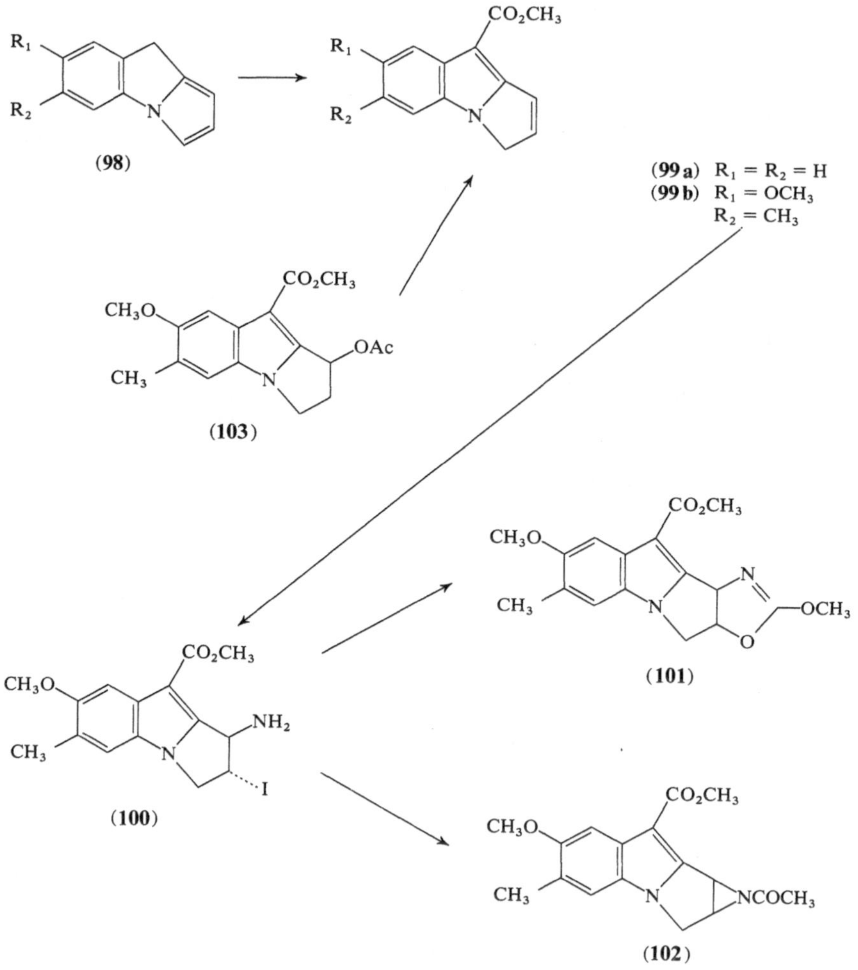
CO_2CH_3
R_1
R_2
N
(98)
(99a) $R_1 = R_2 = H$
(99b) $R_1 = OCH_3$
$R_2 = CH_3$
CH_3O
CH_3
OAc
(103)
(100)
NH_2
I
(101)
N
O
OCH_3
(102)
$NCOCH_3$

Scheme XII

Simultaneous to the reports by the Kyowa group, the Fordham laboratory reported Scheme XIII for synthesis of an aziridinomitosene (**107**) (*64, 65*). This scheme was later extended to the preparation of mitosenes such as (**108**) which have two of the functions considered to be important for the mitomycins, the aziridine and the C-10 hydroxymethyl group (*66*).

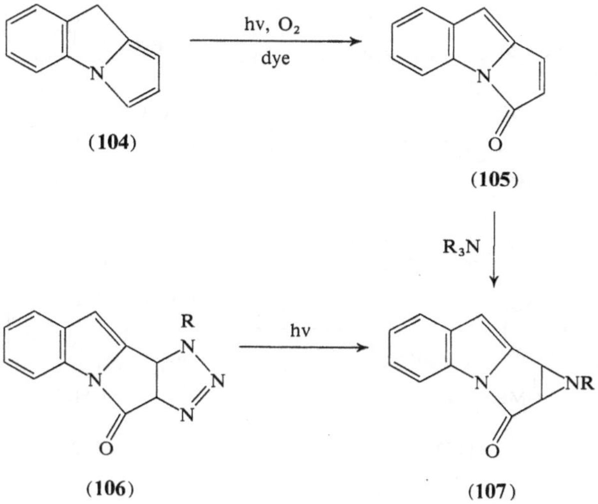

(104) (105) (106) (107)

Scheme XIII

The earliest attempt at the synthesis of a mitosane was reported by MANDELL (*67*) who attempted an intramolecular Michael addition of the ester enolate of quinone (**109**) but failed to obtain (**110**). The stereoelectronic requirements for the desired cyclization impose a great deal of strain on the transition-state for this particular ring-formation. In BALDWIN's terminology, the reaction is "5-*endo*-trig" and is an example of a disfavored pathway (*68*). A mitosane similar to (**110**) has been obtained by DANISHEFSKY *et al.* as shown in the following Scheme XIV.

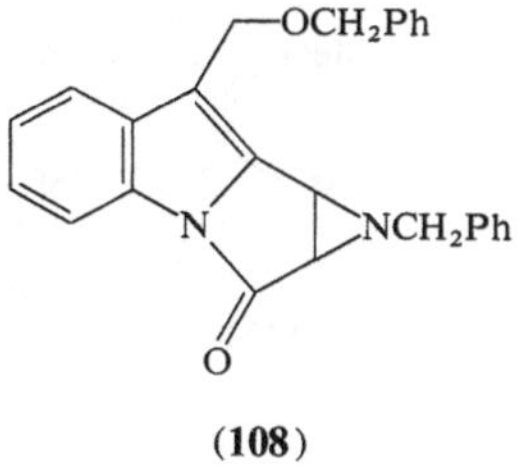

(**108**)

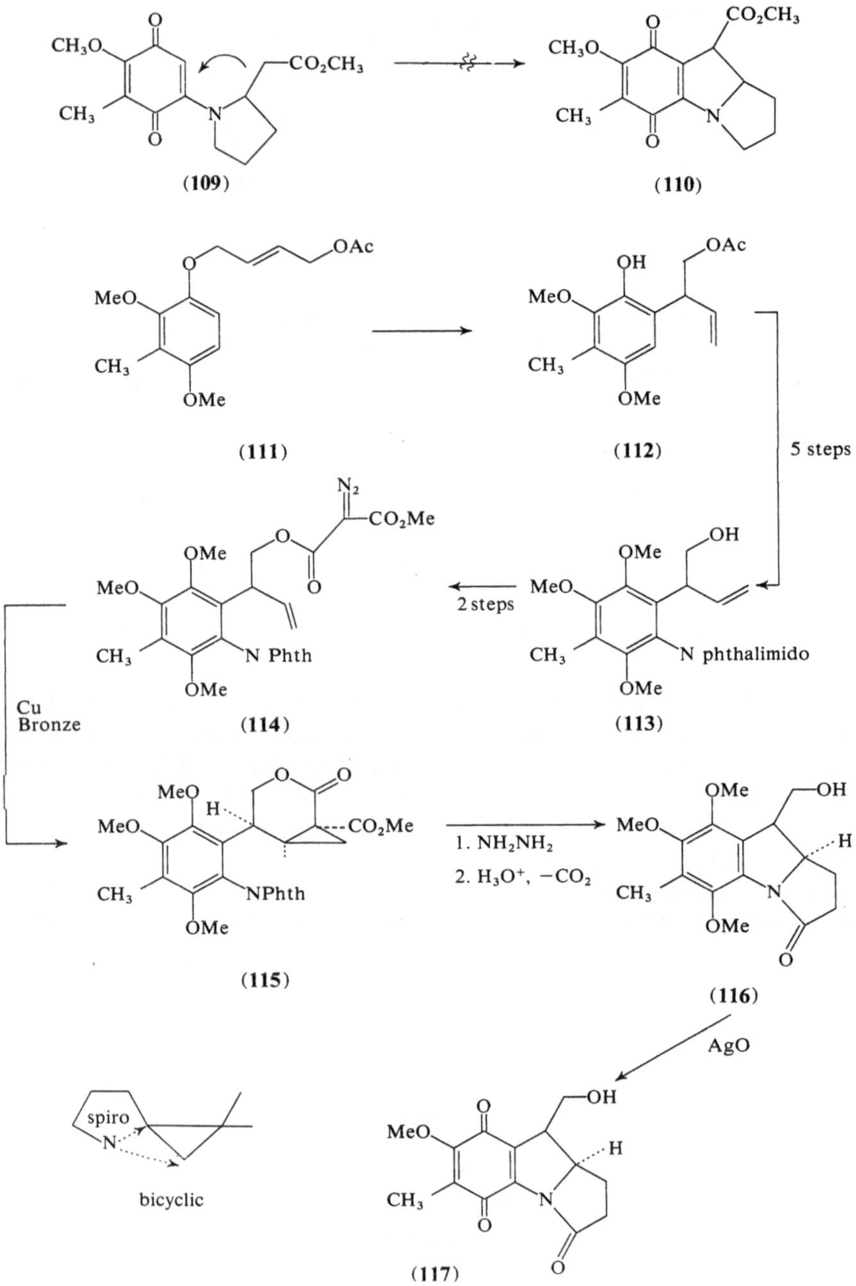

Scheme XIV

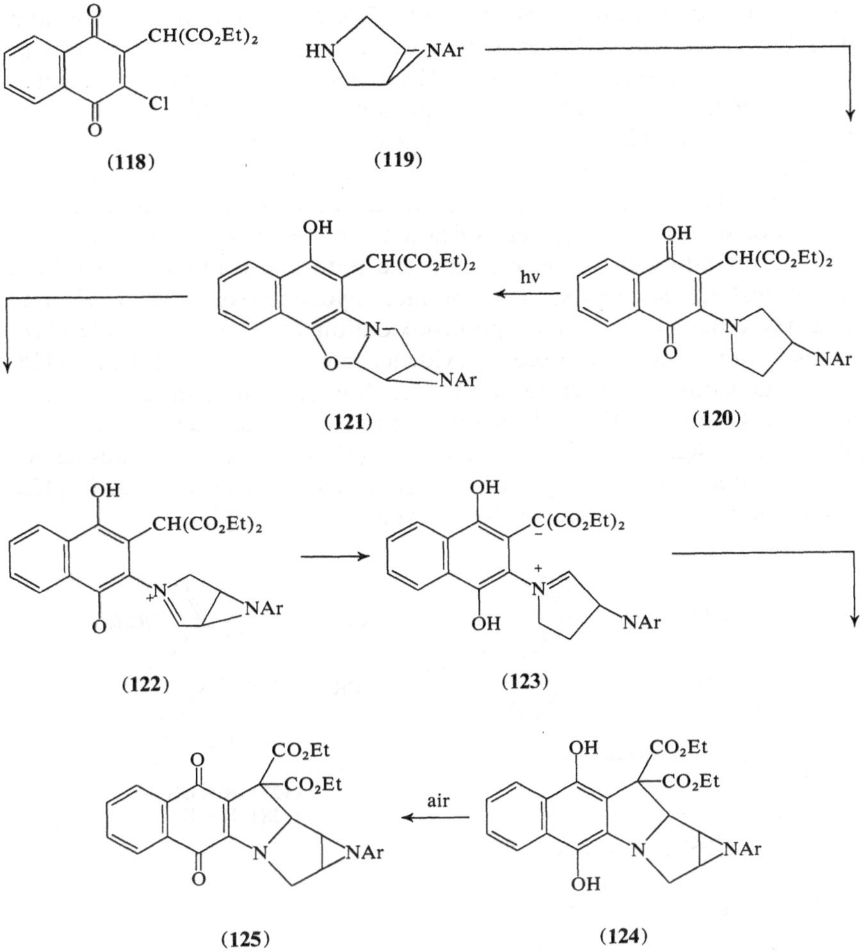

Scheme XV

The carbon framework was created using two C-C bond-forming steps. First, a Claisen rearrangement converted (111) to (112). Second, after introduction of a blocked amino function on the aromatic ring, cyclopropanation was effected to form the bonds that completed the required carbon sequence. With (115) in hand, the amino group was liberated by hydrazinolysis. The key step was the subsequent nucleophilic attack of the amine on the activated cyclopropane *via* the spiro mode to form (116) (after decarboxylation of the extra carbomethoxy) (69).

A very convergent synthesis of an aziridinobenzomitosane lacking only the angular oxygen at C-9a of the natural product has been reported by AKIBA and is shown in Scheme XV (70). Naphthoquinone (118) and diazabicyclohexane (119) reacted to afford quinone (120). Photolysis of (120) afforded insertion product (121). Upon standing (121) dissociated to intermediate (122) which upon proton-transfer yielded (123) which then cyclized to (124). Although the closure of (123) to (124) is "5-endo-trig" (68), it also involves a fully conjugated 6 π electron system and this can also be viewed as an allowed disrotatory electrocyclization of the WOODWARD-HOFFMAN classification $\pi\,6\,s \rightarrow \pi\,4\,s + \sigma\,2\,s$ (71).

The introduction of oxygen into the mitosane framework has been accomplished. KAMETANI has obtained hydroperoxy ether (127) and hydroxy ether (128) (72) by photo-oxidation of tricyclic pyrrole (126) in methanol. The spectroscopic evidence presented for (127) and (128) rules out isomeric structure (129). The downfield shift of a methine H upon acetylation of the hydroxyl is used in support of (128) as the proper structural assignment. The methine H in (129) should not be deshielded upon acetylation of the hydroxyl group across the ring whereas in (128) the methine H is geminal to the acetylatable hydroxyl.

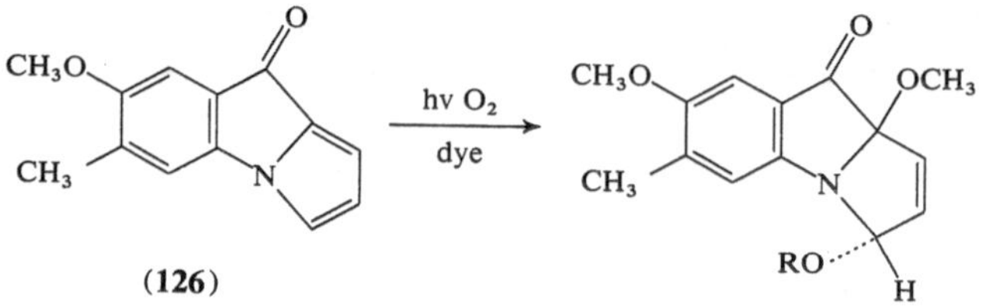

(126)

hv O₂
dye

(127) R = OH
(128) R = H

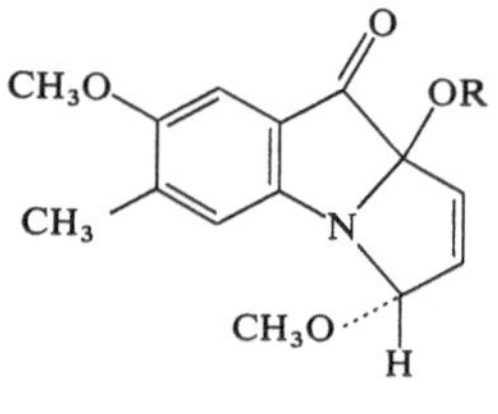

(129)

The Fordham group has also used photo-oxidation for introducing angular oxygen into mitosane-like molecules as shown in Scheme XVI (73). The carbon skeleton was first established by condensing the pyrrole Grignard reagent, whose most nucleophilic site is the α-carbon, with styrene oxide (130). The resultant pyrrole (131), containing all the carbons required for a mitomycin synthesis, was photo-oxidized to yield the lactam (132). The relative configuration of (132) was determined by X-ray crystallography and shown to correspond to Mitomycin B. Equilibration of (132) with acidic methanol afforded a mixture from which (133), whose stereochemistry corresponds to that of mitomycin A was obtained. (132) and (133) were converted to quinones (134) and (135) respectively. Neither quinone nor various derivatives of the quinones could be cyclized to tricyclics such as (136). Clearly, this cyclization attempt, as the earlier MANDELL attempt (109→110), was "5-endo-trig" and disfavored.

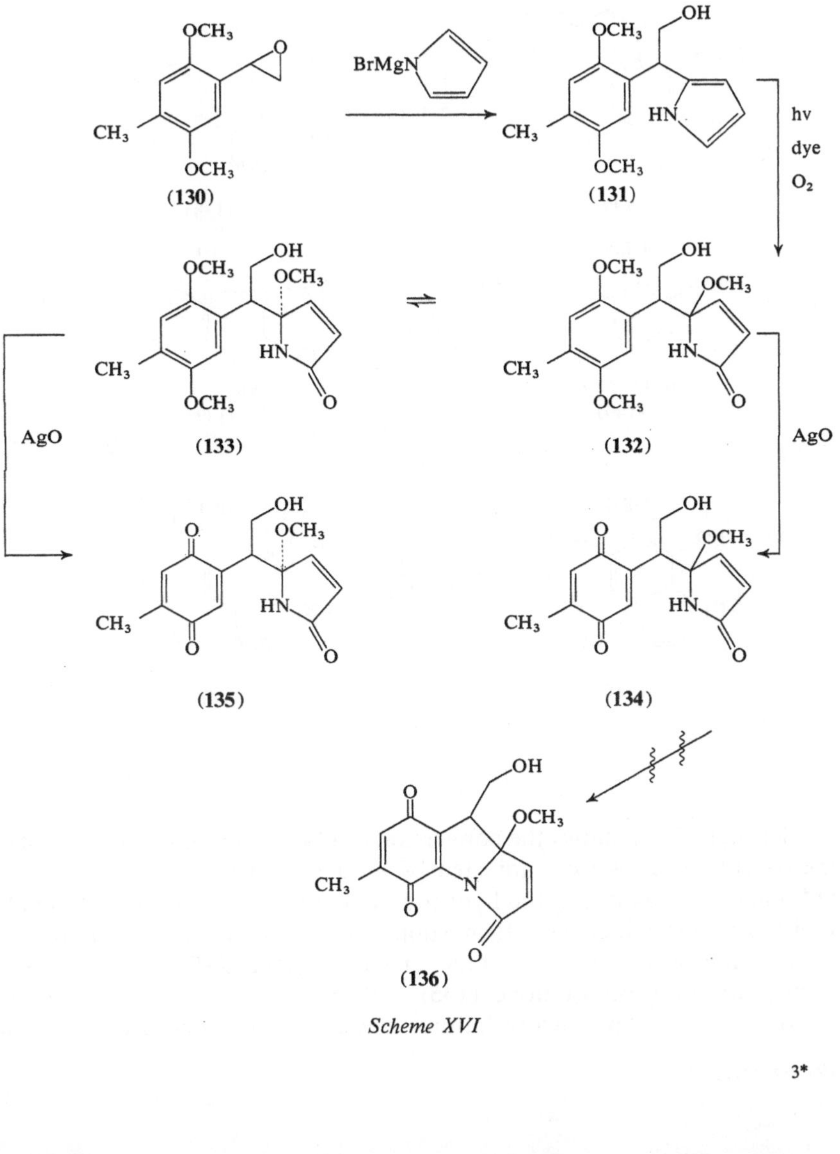

Scheme XVI

A group at Harvard has successfully completed total syntheses of
mitomycins A, C and porfiromycin as well as of a desaziridinomito-
mycin. Scheme XVII illustrates assembly of the carbon framework.
Carbon-carbon bond forming steps were accomplished by Claisen rearrangement of ether (137) (cf. Scheme XIV), opening of epoxide (139)
which was derived from the Claisen product (138), with the anion of
acetonitrile, and lastly, a crossed aldol reaction of formaldehyde with
ketone (141).

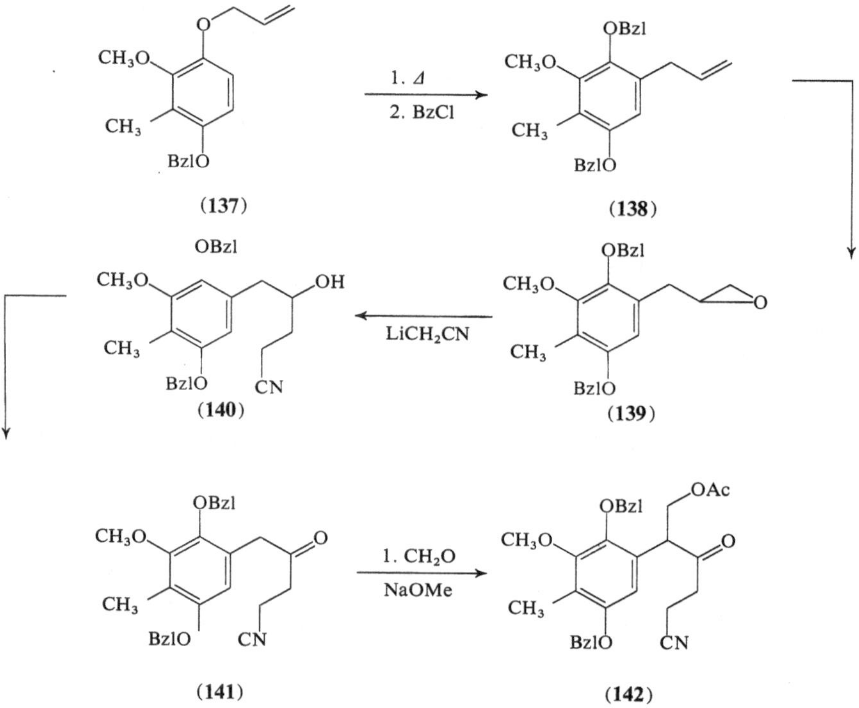

Scheme XVII

Scheme XVIII shows the conversion of (142) to (150) and (151). First,
the oxidation levels were adjusted by reducing nitrile (143) to an amine
(144) and then oxidizing the hydroquinone, derived by deblocking (144),
to its quinone which was then subject to spontaneous intramolecular
nucleophilic attack by the amino group to yield (145). The successful
cyclization to form quinone (145) with its 8-membered ring may be
contrasted with two previous failures where the desired cyclization would

have formed 5-membered rings (Scheme XV and **109→110**). The key step in the deiminomitomycin synthesis was the transannular cyclization of **(147)** to **(148)** and **(149)** where, in contrast to earlier work *(vide supra* **90→91)** the angular oxygen was retained rather than spontaneously eliminated to form an indole. KISHI noted, however, that isomer **(148)** did form an indole simply upon chromatography on Al$_2$O$_3$ while **(149)** survived Al$_2$O$_3$ but was rapidly converted to an indole on silica gel. It should be noted that the elimination products from **(150)** and **(151)** had been prepared previously [see **(60)** above] *(74)*.

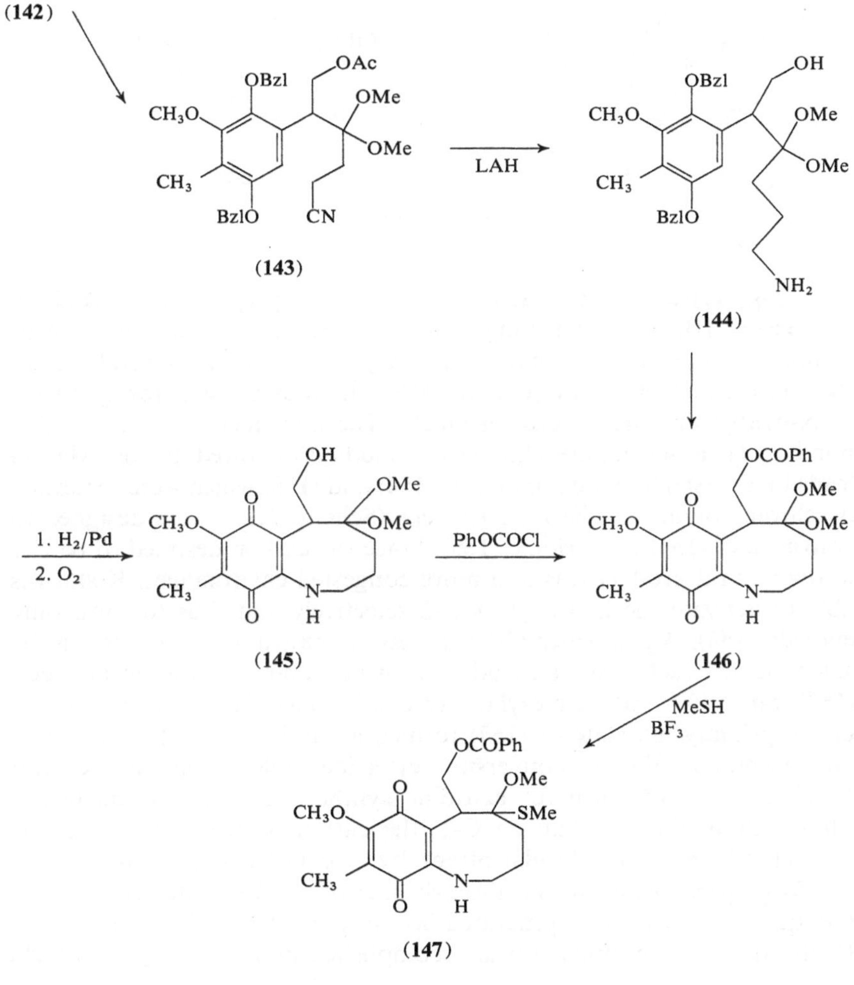

(142)

(143)

LAH

(144)

1. H$_2$/Pd
2. O$_2$

(145)

PhOCOCl

(146)

MeSH
BF$_3$

(147)

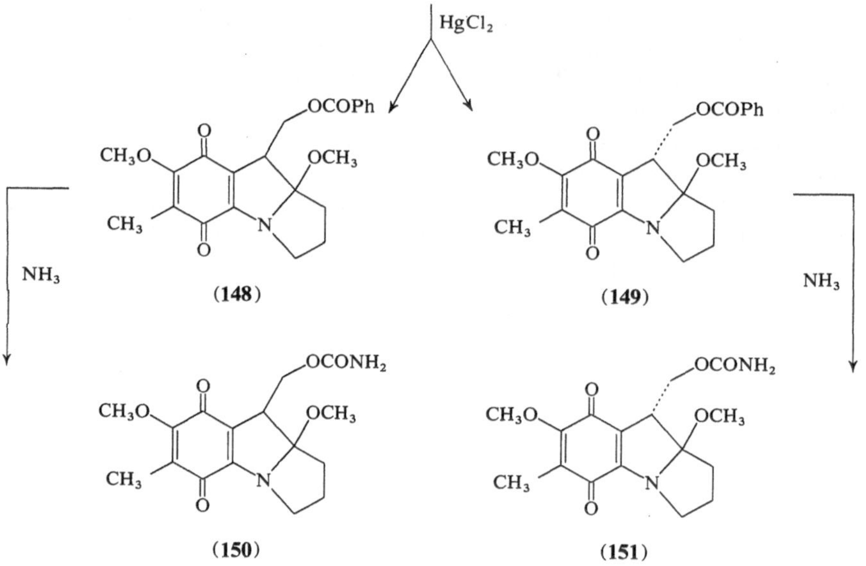

Scheme XVIII

Using derivative (152) (Scheme XIX), easily prepared from (142) in Scheme XVIII, as a branching point, the Harvard group was able to complete syntheses of the natural mitomycins. There are several crucial steps in the conversion of (152) to (159), the branch point for syntheses of N-methyl or N-H natural products. The introduction of the double bond using a selenoxide elimination method followed by osmylation leads to a diastereomeric mixture of (154) and (155) which were separated by chromatography. The next 10 steps (Scheme XIX) were designed to convert diol (154) to aziridine (159). Since the carbon destined to be C-1 in the natural product was in a more congested environment, KISHI was able to activate the hydroxyl at C-2 selectively and thus to form only epoxide (156). Again, since C-2 was less congested, epoxide opening by azide yielded exclusively an azide diol which upon mesylation afforded (157). Since C-1 and its mesylate were still congested, the displacement of the primary mesylate of (157) formed an amine at the future C-3 of the antibiotic. Finally, conversion of azide (158) to an amine with P(OCH₃)₃ set the stage for aziridine synthesis by intramolecular displacement of the mesylate on C-1, the carbon which had been safely protected from intermolecular attack by its congested environment.

To prepare porfiromycin, the aziridine of (159) was methylated and the quinone (160) was generated as in the desaziridino series (75). Cyclization to a mitomycin was accomplished by acid catalysis to yield

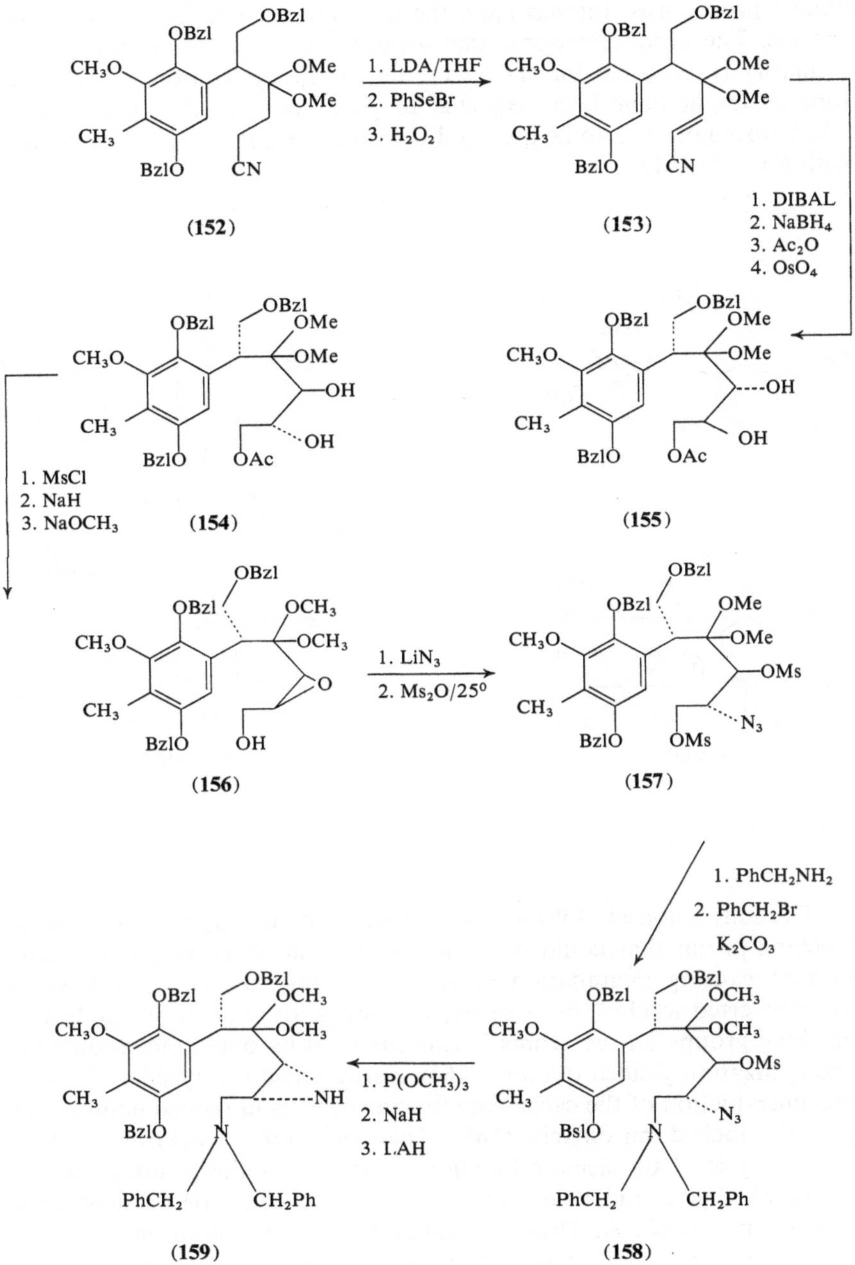

Scheme XIX

(161) which was then converted to the racemic natural product (3) by known procedures. Interestingly, the transannular cyclization is stereo-specific. The authors propose that ground-state conformation (160a) is primarily responsible for the specificity of the cyclization. Other con-formers might have been expected to yield some of the epimer (162) which does not seem to be greatly disfavored when its model is compared with that of (161).

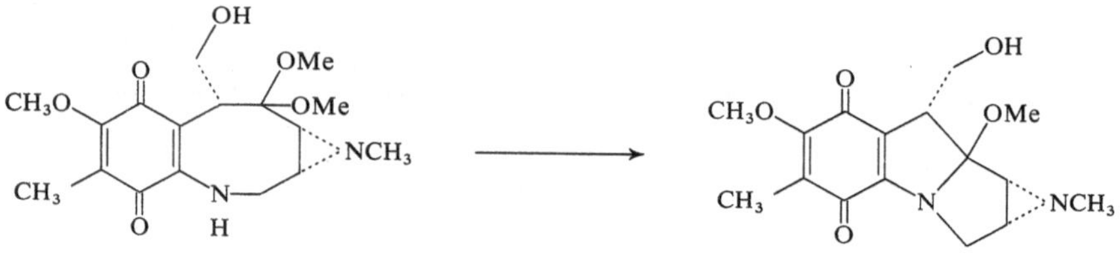

(160)

(161)

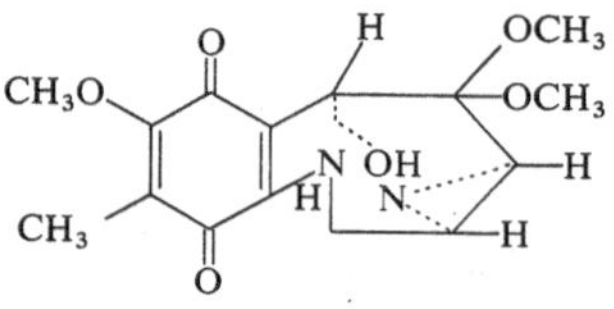

(160a)

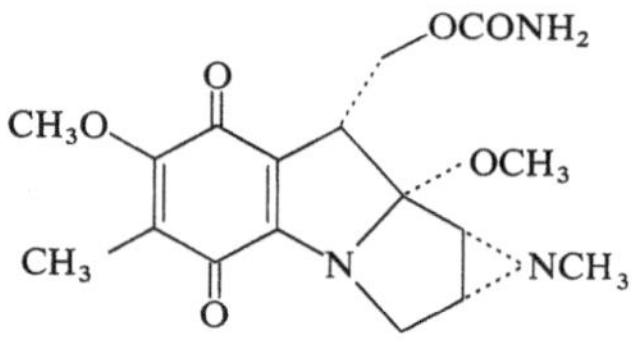

(162)

The conversion of (159) to mitomycins A and C required an aziridine blocking group that could be removed at a late stage of the synthesis without causing degradation of the entire molecule (76). Thus (159) was converted to (163) in three steps. Subsequent removal of the benzyl blocking groups on the aromatic ring followed by oxidation to quinone and cyclization yielded quinone (164). Further acid-catalyzed cyclization and introduction of the carbamate functionality as in related compounds yielded blocked mitomycin (165). The deblocking sequence involved methanolysis of the acetate function, oxidation of the primary alcohol to the aldehyde, and acid-catalyzed reverse Michael reaction to yield racemic mitomycin A. Thus, 21 years after their isolation and 15 years after their structural characterization, a laboratory synthesis of the mitomycins has been accomplished.

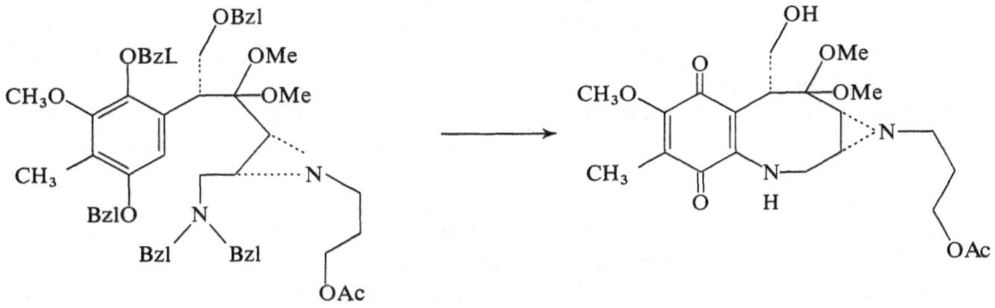

(163)

(164)

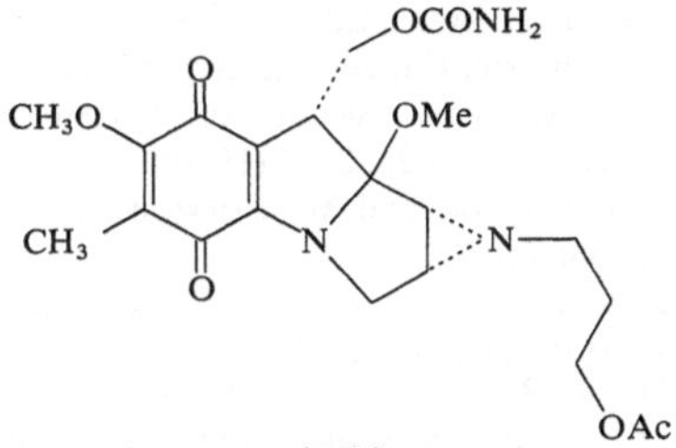

(165)

References

1. HATA, T., Y. SANO, R. SUGAWARA, A. MATSUMAE, K. KANAMORI, T. SHIMA, and T. HOSHI: Mitomycin, a New Antibiotic from Streptomyces. I. J. Antibiotics, Ser. A. **9**, 141 (1956).
2. SUGAWARA, R., and T. HATA: Mitomycin, a New Antibiotic from Streptomyces. II. Description of the strain. J. Antibiotics, Ser. A. **9**, 147 (1956).
3. WAKAGI, S., H. MARUMO, K. TOMOKA, G. SHIMIZU, E. KATO, H. KAMADA, S. KUDO, and Y. FUJIMOTO: Isolation of News Fractions of Antitumor Mitomycins. Antibiot. and Chemoth. **8**, 228 (1958).
4. WAKAGI, S.: Identification and Classification of Antitumor Mitomycin Group. Gann. **49** (Suppl.), 10 (1958).
5. HERR, R. R., M. E. BERGY, T. E. EBLE, and H. K. JAHNKE: Porfiromycin, a New Antibiotic. II. Isolation and Characterization. Antimicrobial Agents Ann. **1960**, 23 (1961).
6. LEFEMINE, D. V., M. DANN, F. BARBATSCHI, W. K. HAUSMANN, V. ZBINOVSKY, P. MONNIKENDAM, J. ADAM, and N. BOHONOS: Isolation and Characterization of Mitiromycin and Other Antibiotics Produced by Streptomyces Verticillatus. J. Amer. Chem. Soc. **84**, 3184 (1962).
7. UZU, K., Y. HARADA, and S. WAKAKI: Mitomycins, Carcinostatic Antibiotics. I. Derivatives and Acid Hydrolysis of Mitomycin A and C. Agr. Biol. Chem. **28**, 388 (1964).
8. UZU, K., Y. HARADA, S. WAKAKI, and Y. YAMADA: Mitomycins, Carcinostatic Antibiotics. II. Mitomycinone. Agr. Biol. Chem. **28**, 394 (1964).

9. WEBB, J. S., D. B. COSULICH, J. H. MOWAT, J. B. PATRICK, R. W. BROSCHARD, W. E. MEYER, R. P. WILLIAMS, C. F. WOLF, W. FULMOR, C. PIDACKS, and J. E. LANCASTER: The Structures of Mitomycins A, B, and C and Porfiromycin-Parts I and II. J. Amer. Chem. Soc. **84**, 3185 (1962).

10. STEVENS, C. L., K. G. TAYLOR, M. E. MUNK, W. S. MARSHALL, K. NOLL, G. D. SHAH, L. G. SHAH, and K. UZU: Chemistry and Structure of Mitomycin C. J. Med. Chem. **8**, 1 (1965).

11. TULINSKY, A.: The Structure of Mitomycin A. J. Amer. Chem. Soc. **84**, 3188 (1962).

12. TULINSKY, A., and J. H. VAN DEN HENDE: The Crystal and Molecular Structure of N-Brosylmitomycin A. J. Amer. Chem. Soc. **89**, 2905 (1967).

13. YAHASHI, R., and I. MATSUBARA: The Molecular Structure of 7-Demethoxy-7-p-bromo-anilinomitomycin B. J. Antibiotics **29**, 104 (1976).

14. WEBB, J. S.: Private Communication.

15. LOWN, J. W., and A. BEGLEITER: Studies Relating to Aziridine Antitumor Antibiotics. Part II. ^{13}C and ^{1}H Nuclear Magnetic Resonance Spectra of Mitomycin C and Structurally Related Streptonigrin. Canad. J. Chem. **52**, 2331 (1974).

16. VAN LEAR, G.: Mass Spectrometric Studies of Antibiotics — I. Mass Spectra of Mitomycin Antibiotics. Tetrahedron **26**, 2587 (1970).

17. MORTON, G. O., G. E. VAN LEAR, and W. FULMOR: The Structure of Mitiromycin. J. Amer. Chem. Soc. **92**, 2588 (1970).

18. NOMURA, S., M. YAMAMOTO, I. UMESWARA, A. MATSUMAE, and T. HATA: Studies on G-253 Substances, New Antibiotics from Streptomyces. I. J. Antibiotics **20**, 55 (1967).

19. TAYLOR, W. G., and W. A. REMERS: Structure and Stereochemistry of Some 1,2-Di-substituted Mitosenes from Solvolysis of Mitomycin C and Mitomycin A. J. Med. Chem. **18**, 307 (1975).

20. CHENG, L., and W. A. REMERS: Comparative Stereochemistry in the Aziridine Ring Openings of N-Methylmitomycin A and 7-Methoxy-1,2-(N-methylaziridino)mitosene. J. Med. Chem. **20**, 767 (1977).

21. KINOSHITA, S., K. UZU, K. NAKANO, and T. TAKAHASHI: Mitomycin Derivatives. 2. Derivatives of Decarbamoylmitosane and Decarbamoylmitosene. J. Med. Chem. **14**, 109 (1971).

22. — — — — New Derivatives of Mitomycins: Decarbamoylmitomycin, Demethoxy-mitomycin, and N-Methylmitomycin B. Progress in Antimicrobial and Anticancer Chemotherapy, Vol. II (Proceedings of the 6th International Congress of Chemotherapy), p. 112—115. Baltimore: University Park Press. 1970.

23. KISHI, Y.: Harvard University, unpublished results.

24. NAKANO, K.: Kyowa Hakko Kogyo, unpublished results.

25. KINOSHITA, S., K. UZU, K. NAKANO, M. SHIMIZU, T. TAKAHASHI, and M. MATSUI: Mitomycin Derivatives. 1. Preparation of Mitosane and Mitosene Compounds and Their Biological Activities. J. Med. Chem. **14**, 103 (1971).

26. PATRICK, J. B., R. P. WILLIAMS, W. E. MEYER, W. FULMOR, D. B. COSULICH, R. W. BROSCHARD, and J. S. WEBB: Aziridinomitosenes; A New Class of Antibiotics Related to the Mitomycins. J. Amer. Chem. Soc. **86**, 1889 (1964).

27. MOORE, H. W.: Bioactivation as a Model for Drug Design: Bioreductive Alkylation. Science **197**, 527 (1977).

28. IYER, V. N., and W. SZYBALSKI: Mitomycins and Porfiromycin: Chemical Mechanism of Activation and Cross-Linking of DNA. Science **145**, 55 (1964).

29. RAO, G. M., J. W. LOWN, and J. A. PLAMBECK: Electrochemical Studies of Antitumor Antibiotics. I. Cyclic Voltammetric Study of Mitomycin B. J. Electrochem. Soc. **124**, 195 (1977).

30. RAO, G. M., A. BEGLEITER, J. W. LOWN, and J. A. PLAMBECK: Electrochemical Studies of Antitumor Antibiotics. II. Polarographic and Cyclic Voltammetric Studies of Mitomycin. C. J. Electrochem. Soc. **124**, 199 (1977).

31. Tomasz, M., C. M. Mercado, J. Olson, and N. Chatterjie: The Mode of Interaction of Mitomycin C with Deoxyribonucleic Acid and Other Polynucleotides in Vitro. Biochemistry 13, 4878 (1974).

32. Lown, J. W., A. Begleiter, D. Johnson, and A. R. Morgan: Studies Related to Antitumor Antibiotics. Part V. Reaction Mitomycin C with DNA Examined by Ethidium Fluorescence Assay. Canad. J. Biochem. 54, 110 (1976). — Hsiung, H., J. W. Lown, and D. Johnson: Effects of Alkylation by Dimethyl Sulfate, Nitrogen Mustard, and Mitomycin C on DNA Structure as Studied by the Ethidium Binding Assay. Canad. J. Biochem. 54, 1047 (1976).

33. Moriguchi, I., and K. Komatsu: Adaptive Least Squares Classification Applied to Structure-Activity Correlation of Antitumor Mitomycin Derivatives. Chem. Pharm. Bull. 25, 2800 (1977).

34. Weissbach, A., and A. Lisio: Alkylation of Nucleic Acids by Mitomycin C and Porfiromycin. Biochemistry 4, 196 (1965).

35. Lipsett, M. N., and A. Weissbach: The Site of Alkylation of Nucleic Acids by Mitomycin C. Biochemistry 4, 206 (1965).

36. Tomasz, M.: Novel Assay of 7-Alkylation of Guanine Residues in DNA. Application to Nitrogen Mustard, Triethylenemelamine and Mitomycin C. Biochim. Biophys. Acta 213, 288 (1970).

37. Crooke, S. T., and W. T. Bradner: Mitomycin C: A Review. Cancer Treat. Rev. 3, 121 (1976).

38. Hornemann, U., and J. C. Cloyd: Studies on the Biosynthesis of the Mitomycin Antibiotics by Streptomyces Verticillatus. Chem. Commun. 1971, 301.

39. Hornemann, U., and M. J. Aikman: Mitomycin Biosynthesis by Streptomyces Verticillatus. Incorporation of the Amino-group of D-[^{15}N] Glucosamine into the Aziridine Ring of Mitomycin B. Chem. Commun. 1973, 88.

40. Hornemann, U., J. P. Kehrer, C. S. Nunez, and R. L. Ranieri: D-Glucosamine and L-Citrulline, Precursors in Mitomycin Biosynthesis by Streptomyces Verticillatus. J. Amer. Chem. Soc. 96, 320 (1974).

41. Kirsch, E. J., and J. D. Korshalla: Influence of Biological Methylation on the Biosynthesis of Mitomycin A. J. Bacteriol. 87, 247 (1964).

42. Bezanson, G. S., and L. C. Vining: Studies on the Biosynthesis of Mitomycin C. Canad. J. Biochem. 49, 911 (1971).

43. Hornemann, U., J. P. Kehrer, and J. H. Eggert: Pyruvic Acid and D-Glucose as Precursors in Mitomycin Biosynthesis by Streptomyces Verticillatus. Chem. Commun. 1974, 1045.

44. Kametani, T., and K. Takahashi: Synthesis of Pyrrolo[1,2-a]indoles and Related Systems. Heterocycles 9, 293 (1978).

45. Allen, Jr., G. R., J. F. Poletto, and M. J. Weiss: The Mitomycin Antibiotics. Synthetic Studies. V. Preparation of 7-Methoxymitosene. J. Organ. Chem. (USA) 30, 2897 (1965).

46. Leadbetter, G., D. L. Fost, N. M. Ekwuribe, and W. A. Remers: Mitomycin Antibiotics. Synthesis of 1-Substituted 7-Methoxymitosenes. J. Organ. Chem. (USA) 39, 3580 (1974).

47. Taylor, W. G., G. Leadbetter, D. L. Fost, and W. A. Remers: Mitomycin Antibiotics. Synthesis and Activity of 1,2-Disubstituted Mitosenes. J. Med. Chem. 20, 138 (1977).

48. Kametani, T., K. Takahashi, M. Ihara, and K. Fukumoto: Synthesis of 2,3-Dihydro-1H-pyrrolo[1,2-a]indoles by Intramolecular Nucleophilic Aromatic Substitution. J. Chem. Soc., Perkin I 1976, 389.

49. Yamada, Y., and M. Matsui: Study on the Synthesis of 7-Hydroxy-9-carbethoxy-2,3-dihydro-1H-pyrrolo[1,2-a]indole. Agr. Biol. Chem. 35, 282 (1971).

50. Takada, T., and M. Akiba: Synthesis of 1H-Pyrrolo[1,2-a]indole Derivatives. III. Synthesis of 2,3-Dihydro-7-hydroxy-6,9-dimethyl-5,8-dioxo-1H-pyrrolo[1,2-a]indole. Chem. Pharm. Bull. (Japan) 20, 1785 (1972).

51. TAKADA, T., Y. KOSUGI, and M. AKIBA: Reactions of Acylaminoquinone Tosyl-hydrazones 3. Simple Syntheses of 7-Substituted Pyrrolo[1,2-a]indole quinones and Related Compounds. Chem. Pharm. Bull. (Japan) **25**, 543 (1977).

52. AKIBA, M., Y. KOSUGI, and T. TAKADA: Reactions of Acylaminoquinone Tosyl-hydrazones. Synthesis of Pyrrolo[1,2-a]indoloquinones via Benzoxazoline by Thermolysis and Photolysis. Heterocycles **6**, 1125 (1977).

53. REBEK, JR., J., and J. C. E. GEHRET: A Synthetic Approach to the Mitosenes. Tetrahedron Letters **1977**, 3027.

54. ANDERSON, W. K., and P. F. COREY: 1,3-Dipolar Cycloaddition Reactions with Isatin-N-acids. Synthesis of Dimethyl 9-Oxo-9H-pyrrole[1,2-a]indole-1,2-dicarboxylates. J. Organ. Chem. (USA) **42**, 559 (1977).

55. GERMERAAO, P., and H. W. MOORE: Rearrangements of Azidoquinones. XII. Thermal Conversion of 2-Azido-3-vinyl-1,4-quinones to Indolequinones. J. Organ. Chem. (USA) **39**, 774 (1974).

56. LOWN, J. W., and T. ITOH: Studies Related to Antitumor Antibiotics. Part III. Synthesis of 1,2,3,4,5,6-Hexahydro-2,3-benzazocin-5-ones as Possible Intermediates in the Biosynthesis of Mitomycins. Canad. J. Chem. **53**, 960 (1975).

57. ITOH, T., T. HATA, and J. W. LOWN: Synthetic Studies on Mitomycins. An Alternative Synthesis of 2,3-Dihydro-1H and 9H-Pyrrolo[1,2-a]indoles by Transannular Ring Closure. Heterocycles **4**, 47 (1976).

58. KAMETANI, T., K. TAKAHASHI, M. IHARA, and K. FUKUMOTO: Interconversion Between Pyrrolo[1,2-a]indoles and 2,3-Benzazocin-5-ones — A Synthetic Approach to Mitomycins. Heterocycles **6**, 1371 (1977).

59. REMERS, W. A., R. H. ROTH, and M. J. WEISS: The Mitomycin Antibiotics. Synthetic Studies. VII. An Exploration of Pyrrolo[1,2-a]indole A-Ring Chemistry Directed toward the Introduction of the Aziridine Function. J. Organ. Chem. (USA) **30**, 2910 (1965).

60. FRANCK, R. W., and K. F. BERNADY: A Study of the Acylation of a Tridentate Carbanion. J. Organ. Chem. (USA) **33**, 3050 (1968).

61. HIRATA, T., Y. YAMADA, and M. MATSUI: Synthetic Studies on Mitomycins. Synthesis of Aziridino-Pyrrolo[1,2-a]indoles. Tetrahedron Letters **1969**, 20.

62. — — — Synthetic Studies on Mitomycins. Part II. Synthesis of Aziridino-pyrrolo-[1,2-a]indoles. Tetrahedron Letters **1969**, 4107.

63. KAMETANI, T., K. TAKAHASHI, Y. KIGANA, M. IHARA, and K. FUKUMOTO: Studies on the Synthesis of Heterocyclic Compounds. Part 676. Synthesis of 1-Substituted 7-Methoxymitosenes. J. Chem. Soc. Perkin I **1977**, 28.

64. AUERBACH, J., and R. W. FRANCK: A Synthesis of the Tetracyclic Mitomycin Nucleus. Chem. Commun. **1969**, 991.

65. FRANCK, R. W., and J. AUERBACH: The Singlet Oxygen Oxidation of N-Phenylpyrroles. Its Application to the Synthesis of a Model Mitomycin. J. Organ. Chem. (USA) **36**, 31 (1971).

66. SIUTA, G. J., R. W. FRANCK, and R. J. KEMPTON: Studies Directed Toward a Mitomycin Synthesis. J. Organ. Chem. (USA) **39**, 3739 (1974).

67. MANDELL, L., and E. C. ROBERTS: The Synthesis of 2-Methoxy-3-methyl-5-(2-carbethoxymethylpyrrolino)-p-benzoquinone, I. J. Hetero. Chem. **2**, 479 (1965).

68. BALDWIN, J. E., J. CUTTING, W. DUPONT, L. KRUSE, L. SILBERMAN, and R. C. THOMAS: 5-Endo-Trigonal Reactions: a Disfavoured Ring Closure. Chem. Commun. **1976**, 736.

69. DANISHEFSKY, S., and R. DOEHNER: A Route to Functionalized Mitosanes. Tetrahedron Letters **1976**, 3031.

70. AKIBA, M., Y. KOSUGI, M. OKUYAMA, and T. TAKADA: A Convenient Photosynthesis of Aziridinopyrrolo[1,2-a]benz[f]indoloquinones and Heterocyclic Quinones as Model Compounds of Mitomycins by a One-Pot Reaction. J. Organ. Chem. (USA) **43**, 181 (1978).

71. WOODWARD, R. B., and R. HOFFMANN: The Conservation of Orbital Symmetry, pp. 38—64. Weinheim: Verlag Chemie, GmbH. 1970.

72. KAMETANI, T., T. OHSAWA, M. IHARA, and K. FUKUMOTO: Studies on the Synthesis of Heterocyclic Compounds. Part 738. Photo-oxygenation of 9-Oxo-9 H-pyrrolo[1,2-a]-indoles. J. Chem. Soc. Perkin I **1978**, 460.

73. FRANCK, R. W., K. MIYANO, and J. F. BLOUNT: Approaches to the Mitomycins 4 4a-Secodeiminoquinones. Heterocycles **9**, 807 (1978).

74. NAKATSUBO, F., A. J. COCUZZA, D. E. KEELEY, and Y. KISHI: Synthetic Studies toward Mitomycins. 1. Total Synthesis of Deiminomitomycin A. J. Amer. Chem. Soc. **99**, 4835 (1977).

75. NAKATSUBO, F., T. FUKUYAMA, A. J. COCUZZA, and Y. KISHI: Synthetic Studies toward Mitomycins. II. Total Synthesis of Porfiromycin. J. Amer. Chem. Soc. **99**, 8115 (1977).

76. FUKUYAMA, T., F. NAKATSUBO, A. J. COCUZZA, and Y. KISHI: Synthetic Studies Toward Mitomycins. III. Total Syntheses of Mitomycins A and C. Tetrahedron Letters **1977**, 4295.

77. HORNEMANN, U., Y.-K. HO, J. K. MACKEY, JR., and S. C. SRIVASTAVA: Studies on the Mode of Action of the Mitomycin Antibiotics. Reversible Conversion of Mitomycin C into Sodium 7-Aminomitosane-9a-sulfonate. J. Amer. Chem. Soc. **98**, 7069 (1976).

(Received July 20, 1978)

The Biogenesis and Chemistry
of Sesquiterpene Lactones

By N. H. FISCHER, E. J. OLIVIER, and H. D. FISCHER,
Department of Chemistry, Louisiana State University,
Baton Rouge, Louisiana, U.S.A.

With 4 Figures

Contents

I. Introduction

It is the major goal of this review to present the presently known naturally occurring sesquiterpene lactones and their distribution in the plant kingdom. Furthermore, biogenetic hypotheses indicating the possible biosynthetic relationships among the different skeletal types of sesquiterpene lactones will be discussed and selected modifications and transformation of the various classes of compounds are presented.

With the exception of some earlier reports, most publications related to the isolation, structure elucidation and chemistry of sesquiterpene lactones have appeared during the last two decades. Early studies in the field were summarized by Korte (*607a*). The germacranolides and guaianolides were reviewed by Sorm (*996, 998*), the pseudoguaianolides by Romo and Romo de Vivar (*879*), and more recently, by Herz (*394, 395*). The eremophilanolides and bakkenolides were recently discussed by Pinder (*834*). A book by Yoshioka, Mabry and Timmermann (*1172*) presents the major skeletal types of sesquiterpene lactones with NMR spectra of over 200 naturally occurring sesquiterpene lactones, besides brief introduction to some typical chemical transformations of the various types of lactones. Reviews dealing with biogenesis of the various skeletal types of sesquiterpene lactones have appeared over the years (*265, 273, 288, 290, 393, 395*).

Reports dealing with isolation and structure elucidation of sesquiterpene lactones have increased dramatically during the last decade.

Table I-1. *Number of Skeletal Types of Naturally Occurring Sesquiterpene Lactones*

Skeletal Type		Number of Compounds	Remarks
Germacranolides:		273	Includes one 2,3-*seco*-germacrolide
A. Germacrolides	123		
B. Melampolides	26		
C. Heliangolides	56		
D. *cis,cis*-Germacranolides	6		
E. Germafurenolides	12		
F. Others	62		
Eudesmanolides:		143	Includes 3 *seco*- and 6 *ent*-compounds
Guaianolides:		196	
Xanthanolides:		23	
Pseudoguaianolides:		138	
A. Ambrosanolides	54		Includes 7 *seco*-dilactones
B. Helenanolides	77		Includes 10 *seco*-compounds
C. Others	7		
Elemanolides:		30	
Eremophilanolides:		23	
Fukinanolides (Bakkenolides):		12	
Drimanolides:		22	
Bisabenolides:		3	
Tutinanolides:		20	
Special Structural Types:		41	
Total Number:		924	

Two reasons can be given for the strongly increasing interest in this group of natural products. First, sesquiterpene lactones have been successfully used as markers in biochemical systematic (chemotaxonomy) studies, mainly in the Compositae (*299, 384, 386, 392, 394, 395, 397, 662, 663, 1172*). Second, more recently a number of compounds received considerable attention due to their various biological activities which resulted in an additional increase in biological activity related publications [major references include: (36), (183), (371a), (375), (575), (616—634), (636—654), (658), (859), (860), (862), (864), (1099)]. At the end of 1977, nearly 950 sesquiterpene lactones were known. The numbers of naturally occurring compounds that belong to each of the major skeletal types of sesquiterpene lactones are recorded in Table I-1.

II. Germacradiene-Derived Sesquiterpene Lactones

1. Skeletal Types of Sesquiterpene Lactones

The biogenetic theory of terpenoids advanced first by Ruzicka (911, 912) and later by Hendrickson (382) and others is now well established, in particular due to steroid and triterpenoid biosynthetic work (765). An increasing number of experimental data have become available which also permit acceptance of the idea that sesquiterpenoid biosynthesis proceeds via the mevalonate-isopentyl pyrophosphate-farnesyl pyrophosphate pathway (765).

It is an interesting fact in terpenoid biosynthesis that the earlier biosynthetic stages involve reductive processes whereas subsequent biomodifications mainly represent oxidative reactions. Recent isolations of sesquiterpene lactone hydroperoxides by Doskotch (230) and El-Feraly (249) suggest that hydroperoxides may be intermediates in the biosyntheses of the various skeletal types of hydroxylated terpenoids which generally involve allylic oxidations. Furthermore, isolation of many sesquiterpene lactone epoxides from plants indicates their involvement in major biosynthetic sequences of natural products, in particular in cyclization reactions. Based on biogenetic assumptions it is now generally accepted that sesquiterpene lactones are derived from farnesyl- or nerolidyl pyrophosphates (182, 288, 290, 392, 393, 395, 818).

The large number of skeletal types of sesquiterpenoids in general (217) is contrasted by a relatively small group of different ring systems among the sesquiterpene lactones. Their classification is based on their carbocyclic skeleton in which the suffix "olide" refers to the lactonic function. In sesquiterpene lactones formed by oxidation of the "head" methyl group of farnesol (393) the lactonic function commonly represents an α-methylene γ-lactone moiety (1), or a biomodified functionality derived from (1).

The majority of sesquiterpene lactones belongs to this category which can be considered biogenetic derivatives of the largest class, the germacranolides (2). The presently known structural classes and names of the various carbocyclic ring systems are shown in Chart II-1 and the presumed biogenetic relationships are indicated by arrows; their detailed biogenetic considerations will be presented in the chapters on

(1)

the different types of sesquiterpene lactones. Other minor groups of sesquiterpene lactones, formed by biosynthetic routes distinctly different from the above skeletal types, will be discussed separately in later chapters.

Although both 7,6- and 7,8-lactonic ring closure is found in naturally occurring compounds, for the sake of simplicity, only the 7,6-lactonized types are presented in Chart II-1.

Numbering of the basic carbocyclic ring systems is generally found to be consistent in the literature with the exception of C-14 and C-15 which are frequently interchanged. To avoid future confusion the adoption of one system, possibly the one used in Chart II-1, is strongly recommended.

2. Biogenesis of the Germacradiene and the Lactone Ring

As outlined in Chart II-2, cyclization of *trans, trans*-farnesyl pyrophosphate (15) results in the *trans, trans*-germacradiene intermediate (16) which by enzymatic oxidative modifications provides the germacranolides represented by its simplest member, costunolide (17). From the germacradiene the different other skeletal types of sesquiterpene lactones shown in Chart II-1 can be derived.

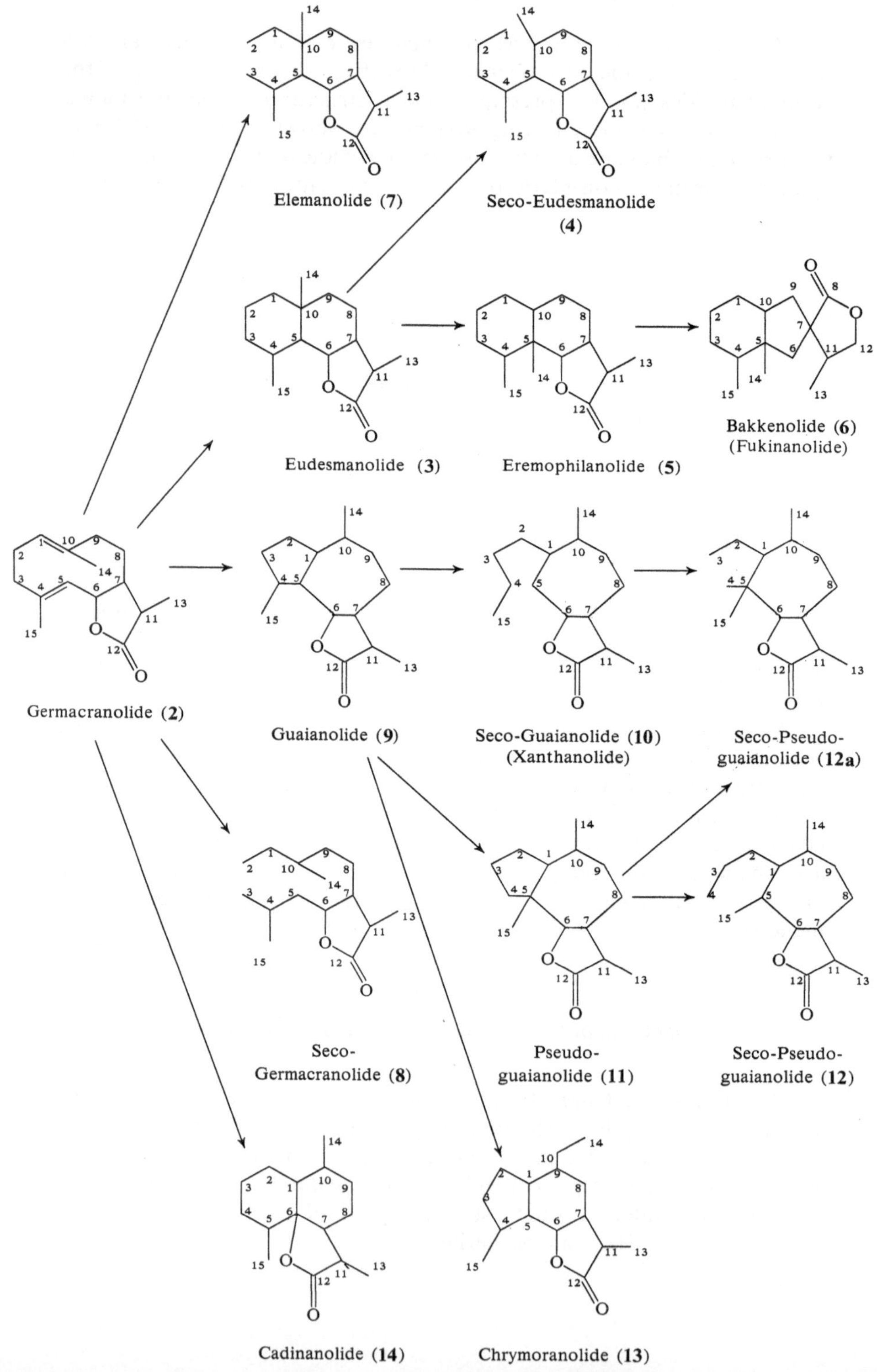

Elemanolide (7)

Seco-Eudesmanolide
(4)

Eudesmanolide (3)

Eremophilanolide (5)

Bakkenolide (6)
(Fukinanolide)

Germacranolide (2)

Guaianolide (9)

Seco-Guaianolide (10)
(Xanthanolide)

Seco-Pseudo-
guaianolide (12a)

Seco-
Germacranolide (8)

Pseudo-
guaianolide (11)

Seco-Pseudo-
guaianolide (12)

Cadinanolide (14)

Chrymoranolide (13)

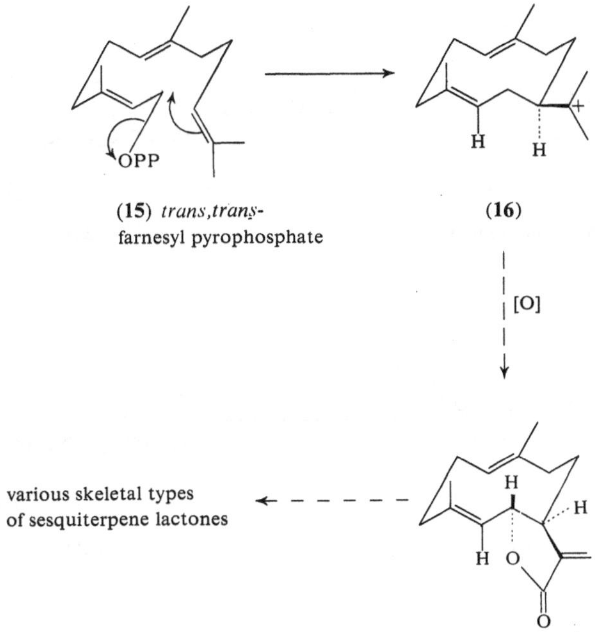

(15) *trans,trans*-
farnesyl pyrophosphate   (16)

various skeletal types
of sesquiterpene lactones

(17) costunolide

Chart II-2. Biogenesis of the germacranolide skeleton

Two possible biogenetic routes have been suggested for the formation
of the lactone ring of these sesquiterpenoids. The various schemes of
formation of the α,β-unsaturated γ-lactone of type (1) have been discussed by GEISSMAN (*288, 290*) and HERZ (*393*). Possible steps involved
in the biogenesis of costunolide (17) and inunolide (24) are outlined in
Chart II-3. The overall process requires oxidative modifications at C-12
and C-6 or C-8, respectively.

One hypothetical intermediate en route from cation (16) to the
lactones (17) and (24) is germacrene A (18) a naturally occurring
hydrocarbon in which all non-olefinic carbons are allyllically activated for
hydroxylation except C-8. Introduction of an oxygen function at C-12
in (18) to give alcohol (21) could either proceed *via* epoxide intermediate
(19) (*288*) or could involve the hydroperoxide (20), the latter being formed
by an enzymatically-mediated reaction mimicking the reaction of singlet
oxygen with olefins. In either case the process involves migration of
a double bond from what was originally C-11, C-13 to C-11, C-12. Further
oxidative modifications of (21) *via* aldehyde (22), acid (23) and hydroxylations at C-6 or C-8 would after lactonization give costunolide (17) or
inunolide (24), respectively. Although, the question regarding the

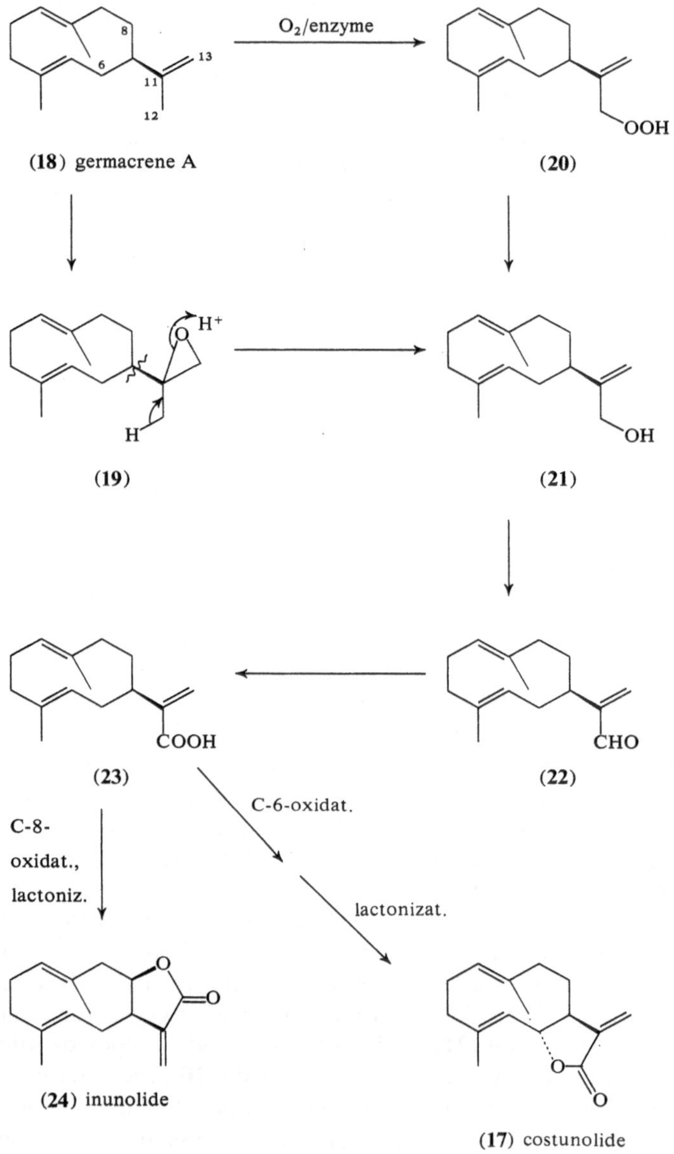
O2/enzyme
8
6
13
11
12
(18) germacrene A
OOH
(20)
H+
O
H
(19)
OH
(21)
COOH
(23)
CHO
(22)
C-8-
oxidat.,
lactoniz.
C-6-oxidat.
lactonizat.
O
O
(24) inunolide
O
O
(17) costunolide

Chart II-3. Biogenesis of the lactone ring

sequence of oxidations and the detailed mechanism remains open, the general routes outlined in Chart II-3 appear reasonable, since sesquiterpenes with oxidation patterns of the isopropenyl side chain indicated in (21) to (23) occasionally accompany the lactonic plant constituents (*288*).

Germacrene B (25) should not be excluded as a possible precursor in lactone biosynthesis, since C-8 hydroxylation in a sesquiterpene lactone precursor of type (25) would now be favored due to allylic activation of C-8.

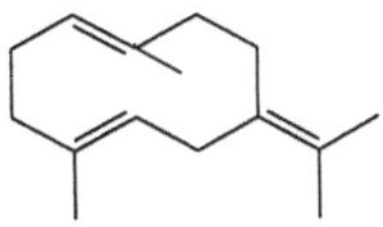

(25)

Furthermore, C-6 in (25) represents a doubly allylic carbon center favoring hydroxylation at this position over all other allylic carbons. This could possibly be the reason for predominant formation of C-6-oxygenated sesquiterpenoids. Sesquiterpene lactones of type (27) commonly cooccur with and are derived from furanosesquiterpenes (26) by autoxidation, suggesting that the lactones are also biogenetically derived from the furan ring adumbrated in Chart II-4:

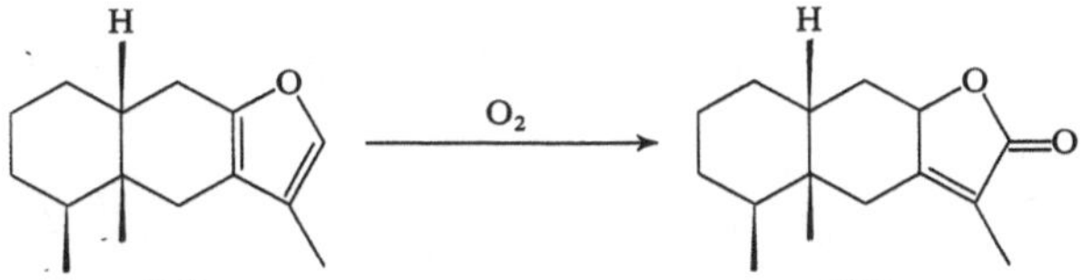

(26)   (27)
                        an eremophilanolide

Chart II-4. Biogenesis of the lactone ring *via* furanosesquiterpenes

A number of laboratory analogies (for a recent review see Ref. *834*) strengthen the furan route for these types of lactones which possibly involves hydroperoxide intermediates in the biogenetic oxidation process as suggested by *in vitro* photosensitized oxygenations (*756, 759*).

3. Common Ester Side Chains in Sesquiterpene Lactones

Sesquiterpene lactones frequently occur as acetates or carry other ester side chains. In order to save space, abbreviations will be used for most side chains throughout this review. The structures of common side chains and their abbreviations are summarized in Table II-1.

Table II-1. *Common Side Chains in Sesquiterpene Lactones*

Structure of Side Chain	Type of Ester	Abbreviation
	acetate	Ac
	propionate	Pro
	isobutyrate	i-But
	methacrylate	Mac
	epoxymethacrylate	Epoxymac
	4-hydroxymethacrylate	Mac-4-OH
	isovalerate	i-Val
	senecioate	Sen

Table II-1 *(continued)*

Structure of Side Chain	Type of Ester	Abbreviation
	2-methylbutanoate	2-Mebut
	tiglate	Tig
	angelate	Ang
	epoxyangelate	Epoxyang
	sarracinate	Sar
	acetylsarracinate	Sarac

Besides typical NMR parameters the ester groups (R-COO-R′) generally show diagnostic mass spectral peaks. Commonly, the M-RCOOH and/or M-ketene are detected as reasonably strong peaks and the peak due to the acylium ion (RCO⁺) frequently represents the base peak.

III. Germacranolides

1. Structural Types of Germacranolides and Biogenetic Considerations

The germacranolides represent the largest group of sesquiterpene lactones with nearly 300 known naturally occurring members. Several reviews related to various aspects of this interesting group of compounds have appeared (*134, 273, 288, 384, 386, 392, 393, 995—998, 1042, 1043, 1172*). Considerable structural variety is recognized within this class of compounds. The variety is mainly due to the unique configurational and conformational features and the reactivity of the cyclodecadiene skeleton.

Recent recognition of configurationally isomeric germacranolides (*274*) has led to a reclassification into four subgroups (*764, 1128*), which are characterized by a cyclodecadiene skeleton with double bonds in the C-1,10- and C-4,5-positions. In Chart III-1 the basic configurational types [(**28**) to (**31**)] are shown.

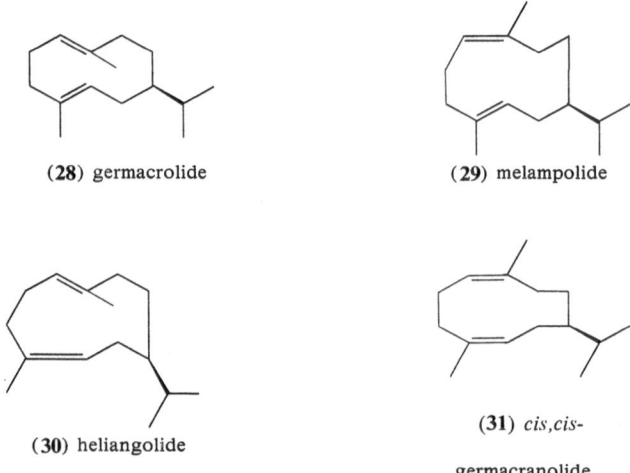

(**28**) germacrolide   (**29**) melampolide

(**30**) heliangolide

(**31**) *cis,cis*-

germacranolide

Chart III-1. Configurational types of germacranolides

Among the four germacranolide subgroups the majority of the medium ring sesquiterpene lactones respresents germacrolides (**28**), although an increasing number of melampolides (**29**) (*954*) and heliangolides (**30**) (*482*) have been isolated. The smallest group within the germacranolides, members of which have been found most recently (*98, 120, 270*), is that of the *cis, cis*-cyclodecadienes (**31**). The question remains unanswered whether the biosynthesis of the four cyclodecadiene subgroups follows independent biogenetic routes from the four possible configurationally isomeric farnesols or whether the configurational isomers (29) to (31) are formed from the germacrolide skeleton (28) by double bond isomerizations at a later stage of biosynthesis. Co-occurrence of more than one skeletal type in the same plant species or genus (270, 954) and the fact that, at least in the melampolide series, all presently known compounds have an oxidized C-14 such that they are either aldehydes or carboxylic acid derivatives (esters or lactones) could be an indication of interconversion from one type of skeletal system to another during or after oxidation of C-14 and/or C-15. Besides enzymatically controlled processes, spontaneous or photochemical double bond isomerizations could be involved.

The only known seco-germacranolide, pycnolide (302) (438, 440) and the structures of the various types of presently known naturally occurring germacranolides are summarized in Charts III-2 to III-8 (pp. 61—81). Names (in alphabetical order), physical properties (empirical formula, m. p., $[\alpha]_D$), plant sources and literature references are found in Table III-3 on pages 116—133.

The representation of the stereochemistries and conformations of germacranolides in two dimensions has in the past caused problems and continues to do so. KUPCHAN, KELSEY, and SIM (630a) proposed a convention in which the ring is drawn with H-7 in the α-configuration with the numbering of the ring running anticlockwise. More recently, ROGERS, MOSS, and NEIDLE (868) pointed out that, since the germacranolide skeleton is symmetric showing a C-2, C-7 axis, the numbering is ambiguous until the absolute configuration of a compound is known. They put forward four new rules: 1) The distinction between α and β faces is to be based on any evidence related to the asymmetry of the molecules' mode of biogenesis, e. g. positions of double bonds or their equivalents (such as epoxides), pattern of oxygen functions that are indicative of the former positions of the double bonds etc., the actual or masked double bonds being standardized in the Δ^4 and $\Delta^{1(10)}$ positions and the numbering running anticlockwise. 2) The chiralities of all dissymmetric ring carbon atoms are to be given in the R, S-notation. If only relative configurations are known one can arbitrarily set H-7α and use the notation R_a and S_a. 3) At tetrahedral ring carbon atoms the substituents are defined by the conventional symbols (wedges and broken lines) with respect to the β and α faces of the ring at the vertex concerned [see (A)] as in the sterols. Reentrant angles [as in (B)] are to be avoided at tetrahedral carbon atoms unless they correspond to reality and then (and only then) the substituents are to be drawn inside the ring. Thus, while the absolute configuration of (B) corresponds to that of (A) (rotation of C_n by 180° through the plane) it may not reflect reality as the

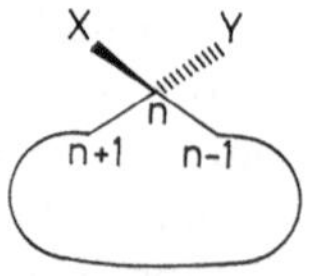

(A) 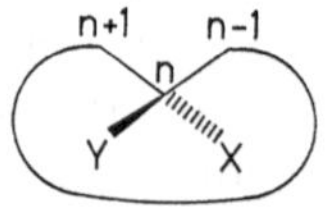   (B)

substituent X now would be interpreted as lying below the plane of the ring. 4) The perimeter of the ring may be drawn in any way that clearly portrays the *cis-trans*-character of the double bond, but keeping in mind the caution about reentrant angles. If the plane of the double bonds is known to be roughly perpendicular to the plane of the macrocycle, the wedge and broken line symbols are employed to indicate the orientation of the double bond substituents β or α to the plane of the macrocycle.

Examples of the use of these rules are given below for costunolide (**17**) and parthenolide (**80**).

An example of the confusion which may arise with regard to chiralities if the formula of a germacranolide is not written in accord with the above rules is the following. In the heliangolide nobilin (**173**) with a *trans*-7,6-

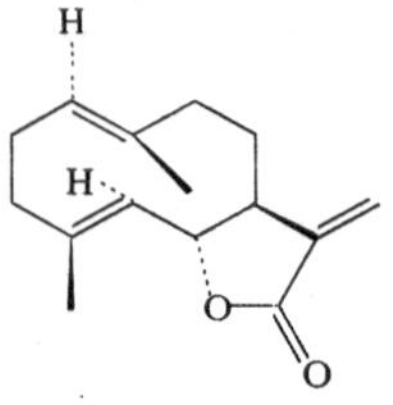

(**17b**) costunolide 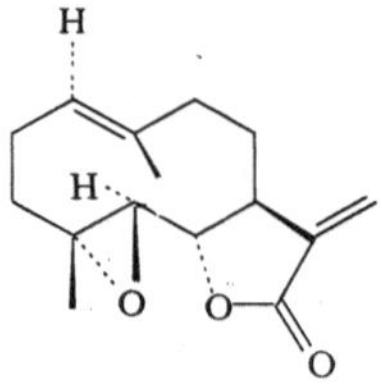   (**80**) parthenolide

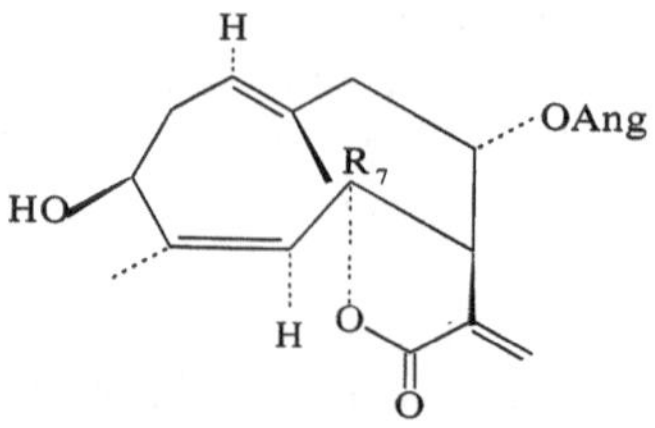

(A)   (B)

(**173**) nobilin

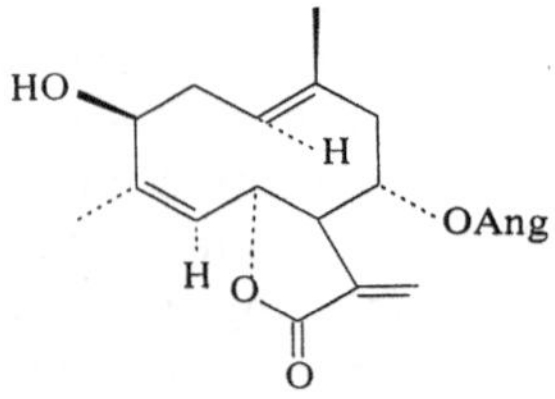

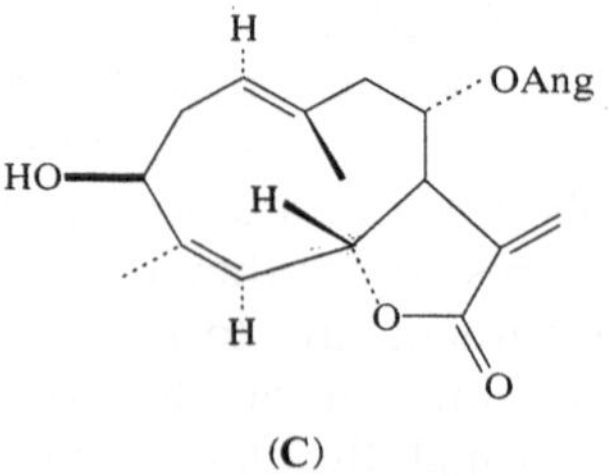

(C)

lactone ring fusion, the C-6-oxygen bond is α with respect to H-7α. An appropriate representation would be (173C) from which R-chirality can be derived for C-6. However, nobilin has also been represented by (173A) and (173B) from which an uninitiated reader would derive S-chirality for C-6, which is opposite to actual fact. The ambiguity derives from the change in depiction of C-6 [reentrant in (173A) and (173B), apical in (173C)] which has involved a change in rotation by 180° without a change in the configurational symbols (wedges or broken lines). A similar problem will exist at any chiral reentrant carbon atom in a two dimensional formula unless the substituents are written inside the ring in accord with the above rule or the chiralities specified. If the representations shown in Chart III-1 are employed, chiralities at the following carbon centres should be more clearly specified: C-5 and C-10 in the germacrolides (28), C-5 and C-9 in the melampolides (29), C-6 and C-10 in the heliangolides (30) and C-6 and C-9 in the cis, cis-germacranolides (31).

If authors wish to retain the commonly used two-dimensional representations of germacranolides it is advisable to present the stereochemical symbol in such a way that an uninitiated reader could derive the correct absolute or relative configuration at any chiral center. If only the relative stereochemistry is known the configurational symbols R_a and S_a (rule 2) in relation to an assumed H-7α, as indicated in formula (A) of (173) should be used. In rare cases where germacranolides contain a nonchiral C-7 any other chiral center could be used as a reference point.

Chart III-2. Naturally occurring 7,6-lactonized germacrolides

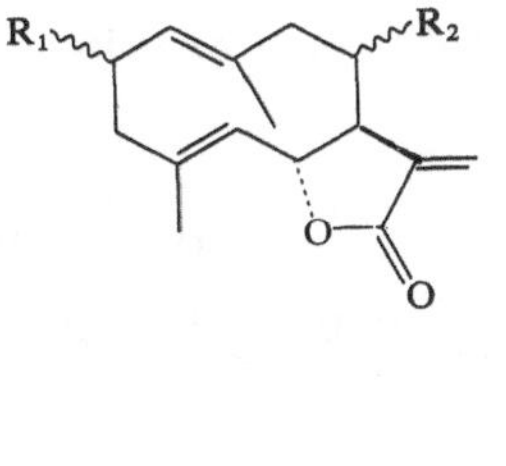

(17) Costunolide; $R_1 = R_2 = H$
(32) Tamaulipin A; $R_1 = α$-OH, $R_2 = H$
(34) Eupaserrin, desacetyl; $R_1 = α$-OH; $R_2 = β$-OSar
(35) Eupaserrin; $R_1 = α$-OH, $R_2 = β$-OSarac
(36?) Costunolide, 8-hydroxy; $R_1 = H_1$, $R_2 = α$-OH
(37) Eupatolide; $R_1 = H$, $R_2 = β$-OH
(38) Tulipinolide; $R_1 = H$, $R_2 = α$-OAc
(38a) Tulipinolide, desacetyl; $R_1 = H$, $R_2 = α$-OH
(39) Tulipinolide, epi; $R_1 = H$, $R_2 = β$-OAc
(40) Eupatoriopicrin; $R_1 = H$, $R_2 = β$-OA

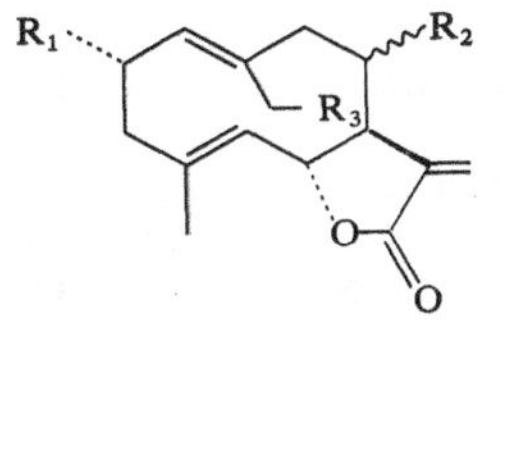

(41) Eriofertopin; $R_1 = OH$, $R_2 = β$-OMac, $R_3 = OH$
(42) Eriofertin; $R_1 = OH$, $R_2 = α$-OAng, $R_3 = OH$
(43) Eriofertopin, 2-O-Acetyl; $R_1 = OAc$, $R_2 = β$-OMac, $R_3 = OH$
(44) Costunolide, 14-hydroxy; $R_1 = R_2 = H$, $R_3 = OH$
(45) Budlein B; $R_1 = H$, $R_2 = α$-OH, $R_3 = OH$ (but see Table III-3)
(46) Ovatifolin; $R_1 = H$, $R_2 = β$-OH, $R_3 = OAc$

Chart III-2 (continued)

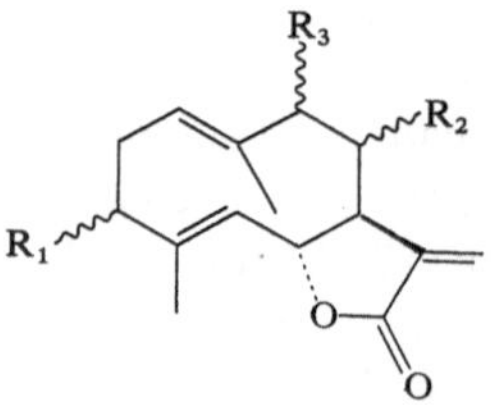

A = 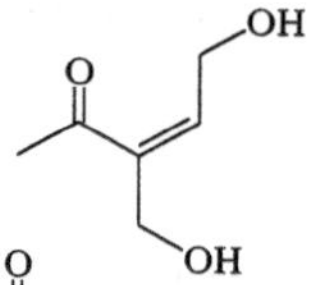

B = 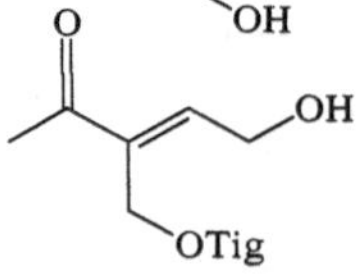

(47) Hanphyllin; R_1 = β-OH, R_2 = R_3 = H
(48) Tamaulipin B; R_1 = α-OH, R_2 = R_3 = H
(49) Novanin; R_1 = β-OAc; R_2 = R_3 = H
(50) Chihuahuin, R_1 = α-OH, R_2 = α-OAc, R_3 = H
(51) Chromolaenide, 4,5-*trans*, 3-desacetyl, 20-tiglyl R_1 = β-OH, R_2 = β-O**B**, R_3 = H
(52) Costunolide, 3β, 9α-dihydroxy, 8β-angeloyloxy; R_1 = β-OH, R_2 = β-OAng, R_3 = α-OH
(53) Haageanolide; R_1 = R_2 = H, R_3 = β-OH
(55) Costunolide, 9β-propionyloxy; R_1 = R_2 = H, R_3 = β-OPro
(56) Costunolide, 9β-isobutyryloxy; R_1 = R_2 = H, R_3 = β-O-i-But
(57) Costunolide, 9β-isovaleryloxy; R_1 = R_2 = H, R_3 = β-O-i-Val
(58) Costunolide, 9β-(2-methylbutyryloxy; R_1 = R_2 = H, R_3 = β-O-2-Mebut
(58a) Tomentosin; R_1 = β-OH, R_2 = α-OAng, R_3 = α-OH

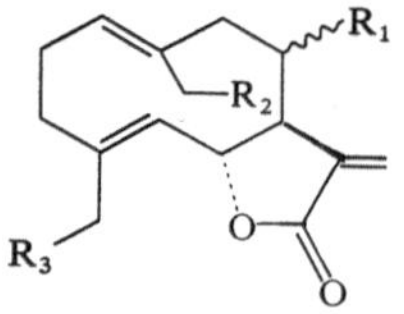

(59) Salonitenolide; R_1 = α-OH, R_2 = H, R_3 = OH
(60) Arctiopicrin; R_1 = α-O**A**, R_2 = H, R_3 = OH
(61) Onopordopicrin; R_1 = α-OMac-4-OH, R_2 = H, R_3 = OH
(62) Costunolide, 14-hydroxy-8β-(4-hydroxy-tiglyloxy); R_1 = β-O**B**, R_2 = H, R_3 = OH
(63) Cnicin; R_1 = α-O**C**, R_2 = H, R_3 = OH
(64) Alatolide; R_1 = α-O-i-But, R_2 = R_3 = OH
(65) Pectorolide; R_1 = α-OMac, R_2 = R_3 = OH
(66) Jurineolide; R_1 = α-O**D**, R_2 = R_3 = OH
(67) Albicolide; R_1 = H, R_2 = R_3 = OH
(68) Salonitenolide, 8-desoxy; R_1 = R_2 = H, R_3 = OH
(69) Costunolide, 15-isovaleryloxy; R_1 = R_2 = H, R_3 = O-i-Val
(70) Costunolide, 15-senecioyloxy; R_1 = R_2 = H, R_3 = OSen

A = 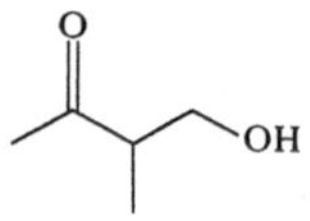 B = 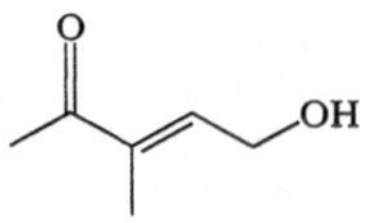 C = 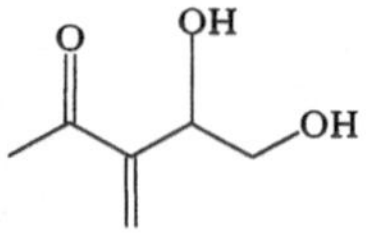 D = 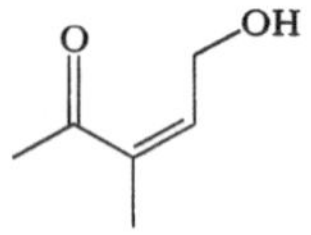

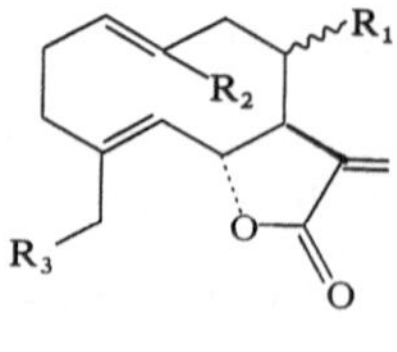

(71?) Germanin A; R_1 = O-2-Mebut, R_2 = COOH, R_3 = H
(72?) Urospermal A; R_1 = α-OH, R_2 = CHO, R_3 = OH
(73?) Urospermal B; R_1 = α-OH, R_2 = CHO, R_3 = OH
(74?) Vernopectolide B; R_1 = α-OMac, R_2 = CHO, R_3 = OH

(75) Elephantopin, deoxy; $R_1 = \alpha$-H, $R_2 = \alpha$-OMac
(76) Elephantopin, isodeoxy; $R_1 = H$, $R_2 = $ OMac

(77) Vernopectolide A

(78) Vernolide, $R = $ OMac
(79) Vernolide, hydroxy; $R = $ OMac-4-OH

(80) Parthenolide; $R_1 = R_2 = R_3 = H$
(81) Stizolin; $R_1 = $ OH, $R_2 = R_3 = H$
(82) Lanuginolide, 11,13-dehydro; $R_1 = \alpha$-OAc,
     $R_2 = R_3 = H$
(83) Lipiferolide; $R_1 = \beta$-OAc, $R_2 = R_3 = H$
(84) Eupassopilin; $R_1 = \beta$-OA, $R_2 = R_3 = H$
(85) Stizolicin; $R_1 = $ OB, $R_2 = R_3 = H$
(86) Eupassopin; $R_1 = \beta$-OA, $R_2 = $ OH, $R_3 = $ OH
(88) Eupassofilin; $R_1 = \beta$-OC, $R_2 = H$, $R_3 = $ OH
(89) Parthenolide, 9α-acetoxy; $R_1 = R_3 = H$,
     $R_2 = \alpha$-OAc
(90) Parthenolide, 9β-acetoxy; $R_1 = R_3 = H$,
     $R_2 = \beta$-OAc

(91) Elephantopin; $R = $ OMac
(92) Elephantin; $R = $ OSen

$A = $ $B = $ $C = $ $(CH_2)_{14}CH_3$

Chart III-2 (continued)

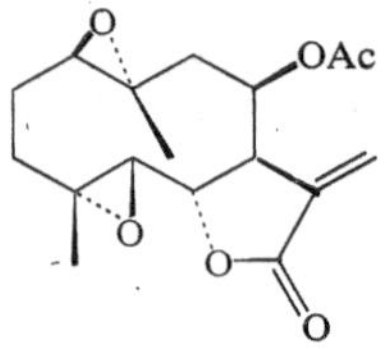

(93) Tulipinolide, epi, diepoxide

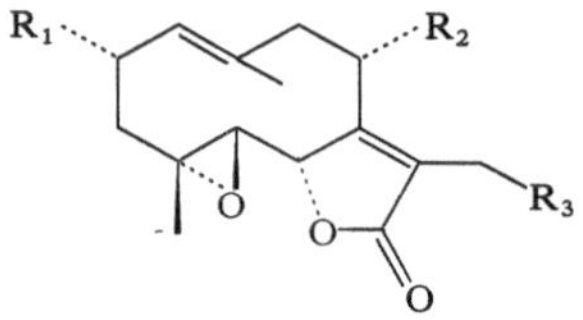

(94) Glaucolide E; R_1=OAc, R_2=OMac, R_3=OAc
(95) Glaucolide D; R_1 = OAc, R_2 = OEpoxymac, R_3 = OAc
(96) Marginatin; R_1 = H, R_2 = OTig, R_3 = OAc
(97) Glaucolide G; R_1 = H, R_2 = OAng, R_3 = OAc

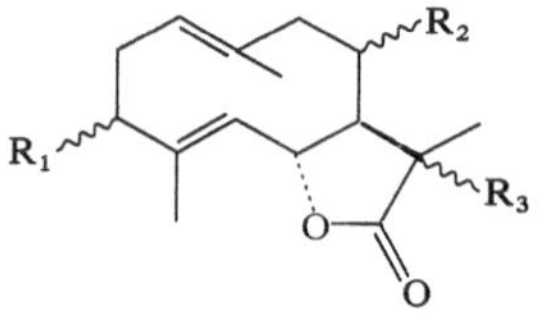

(98) Costunolide, 11,13-dihydro; R_1 = R_2 = H, R_3 = β-H
(99) Artabin; R_1 = OH, R_2 = H, R_3 = α-H
(100) Millefin; R_1 = R_2 = α-OAc, R_3 = H
(101) Carmelin; R_1 = α-OAc, R_2 = β-OAc, R_3 = α-H
(102) Balchanolide; R_1 = H, R_2 = α-OH, R_3 = β-H
(103) Balchanolide acetate; R_1 = H, R_2 = α-OAc, R_3 = β-H
(104) Balchanolide, hydroxy; R_1 = H, R_2 = α-OH, R_3 = β-OH
(105) Laserolide, R_1 = H, R_2 = OAng, R_3 = OAc

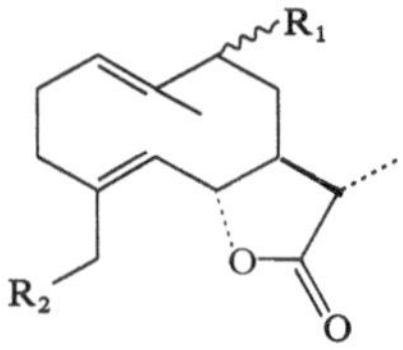

(106) Herbolide A; R_1 = β-OAc, R_2 = H
(107) Salonitenolide, β-11,13-dihydro-8-desoxy; R_1 = H, R_2 = OH

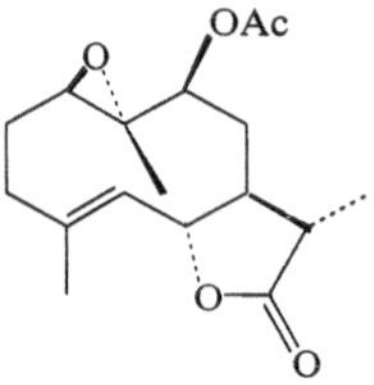

(108) Herbolide B

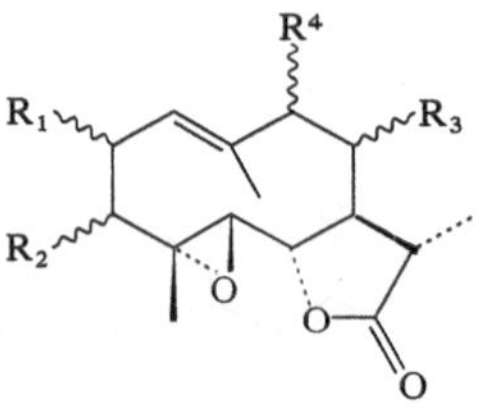

(109) Parthenolide, dihydro; R_1 = R_2 = R_3 = R_4 = H
(110) Euperfolitin; R_1 = α-OH, R_2 = β-OH, R_3 = β-OTig, R_4 = H
(111) Euperfolin; R_1 = R_4 = H, R_2 = β-OH, R_3 = β-OTig
(112) Lanuginolide; R_1 = R_2 = R_4 = H, R_3 = α-OAc
(113) Herbolide C; R_1 = R_2 = R_3 = H, R_4 = β-OAc

Chart III-3. Naturally occurring 7,8-lactonized germacrolides

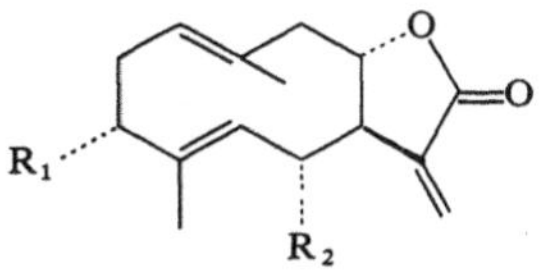

(114) Chamissonin; $R_1 = R_2 = OH$
(114a) Chamissarin, $R_1 = OH, R_2 = OAc$
(115) Laurenobiolide, desacetyl; $R_1 = H, R_2 = OH$
(116) Laurenobiolide; $R_1 = H, R_2 = OAc$

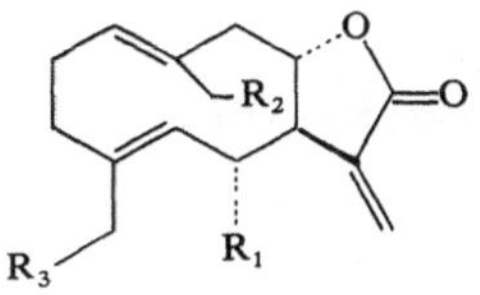

(117) Artemisiifolin; $R_1 = R_3 = OH, R_2 = H$
(118) Artemisiifolin, C-15 acetyl; $R_1 = OH, R_2 = H$, $R_3 = OAc$
(119) Scabiolide; $R_1 = OA; R_2 = H, R_3 = OAc$
(120) Dicomanolide, 14-acetoxy; $R_1 = OH$, $R_2 = R_3 = OAc$

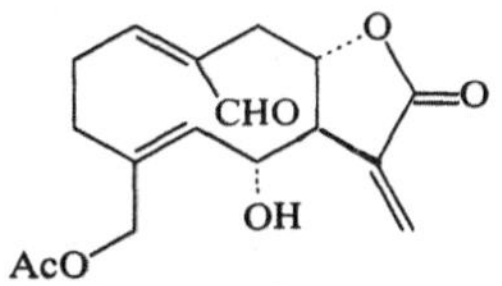

(121) Dicomanolide, 14-oxo

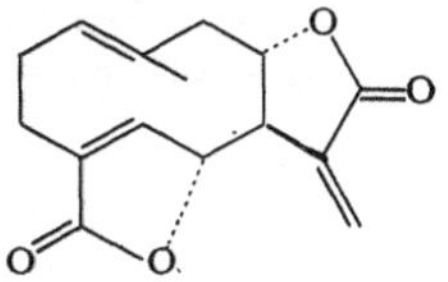

(122) Isabelin

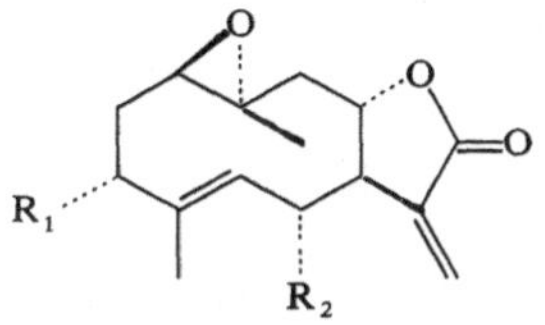

(123) Pyrethrosin; $R_1 = H, R_2 = OAc$
(123a) Chamissonin, 1 (10) epoxy; $R_1 = R_2 = OH$
(124) Tanacin; $R_1 = H, R_2 = OAng$

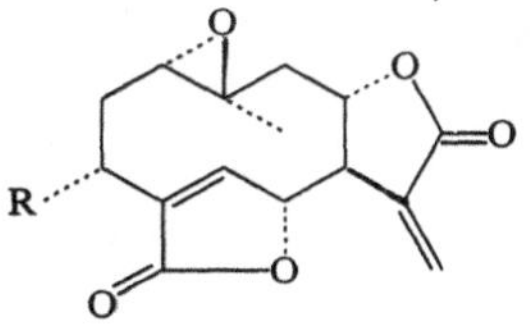

(125) Mikanolide, deoxy; $R = H$
(126) Scandenolide; $R = OAc$

$A =$

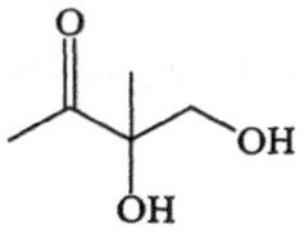

Chart III-3 (continued)

(127) Mikanolide

(128) Spiciformin; R = H
(128a) Chamissonin 4,5-epoxy; R = OH

(129) Baileyin

(130) Laurenobiolide, 6-desacetoxy, dihydro;
       R₁ = R₃ = H, R₂ = β-H
(131) Balchanolide, iso; R₁ = OH, R₂ = β-H, R₃ = H
(132) Salonitolide; R₁ = OH, R₂ = α-H, R₃ = OH

(133) Scandenolide, dihydro

(134) Mikanolide, dihydro

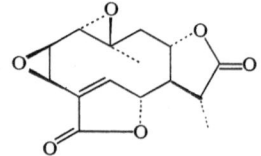

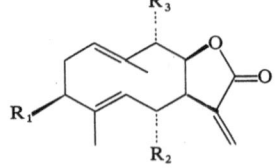

(24)   Inunolide, R₁ = R₂ = R₃ = H
(136) Vernudifloride; R₁ = R₃ = H, R₂ = OSen

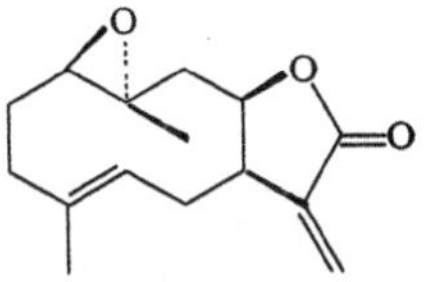

(136a) Inunolide, 1 β,10 α-epoxy -1,10-H

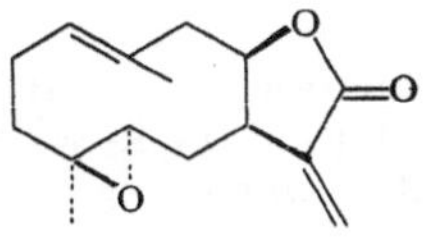

(137) Inunolide, 4 β,5 α-epoxy,4,5-H

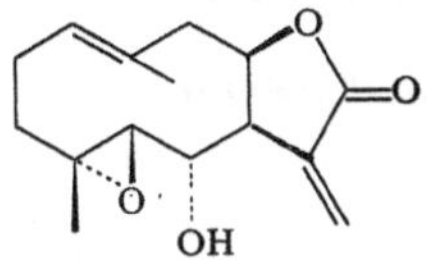

(138) Simsiolide

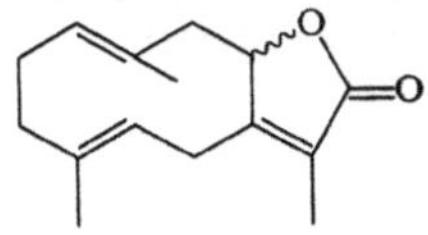

(139) Glechomanolide

Chart III-4. Naturally occurring melampolides and biogenetic derivatives[a]

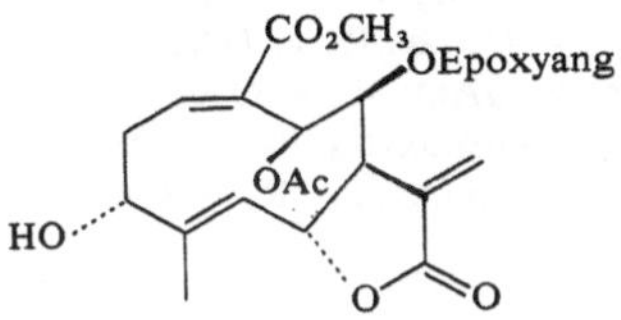

(140) Leucanthinin

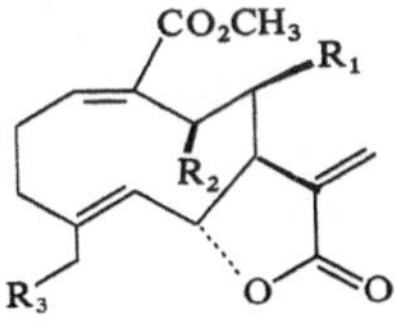

(141) Melampolidin; R_1 = OA, R_2 = R_3 = H
(142) Uvedalin; R_1 = OEpoxyang, R_2 = OAc, R_3 = H
(143) Polydalin; R_1 = OB, R_2 = OAc, R_3 = H
(144) Longipin; R_1 = OEpoxyang, R_2 = H, R_3 = OAc

[a] In relation to C-7 αH, melampolides with a chiral center at C-9 show S-configuration. In several publications (272, 273, 274) from the laboratory of the senior author the stereochemistry at C-9 was erroneously presented as R. The dilactones (166) to (170) have a 7,6-trans lactone with C-6 chirality as indicated (see discussion in section III-1).

Chart III-4 (continued)

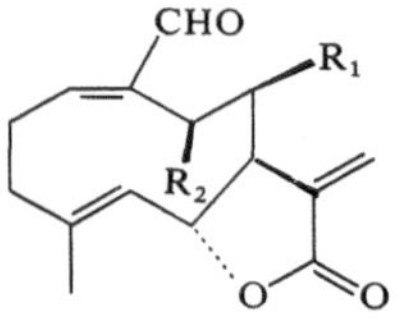

(**145**) Acanthospermal A; R_1 = O-i-But, R_2 = OC
(**146**) Acanthospermal B; R_1 = O-2-Mebut, R_2 = OAc
(**147**) Acanthamolide; R_1 = OH, R_2 = NH-i-But

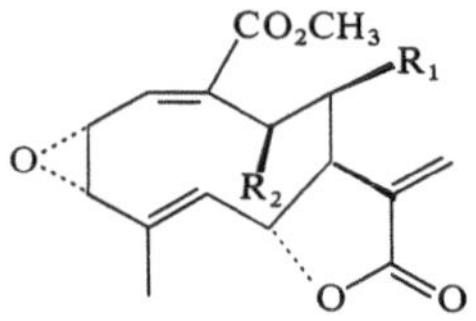

(**148**) Melampodin A; R_1 = OEpoxyang, R_2 = OH
(**149**) Melampodinin, 9-desacetyl; R_1 = OA, R_2 = OH
(**150**) Melampodin A, acetate; R_1 = OEpoxyang,
       R_2 = OAc
(**151**) Melampodin B; R_1 = OEpoxyang,
       R_2 = O-2-Mebut
(**152**) Melampodinin, R_1 = OA, R_2 = OAc

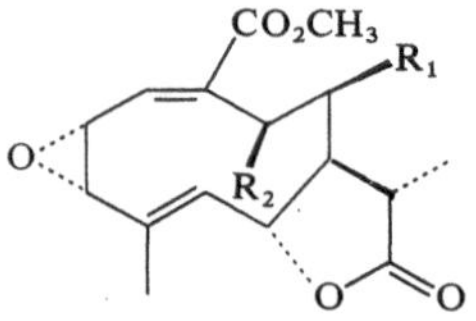

(**153**) Melampodin A, 11,13-dihydro, 9-α-methylbuty-
rate; R_1 = OEpoxyang, R_2 = O-2-Mebut

A =   B =   C =

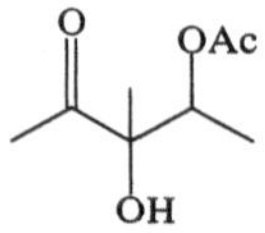

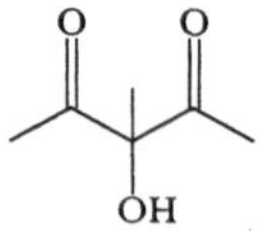

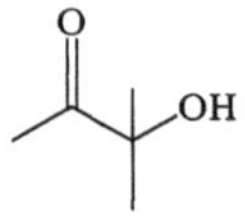

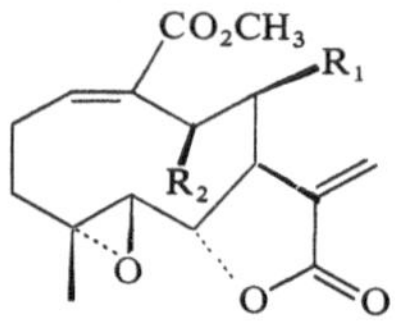

(**154**) Longipilin; R_1 = OAng, R_2 = OH
(**155**) Fluctuadin; R_1 = OMac, R_2 = OAc
(**156**) Fluctuanin; R_1 = OAng, R_2 = OAc
(**157**) Maculatin; R_1 = OEpoxytig, R_2 = OAc
(**158**) Enhydrin; R_1 = OEpoxyang, R_2 = OAc

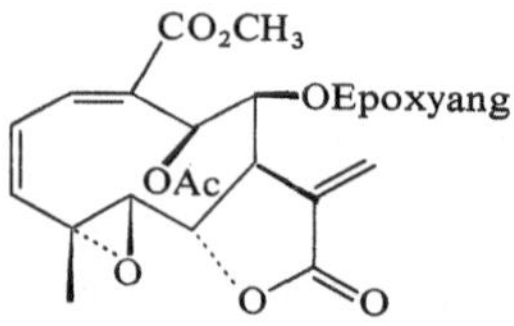

(**159**) Leucanthin A

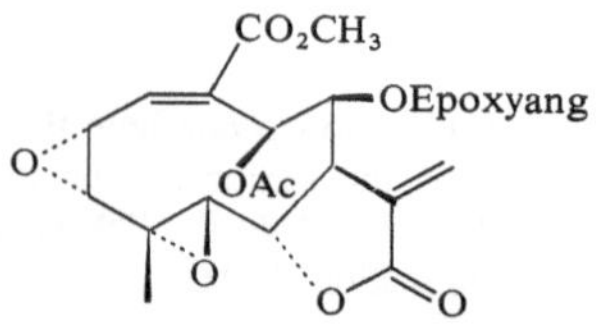

(**160**) Leucanthin B

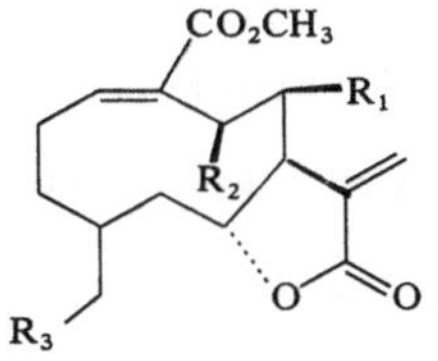

(161) Melnerin A; R_1 = O-i-But, R_2 = H, R_3 = OH
(162) Melnerin B; R_1 = O-2-Mebut, R_2 = H, R_3 = OH
(163) Melnerin A, 9-acetoxy; R_1 = O-i-But,
       R_2 = OAc, R_3 = OH
(164) Melnerin B, 9-acetoxy; R_1 = O-2-Mebut,
       R_2 = OAc, R_3 = OH

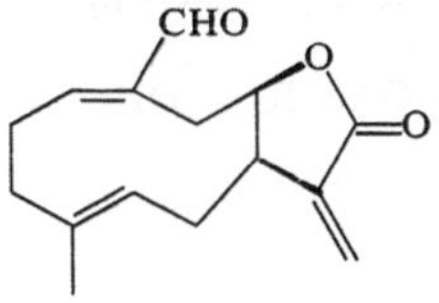

(165) Frutescin

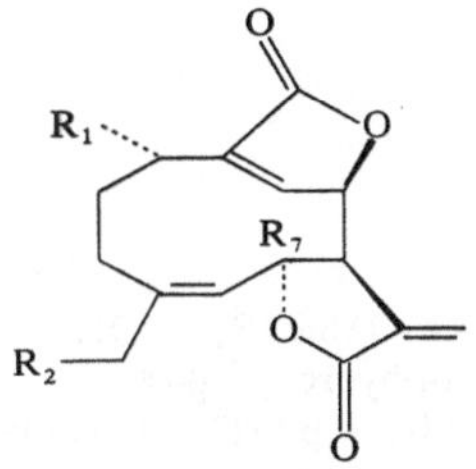

(166) Cinerenin; R_1 = OEt, R_2 = OH
(167) Melampodin B; R_1 = OAc, R_2 = OH
(168) Melampodin C; R_1 = O-i-But, R_2 = OH
(169) Melampodin D; R_1 = O-2-Mebut, R_2 = OH

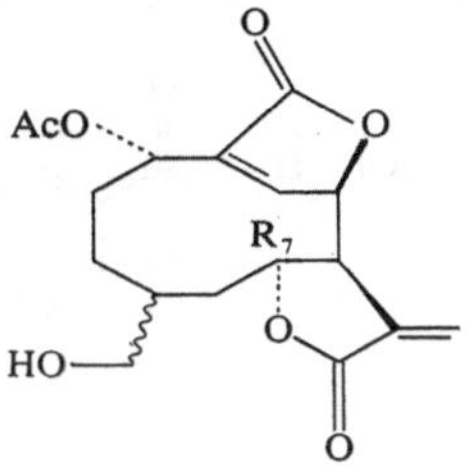

(170) Melampodin B, 4,5-dihydro

Chart III-5. Naturally occurring heliangolides and biogenetic derivatives

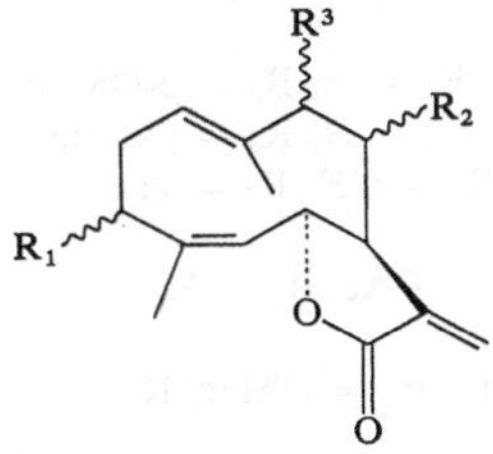

(171) Germacranolide, 4,5-cis, 3-β-hydroxy;
       R_1 = β-OH, R_2 = R_3 = H
(172) Nobilin, 3-epi; R_1 = α-OH, R_2 = α-OAng,
       R_3 = H
(173) Nobilin; R_1 = β-OH, R_2 = α-OAng, R_3 = H
(174) Eupaformonin; R_1 = α-OAc, R_2 = β-OH, R_3 = H
(175) Eupaformosanin; R_1 = α-OAc, R_2 = β-OA,
       R_3 = H
(176) Chromolaenide, 3-epi, 20-acetoxy; R_1 = α-OAc,
       R_2 = β-OB, R_3 = H
(177) Peucephyllin; R_1 = β-OAc, R_2 = β-O-i-But,
       R_3 = H

Chart III-5 (continued)

(178) Chromolaenide; R_1 = β-OAc, R_2 = β-OC,
       R_3 = H
(179) Chromolaenide, 20-tiglinoyloxy; R_1 = β-OAc,
       R_2 = β-OD, R_3 = H
(180) Eucannabinolide; R_1 = β-OAc, R_2 = β-OA,
       R_3 = H
(182) Provincialin; R_1 = β-OAc, R_2 = β-OE, R_3 = H
(183) Eupatocunin; R_1 = β-OAc, R_2 = β-OAng,
       R_3 = OH
(184) Eupatocunoxin; R_1 = β-OAc, R_2 = OH,
       R_3 = OEpoxyang

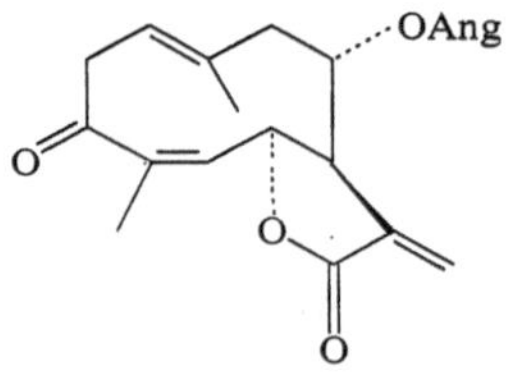

(185) Nobilin, 3-dehydro

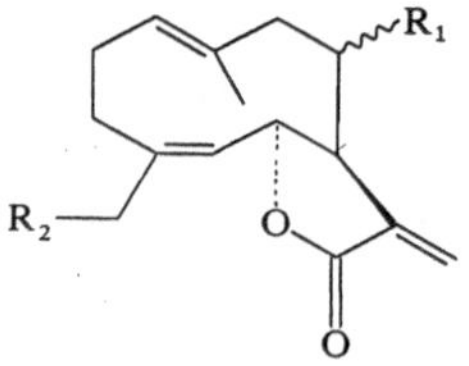

(186) Euparhombin; R_1 = α-OMac, R_2 = OH
(187) Costunolide, 4,5-*cis*, 14-hydroxy-8 β-(4-
       hydroxytiglinoyloxy); R_1 = β-OTig, R_2 = OH
(188) Costunolide, 4,5-*cis*, 14-acetoxy-8 β (4-
       hydroxytiglinoyloxy); R_1 = β-OC, R_2 = OAc

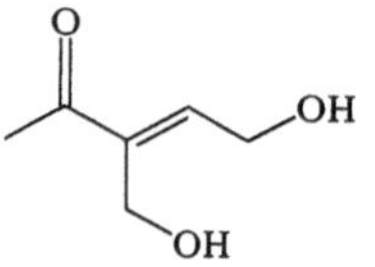

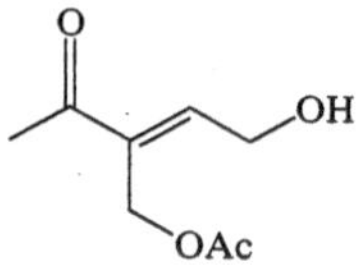

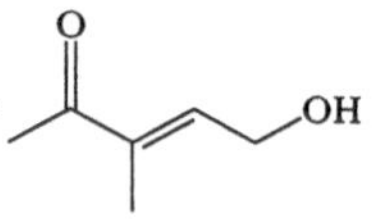

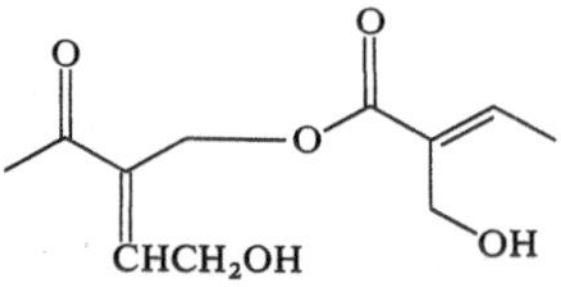

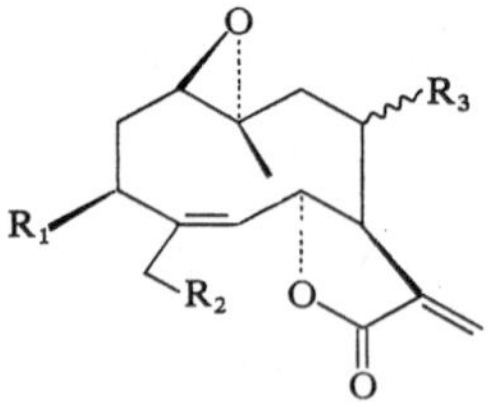

(189) Eriophyllin B; R_1 = R_2 = OH,
       R_3 = β-OMac
(190) Eriophyllin; R_1 = OAc, R_2 = OH,
       R_3 = β-OMac
(191) Erioflorin; R_1 = OH, R_2 = H, R_3 = β-OMac
(192) Heliangin; R_1 = OH, R_2 = H, R_3 = β-OTig
(193) Nobilin, 1,10-epoxy; R_1 = OH, R_2 = H,
       R_3 = α-OAng
(194) Erioflorin acetate; R_1 = OAc, R_2 = H,
       R_3 = β-OMac
(195) Erioflorin methacrylate; R_1 = OMac, R_2 = H,
       R_3 = β-OMac

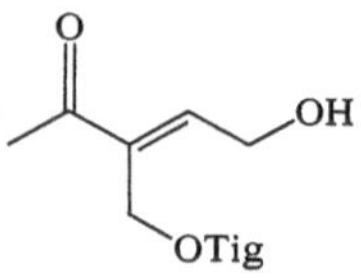

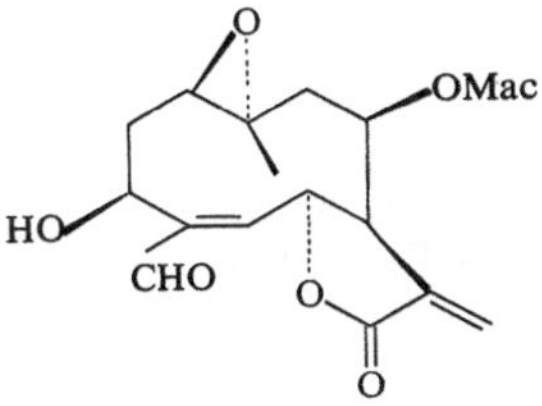

(196) Eriophyllin C

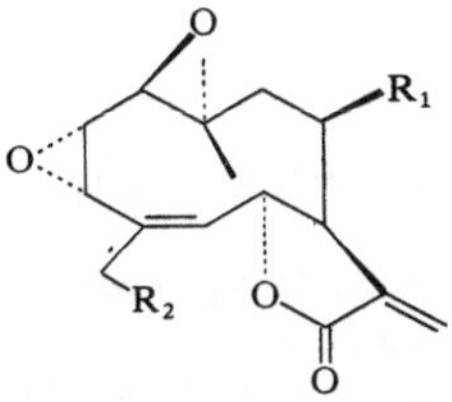

(197) Liscundin; R_1 = OAng, R_2 = OH
(198) Eleganin; R_1 = OA, R_2 = OH
(199) Liscunditrin; R_1 = OSarac, R_2 = OH

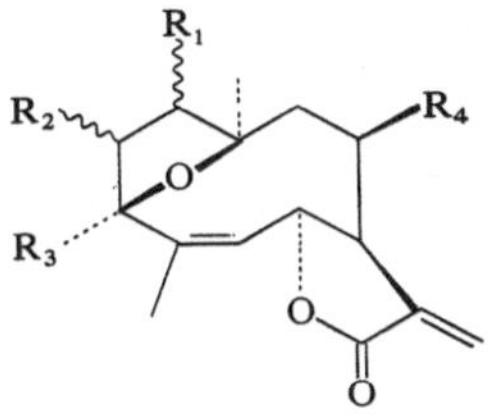

(200) Orizabin; R_1 = α-OH, R_2 = H, R_3 = OH, R_4 = β-O-i-But
(201) Zexbrevin B; R_1 = α-OH, R_2 = H, R_3 = OH, R_4 = β-OMac
(202) Zacatechinolide, 1β-acetoxy; R_1 = β-OAc, R_2 = H, R_3 = OH, R_4 = β-OMac
(203) Tagitinin B; R_1 = H, R_2 = β-OH, R_3 = OH, R_4 = β-O-i-But
(204) Woodhousin; R_1 = H, R_2 = β-OAc, R_3 = OH, R_4 = β-O-i-But

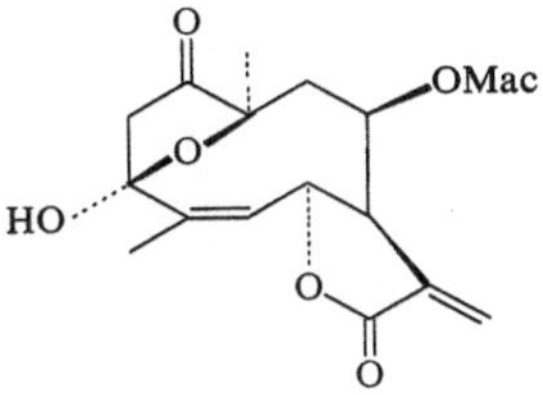

(205) Zacatechinolide, 1-oxo

$A =$

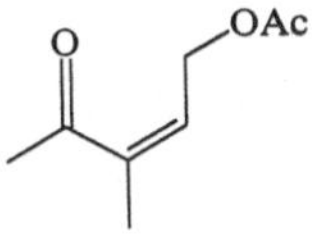

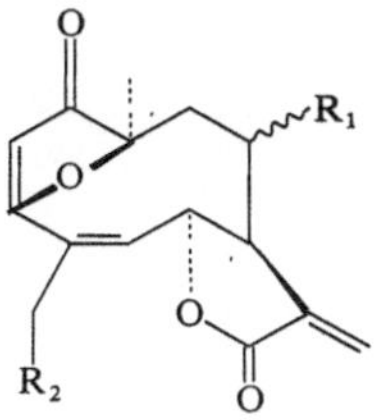

(206) Ciliarin; R_1 = β-O-i-But, R_2 = H
(207) Atripliciolide, isobutyrate; R_1 = β-O-i-But, R_2 = H
(208) Calaxin; R_1 = β-OMac, R_2 = H
(209) Atripliciolide-(2-methylacrylate); R_1 = β-OMac, R_2 = H
(210) Atripliciolide, isovalerate; R_1 = β-O-i-Val, R_2 = H
(211) Atripliciolide, tiglate; R_1 = β-OTig, R_2 = H
(212) Budlein A; R_1 = β-OAng, R_2 = OH

Chart III-5 (continued)

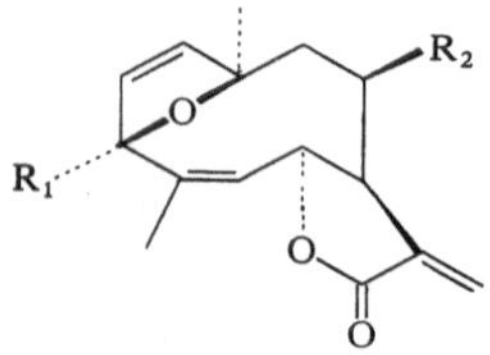

(213) Liatrin; $R_1 = \alpha\text{-OH}$, $R_2 = $ OSarac,
(214) Tagitinin F; $R_1 = \alpha\text{-OH}$, $R_2 = $ O-i-But
C-14 β

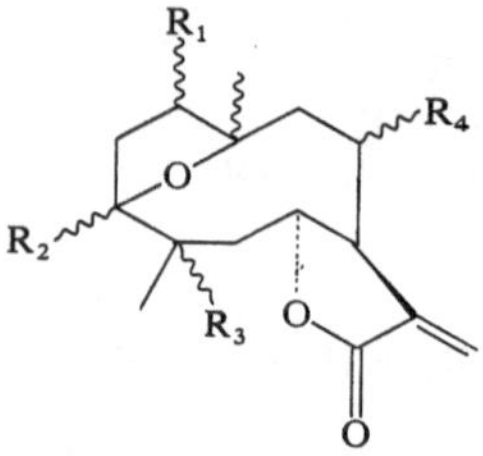

(215) Tagitinin A; $R_1 = R_2 = \alpha\text{-OH}$, $R_3 = \beta\text{-H}$
$R_4 = \beta\text{-O-i-But}$
(216) Tirotundin ethyl ether; $R_1 = $ H, $R_2 = \alpha\text{-OEt}$,
$R_3 = \beta\text{-H}$, $R_4 = \beta\text{-O-i-But}$, C-14α
(217) Tirotundin; $R_1 = $ H, $R_2 = \alpha\text{-OH}$, $R_3 = \beta\text{-H}$,
$R_4 = \beta\text{-O-i-But}$, C-14α

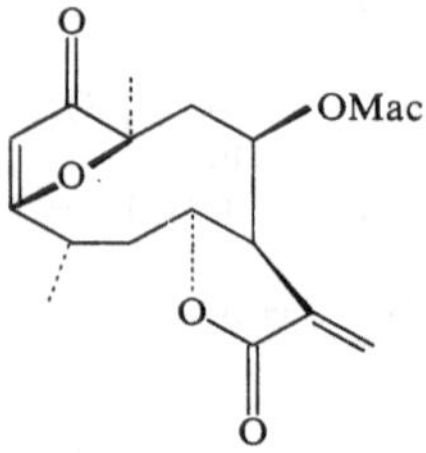

(219) Zexbrevin

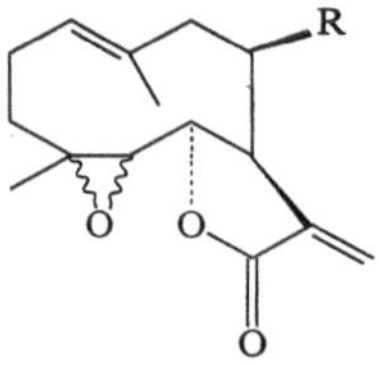

(220) Liatripunctin; R = OA

$A =$

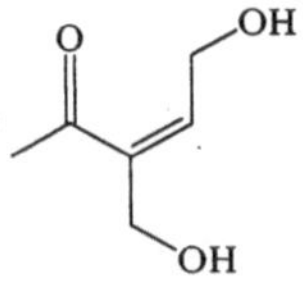

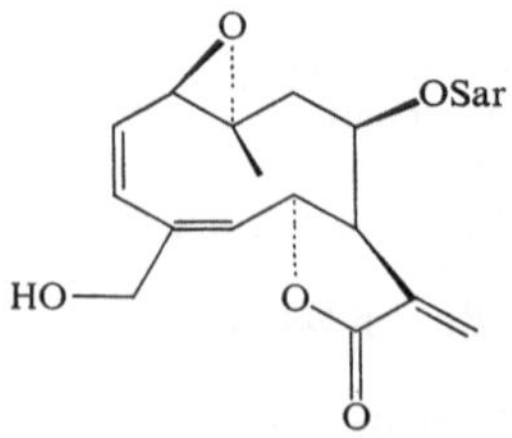

(221) Punctaliatrin

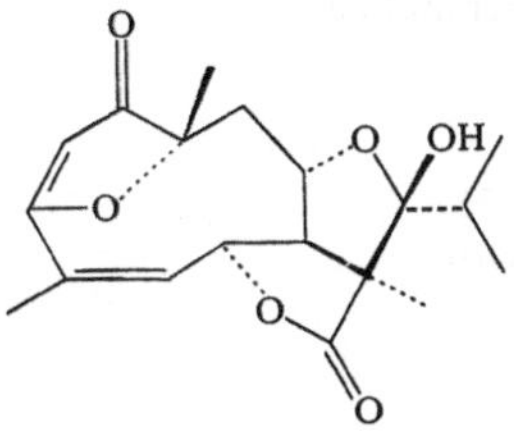

(222) Eremantholide A

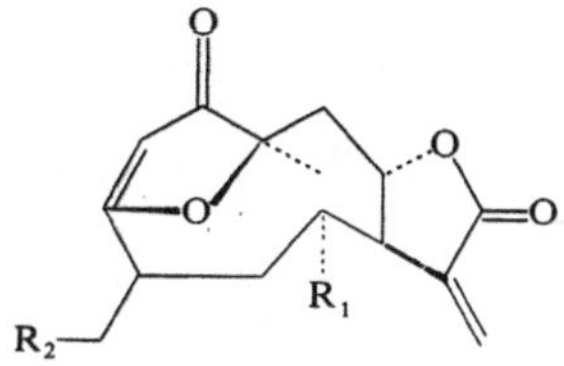

(223) Goyazensolide, 15-deoxy; R_1 = OMac, R_2 = H
(224) Goyazensolide; R_1 = OMac, R_2 = OH

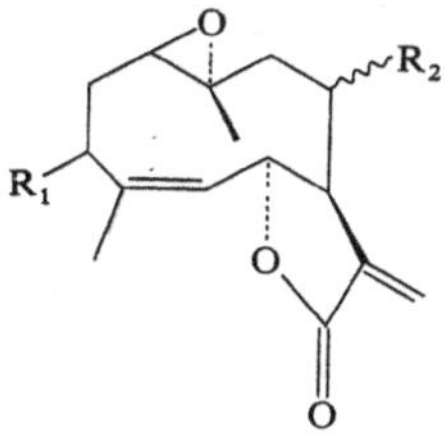

(225) Viguiestenin, desacetyl; R_1 = OH, R_2 = O-i-But
(226) Viguiestenin; R_1 = OAc, R_2 = O-i-But

Chart III-6. Naturally occurring *cis,cis*-germacranolides[a]

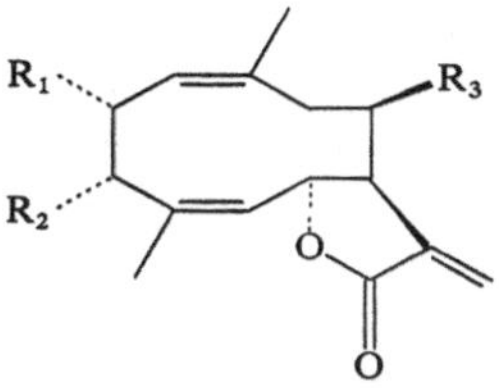

(227) Costunolide, *cis,cis*, 2 α-hydroxy; R_1 = OH, R_2 = R_3 = H
(228) Costunolide, *cis,cis*-3 α-acetoxy-8 β-hydroxy; R_1 = H, R_2 = OAc, R_3 = OH

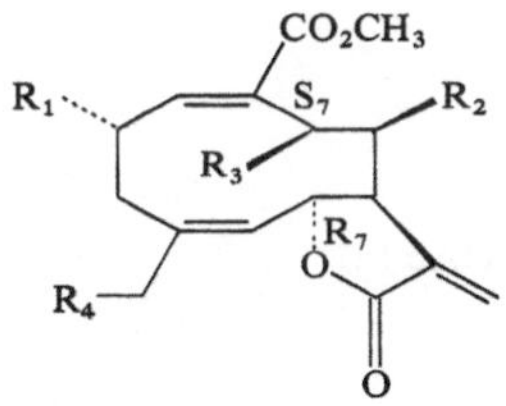

(229) Melcanthin C; R_1 = R_4 = OH, R_2 = O-i-But, R_3 = OAc
(230) Melcanthin B; R_1 = R_4 = OH, R_2 = OAng, R_3 = OAc
(231) Melcanthin A; R_1 = H, R_2 = OAng, R_3 = OAc, R_4 = OH

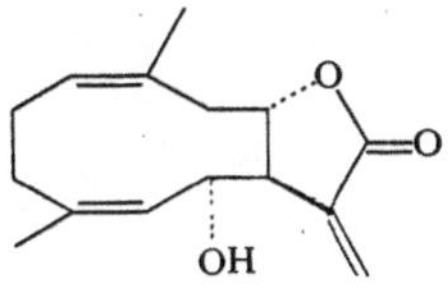

(232) Artemisiifolin, *cis,cis*,15-desoxy

[a] All compounds represent *trans*-lactones.

Chart III-7. Naturally occurring furanogermacranolides

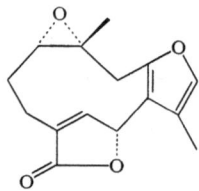

(233) Linderalactone; R = H
(234) Litsealactone; R = OAc

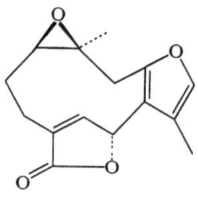

(235) Linderalactone, neo

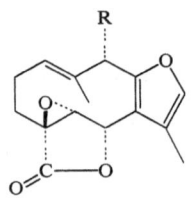

(236) Linderane, pseudo, neo

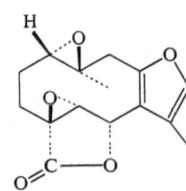

(237) Linderane, neo

(238) Linderane; R = H
(239) Litseaculane; R = OAc

(240) Linderadine

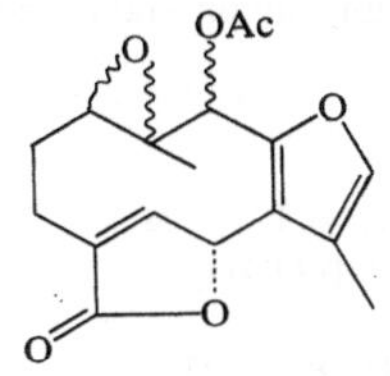

(241)   Zeylanicine

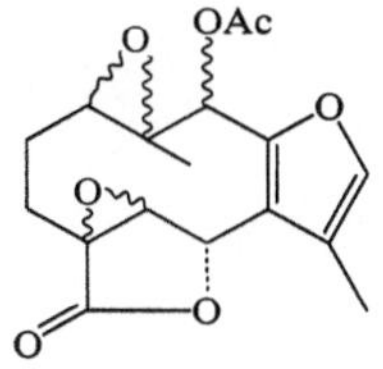

(242)   Zeylanidine

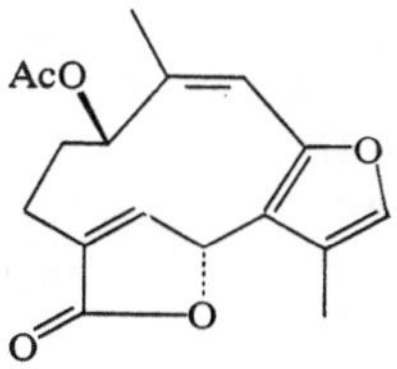

(243)   Zeylanine

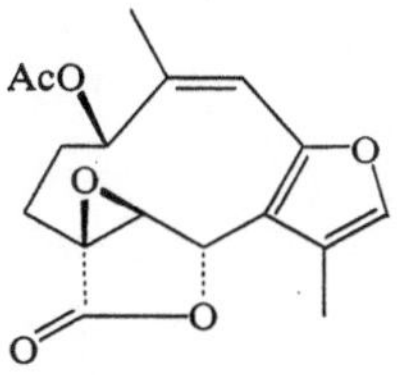

(244)   Zeylanane

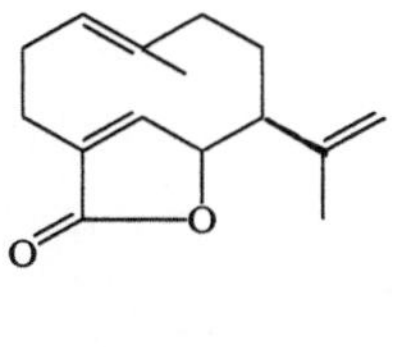

(245)   Aristolactone

Chart III-8. Special structural types of germacranolides and stereochemically unestablished compounds

(246) Artemorin; $R_1 = \beta\text{-OH}$, $R_2 = R_3 = H$
(248) Costunolide, 1-peroxy; $R_1 = \beta\text{-OOH}$, $R_2 = R_3 = H$
(249) Ridentin; $R_1 = R_2 = \beta\text{-OH}$, $R_3 = H$
(250) Dentatin B; $R_1 = \beta\text{-OH}$, $R_2 = H$, $R_3 = \alpha\text{-OH}$
(251) Artevasin; $R_1 = H$, $R_2 = R_3 = OH$

(252) Verlotorin, anhydro

(253) Parthenolide, 1-peroxy; $R = H$
(254) Ferolide, 1-peroxy; $R = OAc$

(255) Ridentin, dihydro

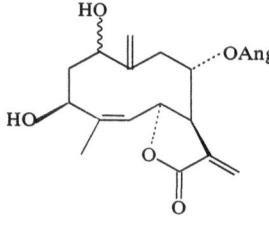

(256) Nobilin, iso, hydroxy

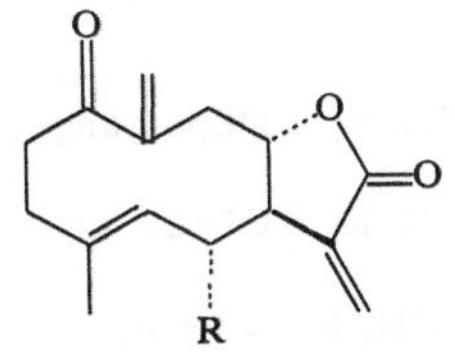

(257) Tamirin; R = OH
(258) Chrysanolide; R = OAc

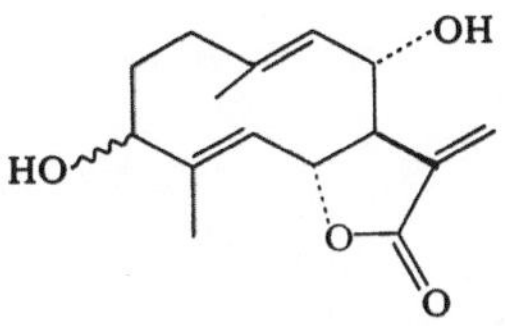

(259) Tatridin A

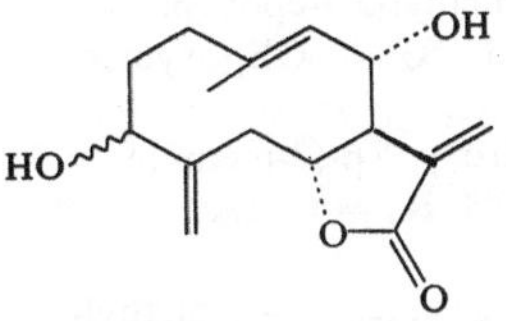

(260) Tatridin B

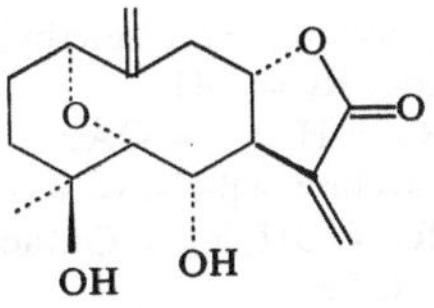

(261) Badgerin

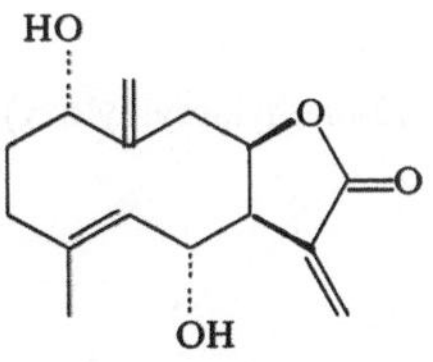

(262) Tanachin

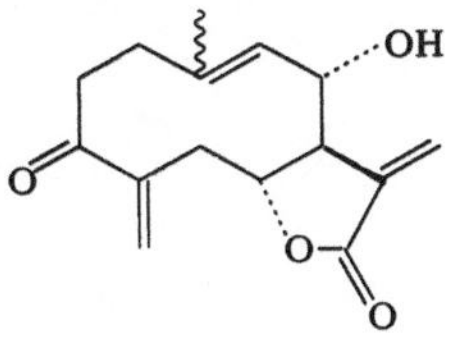

(263) Tatridin C

Chart III-8 (continued)

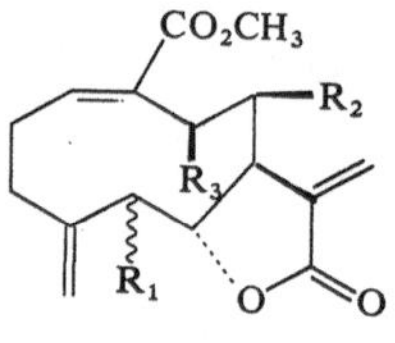

(264) Repandin C; R_1 = β-OH, R_2 = OEpoxyang,
          R_3 = O-i-But
(265) Repandin D; R_1 = β-OH, R_2 = OEpoxyang,
          R_3 = O-2-Mebut
(266) Repandin A; R_1 = β-OH, R_2 = OSar,
          R_3 = O-i-But
(267) Repandin B; R_1 = β-OH, R_2 = OSar,
          R_3 = O-2-Mebut

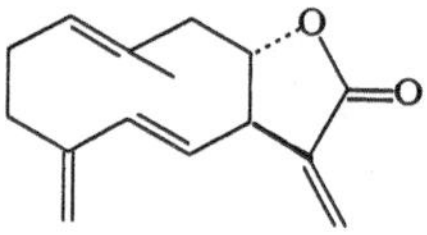

(268) Germacrene D lactone

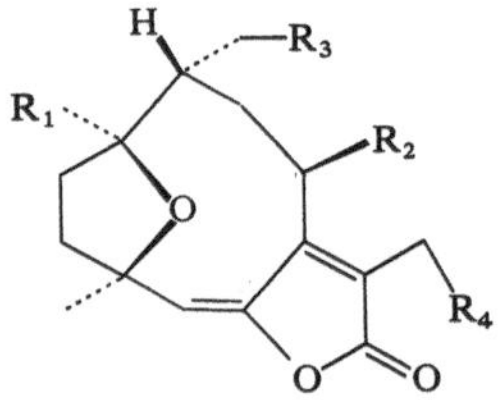

(269) Hirsutinolide, 8 β-(2-methylacryloyloxy);
          R_1 = OH, R_2 = OMac, R_3 = R_4 = H
(270) Hirsutinolide, 8 β-(2-methyl-2,3-epoxyprop-
          ionyloxy); R_1 = OH, R_2 = OEpoxymac,
          R_3 = R_4 = H
(271) Hirsutinolide, 15-hydroxy, 8 β-(2-methyl-
          acryloyloxy); R_1 = OH, R_2 = OMac,
          R_3 = OH, R_4 = OAc
(272) Hirsutinolide, 13(O)-acetate, 8 β-(2-methyl-
          acryloyloxy); R_1 = OH, R_2 = OMac, R_3 = H,
          R_4 = OAc
(273) Hirsutinolide, 13(O)-acetate, 8 β-(2-methyl-
          2,3-epoxypropionyloxy); R_1 = OH,
          R_2 = OEpoxymac, R_3 = H, R_4 = OAc
(274) Hirsutinolide, 13(O)-acetate, 8 β-(2-hydroxy-
          methylacryloyloxy); R_1 = OH, R_2 = OMac-
          4-OH, R_3 = H, R_4 = OAc

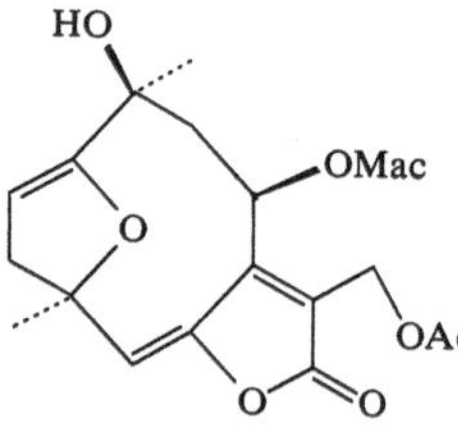

(275) Hirsutinolide, iso, 8 β-(2-methylacryloyloxy)

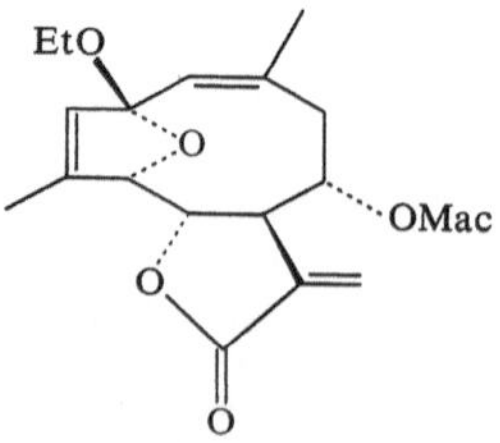

(276) Phantomolin

References, pp. 321—388

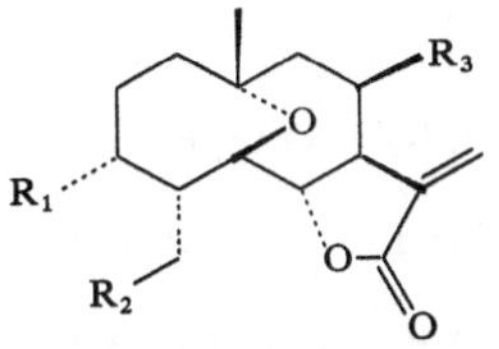

(277) Chapliatrin; R_1 = OH, R_2 = OAc,
       R_3 = OSarac
(278) Chapliatrin, iso; R_1 = OAc, R_2 = OH,
       R_3 = OSarac
(279) Chapliatrin, acetyl; R_1 = OAc, R_2 = OAc,
       R_3 = OSarac

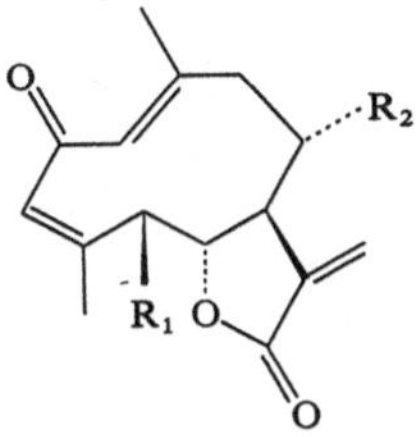

(280) Molephantin; R_1 = OH, R_2 = OMac
(281) Molephantinin; R_1 = OH, R_2 = OTig

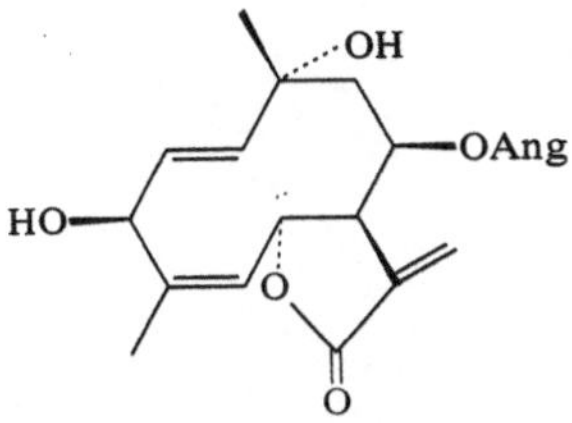

(282) Tifruticin, deoxy

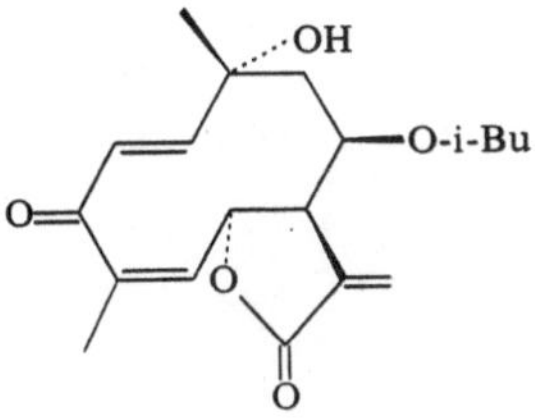

(283) Tagitinin C

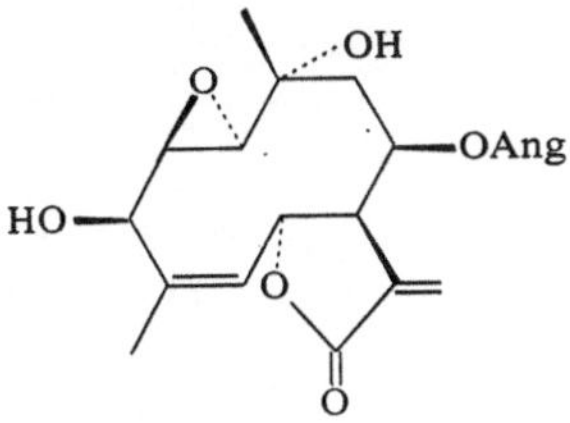

(284) Tifruticin

Chart III-8 (continued)

(285) Eupacunin; R_1 = OH, R_2 = OAc, R_3 = H,
       R_4 = OAng
(286) Eupacunoxin; R_1 = OH, R_2 = OAc, R_3 = H,
       R_4 = OEpoxyang
(287) Eupacunolin; R_1 = OH, R_2 = OAc, R_3 = OH,
       R_4 = OAng

(288) Vernomygdin

(289) Balsamin

(290) Germanin B

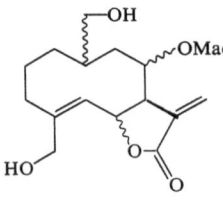

(291) Orientin

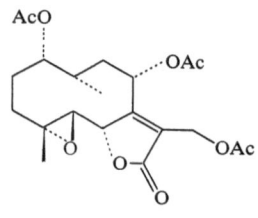

(292) Confertolide

References, pp. 321—388

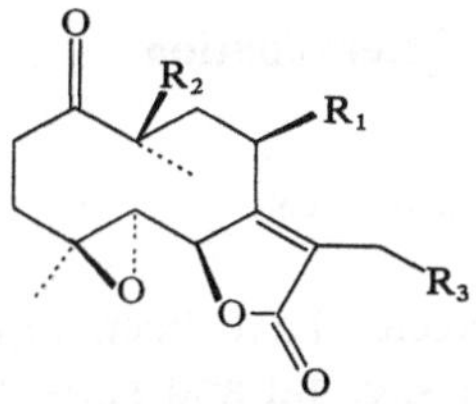

(293) Glaucolide B; $R_1 = R_2 = R_3 = OAc$
(294) Glaucolide A; $R_1 = OMac$, $R_2 = R_3 = OAc$
(295) Glaucolide A, 19-hydroxy;
    $R_1 = OMac$-4-OH, $R_2 = R_3 = OAc$

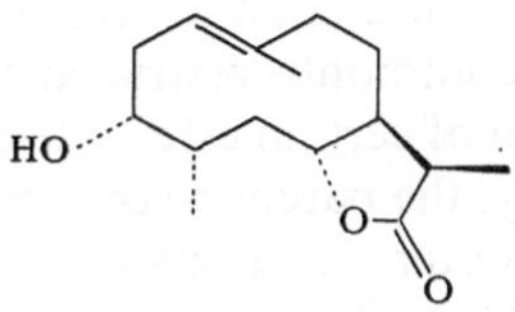

(296) Pelenolide, hydroxy

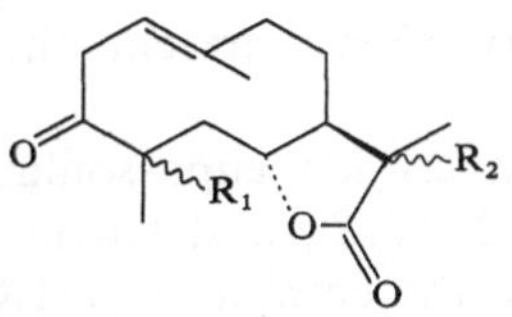

(297) Pelenolide A, keto; $R_1 = \beta$-H, $R_2 = \alpha$-H
(298) Pelenolide B, keto; $R_1 = \alpha$-H, $R_2 = H$

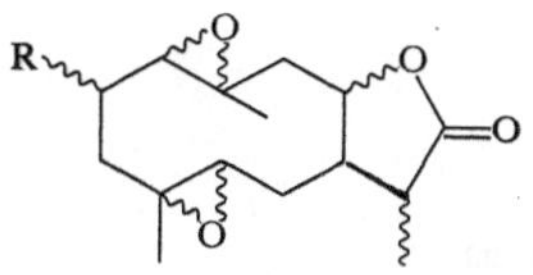

(299) Eriolin; R = H
(300) Eriolin, hydroxy; R = OH

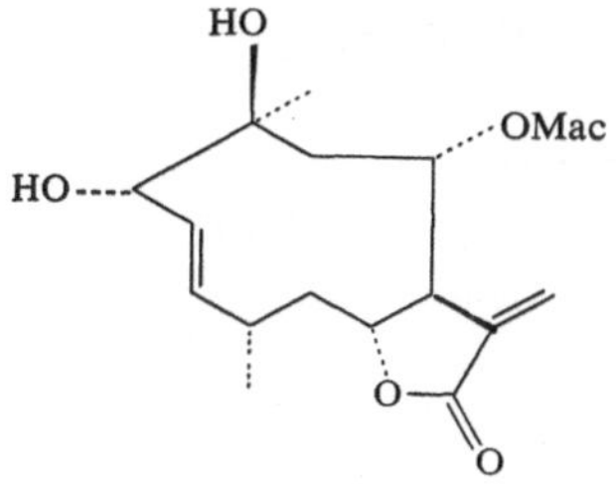

(301) Zexbrevin C

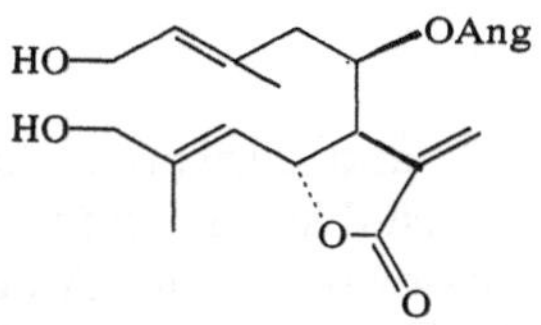

(302) Pycnolide

2. Physical Methods of Structure Determination

2.1 Mass Spectra of Germacranolides

Detailed mass spectra of sesquiterpene lactones have been reported only sporadically. Due to the great variation in skeletal and substitution patterns within the germacranolides the mass spectral fragmentations vary considerably and do not allow the presentation of an unified picture. Use of mass spectra of germacranolides is commonly restricted to the finding of the parent peak and/or the detection of certain ester side chains attached to the ten-membered ring. Frequently, the parent peak is missing in compounds containing hydroxyls and/or ester functions due to the loss of water and/or the side chain by McLafferty rearrangements (M-RCOOH) (828, 829). In such case the acylium ion of the ester group is usually observed as an intense peak and very often represents the base peak (100, 270—272, 417, 436, 623, 829).

In the structure elucidation of the germacrolide eupassofilin (**88**) (441), and the heliangolides provincialin (**182**) (458), and liatrin (**213**) (623) which possess unusual ester side chains, the combined NMR and mass spectral data allowed unambiguous assignment of the side chain as shown for provincialin (Fig. III-1).

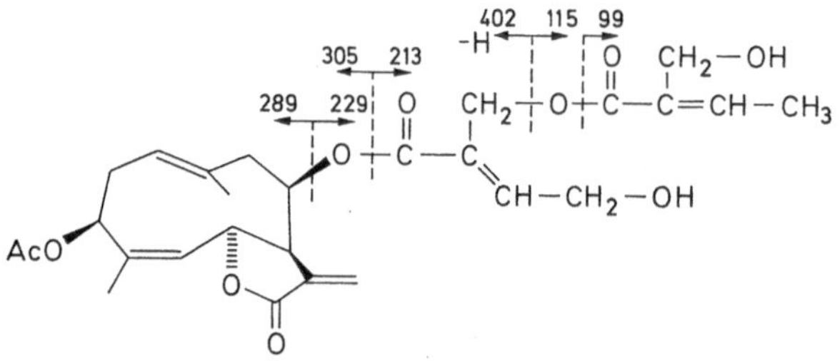

Fig. III-1

The fragmentation modes of costunolide (**17**) and derivatives have been studied by Sathe et al. (942—944) using metastable transitions and deuterium labeling for the interpretation of the major fragments. Among a number of other fragmentations the loss of CH_3 possibly involves cyclization as outlined in Chart III-9. The peak M-55 is due to the loss

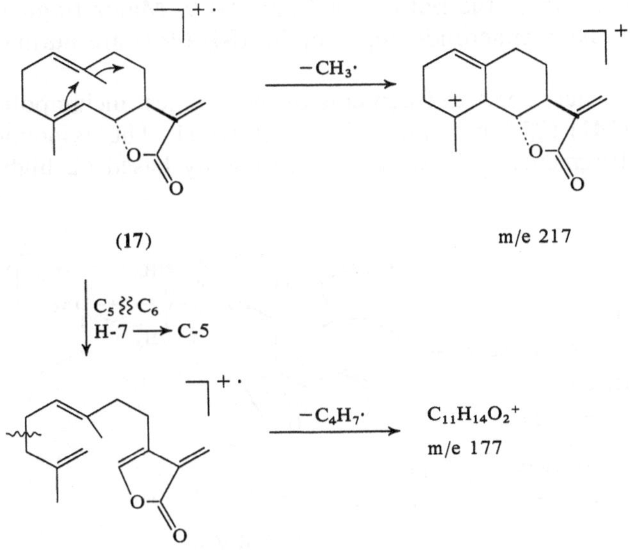
−CH₃·
(17)
m/e 217
C₅ ≀≀ C₆
H-7 → C-5
−C₄H₇·
C₁₁H₁₄O₂⁺
m/e 177

Chart III-9. Mass spectral fragmentation of costunolide

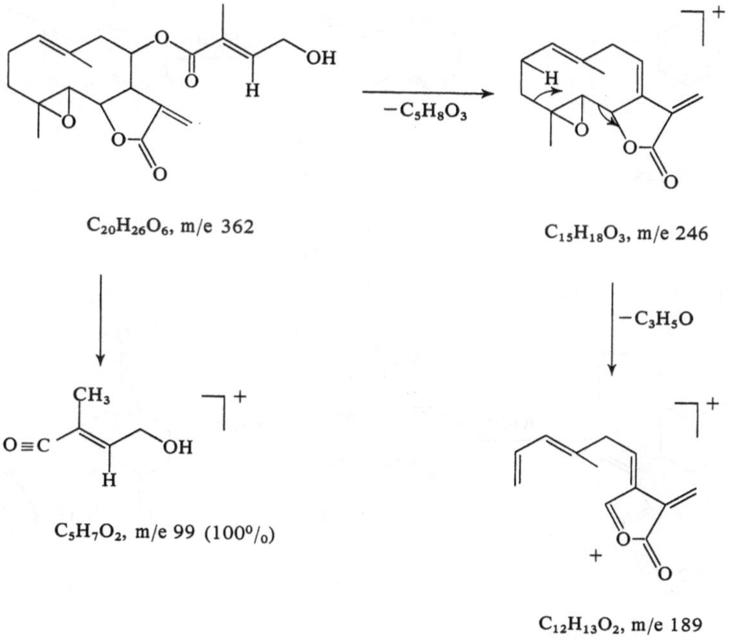
OH
H
−C₅H₈O₃
H
C₂₀H₂₆O₆, m/e 362
C₁₅H₁₈O₃, m/e 246
−C₃H₅O
CH₃
O≡C
OH
H
C₅H₇O₂, m/e 99 (100%)
C₁₂H₁₃O₂, m/e 189

Chart III-10. Major mass spectral fragments of eupassopilin

of C_4H_7 shown at the bottom of Chart III-9. Major fragmentations of the 4,5-epoxygermacrolide eupassopilin (**84**) (*441*) are outlined in Chart III-10.

Major mass spectral fragments of the simplest melampolide, melampolidin (**141**) (*271*) are shown in Chart III-11. The assignments of the various fragments in Chart III-11 are mainly based on high resolution

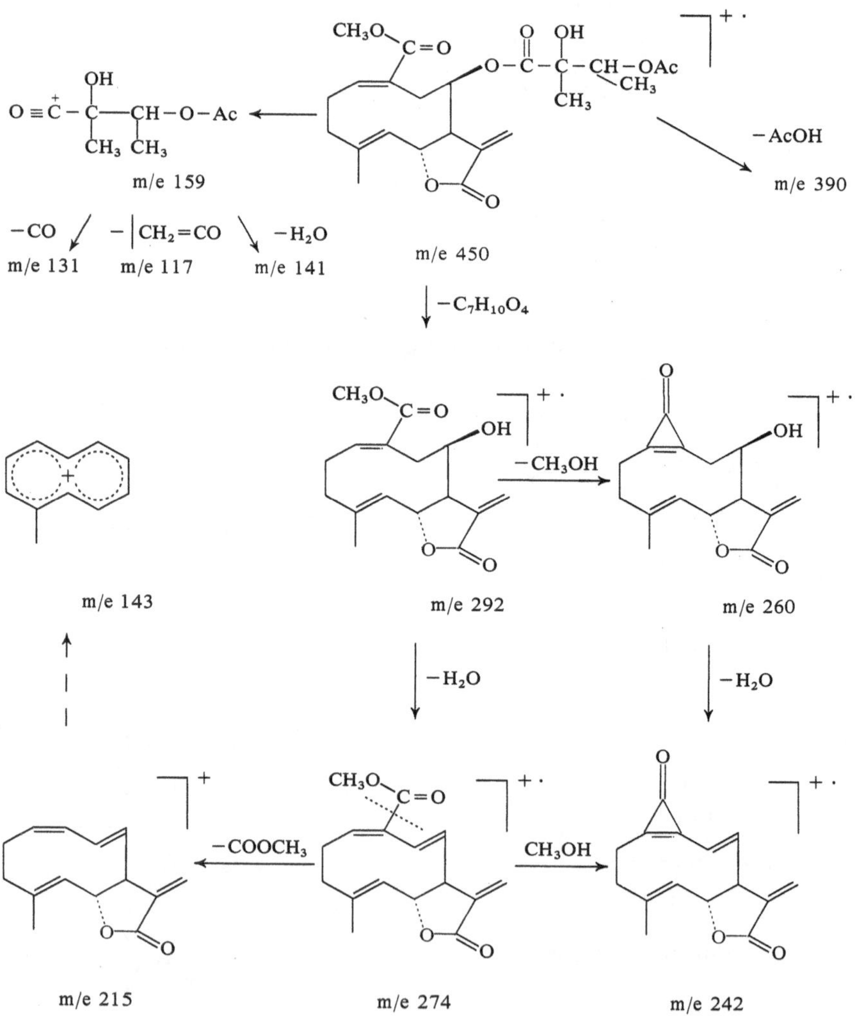

Chart III-11. Mass spectral fragmentation of melampolidin

MS data. Extensive studies on the fragmentation patterns of sesquiterpene dilactones of the melampodin B type (**167**) were carried out more
recently. The cleavage patterns of the medium ring were established by
isotope labeling involving d_3-acetates, d_9-TMSi-derivatives, and D_2O-
exchange products (*828*). Mixtures of hydroxyl-containing sesquiterpene
lactones were analyzed by GC-MS of their TMSi-derivatives (*1129*).
In the case of the *cis,cis*-germacranolide melcanthin B (**230**), the fragmentation patterns are similar to those of the melampolides (*270*). Two
major peaks in the high resolution mass spectrum of melcanthin B, its
acetate and d_3-acetate were interpreted as the fragments (**A**) and (**B**)
formed by cleavages between C-8/C-9 and C-2/C-3, as outlined in
Chart III-12.

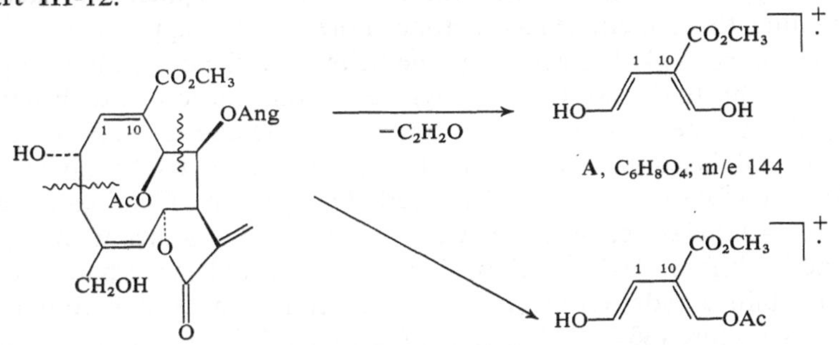

Chart III-12. Mass spectral fragments of melcanthin B

2.2 NMR Spectroscopy

2.21 Proton Spectroscopy

The present discussion will not emphasize general spectral data since
many representative spectra have been published in the book by
YOSHIOKA, MABRY and TIMMERMANN (*1172*). Instead, developments
and trends related to the general NMR spectral behavior of the various
germacranolide subgroups will be considered preferentially. NMR
spectroscopy is the most commonly used technique in the structure
elucidation of sesquiterpene lactones. In general, considerable structural
information is obtained from proton spectra together with the application
of double resonance experiments. In addition, NMR spectra of acetates
and other derivatives of hydroxyl-containing compounds strongly assist
spectral interpretations. Trichloroacetyl isocyanate (TAI) (**303**) has been
introduced for *in situ* reactions with alcohols for NMR studies (*314*),
and has been recently applied to sesquiterpene lactones (*482, 931, 1117*).

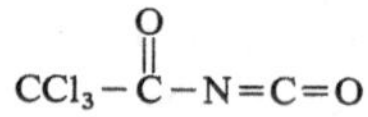

$$CCl_3 - \overset{\overset{\displaystyle O}{\|}}{C} - N = C = O$$

(303)

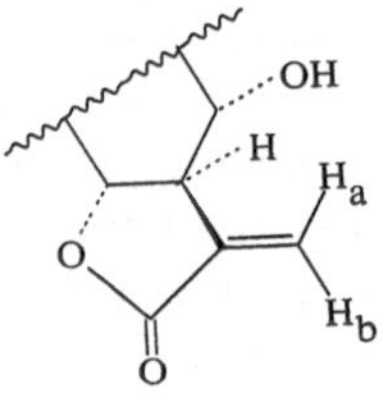

(304)

Various ^{13}C NMR techniques have proved particularly helpful in obtaining structural data on unknown compounds.

The typical feature in the NMR spectra of sesquiterpene lactones containing the α-methylene-γ-lactone moiety is the appearance of two doublets (J = 1—4 Hz), generally one below and the other above 6 ppm, which are due to the two C-13 methylene protons, the low field absorption being due to the proton oriented toward the lactone carbonyl group. It is well established (929, 1008a) that this coupling is due to allylic interactions between H-7 and the C-13 methylene protons. Geminal coupling (J = 0.7—2.0 Hz) between the two C-13 protons together with a paramagnetic shift of H-13a below 6 ppm is observed in germacranolides that contain α-hydroxyl groups at C-8 in 7,6-lactonized compounds [see partial structure (304) (1170), ref. (1172), page 77]. For example, salonitenolide (59) and tatridin A (259) and its TMSi-derivative exhibit doublets of doublets between 6 and 6.5 ppm. This typical paramagnetic shift for H-13a together with the geminal coupling are interpreted as van der Waals proximity effects of the hydroxyl group upon the bonding orbitals of H-13a (1170, 1172). The effects can be helpful in stereochemical and conformational assignments (1172).

A rule for determinating of the stereochemistry of α-methylene γ-lactones which is based on the size of the allylic coupling between the C-13 exocyclic methylene protons and H-7 was proposed by Samek (929). Since there exists a correlation between the allylic dihedral angle and the magnitude of allylic coupling (1008a) certain characteristic J-ranges for $J_{13a,7}$ and $J_{13b,7}$ can be expected with maximal couplings when $\Phi = 90°$ (Fig. III-2-A). On the basis of $J_{7,8}$ values in sesquiterpene

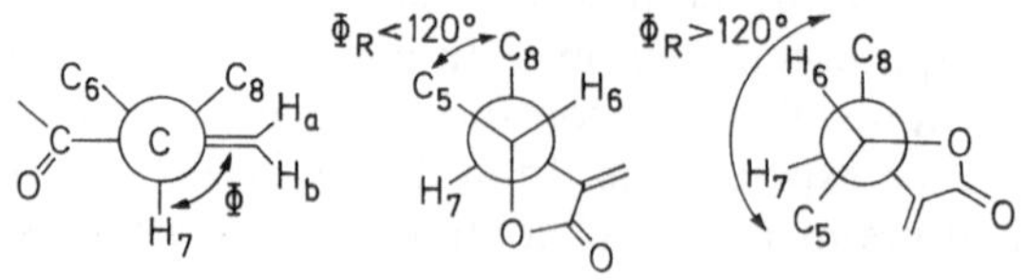

Fig. III-2

lactones with known configurations, SAMEK derived an empirical rule that, *trans*-lactones have larger allylic couplings than *cis*-lactones, the allylic couplings in *trans*-lactones being equal or greater than 3 Hz whereas the allylic J-values for *cis*-lactones are ≤ 3 Hz:

$$J_{trans} \geq 3 \text{ Hz} \geq J_{cis}.$$

The above rule is applicable to a number of sesquiterpene lactones including most 7,6-lactonized germacrolides and the melampolides (*271, 272, 403, 416, 1172*). It was pointed out by HERZ and WAHLBERG (*459*) that heliangolides constitute an exception to SAMEK's rule and show $J_{7,13}$ values between 1.5 and 2.5 Hz (*313, 402, 436, 458, 459, 482, 935*). The differences can be explained as follows.

There exists a direct relationship between the configuration of the endocyclic double bonds and the conformation of germacra-1(10),4-dien-6,12-olides (*1128*). Based on X-ray results, 7,6-*trans*-lactonized germacrolides (*1143*) and melampolides (*1128*) and their 4,5-epoxide derivatives (*558*) preferentially adopt a conformation with $\Phi_R < 120°$ (Fig. III-2-**B**) and H-6 and H-7 in an antiperiplanar relationship with $J_{6,7} = \sim 10$ Hz which is commonly observed for germacrolides and melampolides. In contrast, heliangolides adopt conformations (*704*) with $\Phi_R > 120°$ (Fig. III-2-**C**) narrowing the angle between H_6 and H_7 to nearly 90° thus giving $J_{6,7}$ values < 3 Hz. In all cases the values $J_{6,7}$, $J_{7,13}$ and Φ_R are strongly interrelated and 7,6-*trans*-lactonized germacranolides with $\Phi_R < 120°$ should generally give values $J_{7,6} > 10$ and $J_{7,13} > 3$ (germacrolides and melampolides), and lactones with $\Phi_R > 120°$ would result in $J_{7,6} < 3$ and $J_{7,13} < 3$ (heliangolides) (*313, 482*).

The above considerations represent a part of SAMEK's revised lactone rule (*929a*) which can thus be applied more specifically to the various skeletal germacranolide types. It is particularly interesting that variations in the NMR parameters of the different skeletal types of germacranolides are paralleled by a change in the chirality of the α-methylene γ-lactone function, which results in Cotton effects of opposite sign for germacrolides and heliangolides (see Chapter III., 2.3).

2.22 ^{13}C NMR

Information on the various types of carbon present in sesquiterpene lactones is generally obtained by proton-noise-decoupling experiments which yield chemical shift information. With the aid of the single-frequency-off-resonance decoupling technique, residual carbon-hydrogen splitting allows determination of the number of hydrogens at each carbon.

Selectively decoupled spectra permit assignment of non-proton-bearing carbons (*79a*). In the past, the requirement for relatively large samples has restricted availability of ^{13}C NMR spectral data. ^{13}C NMR data for the germacrolides eupassopilin (**84**), eupassopin (**86**), eupassofilin (**88**) (*441*), euperfolitin (**110**) and euperfolin (**111**) (*417*) have been reported and the ^{13}C-parameters for tamaulipin A (**32**) (*269*) are given in Fig. III-3. Data on two heliangolides, eleganin (**198**) (*438, 439*) and goyazensolide (**224**) (*1111*), and the melampolide melampodinin (**152**) (*271*) and melam-podin B (**167**) (*80*) have been published. A thorough ^{13}C NMR study of melampodin A (**148**) has been carried out using proton-noise-de-coupling, single-frequency-off-resonance-decoupling and selective de-coupling, which allowed unambiguous assignment of all carbons in the molecule (*79a*). Similar techniques permitted ^{13}C NMR spectral assign-ments of the furanogermacranolides linderalactone (**233**), linderane (**238**), and isolinderalactone (**235**) (*1083, 1090*). The use of ^{13}C NMR in confor-mational investigations was reported for laurenobiolide (**116**) (*1086*).

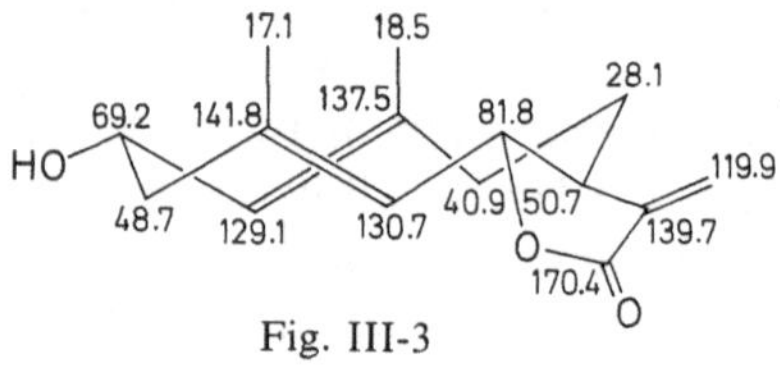

Fig. III-3

2.23 Conformational Considerations

The *cis, trans* nature of the double bonds of the medium ring lactones seems to have a strong influence on their conformation (*273, 1128*). Four major conformations are possible for each of the four geometrically isomeric germacranolides (*265, 1037*) typical representatives of which are given in Chart III-14. In Chart III-13, the four conformations of the germacrolide skeleton (**28**) are presented. In all conformations the double bonds are approximately perpendicular to the plane of the medium ring and thus can adopt two crossed [(**305**) and (**308**)] and two parallel orientations [(**306**) and (**307**)].

Two reviews by TAKEDA (*1042, 1043*) have been concerned with conformational aspects of various types of germacranolides. Of the limited number of compounds that have been studied it was found that 7,6-lactonized germacrolides have a fixed conformation in solution whereas the 7,8-lactonized medium rings seem to be more flexible and can

(305) (306)

(307) (308)

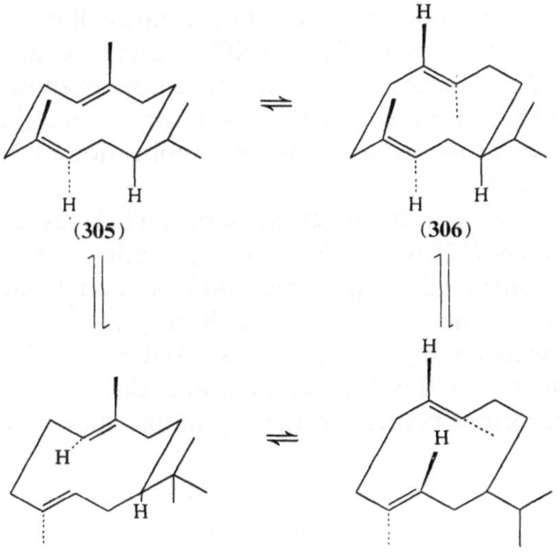

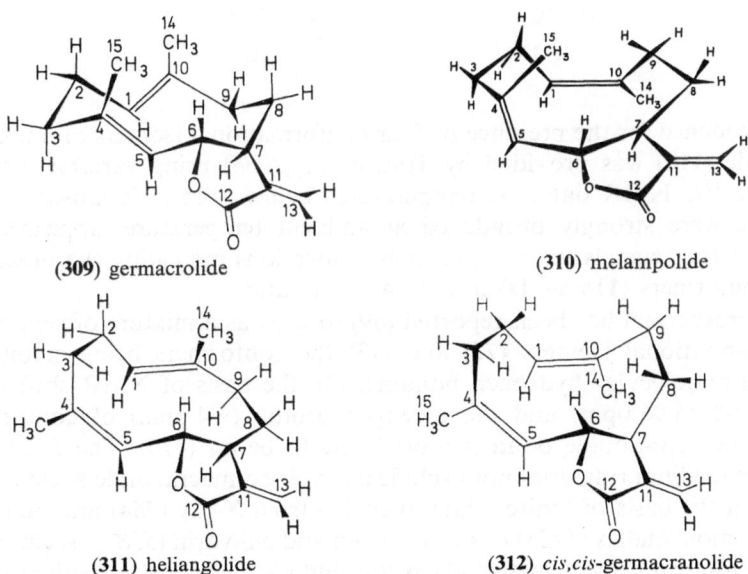

Chart III-13. Four major conformations of the germacrolide skeleton (**28**)

(**309**) germacrolide

(**310**) melampolide

(**311**) heliangolide

(**312**) *cis,cis*-germacranolide

Chart III-14. Conformational representatives of the four subgroups of germacranolides

exist in more than one conformation. For instance, it has been document-
ed by nuclear Overhauser effects (NOE) that the dihydroacetate of
tamaulipin A (**32**) adopts a single conformation as shown in (**309**) (*79*).
The same conformation was established for costunolide (**17**) and its
11,13-dihydroderivative by simultaneous application of NOE and NMR
shift reagents (*1087*).

The NMR spectra of the dilactone isabelin (**122**) revealed the presence
of two conformers (**122 A** and **B**) in a 10 : 7 ratio (*1167*); detailed NOE
studies (*1088*) confirmed the previous conformational assignments of the
two isomers. The minor conformer (**122 B**) appeared as the predominant
isomer when isabelin crystals were dissolved at −50° and the NMR
spectrum measured at this low temperature; this suggested that (**122 B**)
is the only conformer present in the crystalline state (*1167*).

(**122**) isabelin

(**122 A**)　　　　　　　　(**122 B**)

Evidence for the presence of four conformational isomers of laurenobiolide (**116**) was provided by Tori *et al.* (*1086*) using variable-temperature ^{13}C NMR data. At temperatures below −15°, ^{13}C-absorptions,
which were strongly broadened at ambient temperature, appeared as
sets of four signals for most carbon absorptions indicating the presence
of conformers (**116 A—D**) in a 5 : 4 : 3 : 1 ratio.

Urospermal has been reported (*69*) to exist as a mixture of separable
conformational isomers (**72**) and (**73**) the conformers being stabilized
by intramolecular hydrogen bonding. On the basis of NMR shift data
for H-1 (6.96 ppm) and the aldehyde proton (9.4 ppm) of conformer
(**72**) the 1,10-double bond is most likely to be *cis* (*416*). The flexibility
of the medium ring does not exclude a *cis, cis*-germacranolide skeleton.

On the basis of limited data available from X-ray (*764*) and neutron
diffraction studies (*1128*) of melampodin and enhydrin (*558*) it seems that
the *anti*-arrangement, with C-14 below and C-15 above the medium ring,
is typical for the melampolides (**310**). On NMR spectral grounds, no

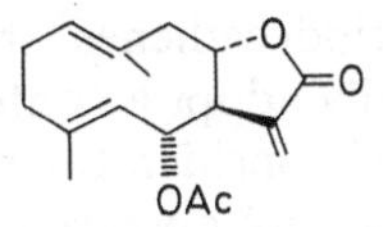

(116) laurenobiolide

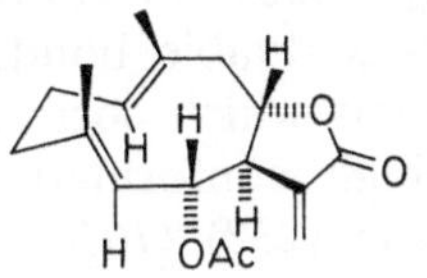

(116A)

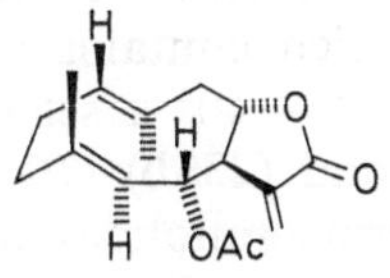

(116B)

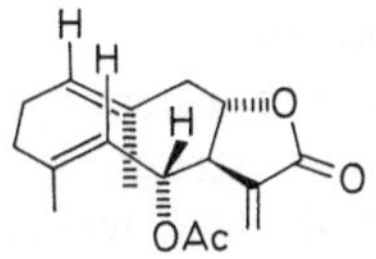

(116C)

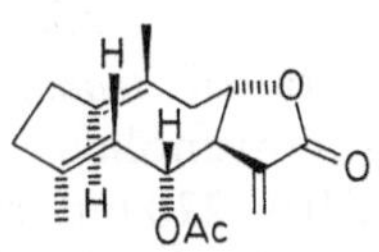

(116D)

conformational changes seem to occur in solution (274). Heliangolides are found to have an *anti*-arrangement of the C-14 and C-15 groups with orientations opposite to those of the melampolides, that is, C-14 is above and C-15 below the plane of the ten-membered ring (311). X-ray analyses of heliangin (192) (776), eupaformonin (174) (704) and peucephyllin (177) (61) show general conformations represented by (311) (Chart III-14). Limited evidence based on chemical data and NMR spectra

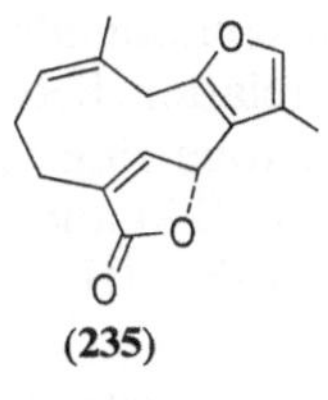

(235)

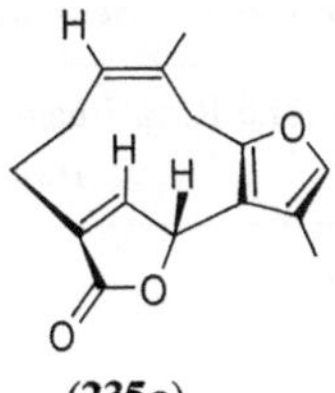

(235a)

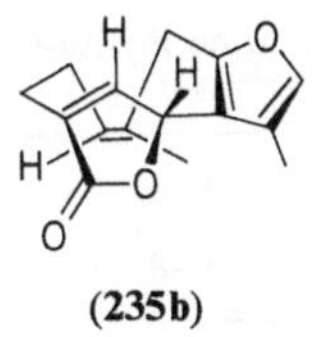

(235b)

indicates that the *cis,cis*-cyclodecadienes are conformationally flexible (*98, 120, 270*) but preferentially adopt a conformation indicated by (**312**) (*270*) as predicted from model considerations (*1128*).

Stable single conformations are observed for a number of the furanogermacranolides (*1043*). With the aid of NOE studies linderalactone (**233**) was shown to adopt a conformation with both C-14 and C-15 below the plane of the cyclodeca-1,5-diene ring (*1056*). In contrast, neolinderalactone (**235**), which contains a 1,10-*cis* double bond, exhibits in the NMR spectrum at −40° a set of signals which correspond to conformations (**235a**) and (**235b**) (*1085*). Single conformations are observed in the furanogermacranolides linderane (**238**) (*1050*), zeylanine (**243**) and zeylanane (**244**) (*1089*).

2.3 UV and CD Spectra

2.31 UV Spectra

The UV spectra of *trans,trans*-cyclodecadienes including germacrolides such as costunolide (**17**) exhibit strong end absorption or maxima in the 210—220 nm region. This low wavelength band was attributed by Sorm et al. (*995, 998*) to homoconjugation (transannular conjugation) of the medium ring double bonds which adopt a crossed conformation that results in transannular interaction of the π orbitals.

2.32 CD of the Lactone Ring

The application of circular dichroism (CD) has been of great value in the determination of the stereochemistry of many sesquiterpene lactones. A large number of sesquiterpene lactones contains α-methylene-γ-lactone groups (**1**) which are optically active; an analysis of the Cotton effect in the range 245—260 nm due to the n-π* transition of this chromophore in a number of known compounds led Geissman and coworkers (*1011, 1124*) to suggest a correlation between sign of the Cotton effect and stereochemistry of the lactone ring. Their rule relating the sign of the *CE* and the position (C-6 or C-8) and stereochemistry of the lactone ring fusion is summarized in Table III-1.

Table III-1. *The Stöcklin-Waddell-Geissman Rule: Predictions of the Type of Lactonization From the Sign of the CE of the n → π* Transition of Sesquiterpene Lactones*

Lactone Position	Lactone Ring Fusion	
	cis	*trans*
C-6	+	−
C-8	−	+

Even at its inception the rule was not uniformly valid; since then a relatively large number of exceptions has been found most significant of which are the heliangolides and some melampolides with 7,6-*trans* lactone ring fusion. An explanation for these observations can be found in BEECHAM's suggestion (*60*) that the sign of the Cotton effect of the α-methylene-γ-lactone chromophore is determined by its chirality.

The conformations of the various types of germacranolides depend mainly on the *cis* or *trans* nature of the 1(10) and 4,5-double bonds (*1128*). Since the exocyclic $C_{13}=C_{11}-C_{12}=O$ torsion angle depends on the endocyclic $C_{11}-C_7-C_6-O$ (or $C_{11}-C_7-C_8-O$) torsion angle, a change in the cyclodecadiene conformation can result in a change in the chirality of the $C=C-C=O$ chromophore and therefore in the sign of the CE (*714*). Hence the absolute configuration of a sesquiterpene lactone can be determined if the CE and the torsion angles of the lactone chromophore are known. McPHAIL and ONAN (*704*) compared the X-ray derived $C=C-C=O$ torsion angles of the germacrolide eupatolide (**37**) with the torsion angles of the heliangolide eupaformonin (**174**) and found that they were of opposite sign (see Fig. III-4), thus leading to opposite Cotton effects in accordance with BEECHAM's rule.

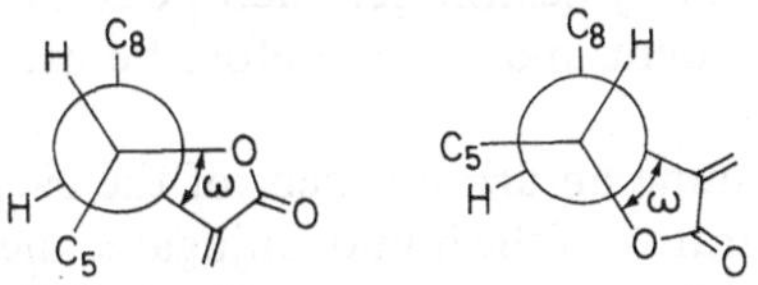

(A), ω positive   (B), ω negative

Fig. III-4

In contrast to the 7,6-lactonized medium rings, the 7,8-lactonic germacrolides show an inconsistent picture which seems to be due to the conformational flexibility of these type of compounds. Since the sign of the CE should be sensitive to deviations from the normal lactone geometry CE-predictions based on models of conformationally flexible compounds appear to be particularly difficult.

Generally, heliangolides show CE's of the same sign (positive) as that of eupaformonin (*402, 439, 458, 482, 935*), indicating that the conformations around the lactone portion of the medium ring are similar. All disobey the Stöcklin-Waddell-Geissman rule but follow BEECHAM's rule (*60*) in accordance with the discussion in the preceding paragraphs.

CD data within the melampolide series are inconsistent. A number of compounds related to melampodin A (**148**) exhibit relatively strong positive CE's near 245 nm as exemplified for melampodin A (*271, 272, 404*). Other melampolides such as uvedalin (**142**) and polydalin (**143**)

(*403*) exhibit negative CE values but reverse their CE sign upon epoxidation of the 4,5-double bond (*404*). Negative CD absorptions are also observed for acanthospermal A and B with C-14 being aldehydic (*416*). There inconsistencies may be due to minor conformational changes within the melampolide skeleton which could invert the sign of the torsional angle of the lactone chromophore and therefore the sign of the CE. Future CD studies on naturally occurring and/or synthetic melampolides lacking C-14 carbonyls may shed light on this question.

2.33 CD of the Cyclodecadiene Ring

A number of germacrolides exhibit two strong bands of opposite sign at low wave lengths. A positive CE is usually observed near 220 nm and a negative band at 200 nm or below, the latter absorption frequently not being detected due to instrumental cut off. These bands are highly characteristic of homoconjugation (*1013*) between the crossed 1,10- and 4,5-*trans* double bonds and indicate the chirality of the diene arrangement. The bands have been ascribed to exiton splitting caused by chiral overlap of the homoconjugated double bonds in the cyclodecadiene skeleton (*992*). In conformationally flexible germacrolides the CD spectrum should be temperature dependent and can therefore be used for conformational investigations (*1084*).

Heliangolides assume an arrangement of the cyclodecadiene skeleton (**311**) in which the chirality of the homoconjugated diene system is opposite to that found in the germacrolides (*482*). Consequently, inverted low wave length CE signs should be expected for heliangolides as shown by a number of compounds in Table III-3. Indeed, a negative CE due to the exiton band is usually observed in the region between 200—210 nm (*482*). In melampolides assignment of the CD bands has so far been hampered by the presence of additional α,β-unsaturated ester or aldehyde chromophores which are part of the medium ring. A strong negative band between 205 and 220 nm is observed in most compounds independent of the presence of a 4,5-double bond or a 4,5-epoxide function. Retention of the strong negative CE in 4,5-epoxides might be used as evidence against the assignment of this absorption to an exiton band caused by homoconjugation.

In a number of cases the CD-bands of the pyrazoline derivatives which are easily obtained from the α-methylene lactones by treatment with diazomethane have been used for the determination of the lactone stereochemistry (*1011, 1013*). Generally, the n → π* band of the azochromophore appears near 330 nm (*1011, 1013*). For instance, in the pyrazoline derivative (**313**) of costunolide (**17**), a positive CE is observed at 327 nm and strong exiton bands appear at 220 and 199 nm (*1013*).

Table III-2. *Germacranolides of Known X-Ray Structure*

Structure Number	Compound	Structure Type[a]	References	Comments
(64)	Alatolide	GG	(*189*)	Monohydrate
(17)	Costunolide	GG	(*132, 999*)	Examined directly and as silver nitrate adduct
(92)	Elephantin	GG	(*715*)	*p*-Bromobenzoate of alkaline hydrolysis product
(91)	Elephantopin	GG	(*715*)	*p*-Bromobenzoate of alkaline hydrolysis product
(158)	Enhydrin	GM	(*558*)	Bromohydrin examined
(222)	Eremantholide A	GH	(*848*)	
(285)	Eupacunin	G	(*633*)	As ortho-bromobenzoate
(286)	Eupacunoxin	G	(*633*)	As meta-bromobenzoate
(174)	Eupaformonin	GH	(*704*)	
(86)	Eupahyssopin	GG	(*652*)	
(37)	Eupatolide	GG	(*708*)	
(294)	Glaucolide A	G	(*188, 811, 1132*)	
(95)	Glaucolide D	GG	(*77, 338, 1071*)	Mixture with Glaucolide E
(94)	Glaucolide E	GG	(*338*)	Mixture with Glaucolide D
(97)	Glaucolide G	GG	(*78*)	
(192)	Heliangine	GH	(*776*)	Monochloroacetate of dihydro derivative examined
(213)	Liatrin	GH	(*624*)	Monobromoacetate of reduction product
(233)	Linderalactone	GF	(*610*)	
(148)	Melampodin A	GM	(*75, 764, 1128*)	By neutron diffraction
(167)	Melampodin B	G	(*1127, 270*)	
(161)	Melnerin A	G	(*1129*)	Mixture with Melnerin B
(162)	Melnerin B	G	(*1129*)	Mixture with Melnerin A
(134)	Mikanolide, dihydro	GG	(*192*)	
(280)	Molephantin	G	(*640*)	As *p*-bromobenzenesulfonate
(46)	Ovatifolin	GG	(*339*)	
(80)	Parthenolide	GG	(*845*)	
(296)	Pelenolide, hydroxy	G	(*57*)	As *p*-bromobenzoate
(177)	Peucephyllin	GH	(*61*)	
(276)	Phantomolin	G	(*713*)	As epoxide
(123)	Pyrethrosin	GG	(*282*)	As 3-o-chlorophenyloxazoline derivative
(32)	Tamaulipin A	GG	(*1143*)	
(216)	Tirotundin ethyl ether	GH	(*408 a*)	
(78)	Vernolide	GG	(*820*)	
(204)	Woodhousin	GH	(*408 b*)	

[a] G = Germacranolides; GG = Germacrolides; GM = Melampolides; GH = Heliangolides; GF = Furangermacranolides.

(313)

2.4 X-ray Diffraction

In recent years, an increasing number of sesquiterpene lactones has been analyzed by X-ray diffraction methods (*988, 989*). The germacranolides have attracted special attention due to their configurational and conformational variations. Presently, X-ray data are available for three germacranolide subgroups: the germacrolides, melampolides and heliangolides. The compounds whose structure has been established by X-ray analysis are summarized in Table III-2 on page 95.

3. Chemical Transformations of Germacranolides

3.1 Hydrolysis, Relaconizations and Esterifications

In the course of structure elucidation alkaline hydrolysis of a sesquiterpene lactone is frequently desired. A distinct difference in ease of hydrolysis of various ester side chains attached to a germacrolide was described by Herz and Sharma (*441*). The authors demonstrated that a solution of K_2CO_3 in methanol-water at room temperature hydrolyzes 3-hydroxy-2-methylbut-2-enoate in 20 to 30 minutes whereas the iso-butyrate, α-methyl-n-butyrate, tiglate and angelate give no reaction under the same conditions.

Mild hydrolysis with K_2CO_3 of cnicin (63) provides salonitenolide (59) without opening of the γ-lactone group (*1175*). Treatment under stronger alkaline conditions followed by acidification results in the formation of the 7,8-lactonized germacrolide artemisiifolin (117). Upon strong alkaline treatment and subsequent acidification, similar relactonizations of 7,6-lactonized germacranolides to C-8-lactonic compounds have been reported.

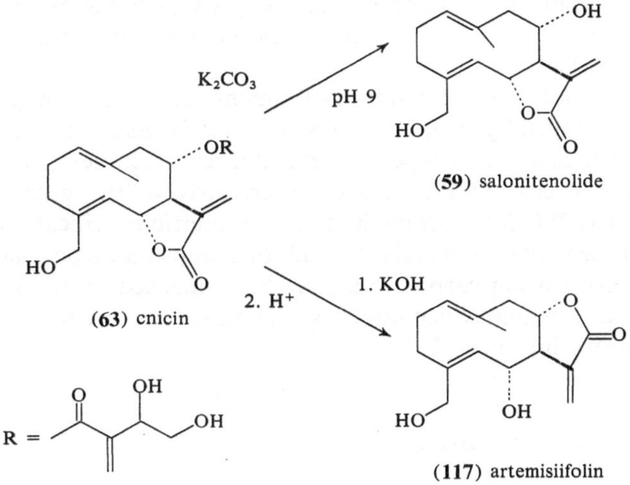

(59) salonitenolide

(63) cnicin

(117) artemisiifolin

$$R =$$

Chart III-15. Alkaline hydrolysis of cnicin

Chihuahuin (50) can be converted to chamissonin (114) (*1175*) and, as outlined in Chart III-16, the dilactones elephantin (92) and elephantopin (91) provide the C-8 lactonic elephantol (314) (*620*). Contrary to the previous examples, the C-8-β-oxygenated eupatoriopicrin (40) reforms the 7,6-lactonic skeleton upon acidification providing eupatolide (37) (*220*).

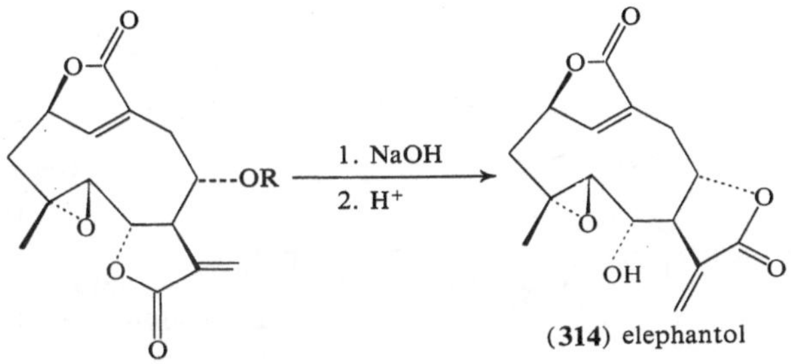

(314) elephantol

(91) elephantopin; R = methacryl
(92) elephantin; R = β-dimethylacryl

Chart III-16. Hydrolysis of elephantopin and elephantin

The data presented above led to a general relactonization rule for germacrolides containing C-6 and C-8 α-oxygen functions which states that upon strong alkaline treatment followed by acidification, this type of germacranolide always relactonizes to C-8 (*1175*). Formation of 7,8-

cis-lactonized germacrolides containing C-8-β-oxygens seems to be less favored, possibly due to interactions of the lactone ring with the C-14 and C-15 methyl groups.

The partial resolution method for assignment of absolute configuration to alcohols developed by HOREAU and KAGAN (*493*) has been successfully applied to various skeletal types of sesquiterpene lactones (*269, 414a*) but failed in the case of the desacetyl derivative of the germacrolide epitulipinolide (**39**) (*228*). This method of asymmetric esterification seems to be not only sensitive to the relative bulk of groups on carbons adjacent to the hydroxyl-bearing carbon, but may be influenced by the shape of the molecule as a whole which could be the reason for the failure of the rule in the above instance (*228*).

3.2 Reductions and Oxidations

Generally, catalytic hydrogenation of sesquiterpene lactones with Pd-C as a catalyst as well as chemical reduction with $NaBH_4$ in methanol proceed with ease under saturation of the lactonic exocyclic methylene group to form the 11,13-dihydro derivatives. As outlined in Chart III-17, catalytic hydrogenation of costunolide (**17**) provides the 11,13-dehydro derivatives, the predominant product (**315**) being the isomer with an α-oriented C-11-methyl group (*849*). In the presence of platinum oxide at 60 atm the hexahydroderivative (**316**) is formed (*849*).

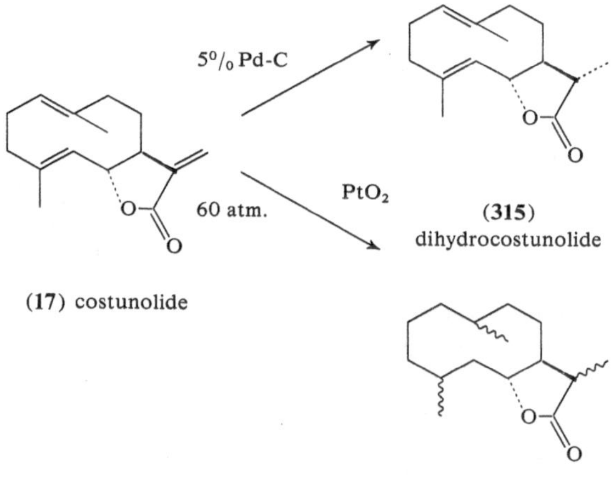

(**17**) costunolide

(**315**) dihydrocostunolide

(**316**)

Chart III-17. Hydrogenation of costunolide

Other examples of Pd-C catalyzed hydrogenations include tamaulipin A (**32**) (*269*), tamaulipin B (**48**) (*268*), artemisiifolin (**117**) (*839*), and isabelin (**122**) (*1171*). Reductions using NaBH₄ have been reported for chihuahuin (**50**) (*851*), pectorolide (**65**) (*733*) and many other compounds.

Catalytic reduction of the melampolide maculatin (**157**) gives the 11,13-dihydro derivative (**317**) whereas NaBH₄ results in the 11-epimer (**318**) (*404*) (Chart III-18).

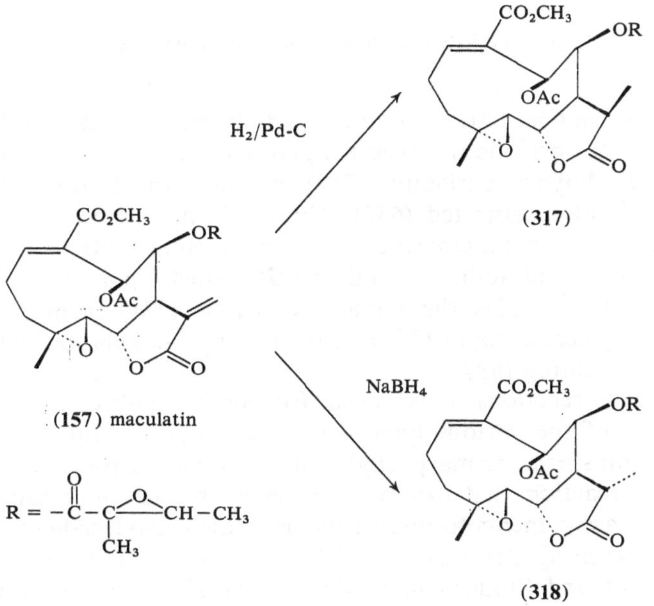

Chart III-18. Reductions of maculatin

Upon NaBH₄ reduction, uvedalin (**142**) provides the 11,13-dihydro product, but catalytic hydrogenation with 10% Pd-C gives the 4,5-11,13-tetrahydro derivative and PtO₂-reduction results in hexahydrouvedalin, in which the 1,10-double bond is additionally saturated (*403*). Enhydrin (**158**), when hydrogenated with 10% Pd-C, affords a mixture of the two epimeric 11,13-dihydro derivatives which can be separated chromatographically (*19*). Examples of the reduction of heliangolides by various methods are: heliangin (**192**) (*502a, 738, 764*), erioflorin (**191**) (*482, 1082*), and provincialin (**182**) (*458*).

Reductive transformations of epoxides to alkenes have in some instances been carried out by using either zinc-copper couple (*631, 634*) or CrCl₂ (*542*) as reducing reagents. Linderane (**238**) was correlated with linderalactone (**233**) by reduction of (**238**) with CrCl₂ in 50% yield (*542*) (Chart III-19).

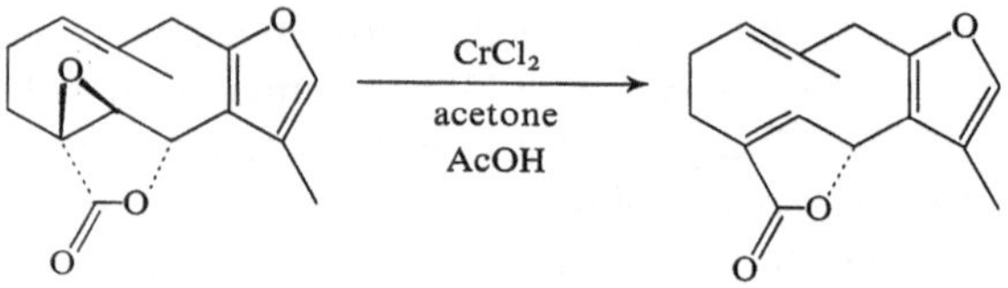

(238) linderane   (233) linderalactone

Chart III-19. Reduction of linderane with CrCl₂

As shown in Chart III-20, reduction of elephantopin (91) with Zn-Cu couple does not provide the expected deoxyelephantopin (75) but results in the deoxydihydro derivative (319) in which the lactonic exocyclic methylene is also saturated (631). This problem can be circumvented by protection of the α-methylene lactone and methacrylate moieties (626) of (91) via the thiol addition product (320) which upon reduction with Zn-Cu couple provides the reduced dilactone (321) from which the desired deoxyelephantopin (75) is generated by S-alkylation with CH₃I followed by heating (634).

Oxidation reactions have been frequently applied in structural elucidations of the various type of germacranolides. MnO₂ oxidations generally transform primary allylic alcohols into α,β-unsaturated aldehydes, a reaction that can be of considerable use for making configurational assignments to double bonds in the cyclodecadiene skeleton as well as for siting OH groups at C-14 and/or C-15. It has been pointed out by Herz and Sharma (440) that the NMR chemical shift of the aldehyde proton in α,β-unsaturated medium ring aldehydes depends on the configuration of the carbon carbon double bond. Compounds with trans-double bonds generally exhibit NMR absorptions near 10 ppm or above whereas values near 9.5 ppm are found in cis-configurated α,β-unsaturated aldehydes. The aldehydic chemical shifts of the germacrolide derivatives (322) and (323) obtained from 14-hydroxycostunolide (44) (200 a) and from 8-desoxysalonitenolide (68) (124) are in good agreement with the predicted values for a trans, trans-cyclodecadiene and the melampolide (324) (416) also shows aldehyde proton absorptions expected for this skeletal type (Chart III-21).

In the case of melampodin B (167) and analogues (80) preparation of the aldehyde (325) (829) with a proton shift near 9.5 ppm indicated a cis-4,5-double bond. This led to a reexamination of melampodin B by X-ray diffraction (1127) which provided unambiguous proof for the presence of a 4,5-cis double bond in the dilactone skeleton of melampodin B (829).

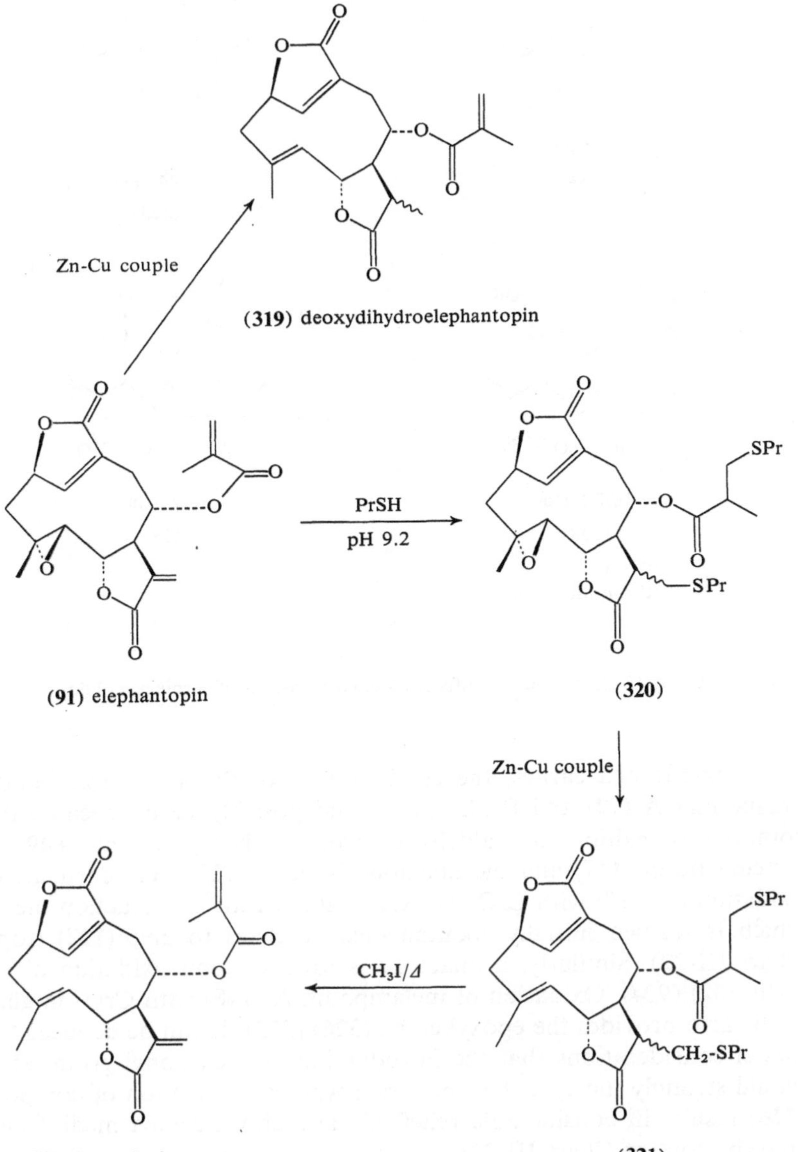

Zn-Cu couple

(319) deoxydihydroelephantopin

(91) elephantopin

$$\xrightarrow[\text{pH 9.2}]{\text{PrSH}}$$

(320)

Zn-Cu couple

(75) deoxyelephantopin

$$\xleftarrow{\text{CH}_3\text{I}/\Delta}$$

(321)

Chart III-20. Zn-Cu-couple reduction of elephantopin and derivatives

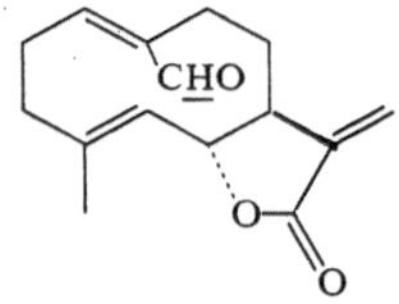

9.95 ppm

(322)

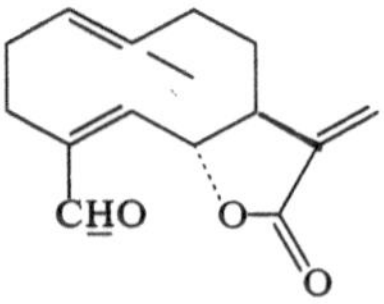

9.98 ppm

(323)

9.48 ppm

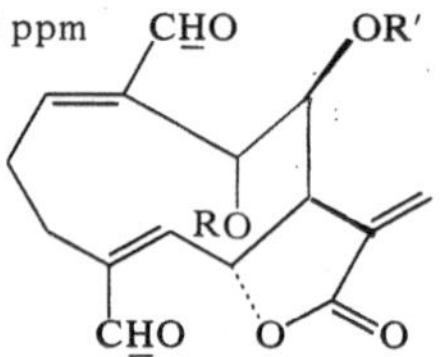

10.22 ppm

(324)

9.49 ppm

(325)

R = Ac
R' = α-Me-butyr.

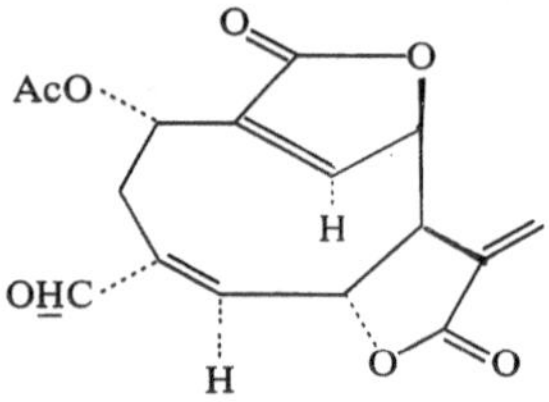

Chart III-21. Chemical shifts of aldehydic protons of germacranolides

As mentioned earlier, the configuration of the 4,5-double bond of urospermal A (**72**) and B (**73**) (*69*) could possibly be *cis* because these compounds exhibit an aldehyde proton absorption at 9.49 ppm. Artemisiifolin (**117**) and the dilactone isabelin (**122**) were correlated by oxidation of (**117**) with CrO_3/H_2SO_4 (*839*). During this reaction the acid which is formed initially spontaneously cyclizes to give (**122**) (top of Chart III-22). Similarly, a dilactone is formed upon oxidation of erio-fertin (**42**) (*924*). Oxidation of melampodin A (**148**) with CrO_3 in glacial acetic acid provides the epoxyketone (**326**) (*272*). It can be deduced from model considerations that the introduction of a carbonyl group at C-9 would strongly increase the ring strain whereas formation of compound (**326**) results in considerable relief of the highly strained medium-sized ring (bottom of Chart III-22).

When 7,6-lactonic germacrolides are oxidized with peracids the 1,10-epoxides are formed preferentially. For instance, costunolide (**17**) is converted to the acid-labile 1,10-epoxide (**327**) with *m*-chloroperbenzoic acid in $CHCl_3$ in the presence of sodium acetate as a buffer (*858*) (Chart III-23). Without buffer, the epoxide cyclizes under the reaction conditions

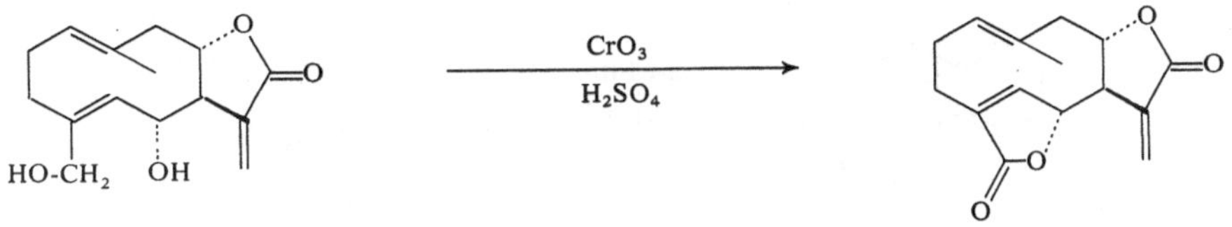

(117)
artemisiifolin

(122) isabelin

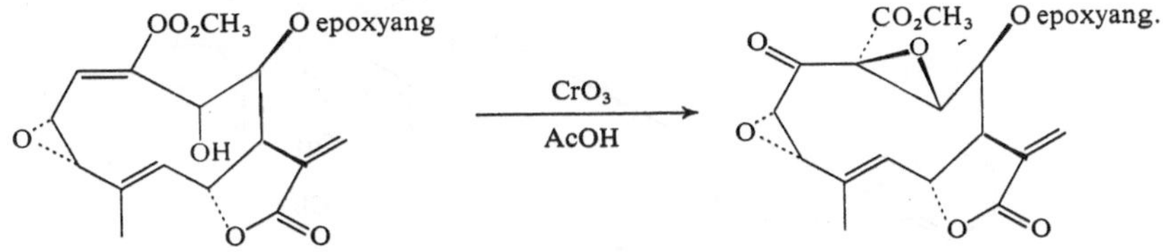

(148) melampodin A

(326)

Chart III-22. CrO₃-oxidation of germacranolides

forming a mixture of eudesmanolides (821). In another example, the 11,13-dihydro derivative of pectorolide (65) can first be epoxidized to the 1,10-epoxide which with an excess of reagent provides the 1,10,4,5-di-epoxide (733). Epoxidation of the 7,8-lactonized desacetyllaurenobiolide (115) which exists as an equilibrating mixture of conformers (1084) results in a mixture of two 4,5-epoxide derivatives (978). As outlined in Chart III-23, epoxide (328) must be derived from a conformation of (115) in which the C-4 methyl group adopts a β-orientation whereas the second isomer, spiciformin (128), results from epoxidation of a conformer of (115) with a C-4-α-oriented methyl group. Preferential 4,5-epoxidation of desacetyllaurenobiolide might be due to anchimeric assistance of the C-6-hydroxyl group in (115). Epoxidation of the conformationally flexible dilactone isabelin (122) provides in excellent yield deoxymikanolide (125) which must be derived from the C-10-α-methyl conformer of (122) (451). Reaction of melampodin A acetate (150) with m-CPBA results in leucanthin B (160) (272) thus establishing its absolute configuration since (150) had been correlated with melampodin A (148) of known absolute configuration (764, 1128).

Photooxidation of germacranolides will be discussed later in the photochemical section.

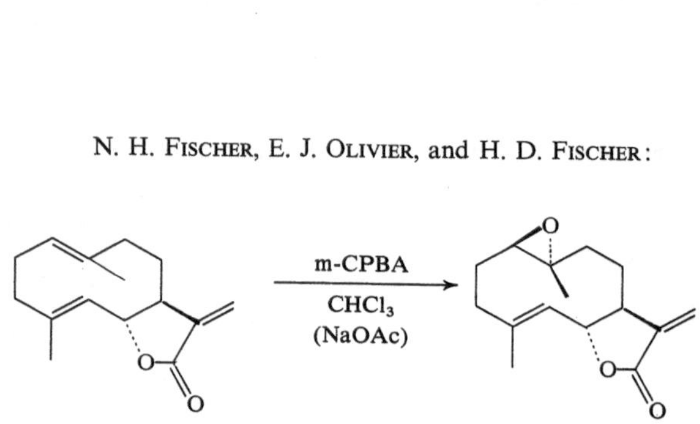

(17) costunolide

(327)

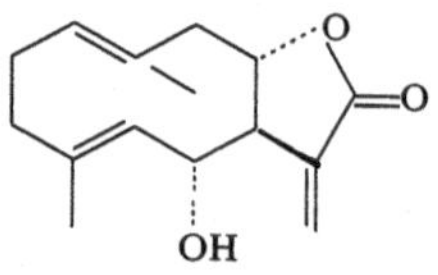

(328)

(115) desacetyl-
laurenobiolide
(2 conformers)

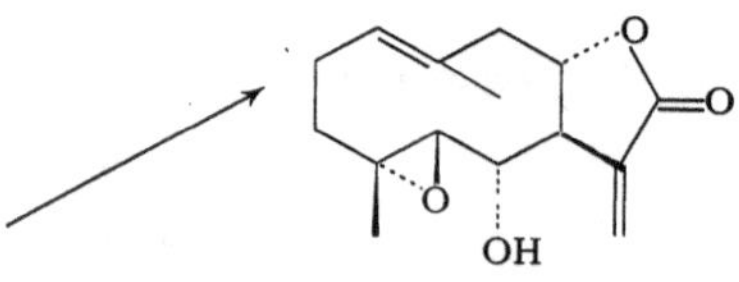

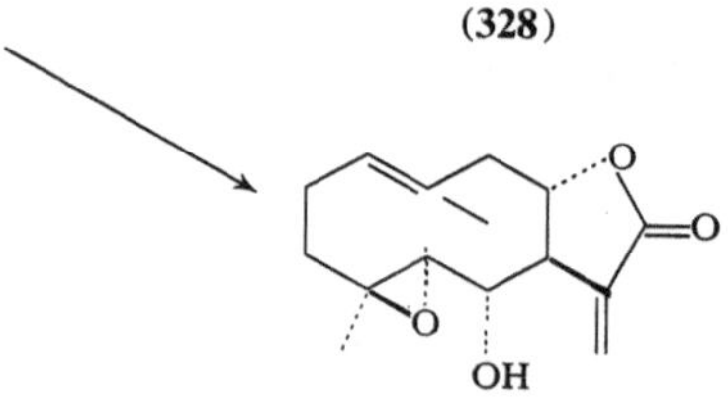

(128) spiciformin

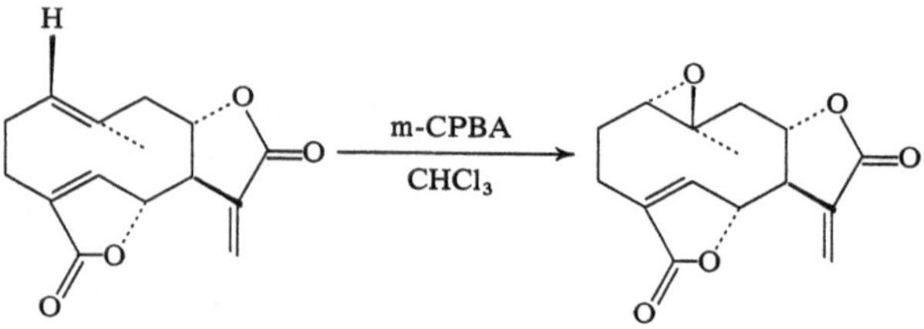

(122) isabelin

(125)
deoxymikanolide

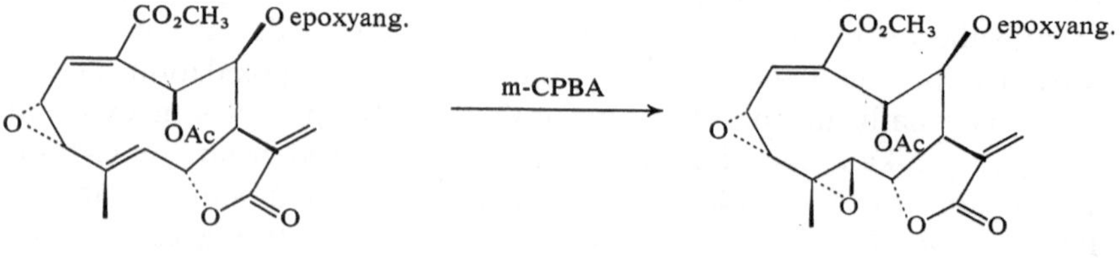

(150) melampodin A acetate

(160) leucanthin B

Chart III-23. Epoxidations of germacranolides

3.3 Cyclization Reactions of Germacranolides

In the process of structure elucidation many Lewis acid-catalyzed cyclization reactions of the cyclodecadiene system or of 1,10- and 4,5-epoxide derivatives have been studied (*1037*). In general, cyclization of germacra-1,5-dienes provides eudesmanolides. As outlined in Chart III-24, costunolide (**17**), when treated with a cation exchange resin (*524, 525*) or HClO₄/AcOH (*615*), undergoes an acid-initiated cyclization *via* cation (**329**) to give a mixture of the eudesmanolides (**330**) and (**331**) (*525*). Cyclizations of costunolide *via* amine adducts are reported to give higher yields (*520, 523*).

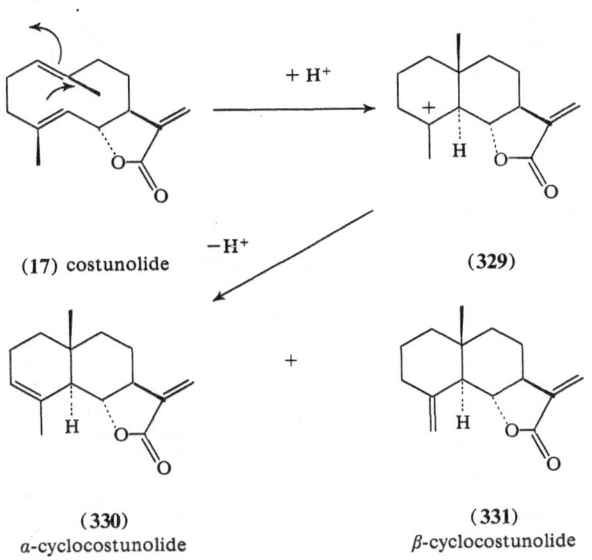

(**17**) costunolide   (**329**)

(**330**)
α-cyclocostunolide

(**331**)
β-cyclocostunolide

Chart III-24. Cyclization of costunolide

Analogous reactions with various Lewis acid catalysts are known for arctiopicrin (**60**) (*1016*), chamissonin (**114**) (*657*), novanin (**49**) (*509*), tulipinolide (**38**) and epitulipinolide (**39**) (*228*), laserolide (**105**) (*488*), laurenobiolide (**116**) (*1041*), and other germacrolides. Chart III-25 shows that treatment of eriofertin diacetate (**332**) with *p*-toluenesulfonic acid in benzene results in the acid (**335**). The conversion most likely proceeds *via* the cyclopropane intermediate (**333**) which upon ring opening and loss of acetic acid gives (**334**) from which the naphthalene skeleton (**335**) is formed by double elimination (*924*).

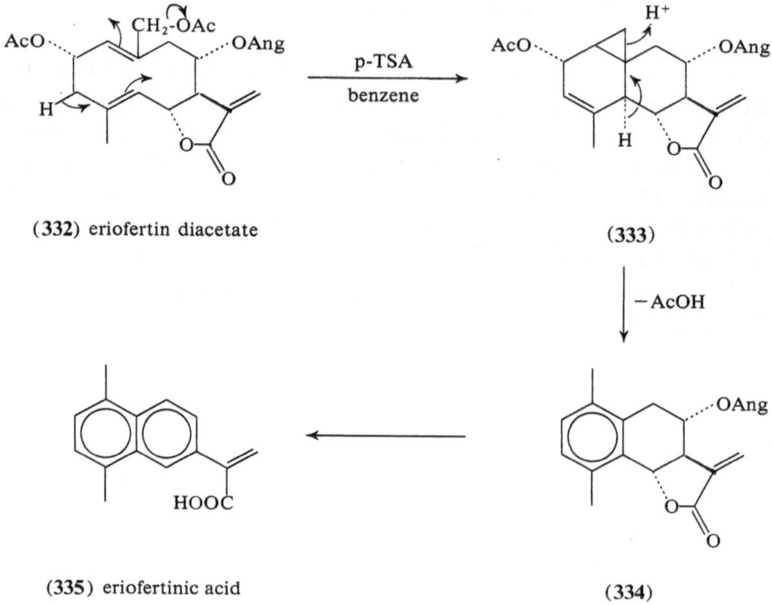

(332) eriofertin diacetate

(333)

(335) eriofertinic acid

(334)

Chart III-25. Cyclization of eriofertin diacetate

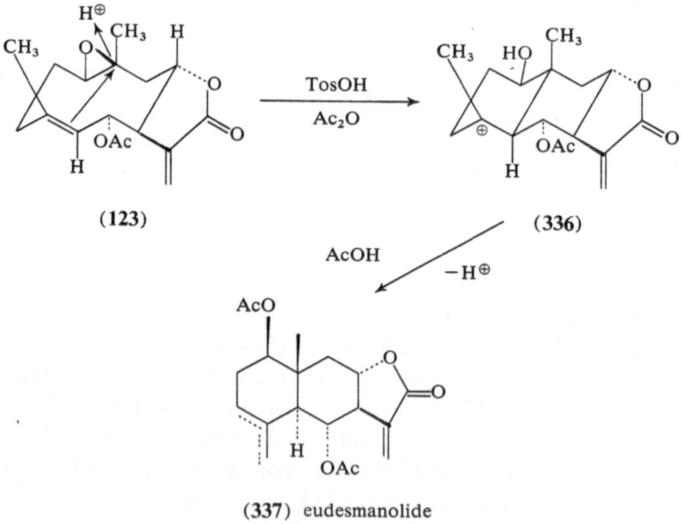

(123)

(336)

(337) eudesmanolide

Chart III-26. Cyclization of pyrethrosin

The conversion of a double bond into an epoxide function either at the 1,10- or the 4,5-position of a germacrolide increases the reactivity, the regiospecificity and stereospecificity of a cyclization reaction (265). As shown above, most acid-catalyzed germacrolide cyclizations provide eudesmanolides. Chart III-26 outlines the cyclization of pyrethrosin (123), a 1,10-epoxide which in an acid-catalyzed Markovnikov type attack of C-5 at C-10 forms the cation (336). This is stabilized by the uptake of acetic acid to give the eudesmanolide (337) (48, 503, 504).

In contrast, BF$_3$-catalyzed reaction of the 4,5-epoxide dihydroparthenolide (109), again proceeds highly specifically in a Markovnikov type attack of C-1 at C-5 to give cation (338) which after the loss of a proton from C-1 now forms the guaianolide (339) (340) (Chart III-27).

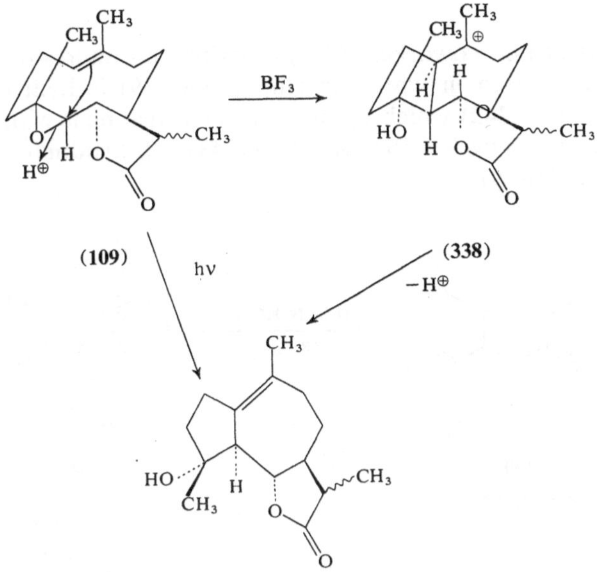

(339) guaianolide

Chart III-27. Cyclization of dihydroparthenolides

Further examples of the conversion of 1,10-epoxygermacrolides into eudesmanolides include the derivatives of costunolide (17) (615, 858), epitulipinolide (39) (235), tanacin (124) (229, 231, 1184), herbolide B (108) (993), and eriofertin (42) (924). Cyclization reactions of 4,5-epoxides which result in formation of guaianolides are also described for parthenolide (80) (687), lanuginolide (112) (1058, 1059) and derivatives.

As shown in Chart III-28, treatment of the cyclodecadiene ketone (340) with silica gel for one week provides the guaianolide (341) (280).

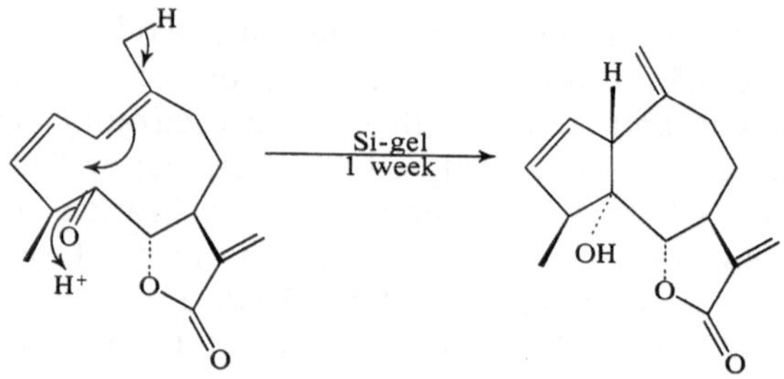

(340)   (341)

Chart III-28. Conversion of a cyclodecadiene ketone to a guaianolide

Doskotch and coworkers (*233*) described an interesting hydrolysis-cyclization reaction in which epitulipinolide (**39**) is hydrolyzed to the carboxylate ion (**342**) which upon controlled treatment with acid undergoes transannular cyclization *via* cation (**343**) to the cadinane type lactone (**344**) (Chart III-29).

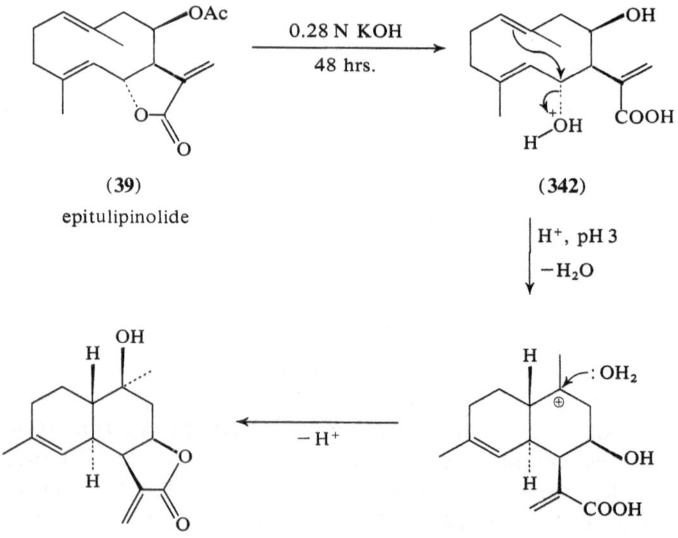

(39)   (342)
epitulipinolide

(344)   (343)

Chart III-29. Conversion of epitulipinolide into a cadinane lactone

Many sesquiterpene lactones, when treated with conc. mineral acids, form deep red or blue solutions (*361*). The reactions involved in the color-forming process have been studied by GEISSMAN and coworkers (*293, 296, 361, 362, 512, 1123*). The proposed mechanism for the color-

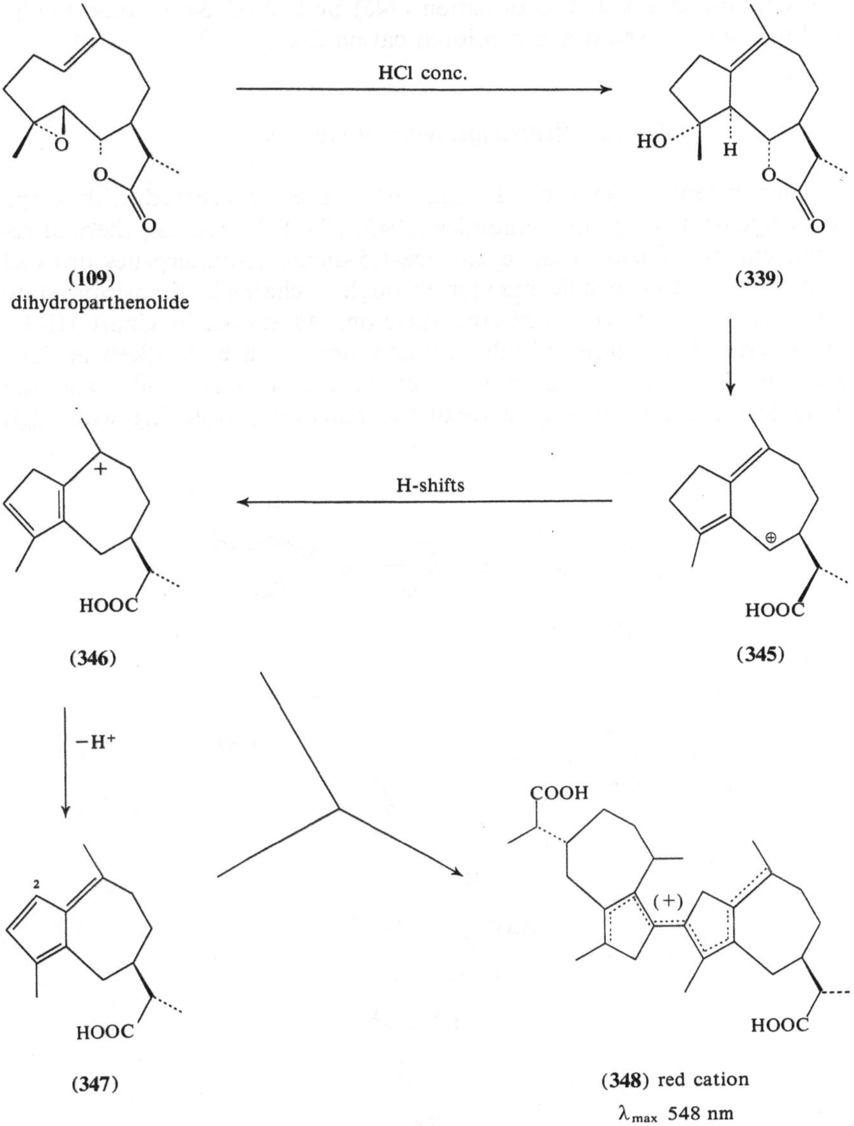

Chart III-30. Formation of a red cation from dihydroparthenolide

forming reactions of dihydroparthenolide (**109**) is outlined in Chart III-30. Treatment of dihydroparthenolide (**109**) with conc. HCl gives first the guaianolide (**339**) from which by acid-catalyzed opening of the lactone ring cation (**345**) can be formed. Rearrangement of (**345**) to the allylic cation (**346**) followed by the loss of a proton could provide (**347**). Nucleophilic attack at C-3 of cation (**346**) by C-2 of (**347**) would finally lead to the proposed dimeric colored cation (**348**) (*361*).

3.4 Cope Rearrangements of Germacranolides

Two recent reviews by Takeda have been concerned with Cope rearrangements of germacranolides (*1042, 1043*). In general, thermal rearrangements of *trans,trans*-cyclodeca-1,5-dienic sesquiterpenes proceed in a highly stereospecific manner through a chair-like transition state resulting in a divinylcyclohexane skeleton. As shown in Chart III-31, short term thermolysis of dihydrotamaulipin A acetate (**309**) at 220° gives an 2:3 equilibrium mixture of starting material (**309**) and the divinylcyclohexane derivative (**350**) the reaction proceeding with high

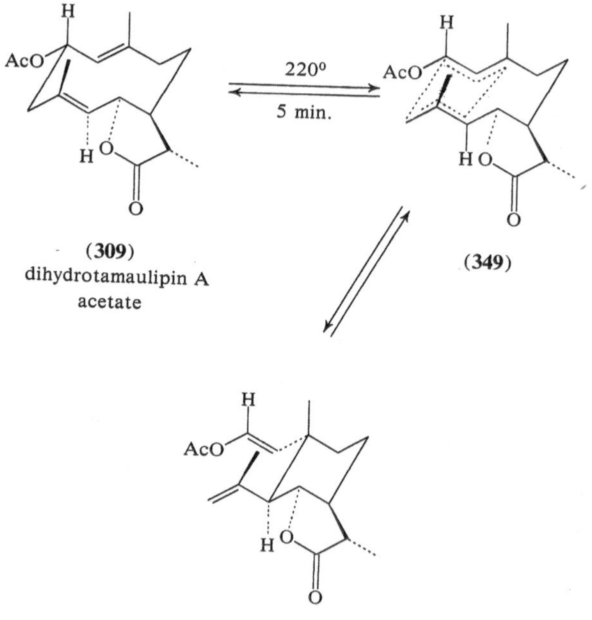

(**309**)
dihydrotamaulipin A
acetate

(**349**)

(**350**)

Chart III-31. Cope rearrangement of dihydrotamaulipin A acetate

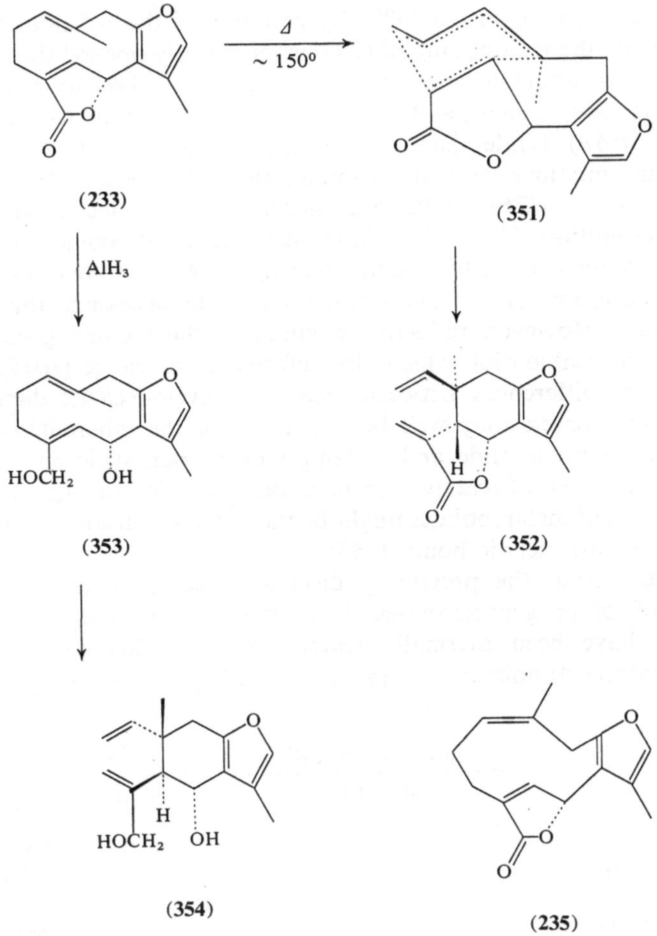

Chart III-32. Cope rearrangements of linderalactone and derivative

stereospecificity *via* the chair-like transition state (349) (269). The reverse of the above Cope rearrangement has recently been applied by GRIECO and NISHIZAWA toward the total synthesis of costunolide (17) from an elemene precursor (355).

Unlike dihydrotamaulipin A acetate, whose stable conformation (309) in solution was described earlier in this chapter (79), linderalactone (233) gives an elemadiene compound (352) with an α-methyl group at the C-10 angular position (1046a) (Chart III-32). This change in the stereochemistry of the Cope rearrangement was ascribed (1046a) to the presence of the C-5, C-7 lactone ring which prohibits conversion into a conformation analogous to (309) by rotation of the 4,5-double bond. Indeed, when the lactone ring of (233) is reductively opened the resulting diol (353) can adopt a transition state similar to (349) to provide the elemadiene (354) which has the same stereochemistry at C-5 and C-10 as (350) (*1046a*). Linderalactone as well as litsealactone (234) give 2 : 3 equilibrium mixtures of starting materials and products (*1046a*) but neolinderalactone (235), a 1,10-*cis* isomer of linderalactone is stable under the same conditions (*1045*). The difference in reactivity could possibly be due to the rigidity of the ten-membered ring the 4,6-lactone grouping not allowing adoption of the chair transition state necessary for the rearrangement. However, reductive opening of the lactone grouping in (235) gave a flexible diol, which also refused to rearrange (*1045*). These experimental differences between *trans, trans*-cyclodeca-1,5-dienes and the *cis, trans*-isomers may well be diagnostic of the inherent difference between the germacrolide and melampolide carbon skeleton. A major reason for the lack of reactivity in reactions involving the 4,5- and 1,10-double bonds of melampolides might be the greater transannular distance between the two olefinic bonds (*265*).

In addition to the previously discussed examples, a considerable number of other germacrolides, lactonized toward the 7,6- or 7,8-positions, have been thermally rearranged to elemadienes. Whereas 11,13-dihydrocostunolide (98) gives a good yield of rearrangement

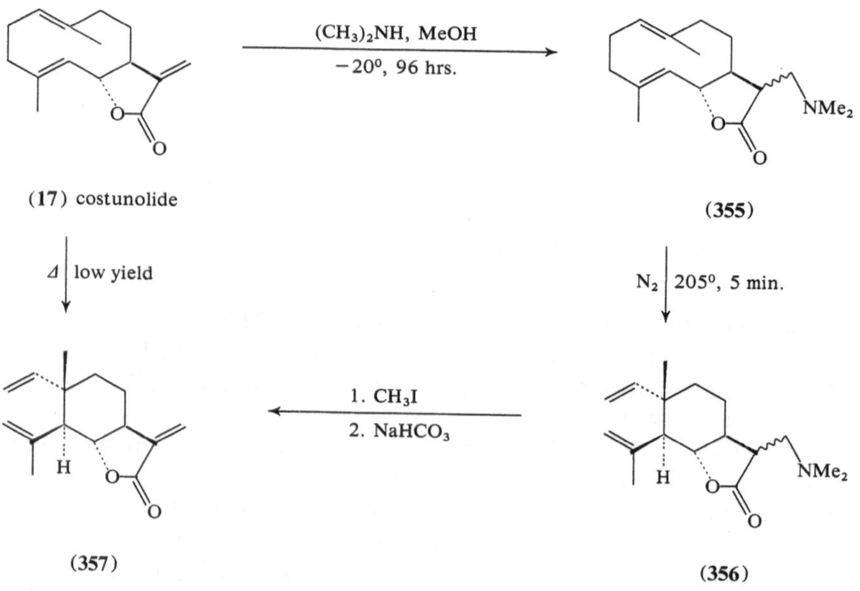

Chart III-33. Cope rearrangement of costunolide and derivatives

products (*220*) the yield of the 11,13-dehydro compound (**357**) from thermolysis of costunolide (**17**) is very low due to side reactions involving the α-methylene lactone group of (**17**). This problem can be circumvented by protecting of α,β-unsaturated γ-lactone group with dimethylamine to give (**355**) and regenerating the α-methylene lactone function after thermal conversion of (**355**) to (**356**) (*521*) as shown in Chart III-33. Other Cope rearrangements were carried out on the following germacrolides or their derivatives: carmeline (**101**) (*928*), chihuahuin (**50**) (*851*), chamissonin (**114**) (*657*), epitulipinolide (**39**) (*235*), eupaserrin (**35**) (*625*), eriofertin (**42**) (*924*), laurenobiolide (**116**) (*133, 1041, 1044*), laserolide (**105**) (*488*) and tamaulipin B (**48**) (*268*).

3.5 Photochemical Reactions of Germacranolides

Recently, DOSKOTCH and coworkers (*230*) and EL-FERALY et al. (*249*) isolated the first hydroperoxide-containing sesquiterpene lactones peroxyferolide (**254**), peroxycostunolide (**248**) and peroxyparthenolide (**253**), compounds of unexpected high thermal stability. The authors were able to prepare the hydroperoxides (**248**) and (**254**) by a common photooxygenation procedure involving singlet oxygen generated by methylene blue sensitized oxygenation. As outlined in Chart III-34, costunolide (**17**) was converted to peroxycostunolide (**248**). Subsequent reduction of (**248**) with PPh$_3$ provided artemorin (**246**). In a similar oxidation reaction, peroxyferolide (**254**) was prepared from lipiferolide (**83**). These laboratory conversions strongly suggest that sesquiterpene lactones with C-10,14-double bonds and oxygen functions at C-1 such as artemorin (**246**) are produced *in vivo* by similar oxygenation-reduction reactions. Furthermore, the new evidence for the natural existence of sesquiterpene hydroperoxides may indicate their general importance as intermediates in the oxydative steps of sesquiterpene biogenesis.

Irradiation of 11,13-dihydrocostunolide (**98**) with a quartz lamp in xylene-isopropyl alcohol (1 : 9) gave the guaianolide (**358**). The reaction possibly involves an intramolecular C-14 to C-4 H shift as indicated by the arrows in Chart III-35 (*1141*).

Other germacrolides of this type were recovered unchanged upon prolonged irradiation at 254 nm (ref. *1172,* page 19). However, as shown in Chart III-27, dihydroparthenolide (**109**), when irradiated in benzene, provided the guaianolide (**339**) (*340*).

Extensive photolytic studies on the dilactone isabelin (**122**) were carried out by MABRY and coworkers (*1169*). As outlined in Chart III-36,

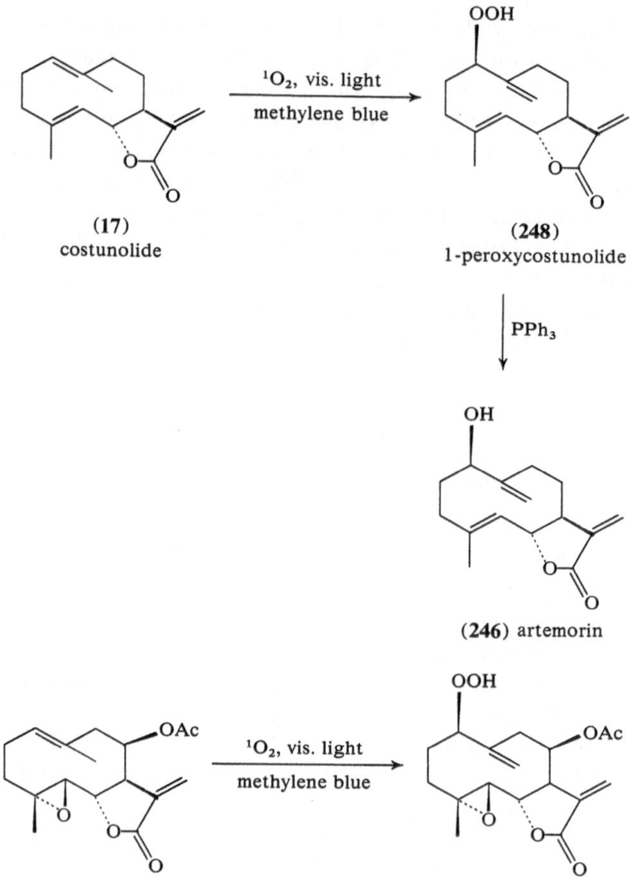

(**17**) costunolide

(**248**) 1-peroxycostunolide

(**246**) artemorin

(**83**) lipiferolide

(**254**) peroxyferolide

Chart III-34. Photooxygenations of germacrolides

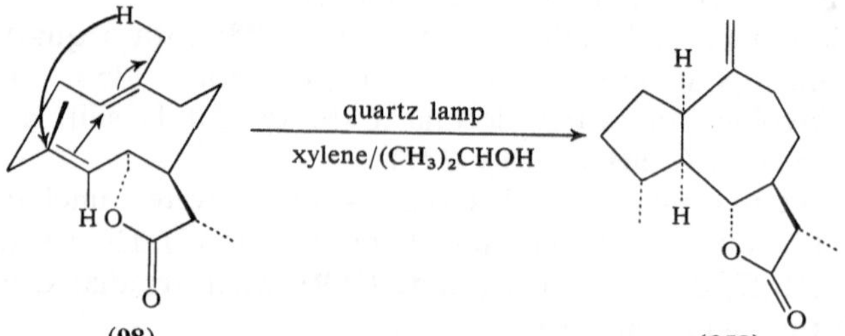

(**98**) dihydrocostunolide

(**358**)

Chart III-35. Photolysis of dihydrocostunolide

that conformer of isabelin reacts in which the C-1,10- and C-4,5-double bonds are parallel to form in a photochemically allowed suprafacial 2 + 2 cycloaddition photoisabelin (**359**) which can be catalytically reduced to the dihydroderivative (**361**). Hydrogenation of isabelin (**122**) provided the 11,13-dihydro derivative (**360**) which upon irradiation gave the 2 + 2 cycloadduct (**361**) as minor product. The major compound from the UV irradiation of dihydroisabelin (**360**) was lumidihydroisabelin (**362**) which is formed from the isabelin conformer with an α-oriented C-14 methyl by an ene reaction analogous to the process shown in Chart III-35.

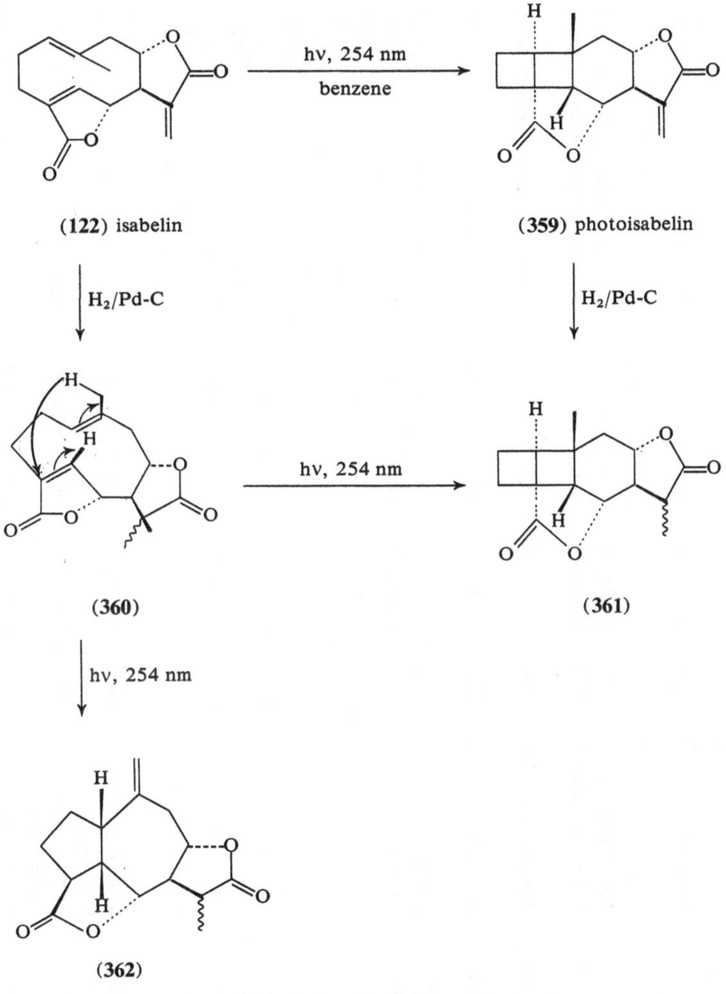

(**122**) isabelin   (**359**) photoisabelin

(**360**)   (**361**)

(**362**)

Chart III-36. Photolysis of isabelin and derivatives

Table III-3. *Naturally Occurring Germacranolides and Biogenetic Derivatives*

Structure Number	Name of Compound	Type[c]	Formula	m.p. °C	$[\alpha]_D$	Plant Source[d]	References	Comments
(147)	Acanthamolide	GM	$C_{19}H_{25}NO_5$	249–51	—	Acanthospermum glabratum (926)		
(145)	Acanthospermal A	GM	$C_{23}H_{30}O_8$	gum	—	Acanthospermum australe (416)		$[\alpha]_{Hg} -54$; $[\theta]_{224} -54400$
(146)	Acanthospermal B	GM	$C_{22}H_{28}O_8$	gum	—	Acanthospermum hispidum (416)		$[\alpha]_{Hg} -33$; $[\theta]_{224} -40200$ $[\theta]_{262} -4390$; $[\theta]_{220} +120680$
(64)	Alatolide	GG	$C_{19}H_{26}O_6$	59–61	+64.4	Jurinea alata	(242, 189[b], 87, 467, 468)	
(67)	Albicolide	GG	$C_{15}H_{20}O_4$	104–5	+73	Jurinea albicaulis	(1032)	
(60)	Arctiopicrin	GG	$C_{19}H_{26}O_6$	115–7	+133	Arctium minus	(152, 1028[a], 1031[a], 1017)	
						A. lappa	(283)	
						A. nemorosum	(283)	
						A. tomentosum	(283, 241)	
(99)	Artabin	GG	$C_{15}H_{22}O_3$	162–4	+220	Artemisia absinthium	(18, 16)	
						Ambrosia artimisiifolia	(839)	
(117)	Artemisiifolin	GG	$C_{15}H_{20}O_4$	131	+54.6	Centaurea seridis	(316)	
						Ambrosia artemisiifolia	(839)	
						A. castonensis	(463)	
						A. confertiflora	(851)	
(118)	Artemisiifolin, C-15-acetyl	GG	$C_{17}H_{22}O_5$	102–4	+48	Centaurea seridis	(316)	
(232)	Artemisiifolin, cis,cis-15-desoxy	GC	$C_{15}H_{20}O_3$	oil	+35	Inula britannica	(115)	
(245)	Aristolactone	G	$C_{15}H_{20}O_2$	110–1	−156	Aristolochia serpentaria	(1005, 675[a])	
(251)	Artevasin	G	$C_{15}H_{20}O_4$	209–10	+244	Artemisia tridentata	(39)	
						A. cana	(81)	
						A. tripartita	(82)	

No.	Name	Class	Formula	m.p.	$[\alpha]$	Plant source	Refs.	Comments
(246)	Artemorin	G	$C_{15}H_{20}O_3$	120–1	+89	Artemisia verlotorum; A. mexicana; Magnolia grandiflora	(285, 300) (887) (249a)	Magnoliaceae
(207)	Atripliciolide, isobutyrate	G	$C_{19}H_{22}O_6$	oil	—	Isocarpha atriplicifolia	(113)	
(210)	Atripliciolide, isovalerate	G	$C_{20}H_{24}O_6$	oil	–41	Isocarpha atriplicifolia	(113)	
(209)	Atripliciolide, (2-methylacrylate)	G	$C_{19}H_{20}O_6$	oil	—	Isocarpha atriplicifolia	(113)	
(211)	Atripliciolide, tiglate	G	$C_{20}H_{22}O_6$	oil	—	Isocarpha atriplicifolia	(113)	
(261)	Badgerin	G	$C_{15}H_{20}O_5$	207–8	+8.5	Artemisia arbuscula	(979)	Structure redrawn by these authors
(289)	Balsamin	G	$C_{20}H_{26}O_6$			Stizolophus balsamita	(920)	
(129)	Baileyin	G	$C_{15}H_{20}O_4$	189		Baileya pleniradiata	(1123)	acetate: m. p. 176–7°, $[\alpha]_D$ +17.6
(102)	Balchanolide	GG	$C_{15}H_{22}O_3$	154	+183	Artemisia balchanorum	(391, 228, 1020)	
(103)	Balchanolide acetate	GG	$C_{17}H_{24}O_4$	125	+128.1	Achillea millefolium	(472)	
(104)	Balchanolide, hydroxy	GG	$C_{15}H_{22}O_4$	163	+105	Artemisia balchanorum	(391, 1020)	
(131)	Balchanolide, iso-	GG	$C_{15}H_{22}O_3$	133	+122	Artemisia balchanorum	(391, 1020)	
(212)	Budlein-A	GH	$C_{20}H_{22}O_7$	106–8	–82.3	Viguiera buddleiaeformis; V. angustifolia	(898) (368)	
(45)	Budlein-B	G	$C_{15}H_{20}O_4$	162	+3.1	Viguiera buddleiaeformis	(898, 804)	$J_{7,13}$-Values indicate that this substance is a heliangolide rather than a germacrolide
(208)	Calaxin	GH	$C_{19}H_{20}O_6$	180–2	–115	Calea axillaris	(806, 1200)	
(101)	Carmelin	GG	$C_{19}H_{26}O_6$	176–7	+146.1	Stevia serrata	(928)	
(114a)	Chamissarin	GG	$C_{17}H_{22}O_5$	gum	—	Ambrosia chamissonis	(303)	
(115)	Chamissellin	GG	$C_{15}H_{20}O_3$			Ambrosia chamissonis	(303)	See Laurenobiolide, desacetyl
(114)	Chamissonin	GG	$C_{15}H_{20}O_4$	124–5	–19.8	Ambrosia chamissonis; A. acanthicarpa; A. dumosa	(310, 657a, 303) (302) (463)	

[a] Reference containing structural revision; [b] Reference gives X-ray data; [c] G = Germacranolides, GC = cis-l(10), cis-4,5-Germacranolides, GF = Furangermacranolides, GG = Germacrolides, GH = Heliangolides, GM = Melampolides; [d] When plant sources are from a family other than Compositae, the family name is listed under Comments.

Table III-3 *(continued)*

Structure Number	Name of Compound	Type[c]	Formula	m. p. °C	$[\alpha]_D$	Plant Source[d]	References	Comments
(123a)	Chamissonin, 1(10)-epoxy	GG	$C_{15}H_{20}O_5$			Ambrosia chamissonis	(303)	Mixture with (128a)
(128a)	Chamissonin, 4,5-epoxy	GG	$C_{15}H_{20}O_5$			Ambrosia chamissonis	(303)	Mixture with (123a)
(277)	Chapliatrin	G	$C_{24}H_{32}O_{10}$	gum	—	Liatris gracilis L. chapmanii L. tenuifolia	(460) (460) (438)	$[\alpha]_{Hg}$ −35
(279)	Chapliatrin, acetyl-	G	$C_{26}H_{34}O_{11}$	143		Liatris gracilis	(460)	$[\alpha]_{Hg}$ −53
(278)	Chapliatrin, iso-	G	$C_{29}H_{32}O_{10}$	162	−74	Liatris chapmanii	(460)	
(50)	Chihuahuin	GG	$C_{17}H_{22}O_5$	175	+112	Ambrosia confertiflora	(851)	
(178)	Chromolaenide	GH	$C_{22}H_{28}O_7$	138	−146	Chromolaena glaberirma Isocarpha oppositifolia	(100) (113)	
(176)	Chromolaenide, 3-epi-20-acetoxy	GH	$C_{24}H_{30}O_9$	oil	—	Isocarpha oppositifolia	(113)	
(180)	Chromolaenide, 20-hydroxy	GH	$C_{22}H_{28}O_7$	oil	−118	Isocarpha oppositifolia	(113)	See Eucannabinolide
(179)	Chromolaenide, 20-tiglinoyloxy	GH	$C_{27}H_{34}O_9$	oil	−95	Isocarpha oppositifolia	(113)	
(51)	Chromolaenide, 4,5-trans-3-desacetyl-20-tiglinoyloxy	GG	$C_{25}H_{32}O_8$	149	—	Isocarpha oppositifolia	(113)	
(258)	Chrysanolide	G	$C_{17}H_{20}O_5$	204–5	−52	Chrysanthemum cinerariaefolium	(231)	
(206)	Ciliarin	GH	$C_{19}H_{22}O_6$	148	−143	Helianthus ciliaris	(806, 1200)	
(166)	Cinerenin	G	$C_{17}H_{20}O_6$	161–3	—	Melampodium cinereum M. argophyllum	(829, 1127a) (829)	

						Species	References	Notes
(63)	Cnicin	GG	$C_{20}H_{26}O_7$	143	+58	Cnicus benedictus	(936, 1175[a])	
						Centaurea diffusa	(236)	
						C. calcitrapa	(237)	
						C. iberica	(237)	
						C. micranthos	(238)	
						C. ovina	(237)	
						C. stoebe	(236, 1015)	
(292)	Confertolide	G	$C_{21}H_{28}O_9$	185–8	−82	Vernonia conferta	(1097, 1096, 1098[a])	$[\theta]_{249} + 2{,}420$; $[\theta]_{222} − 15{,}676$; $[\theta]_{262} − 6600$; $[\theta]_{220} + 110{,}000$
(17)	Costunolide	GG	$C_{15}H_{20}O_2$	106–7	+128	Saussurea lappa	(389, 849, 391, 957, 999[b])	
						Ambrosia chamissonis	(132[b], 228, 1087, 355, 93)	
						Artemisia balchanorum	(1141, 547, 942, 944, 391)	
						Centaurea kurdica	(1012, 93)	
						Chrysanthemum achillea	(93)	
						Cosmos hybridus	(93)	
						C. sulphuxus	(93)	
						Critonia morifolia	(93)	
						C. sexangulatis	(114)	
						Frullania tamarisci	(173)	
						Hymenoclea monogyra	(1091)	
						Inula helenium	(93)	
						Lauris nobilis	(1040)	
						Matricaria nigellaefolia	(93)	
(228)	Costunolide, cis, cis, 3α-acetoxy-8β-hydroxy	GC	$C_{17}H_{22}O_5$	oil	+60.7	Anthemis cretica	(120)	
(188)	Costunolide, 4,5-cis, 14-acetoxy-8β-(4-hydroxy-tiglinoyloxy)	GH	$C_{22}H_{28}O_7$	oil	−101.5	Eupatorium hyssopifolium	(112)	acetate: m. p. 142–3°
(98)	Costunolide, 11,13-dihydro	GG	$C_{15}H_{22}O_2$	77	+113	Saussurea lappa	(957, 545, 279, 185, 1087, 943, 522)	

Table III-3 *(continued)*

Structure Number	Name of Compound	Type[c]	Formula	m.p.°C	$[\alpha]_D$	Plant Source[d]	References	Comments
(52)	Costunolide, 3β,9α-dihydroxy-8β-angeloyloxy	GG	$C_{20}H_{27}O_6$	oil	—	*Eupatorium rotundifolium*	(112)	
(187)	Costunolide, 4,5-cis, 14-hydroxy-8β-(4-hydroxytiglinoyloxy)	GH	$C_{20}H_{26}O_6$	oil	−80	*Eupatorium hyssopifolium* *E. mohrii*	(112) (112)	
(227)	Costunolide, *cis, cis,* 2α-hydroxy	GC	$C_{15}H_{20}O_3$	oil	—	*Chrysanthemum poterifolium*	(98)	identified as acetate
(36)	Costunolide, 8-hydroxy	GG	$C_{15}H_{20}O_3$	gum	—	*Artemisia balchanorum*	(1021, 228)	acetate m. p. 98°, $[\alpha]_D$ + 37.4
(53)	Costunolide, 9β-hydroxy	GG	$C_{15}H_{20}O_3$			*Inula helenium* *I. royleana*	(113) (113)	See Haageanolide
(44)	Costunolide, 14-hydroxy	GG	$C_{15}H_{20}O_3$	—	—	*Clibadium surinamense*	(200a, 954)	
(62)	Costunolide, 14-hydroxy-8β-(4-hydroxytiglinoyloxy)	GG	$C_{20}H_{26}O_6$	oil	+20	*Eupatorium mohrii*	(112)	
(56)	Costunolide, 9β-isobutyroyloxy	GG	$C_{19}H_{26}O_4$	154	+104.8	*Inula helenium*	(111)	
(57)	Costunolide, 9β-isovaleroyloxy	GG	$C_{20}H_{28}O_4$	—	—	*Inula helenium*	(111)	
(69)	Costunolide, 15-isovaleroyloxy	GG	$C_{20}H_{28}O_4$	122	+67.4	*Vernonia hirsuta*	(94)	
(58)	Costunolide, 9β-(2-methylbutyroyloxy)	GG	$C_{20}H_{28}O_4$	oil	—	*Inula helenium* *I. royleana*	(111) (111)	
(248)	Costunolide, 1-peroxy	G	$C_{15}H_{20}O_4$	141	+171	*Magnolia grandiflora*	(249, 285)	Identical with Verlotorin; Magnoliaceae

No.	Name	Code	Formula	m.p.	$[\alpha]$	Species	Ref.	Notes
(55)	Costunolide, 9β-propionyloxy	GG	$C_{18}H_{24}O_4$	oil	+54.3	*Inula royleana*	(111)	
(70)	Costunolide, 15-senecioyloxy	GG	$C_{20}H_{26}O_4$	146	—	*Vernonia hirsuta*	(94)	
(250)	Dentatin B	G	$C_{15}H_{20}O_4$	—	—	*Artimisia tridentata*	(505)	identified as diacetate m.p. 142.5–143.0
(120)	Dicomanolide, 14-acetoxy	GG	$C_{19}H_{24}O_7$	oil	−35.3	*Dicoma anomala*	(108)	
(121)	Dicomanolide, 14-oxo	GG	$C_{17}H_{20}O_6$	oil	+68	*Dicoma anomala*	(108)	
(198)	Eleganin	GH	$C_{22}H_{26}O_9$	142–3	−108	*Liatris elegans* / *L. sabra*	(439, 438)	
(92)	Elephantin	GG	$C_{20}H_{22}O_7$	242–4	−380	*Elephantopus elatus*	(620, 619, 715[b])	
(91)	Elephantopin	GG	$C_{19}H_{20}O_7$	262–4	−398	*Elephantopus elatus*	(620, 619, 715[b])	
(75)	Elephantopin, deoxy	GG	$C_{19}H_{20}O_6$	198–200	−54.6	*Elephantopus scaber* / *E. carolinianus*	(341, 634, 715[b]) (637)	
(76)	Elephantopin, iso-deoxy	GG	$C_{19}H_{20}O_6$	150–3	+188.4	*Elephantopus scaber*	(342)	
(158)	Enhydrin	GM	$C_{23}H_{28}O_{10}$	183–4	−55.6	*Enhydra fluctans* / *Melampodium perfoliatum* / *M. longipilum* / *Polymnia uvedalia*	(540, 544[a], 558[b], 19, 613) (126) (952) (265)	$[\theta]_{281}$ −442; $[\theta]_{243}$ +8980; $[\theta]_{216}$ −85,400
(222)	Eremantholide A	GH	$C_{19}H_{24}O_6$	181–3	+65	*Eremanthus elaeagus*	(848[b])	diacetate: m.p. 161–3°; $[\alpha]_D$ +770
(42)	Eriofertin	GG	$C_{20}H_{26}O_6$	118–23	—	*Eriophyllum confertiflorum*	(924)	
(43)	Eriofertopin, 2-O-Acetyl	GG	$C_{21}H_{26}O_7$	—	+29	*Eriophyllum confertiflorum*	(618)	
(41)	Eriofertopin	GG	$C_{19}H_{24}O_6$	—	+89	*Eriophyllum confertiflorum*	(618)	
(191)	Erioflorin	GH	$C_{19}H_{24}O_6$	236–8	−100	*Eriophyllum confertiflorum*	(1082)	
(194)	Erioflorin acetate	GH	$C_{21}H_{26}O_7$	193–6	−114	*Podanthus ovatifolius*	(313, 1082)	diacetate: m.p. 132–3°
(195)	Erioflorin methacrylate	GH	$C_{23}H_{28}O_7$	155–8	−87	*Podanthus ovatifolius*	(313, 313)	

Table III-3 (continued)

Structure Number	Name of Compound	Type[c]	Formula	m.p. °C	$[\alpha]_D$	Plant Source[d]	References	Comments
(299)	Eriolin	G	$C_{15}H_{22}O_4$	238–40	−42	Eriophyllum confertiflorum	(1082)	
(300)	Eriolin, hydroxy	G	$C_{15}H_{22}O_5$	256–60	−13	Eriophyllum confertiflorum	(1082)	
(190)	Eriophyllin	GH	$C_{22}H_{28}O_8$	220–2	−118	Eriophyllum confertiflorum	(1082, 313[a])	
(189)	Eriophyllin B	GH	$C_{19}H_{24}O_7$	oil	—	Eriophyllum confertiflorum	(1082, 313[a])	$[\alpha]_{546}$ −72.6
(196)	Eriophyllin C	GH	$C_{19}H_{22}O_7$	166–9	—	Eriophyllum confertiflorum	(1082, 313[a])	$[\alpha]_{546}$ −65.8
(180)	Eucannabinolide	GH	$C_{22}H_{28}O_8$	gum	−121	Eupatorium cannabinum	(240, 482[a], 458[a])	Reported before Chromolaenide, 20-hydroxy
(285)	Eupacunin	G	$C_{22}H_{28}O_7$	166–7	+55	Eupatorium cuneifolium	(632, 633)	
(287)	Eupacunolin	G	$C_{22}H_{28}O_7$	164–5	+46	Eupatorium cuneifolium	(632)	
(286)	Eupacunoxin	G	$C_{22}H_{28}O_8$	171–2	+27	Eupatorium cuneifolium	(632)	
(174)	Eupaformonin	GH	$C_{17}H_{22}O_5$	216–8	—	Eupatorium formosanum	(712, 704[b])	
(175)	Eupaformosanin	GH	$C_{22}H_{28}O_8$	91	−99.5	Eupatorium formosanum	(651)	
(86)	Eupahyssopin	GG	$C_{20}H_{26}O_7$	125	−138.9	Eupatorium hyssopifolium	(652[b])	See Eupassopin
(186)	Euparhombin	GH	$C_{19}H_{24}O_5$	140–1	−149.6	Eupatorium rhomboideum	(363)	
(35)	Eupaserrin	GG	$C_{22}H_{28}O_7$	153–4	+71.2	Eupatorium semiserratum	(632)	
						E. cuneifolium	(625)	
(34)	Eupaserrin, desacetyl	GG	$C_{20}H_{26}O_6$	gum	+75	Eupatorium semiserratum	(625)	
				134–5		Helianthus pumilus	(409a)	
(88)	Eupassofilin	GG	$C_{38}H_{60}O_9$	gum	−143	Eupatorium hyssopifolium	(441)	$[\theta]_{250}$ +954
(84)	Eupassopilin	GG	$C_{20}H_{26}O_6$	gum	−161	Eupatorium hyssopifolium	(441)	$[\theta]_{235}$ −7590
(86)	Eupassopin	GG	$C_{20}H_{26}O_7$	gum	−137.5	Eupatorium hyssopifolium	(441)	See Eupahyssopin; $[\theta]_{235}$ −6038
(183)	Eupatocunin	GH	$C_{22}H_{28}O_7$	163–4	−129	Eupatorium cuneifolium	(632)	
(184)	Eupatocunoxin	GH	$C_{22}H_{28}O_8$	200–1	−209	Eupatorium cuneifolium	(632)	
(37)	Eupatolide	GG	$C_{15}H_{20}O_3$	186–8	+41.3	Eupatorium cannabinum	(220, 228[a], 239, 240)	8-β-Hydroxy-costunolide
						E. formosanum	(644, 708[b])	

No.	Name		Formula	M.p.	$[\alpha]$	Species	Refs.	Notes
(40)	Eupatoriopicrin	GG	$C_{20}H_{26}O_6$	157–61	+95	*Eupatorium cannabinum*	(220, 228[a], 239, 240)	$[\theta]_{260}$ +554; $[\theta]_{250}$ +998; $[\theta]_{240}$ +1164; $[\theta]_{231}$ +44
(111)	Euperfolin	GG	$C_{20}H_{28}O_6$	173	−13.9	*Eupatorium perfoliatum*	(417)	
(110)	Euperfolitin	GG	$C_{20}H_{28}O_7$	190–2	−5.8	*Eupatorium perfoliatum*	(417)	
(254)	Ferolide, 1-peroxy	G	$C_{17}H_{22}O_7$	190	+20	*Liriodendron tulipifera*	(230)	Magnoliaceae
(155)	Fluctuadin	GM	$C_{22}H_{26}O_9$	202–5	−18.4	*Enhydra fluctuans*	(19)	
(156)	Fluctuanin	GM	$C_{23}H_{26}O_9$	161–3	−23.5	*Enhydra fluctuans*	(19)	
(165)	Frutescin	GM	$C_{15}H_{18}O_3$	158–60	—	*Iva frutescens*	(408)	$[\theta]_{313}$ −1951; $[\theta]_{247}$ −7900
(171)	Germacranolide, 4,5-cis-3β-hydroxy	GH	$C_{15}H_{20}O_3$	136–7	−80	*Tanacetum tenacetioides*	(114)	
(268)	Germacrene D Lactone	G	$C_{15}H_{18}O_2$	142	−200.9	*Inula helenium*	(111)	
(71)	Germanin A	G	$C_{20}H_{26}O_6$	150–3	+17.8	*Inula germanica*	(161, 605)	NMR: H-1 (6.8 ppm) suggests 1,10-*cis* double bond
(290)	Germanin B	G	$C_{20}H_{24}O_6$	—	—	*Inula germanica*	(161, 605)	NMR: H-1 (6.8 ppm) suggests 1,10-*cis* double bond
(294)	Glaucolide A	G	$C_{23}H_{28}O_{10}$	153–4.5	−29.0	*Vernonia glauca*	(811[b], 188[b], 5, 1132[b])	For occurrence in other *Vernonia* species see refs. (5) and (663a)
(295)	Glaucolide A, 19-hydroxy	G	$C_{23}H_{28}O_{11}$	gum	−22.9	*Erlangea remifolia*	(95)	
(293)	Glaucolide B	G	$C_{21}H_{26}O_{10}$	75–7	−50	*Vernonia baldwinii*	(811, 5, 188[b])	For occurrence in other *Vernonia* species see refs. (5) and (663a)
(95)	Glaucolide D	GG	$C_{23}H_{28}O_{10}$	187–8	−9.0	*Vernonia uniflora*	(77[b], 338[b], 1071[b])	
(94)	Glaucolide E	GG	$C_{23}H_{28}O_9$	150–1	—	*Vernonia uniflora*	(77, 338[b])	

Table III-3 (continued)

Structure Number	Name of Compound	Type[c]	Formula	m.p. °C	$[\alpha]_D$	Plant Source[d]	References	Comments
(97)	Glaucolide G	G	$C_{22}H_{28}O_7$	—	—	Vernonia leiocarpa	(78[b])	
(139)	Glechomanolide	G	$C_{15}H_{20}O_3$	110	+ 120.5	Glechoma hederacea	(1004)	Labiatae
(224)	Goyazensolide	G	$C_{19}H_{20}O_7$	175–7	− 22.5	Eremanthus goyazensis	(1111)	
(223)	Goyazensolide, 15-deoxy	G	$C_{19}H_{20}O_6$	132–4	—	Vanillosmopsis erythropappa	(1110)	$[\alpha]_{Hg}^{24} - 38$
(53)	Haageanolide	GG	$C_{15}H_{20}O_3$	gum	—	Zinnia haageana	(586)	Reported before costunolide, 9β-hydroxy; acetate m.p. 196–7°
(47)	Hanphyllin	GG	$C_{15}H_{20}O_3$	—	+ 155.2	Handelia trichophylla Artemisia ashurbajevii	(1063) (1189)	
(192)	Heliangin	GH	$C_{20}H_{26}O_6$	227–9	− 110	Helianthus tuberosis	(502a, 738, 712[a], 776[b], 764[a])	$[\theta]_{214} + 110{,}020$
(106)	Herbolide A	GG	$C_{17}H_{24}O_4$	162	+ 84	Artemisia herba alba	(955)	
(108)	Herbolide B	GG	$C_{17}H_{24}O_5$	209	+ 23	Artemisia herba alba	(955, 993)	
(113)	Herbolide C	GG	$C_{17}H_{24}O_5$	197–8	− 26	Artemisia herba alba	(955)	
(274)	Hirsutinolide-13(O)-acetate, 8β-(2-hydroxy-methacryloyloxy)	G	$C_{21}H_{26}O_9$	72	+ 111.2	Vernonia hirsuta	(94)	
(272)	Hirsutinolide-13(O)-acetate, 8β-(2-methacryloyloxy)	G	$C_{21}H_{26}O_8$	oil	+ 19.5	Vernonia hirsuta	(94)	
(273)	Hirsutinolide-13(O)-acetate, 8β-(2-methyl-2,3-epoxypropionyloxy)	G	$C_{21}H_{26}O_9$	oil	+ 18.3	Vernonia hirsuta V. angulifolia	(94)	

No.	Name		Formula	M.p.	Rotation	Species	Ref.	Notes
(271)	Hirsutinolide, 15-hydroxy-8β-(2-methacryloyloxy)	G	$C_{21}H_{26}O_9$	oil	—	*Vernonia novebaracensis*	*(94)*	
(275)	Hirsutinolide, iso, 8-β-(2-methyl-acryloyloxy)	G	$C_{21}H_{24}O_9$	oil	—	*Vernonia novebaracensis*	*(94)*	
(269)	Hirsutinolide, 8β-(2-methylacryloyloxy)	G	$C_{19}H_{24}O_7$	184	−6.9	*Vernonia angulifolia*	*(94)*	
(270)	Hirsutinolide, 8β-(2-methyl-2,3-epoxypropionyloxy)	G	$C_{19}H_{24}O_8$	170	+0.7	*Vernonia angulifolia*	*(94)*	
(24)	Inunolide	GG	$C_{15}H_{20}O_2$	84–5	+56.3	*Inula racemosa*	*(847, 111)*	
(136a)	Inunolide, 1β,10α-epoxy-1,10-H	GG	$C_{15}H_{20}O_3$	156	−20	*Inula helenium*	*(111)*	
(137)	Inunolide, 4β,5α-epoxy-4,5-H	GG	$C_{15}H_{20}O_3$	131	−14.1	*Inula helenium* / *I. royleana*	*(111)* / *(111)*	
(122)	Isabelin	GG	$C_{15}H_{16}O_4$	169–70	−57.2	*Ambrosia psilostachya* / *A. artemisifolia*	*(1171, 1167, 1169)* / *(839)*	
(66)	Jurineolide	GG	$C_{20}H_{26}O_7$	170–2	+135	*Jurinea cyanoides*	*(1013, 992)*	
(112)	Lanuginolide	GG	$C_{17}H_{24}O_5$	185	−57	*Michelia lanuginosa*	*(1058)*	Magnoliaceae
(82)	Lanuginolide, 11,13-dehydro	GG	$C_{17}H_{22}O_5$	168	−96.5	*Michelia lanuginosa*	*(1095)*	Magnoliaceae
(105)	Laserolide	GG	$C_{22}H_{30}O_6$	140–1	−234	*Laser trilobum*	*(479, 475, 488, 486)*	Umbelliferae;NMR indicates C_8-αOAc
(116)	Laurenobiolide	GG	$C_{17}H_{22}O_4$	101–3	+17.1	*Laurus nobilis*	*(1041, 1084, 233, 1086, 1040)*	Lauraceae; $[θ]_{253}$ −5650
(130)	Laurenobiolide, 6-desacetoxy, dihydro	GG	$C_{15}H_{22}O_2$	132–3	+139	*Callitris columellaris*	*(133, 1084)*	Cupressaceae
(115)	Laurenobiolide, desacetyl	GG	$C_{15}H_{20}O_3$	gum	+34.5	*Artemisia tridentata* / *A. arbuscula* / *Ambrosia chamissonis*	*(978, 1041)* / *(978)* / *(303)*	Chamisellin;
(159)	Leucanthin A	GM	$C_{23}H_{27}O_{10}$	211–3	—	*Melampodium leucanthum*	*(272)*	
(160)	Leucanthin B	GM	$C_{23}H_{26}O_{11}$	217–9	—	*Melampodium leucanthum*	*(272)*	

Table III-3 (continued)

Struc- ture Number	Name of Compound	Type[c]	Formula	m.p. °C	$[\alpha]_D$	Plant Source[d]	References	Comments
(140)	Leucanthinin	GM	$C_{23}H_{28}O_{10}$	163-4	—	Melampodium leucanthum	(271)	
(213)	Liatrin	GH	$C_{22}H_{26}O_8$	130-2	−142.0	Liatris chapmanii	(623, 624[b])	
(220)	Liatripunctin	GH	$C_{20}H_{26}O_7$	gum	−41.7	Liatris punctata	(439)	
(240)	Linderadine	GF	$C_{15}H_{16}O_5$	130-2	−68.7	Neolitsea aciculata	(1051, 1047)	Lauraceae
(233)	Linderalactone	GF	$C_{15}H_{16}O_3$	136-8	117.2	Lindera strychnifolia	(1054, 1050[a], 610[b], 1081, 1090, 1056)	Tissue culture; Lauraceae
(235)	Linderalactone, neo-	GF	$C_{15}H_{16}O_3$	116-8	+100	Lindera strychnifolia	(1050, 1052)	Lauraceae
(238)	Linderane	GF	$C_{15}H_{16}O_4$	190-1	+180.3	Lindera strychnifolia	(1053, 1050, 1090)	Lauraceae
(237)	Linderane, neo-	GF	$C_{15}H_{16}O_4$	180-3	+32.5	Neolitsea zeylanica	(542, 1048, 1047[a], 543)	Lauraceae
(236)	Linderane, pseudo, neo	GF	$C_{15}H_{16}O_4$	200-2	+90.3	Neolitsea aciculata	(1047, 1048, 1051)	Lauraceae
(83)	Lipiferolide	GG	$C_{17}H_{22}O_5$	118-9	+125	Liriodendron tulipifera	(235)	Magnoliaceae
(197)	Liscundin	GH	$C_{20}H_{34}O_7$	gum	−75	Liatris secunda	(439, 67)	
(199)	Liscunditrin	GH	$C_{22}H_{26}O_9$	161	−5.0	Liatris secunda	(439)	
(239)	Litseaculane	GF	$C_{17}H_{18}O_6$	145-6	+76.1	Neolitsea aciculata	(1051, 1048)	Lauraceae
(234)	Litsealactone	GF	$C_{17}H_{18}O_5$	157-9	+57.5	Neolitsea aciculata	(1051, 1048, 1056)	Lauraceae
(144)	Longipin	GM	$C_{23}H_{28}O_9$	206	—	Melampodium longipes	(951)	
(154)	Longipilin	GM	$C_{21}H_{26}O_8$	170-3	—	Melampodium longipilum	(952)	
(157)	Maculatin	GM	$C_{23}H_{28}O_{10}$	226-8	−74.8	Polymnia maculata	(404)	$[\theta]_{283} -294$; $[\theta]_{242} +7060$; $[\theta]_{215} -8100$
(96)	Marginatin	GG	$C_{22}H_{28}O_7$	104-5	−6.0	Vernonia marginata / V. arkansana / V. fasciculata	(810, 5) / (5) / (810, 5)	
(148)	Melampodin A	GM	$C_{21}H_{24}O_9$	210-1	+155	Melampodium leucanthum	(272, 274, 764[b], 75[b], 1128[b])	

No.	Compound		Formula	mp	[α]	Source (refs)	Notes
(150)	Melampodin A acetate	GM	$C_{23}H_{26}O_{10}$	182.5–3.5	—	Melampodium leucanthum (272, 274)	
(153)	Melampodin A, 11,13-dihydro-9-α-methylbutyrate	GM	$C_{26}H_{34}O_{10}$	191–2	—	Melampodium americanum (266)	
(167)	Melampodin B	G	$C_{17}H_{18}O_7$	226–8	—	Melampodium leucanthum (80, 829, 1127b); M. cinereum (829); M. argophyllum (829); M. leucanthum (829)	
(170)	Melampodin B, 4,5-dihydro	G	$C_{17}H_{20}O_7$	204–5	—	Melampodium cinereum (829)	
(168)	Melampodin C	G	$C_{18}H_{22}O_7$	199–201	—	Melampodium argophyllum (829, 1127b)	
(169)	Melampodin D	G	$C_{19}H_{24}O_7$	—	—	Melampodium argophyllum (828, 1127b)	
(152)	Melampodinin	GM	$C_{25}H_{30}O_{12}$	208–10	—	Melampodium americanum (271); M. longipes (951)	
(151)	Melampodinin B	GM	$C_{26}H_{34}O_{10}$	—	—	Melampodium americanum (266)	
(149)	Melampodinin, 9-desacetyl	GM	$C_{23}H_{28}O_{11}$	—	—	Melampodium americanum (266)	
(141)	Melampolidin	GM	$C_{23}H_{30}O_9$	gum	—	Melampodium leucanthum (271)	
(231)	Melcanthin A	GC	$C_{23}H_{28}O_9$	gum	—	Melampodium leucanthum (270)	$[\theta]_{243} + 3 \times 10^3$; $[\theta]_{220} - 1 \times 10^5$
(230)	Melcanthin B	GC	$C_{23}H_{28}O_{10}$	83–4	—	Melampodium leucanthum (270)	$[\theta]_{244} + 4 \times 10^3$; $[\theta]_{218} - 7.4 \times 10^4$
(229)	Melcanthin C	GC	$C_{22}H_{28}O_{10}$	gum	—	Melampodium leucanthum (270)	
(161)	Melnerin A	G	$C_{20}H_{28}O_7$	194–5	—	Melampodium cinereum (1129b)	
(163)	Melnerin A, 9-acetoxy	G	$C_{22}H_{30}O_9$	—	—	Melampodium leucanthum (800)	
(162)	Melnerin B	G	$C_{21}H_{30}O_7$	—	—	Melampodium cinereum (1129b)	
(164)	Melnerin B, 9-acetoxy	G	$C_{23}H_{32}O_9$	—	—	Melampodium leucanthum (800)	
(127)	Mikanolide	GG	$C_{15}H_{14}O_6$	230–3	+53.4	Mikania scandens (451, 192ab); M. monagasensis (683, 682)	
(125)	Mikanolide, deoxy	GG	$C_{15}H_{16}O_5$	198–200	+98.9	Mikania scandens (451)	1,10-epoxide stereochemistry based on ref. (192)
(134)	Mikanolide, dihydro	GG	$C_{15}H_{16}O_6$	240–4	+91.1	Mikania scandens (451, 192ab); M. monagasensis (683)	

Table III-3 (continued)

Structure Number	Name of Compound	Type[c]	Formula	m.p.°C	[α]_D	Plant Source[d]	References	Comments
(100)	Millefin	GG	$C_{19}H_{26}O_6$	209-10	—	Achillea millefolium	(561)	
(280)	Molephantin	G	$C_{19}H_{22}O_6$	214-6	—	Elephantopus mollis	(640[b])	
(281)	Molephantinin	G	$C_{20}H_{24}O_6$	223-5	—	Elephantopus mollis	(645)	
(173)	Nobilin	GH	$C_{20}H_{26}O_5$	178	±0	Anthemis nobilis	(67, 64, 841, 482[a])	
(185)	Nobilin, 3-dehydro	GH	$C_{20}H_{24}O_5$	205	+136.6	Anthemis nobilis	(482)	
(172)	Nobilin, 3-epi	GH	$C_{20}H_{26}O_5$	137	±0	Anthemis nobilis	(482)	
(193)	Nobilin, 1,10-epoxy	GH	$C_{20}H_{26}O_6$	192	±0	Anthemis nobilis	(482)	
(256)	Nobilin, iso, hydroxy	G	$C_{20}H_{26}O_6$	144-6	+34.5	Anthemis nobilis	(935)	
(49)	Novanin	GG	$C_{17}H_{22}O_4$	gum	—	Artemisia nova	(509, 1125)	characterized as 11,13-dihydro cpd.; m.p. 135-6°
						A. tripartita	(509)	
(61)	Onopordopicrin	GG	$C_{19}H_{24}O_6$	55-6	+166.8	Onopordon acanthium	(241)	
						O. algeriensis	(243)	
						O. alexandrinum	(573)	
						O. brackteatum	(243)	
						O. illyricum	(243)	
						O. nervosum	(243)	
						O. tauricum	(243)	
(291)	Orientin	G	$C_{19}H_{26}O_6$	—	—	Sigesbeckia orientalis	(919)	
(200)	Orizabin	GH	$C_{19}H_{26}O_7$	84-5	-140	Tithonia tubaeformis	(805, 402[a], 1200)	
(46)	Ovatifolin	GG	$C_{17}H_{22}O_5$	131-4	-75	Podanthus ovatifolius	(313, 339[b])	
(80)	Parthenolide	GG	$C_{15}H_{20}O_3$	116-7	-81.4	Chrysanthemum parthenium	(1000, 845[b], 59)	
						Michelia champaca	(340[a], 1139, 31)	Magnoliaceae
						M. lanuginosa	(1059)	Magnoliaceae
						Anthemis cretica	(120)	
						Ambrosia confertiflora	(1174)	
						A. dumosa	(304)	
						Arctotis aspera	(107)	
						A. repens	(107)	

No.	Name		Formula	m.p.	[α]	Source	Ref.	Remarks
(89)	Parthenolide, 9α-acetoxy	GG	$C_{17}H_{22}O_5$	oil	−59.7	Matricaria suffructicosa	(122)	
(90)	Parthenolide, 9β-acetoxy	GG	$C_{17}H_{22}O_5$	—	—	Anthemis cretica	(120)	
(109)	Parthenolide, dihydro	GG	$C_{15}H_{22}O_3$	137	−62	Michelia lanuginosa	(1058, 1000, 340)	Magnoliaceae
						Ambrosia artemisiifolia	(265a)	
(253)	Parthenolide, 1-peroxy	G	$C_{15}H_{20}O_5$	190	+27	Magnolia grandiflora	(249)	Magnoliaceae
(65)	Pectorolide	GG	$C_{19}H_{24}O_6$	—	+243	Vernonia pectoralis	(731, 733)	
(297)	Pelenolide-A, keto	G	$C_{15}H_{22}O_3$	112–4	−227	Artemisia absinthium	(385, 1033, 57[b], 215)	Δε₂₉₆ −5.17
(298)	Pelenolide-B, keto	G	$C_{15}H_{22}O_3$	172	+213	Artemisia absinthium	(385, 1033, 57[b])	
						A. anethifolia	(263)	
						A. jacutica	(66)	
(296)	Pelenolide, hydroxy	G	$C_{15}H_{24}O_3$	108	−41	Artemisia absinthium	(385, 1033, 57[b])	
(177)	Peucephyllin	GH	$C_{21}H_{28}O_6$	120.5–1.5	−140.9	Peucephyllum schottii	(61[b])	
(276)	Phantomolin	G	$C_{21}H_{26}O_6$	—	—	Elephantopus mollis	(713[b])	
(143)	Polydalin	GM	$C_{23}H_{28}O_{10}$	181–3	+8.4	Polymnia uvedalia	(403, 274[a], 404)	1,10-epoxide m.p. 172° [θ]₃₁₀ −694; [θ]₂₅₃ −1740
(182)	Provincialin	GH	$C_{27}H_{34}O_{14}$	gum	−85	Liatris provincialis	(458)	
(221)	Punctaliatrin	GH	$C_{20}H_{24}O_7$	163–5	—	Liatris punctata	(459)	
(302)	Pycnolide	GG	$C_{20}H_{28}O_6$	—	+39.8	Liatris pycnostachya	(438, 440)	[θ]₂₂₅ −140,000 (originalnamepunctatin) [θ]₂₄₀ +990, seco-Germacrolide
(123)	Pyrethrosin	GG	$C_{17}H_{22}O_5$	198–202	−31	Chrysanthemum cinerariaefolium	(1075, 49, 282[b], 48, 503)	
						C. coccineum	(93)	
						Anthemis cupaniana	(93)	
(266)	Repandin A	G	$C_{25}H_{32}O_{10}$	132–3	—	Tetragonotheca repanda	(953)	
(267)	Repandin B	G	$C_{26}H_{34}O_{10}$	127–8	—	Tetragonotheca repanda	(953)	
(264)	Repandin C	G	$C_{25}H_{32}O_{10}$	—	—	Tetragonotheca repanda	(953)	
(265)	Repandin D	G	$C_{26}H_{34}O_{10}$	—	—	Tetragonotheca repanda	(953)	
(249)	Ridentin	G	$C_{15}H_{20}O_4$	215–8	−113	Artemisia cana	(512, 655, 513)	
						A. tridentata	(513)	
						A. tripartita	(513)	

Table III-3 (continued)

Structure Number	Name of Compound	Type[c]	Formula	m. p. °C	$[\alpha]_D$	Plant Source[d]	References	Comments
(255)	Ridentin, dihydro	G	$C_{15}H_{22}O_4$	193–4	—	Artemisia rupicola, A. tripartita	(512, 505), (505)	
(59)	Salonitenolide	GG	$C_{15}H_{20}O_4$	137–9	+199.4	Centaurea salonitana, Cnicus benedictus, Jurinea maxima	(1029, 1175[a]), (1108), (192)	
(68)	Salonitenolide, 8-desoxy	GG	$C_{15}H_{18}O_3$	102	+93	Platycarpha glomerata	(124)	
(107)	Salonitenolide, β-11,13-dihydro-8-desoxy	GG	$C_{15}H_{20}O_3$	146	+84	Platycarpha glomerata	(124)	
(132)	Salonitolide	GG	$C_{15}H_{22}O_4$	184	+100	Centaurea salonitana, C. seridis, Jurinea maxima	(1027, 839[a]), (316), (192)	
(119)	Scabiolide	GG	$C_{21}H_{28}O_8$	118–20	+101	Centaurea scabiosa, C. calcitrapa, C. millitensis, C. solstitialis	(1023, 1030), (238), (238), (238, 743)	
(126)	Scandenolide	GG	$C_{17}H_{18}O_7$	230–4	+62.0	Mikania scandens	(451)	1,10-epoxide stereochemistry based on ref. (192)
(133)	Scandenolide, dihydro	G	$C_{17}H_{20}O_7$	278–30	+83.3	Mikania scandens	(451)	1,10-epoxide stereochemistry based on ref. (192)
(138)	Simsiolide	GG	$C_{15}H_{20}O_4$	oil	—	Simsia dombeyana	(125)	
(128)	Spiciformin	GG	$C_{15}H_{20}O_4$	gum	+81.7	Artemisia tridentata, A. arbuscula	(978), (978)	acetate: m. p. 168–170°; $[\alpha]_D$ +64

No.	Name		Formula	m.p.	$[\alpha]$	Plant source	References	Notes
(85)	Stizolicin	GG	$C_{20}H_{26}O_7$	152–3.5	−32.4	Stizolophus coronopifolius / Centaurea solstitialis / Saussurea elongata	(745) (918) (743)	
(81)	Stizolin	GG	$C_{15}H_{20}O_4$	184–6	−30.5	Stizolophus balsamita	(745, 747, 744)	
(215)	Tagitinin A	GH	$C_{19}H_{28}O_7$	168–70	−154	Tithonia diversifolia	(813, 1200)	$[\theta]_{238}$ −910, plant source misnamed T. tagitiflora
(203)	Tagitinin B	GH	$C_{19}H_{26}O_7$	125	−142	Tithonia diversifolia	(814, 408 b, 1200)	$[\theta]_{245}$ +4400, plant source misnamed
(283)	Tagitinin C	G	$C_{19}H_{24}O_6$	—	−204	Tithonia diversifolia	(816, 408 b, 1200)	Plant source misnamed
(218)	Tagitinin D	GH	$C_{19}H_{28}O_6$	138–40	−137	Tithonia diversifolia	(816, 1200)	See Tirotundin; $[\theta]_{275}$ +760; $[\theta]_{252}$ −1300; plant source misnamed
(214)	Tagitinin F	G	$C_{19}H_{24}O_6$	128–30	−144	Tithonia diversifolia	(816, 1200)	$[\theta]_{235}$ +7400; plant source misnamed
(32)	Tamaulipin A	GG	$C_{15}H_{20}O_3$	159–60	+171	Ambrosia confertiflora	(269, 1143[b])	
(48)	Tamaulipin B	GG	$C_{15}H_{20}O_3$	140–2	+99	Ambrosia confertiflora	(1174, 268)	
(257)	Tamirin	G	$C_{15}H_{18}O_4$	167–8	−36	Tanacetum myriophyllum / T. chilliophyllum	(729, 231) (727)	
(262)	Tanachin	G	$C_{15}H_{20}O_4$	—	—	Tanacetum pseudoachillea	(1178, 1180)	
(124)	Tanacin	GG	$C_{20}H_{26}O_5$	128–9	−74	Tanacetum pseudoachillea	(1184, 231, 1179)	
(263)	Tatridin C	G	$C_{15}H_{18}O_4$	178–9	+49	Artemisia tridentata	(925, 505)	
(259)	Tatridin A	GG	$C_{15}H_{20}O_4$	178–9	−49	Artemisia tridentata	(505, 979, 978)	
(260)	Tatridin B	G	$C_{15}H_{20}O_4$	gum	—	Artemisia tridentata	(505, 978)	Two structures proposed
(284)	Tifruticin	G	$C_{20}H_{26}O_7$	141	−22	Tithonia fruticosa	(436, 1200)	$[\theta]_{257}$ −4990
(282)	Tifruticin, deoxy	G	$C_{20}H_{26}O_6$	gum	—	Tithonia fruticosa	(436, 1200)	Structure redrawn by these authors; $[\theta]_{240}$ +6300
(217)	Tirotundin	GH	$C_{19}H_{28}O_6$	141	−77	Tithonia rotundifolia / T. diversifolia	(436, 408 a, 1200) (816)	Reported before Tagitinin D; $[\theta]_{263}$ −1560 Plant source misnamed

9*

Table III-3 (continued)

Structure Number	Name of Compound	Type[c]	Formula	m.p. °C	[α]_D	Plant Source[d]	References	Comments
(216)	Tirotundin ethyl ether	GH	$C_{21}H_{32}O_6$	125	−55	Tithonia rotundifolia	(436, 408a[a,b])	
(58a)	Tomentosin	G	$C_{20}H_{26}O_6$	193–4	—	Montanoa tomentosa	(295)	
(38)	Tulipinolide	GG	$C_{17}H_{22}O_4$	181	+249	Liriodendron tulipifera	(228, 235, 233)	Magnoliaceae; [θ]₂₆₄ −4780; [θ]₂₂₁ +121,000
(38a)	Tulipinolide, desacetyl	GG	$C_{15}H_{20}O_3$	gum	—	Artemisia mexicana Ambrosia chamissonis	(905) (303)	Chamissanthin
(39)	Tulipinolide, epi	GG	$C_{17}H_{22}O_4$	91–2	+76	Liriodendron tulipifera	(228, 235, 233)	Magnoliaceae; [θ]₂₆₄ −7180; [θ]₂₂₂ +146,000
(93)	Tulipinolide, epi, diepoxide	GG	$C_{17}H_{32}O_6$	214–5	−55.7	Liriodendron tulipifera	(235)	Magnoliaceae
(72)	Urospermal A	G	$C_{15}H_{18}O_5$	163–5	−2	Urospermum dalechampii	(69)	NMR: H-1 (6.96 ppm) suggests 1,10-cis double bond; stable conformer of urospermal B
(73)	Urospermal B	G	$C_{15}H_{18}O_5$	191–3	—	Urospermum dalechampii	(69)	NMR: H-1 (6.78 ppm) suggests 1,10-cis double bond; stable conformer of urospermal A
(142)	Uvedalin	GM	$C_{23}H_{28}O_9$	131–3	+12.8	Polymnia uvedalia	(403, 274[a], 404)	[θ]₃₁₆ −4900; [θ]₂₅₅ −1810
(248)	Verlotorin	G	$C_{15}H_{20}O_4$	130–2	+171	Artemisia verlotorum	(285, 300)	See Costunolide, 1-peroxy
						Magnolia grandiflora	(249[a])	Magnoliaceae

No.	Name		Formula	M.p.	$[\alpha]$	Species	Ref.	Notes
(252)	Verlotorin, anhydro	G	$C_{15}H_{18}O_3$	123–4	—	*Artemisia verlotorum*	*(285, 300, 249)*	
(78)	Vernolide	GG	$C_{19}H_{22}O_7$	180–3	+230	*Vernonia colorata*	*(1094, 471, 820^b)*	
(79)	Vernolide, hydroxy	GG	$C_{19}H_{22}O_8$	150	+135	*Vernonia colorata*	*(471, 1093)*	
(288)	Vernomygdin	G	$C_{19}H_{24}O_7$	208–10	+65	*Vernonia amygdala*	*(627)*	
(77)	Vernopectolide A	G	$C_{21}H_{26}O_8$	gum	+78	*Vernonia pectoralis*	*(733)*	NMR: H-1 (6.60 ppm) might suggest 1,10-*cis* double bond
(74)	Vernopectolide B	G	$C_{19}H_{22}O_6$	gum	+93	*Vernonia pectoralis*	*(733)*	
(136)	Vernudifloride	GG	$C_{20}H_{26}O_4$	gum	+163	*Vernonia nudiflora*	*(128)*	
(226)	Viguiestenin	GH	$C_{21}H_{28}O_7$	196–8	−120	*Viguiera stenoloba*	*(367, 1200)*	Structure redrawn
(225)	Viguiestenin, desacetyl	GH	$C_{19}H_{26}O_6$	212–4	—	*Viguiera stenoloba*	*(367, 1200)*	Structure redrawn
(204)	Woodhousin	GH	$C_{21}H_{28}O_8$	183–4.5	−206.3	*Bahia woodhousei*	*(402, 408 b^{ab}, 1200)*	$[\theta]_{240}$ +4000
(202)	Zacatechinolide, 1β-acetoxy	GH	$C_{21}H_{26}O_8$	199	−222	*Calea zacatechichi*	*(322)*	
(205)	Zacatechinolide, 1-oxo	GH	$C_{19}H_{22}O_7$	167	−263	*Calea zacatechichi*	*(322)*	
(244)	Zeylanane	GF	$C_{17}H_{18}O_6$	150–2	+231.2	*Neolitsea aciculata*	*(1051, 1048)*	Lauraceae
(243)	Zeylanine	GF	$C_{17}H_{18}O_5$	175	+271	*Neolitsea zeylanica*	*(542, 1048^a, 541)*	Lauraceae
(241)	Zeylanicine	GF	$C_{17}H_{18}O_6$	235	−153	*Neolitsea zeylanica*	*(541)*	Lauraceae
(242)	Zeylanidine	GF	$C_{17}H_{18}O_7$	226	−174	*Neolitsea zeylanica*	*(541)*	Lauraceae
(219)	Zexbrevin	GH	$C_{19}H_{22}O_6$	217–8	+41	*Zexmenia brevifolia*	*(897, 890, 1200)*	
(201)	Zexbrevin B	GH	$C_{19}H_{24}O_7$	103–4	−145	*Zexmenia brevifolia*	*(805, 1200)*	
(301)	Zexbrevin C	G	$C_{19}H_{26}O_6$	92–3	−165	*Zexmenia brevifolia*	*(808)*	

IV. Eudesmanolides and Biogenetic Derivatives

1. Structural, Biosynthetic, and Biogenetic Considerations

The eudesmanolides (selinanolides) are based on the eudesmane (selinane) skeleton, most members containing *trans*-7,6-α, β-unsaturated γ-lactones or their 11,13-dihydro derivatives. Compounds with 7,8-lactone groups may occur as *cis*- and *trans*-γ-lactones. Many members contain 3,4-, 4,5- and 4,15-double bonds as well as epoxide derivatives thereof and hydroxyl and/or ketonic oxygen functions predominantly at C-1, C-3, and C-8. The C-6 and C-8 lactonized types of naturally occurring eudesmanolides and their biogenetic derivatives are presented in Chart IV-1. The three 1,10-*seco*-eudesmanolides eriolanin (**487**), eriolangin (**488**) and ivangulin (**489**) as well as lumisantonin (**485**) and the phenyl-containing vernodesmine (**486**) represent structural exceptions.

It is of particular interest that various taxa of the Hepaticae produce antipodal eudesmanolides. The liverworts *Frullania tamarisci* (*36, 173, 350, 351, 597*) and *F. nisqualensis* contain eudesmanolides such as (−)-frullanolide (**422**), which are of the skeletal type commonly found in higher plants. However, other liverworts (*36, 794*) produce the *ent*-eudesmanolide (+)-frullanolide (**492**) and related compounds.

Table IV-2 summarizes all known compounds by names in alphabetical order. Also, physical data, plant sources and literature references are listed in Table IV-2 on pages 157—165.

Chart IV-1. Naturally occurring eudesmanolides and biogenetic derivatives

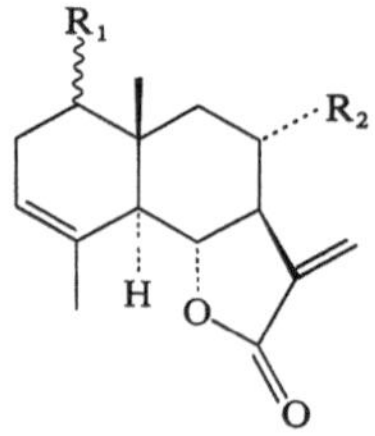

(**330**) α-Cyclocostunolide; $R_1 = R_2 = H$
(**363**) Douglanin; $R_1 = $ α-OH, $R_2 = H$
(**364**) Balchanin; $R_1 = $ β-OH, $R_2 = H$
(**366**) Ludalbin; $R_1 = $ α-OH, $R_2 = $ OAc

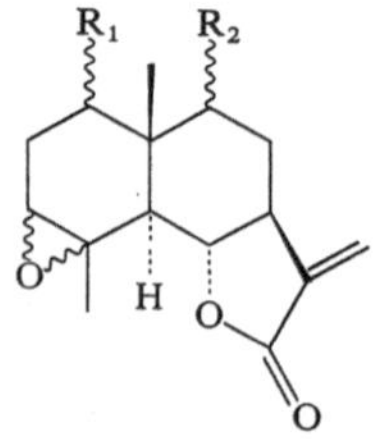

(**367**) Ludovicin A; $R_1 = $ α-OH, $R_2 = H$, epoxide *cis*, α
(**368**) Santamarin, epoxy; $R_1 = $ β-OH, $R_2 = H$, epoxide *cis*, α
(**369**) Pluchea lactone; $R_1 = $ α-OAng, $R_2 = $ α-OH, epoxide *cis*, β

(370) Arbusculin B; R_1 = R_2 = H
(370a) γ-Liriodenolide; R_1 = β-OH, R_2 = β-OAc
(371) Rothin A; R_1 = H, R_2 = α-OH

(372) Ludovicin C; R_1 = α-OH, R_2 = H
(373) Armexifolin; R_1 or R_2 = OH, R_1 or R_2 = H

(331) β-Cyclocostunolide; R_1 = R_2 = R_3 = H
(374) Reynosin; R_1 = β-OH, R_2 = R_3 = H
(375) Ludovicin B; R_1 = R_2 = α-OH, R_3 = H
(376) Ridentin B; R_1 = R_2 = β-OH, R_3 = H
(377) Alantolactone, 1α, 8α-dihydroxy; R_1 = α-OH,
       R_2 = H, R_3 = OH
(378) Alantolactone, 8α-hydroxy-1α(2-hydroxymethyl-
       acryloxy); R_1 = α-OMac-4-OH, R_2 = H, R_3 = α-OH
(379) Alantolactone, 1α-hydroxy-8α-(2-hydroxymethyl-
       acryloxy); R_1 = α-OH, R_2 = H, R_3 = α-OMac-
       4-OH
(380) Dentatin A; R_1 = β-OH, R_2 = H, R_3 = α-OH

(381) Tanacetin; R_1 = R_2 = OH, R_3 = H
(382) Arbusculin C; R_1 = R_3 = H, R_2 = OH
(383) Rothin B; R_1 = H, R_2 = R_3 = OH

(384) Oopodin, dehydro

Chart IV-1 (continued)

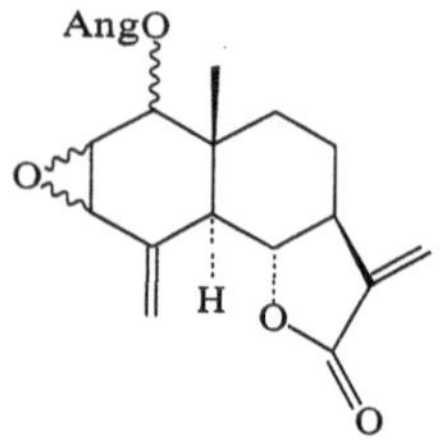

(385) Badkhysinin

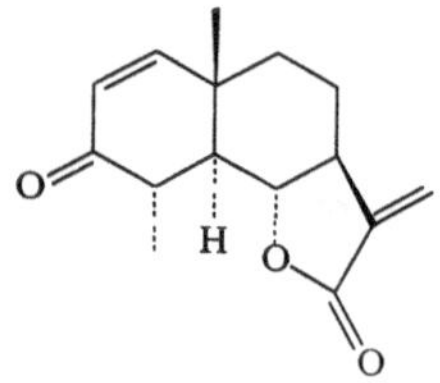

(386) Tuberiferin

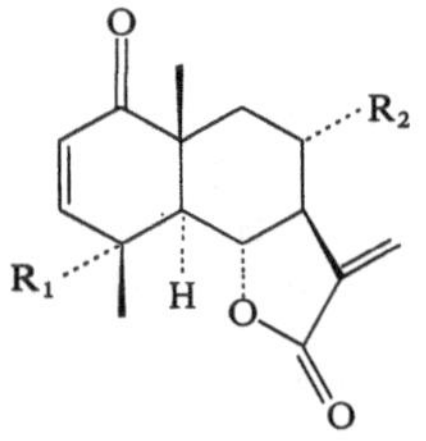

(387) Arglanin; $R_1 = OH, R_2 = H$
(388) Artemexifolin; $R_1 = R_2 = OAc$

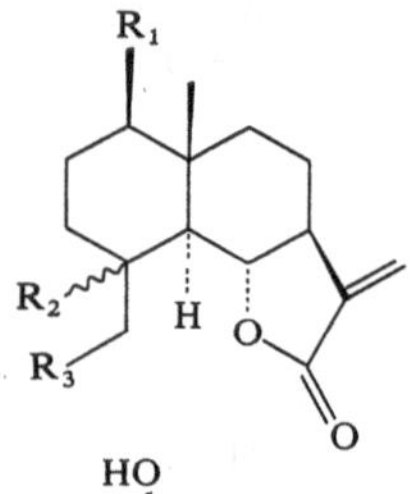

(389) Arbusculin, 1 β-hydroxy; $R_1 = OH, R_2 = \alpha\text{-}OH$, $R_3 = H$
(390) Vahlenin; $R_1 = OH, R_2 = OH, R_3 = OMac$
(391) Arbusculin A; $R_1 = R_3 = H, R_2 = \alpha\text{-}OH$
(392) Arbusculin A, 4-epi; $R_1 = R_3 = H, R_2 = \beta\text{-}OH$

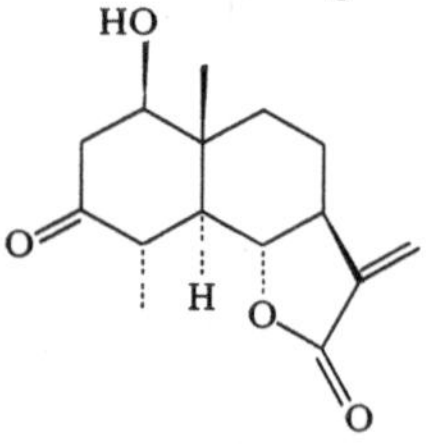

(393) Artecalin

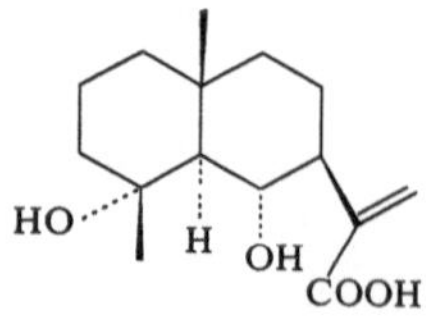

(394) Arbusculin E

(395) Santamarin, dihydro; R_1 = β-OH, R_2 = H,
R_3 = β-H
(397) Decipienin H; R_1 = α-OH, R_2 = OH, R_3 = β-OH
(398) Decipienin G; R_1 = α-OH, R_2 = OH, R_3 = β-OAng

(399) Feropodin

(400) Artesin; R_1 = β-OH, R_2 = α-H
(401) Eudesm-4-en-6,12-olide, 1-hydroxy-6β,7α,11β-H;
R_1 = OH, R_2 = β-H

(402) Taurin, R = H
(403) Eudesm-4-en-6,12-olide, 1-oxo-6β,7α,11β-H;
R = β-H

(404) Santonin, 1,2-dihydro

(405) α-Santonin; R_1 = H, R_2 = β-H
(406) β-Santonin; R_1 = H, R_2 = α-H
(407) Artemisin; R_1 = OH, R_2 = β-H
(408) Santonin, 11-oxy; R_1 = H, R_2 = OH
(408 a) Decipienin A; R_1 = H, R_2 = OAng

Chart IV-1 (continued)

(409) β-Cyclocostunolide, dihydro; $R_1 = R_2 = R_3 = H$, $R_4 = β$-H

(410) 1β-Hydroxysant-4(14)-en-6,12-olide C; $R_1 = β$-OH, $R_2 = R_3 = H$, $R_4 = β$-H

(411) Arsubin; $R_1 = β$-OH, $R_2 = H$, $R_3 = OH$, $R_4 = α$-H

(412) Artemin; $R_1 = β$-OH, $R_2 = H$, $R_3 = OH$, $R_4 = β$-H

(413) Erivanin; $R_1 = R_2 = α$-OH, $R_3 = H$, $R_4 = β$-H

(414) Oopodin

(415) Badkhysidin

(416) Tauremisin; $R_1 = OH$, $R_2 = H$, $R_3 = α$-H

(417) Tabarin; $R_1 = R_2 = OH$, $R_3 = β$-H

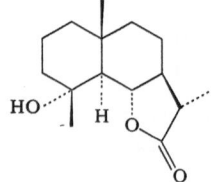

(418) Colartin

(419?) Arsantin; $R_1 = \alpha\text{-OH}$, $R_2 = \beta\text{-H}$, $R_3 = H$
(420) Arsanin; $R_1 = \beta\text{-OH}$, $R_2 = \beta\text{-H}$, $R_3 = H$
(421) Arabsin; $R_1 = H$, $R_2 = \alpha\text{-H}$, $R_3 = \alpha\text{-OH}$

(422) $(-)$ Frullania lactone; $R_1 = R_2 = H$
(423) Armexin diacetate; $R_1 = R_2 = OAc$

(424) Finitin; $R = H$, $C13\alpha$
(425) ψ-Santonin, deoxy; $R = H$, $C13\beta$
(426) ψ-Santonin; $R = OH$, $C13\beta$

(427) Arbusculin D

(428) Silerolide; R_1 or $R_2 = OAc$, R_1 or $R_2 = OAng$

Chart IV-1 (continued)

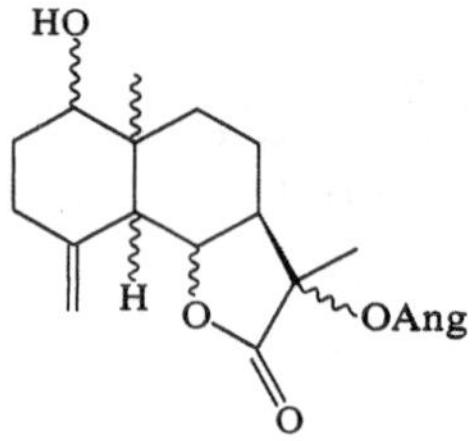

(**429**) Lasolide

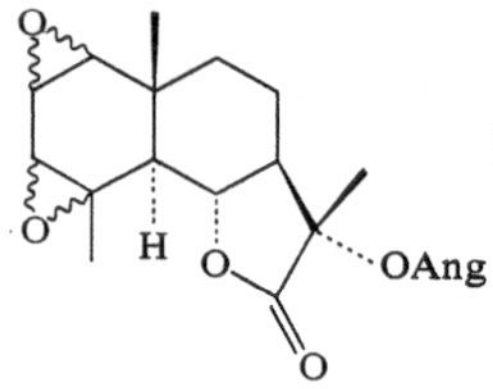

(**430**) Decipienin B

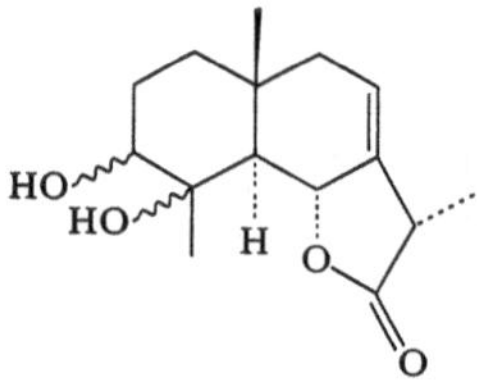

(**431**) Mibulactone (?, possibly identical with artemin, **412**)

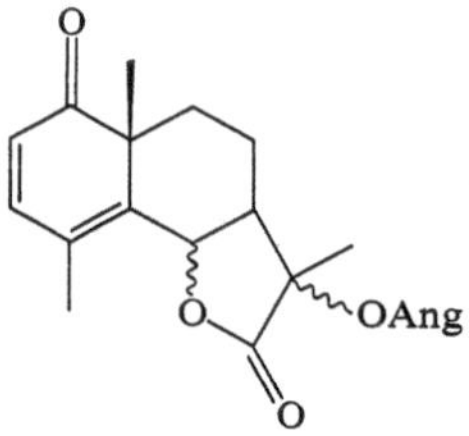

(**432**) Semopodin

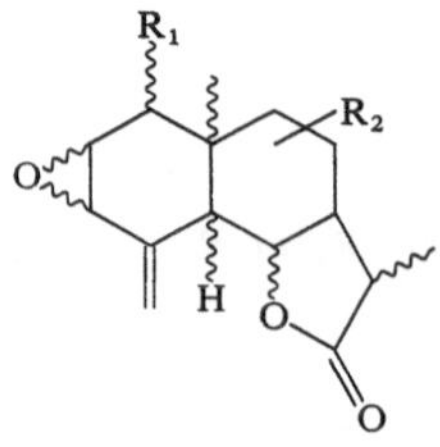

(**433**) Ferula hydroxylactone; R_1 = OH, R_2 = H, C 11 β
(**434**) *Ferula oopoda*, hydrolactone; R_1 or R_2 = H, R_1 or
        R_2 = OH

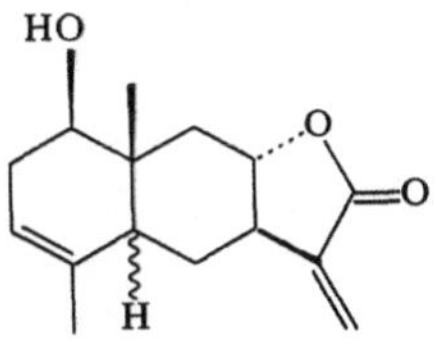

(435) Ivangustin, iso, 8-epi

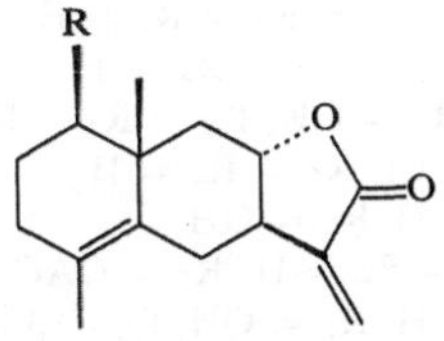

(436) Ivangustin, 1-desoxy, 8-epi; R = H
(437) Ivangustin, 8-epi; R = OH

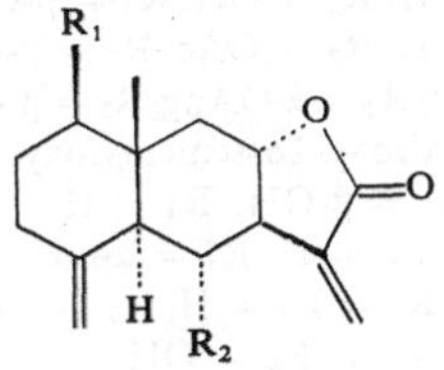

(438) β-Cyclopyrethrosin; R_1 = OH, R_2 = OAc
(439) Chrysanin; R_1 = OH, R_2 = OAng

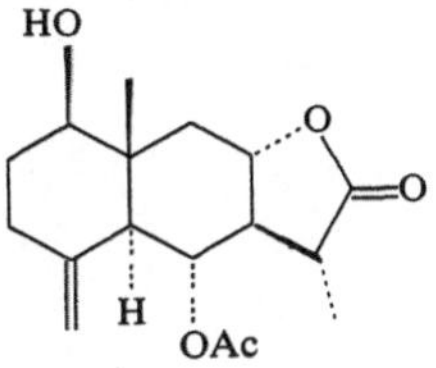

(441) β-Cyclopyrethrosin, dihydro

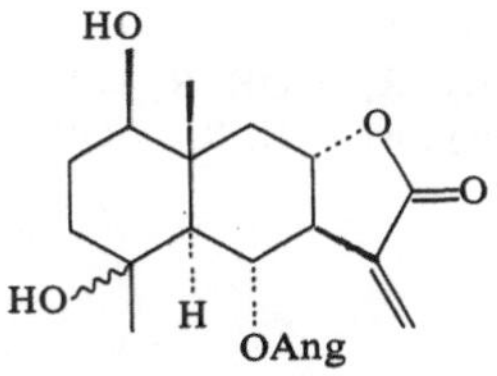

(442) Tanapsin

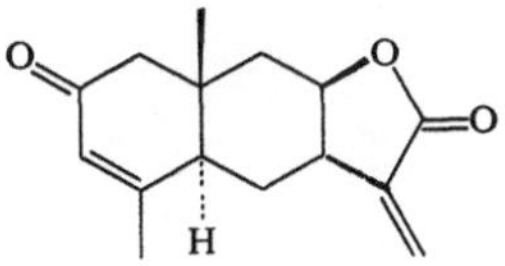

(443) Pinnatifidin

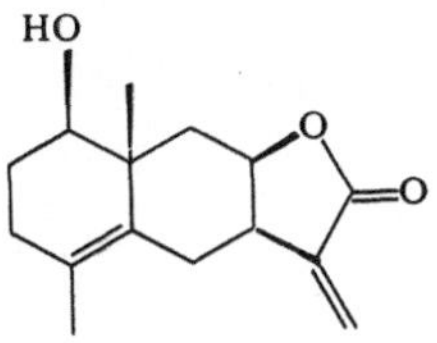

(444) Ivangustin

Chart IV-1 (continued)

(445) Yomogin

(446) Alantolactone, iso; $R_1 = R_2 = R_3 = R_4 = H$
(447) Asperilin; $R_1 = \beta$-OH, $R_2 = R_3 = R_4 = H$
(448) Ivasperin; $R_1 = \alpha$-OH, $R_2 = OH$, $R_3 = R_4 = H$
(449) Granilin; $R_1 = R_3 = \alpha$-OH, $R_2 = R_4 = H$
(450) Ivalin; $R_1 = R_3 = R_4 = H$, $R_2 = OH$
(451) Ivalin acetate; $R_1 = R_3 = R_4 = H$, $R_2 = OAC$
(452) Pulchellin C; $R_1 = R_4 = H$, $R_2 = OH$, $R_3 = \beta$-OH
(453) Pulchellin E; $R_1 = R_4 = H$, $R_2 = OH$, $R_3 = \beta$-OAc
(454) Pulchellin B; $R_1 = R_4 = H$, $R_2 = OAc$, $R_3 = \beta$-OH
(456) Pulchellin F; $R_1 = R_4 = H$, $R_2 = \alpha$-OAng, $R_3 = \beta$-OH
(457) Alantolactone, iso, 3 β-hydroxy-2 α-senicioyloxy;
            $R_1 = H$, $R_2 = OSen$, $R_3 = \beta$-OH, $R_4 = H$
(458) Telekin, iso; $R_1 = R_2 = R_4 = H$, $R_3 = \alpha$-OH
(459) Telekin, 3-epi-iso-; $R_1 = R_2 = R_4 = H$, $R_3 = \beta$-OH
(460) Telekin; $R_1 = R_2 = R_3 = H$, $R_4 = OH$

(461) Encelin

(462) Alantolactone; $R_1 = R_2 = H$
(463) Alantolactone, 1 β-hydroxy; $R_1 = OH$, $R_2 = H$
(464) Alantolactone, 2 α-hydroxy; $R_1 = H$, $R_2 = OH$

(465) Alantolactone, 2-oxo

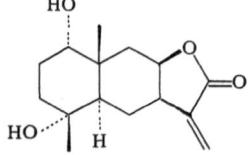

(466) Microcephalin

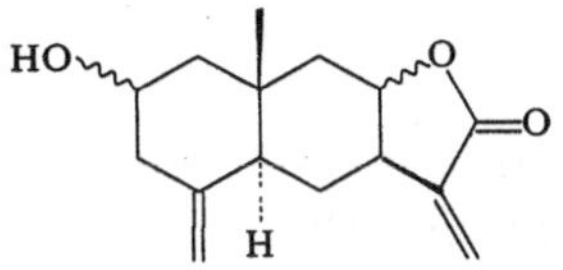

(467) Carpesin

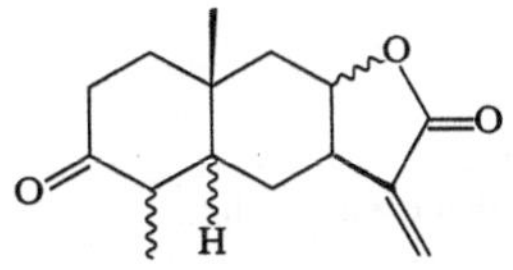

(468) Graveolide

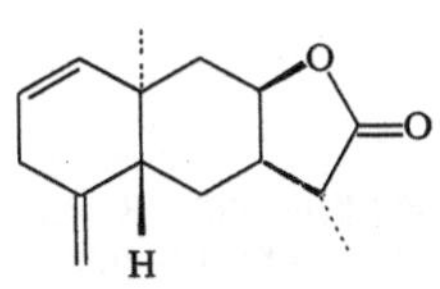

(469) Callitrisin

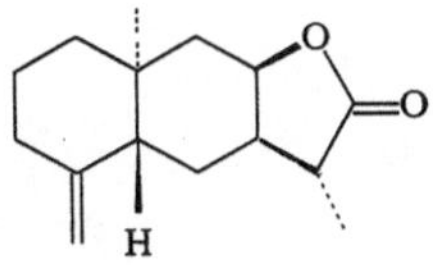

(470) Dihydrocallitrisin

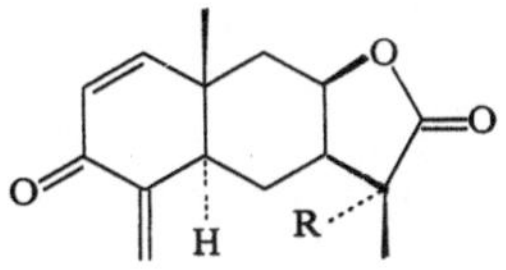

(471) Virginin; R = H
(472) Farinosin; R = OH

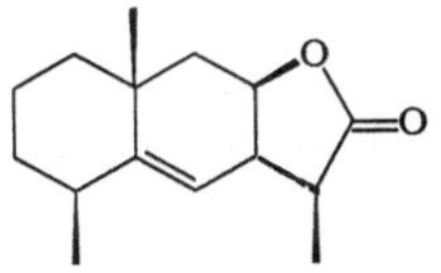

(473) Alantolactone, dihydro

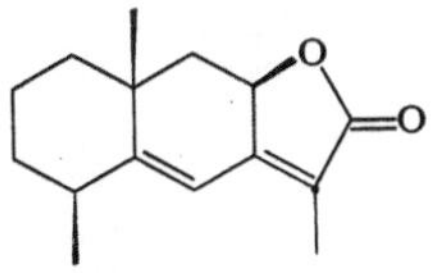

(474) Alantolactone, neo

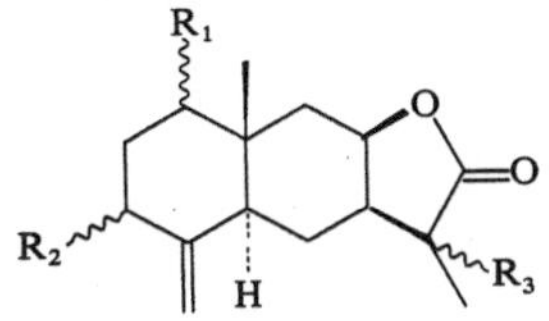

(475) Alantolactone, iso, dihydro; R₁ = R₂ = H, R₃ = α-H
(476) Ashurbin; R₁ = R₂ = α-OH, R₃ = β-H
(477) Hybrifarin; R₁ = H, R₂ = β-OH, R₃ = OH

Chart IV-1 (continued)

(478) Lindestrenolide; R = H
(479) Lindestrenolide, hydroxy; R = OH

(480) Eudesma-4(15),7(11)diene-8 β-12-olide

(481) Eudesma-5,7(11)-diene-8 β,12-olide; R = H
(482) Eudesma-5,7(11)-diene-13-ol-8 β,12-olide; R = OH

(483) Dehydrolindestrenolide

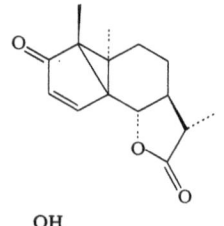

(484) Commiferin

(485) Lumisantonin

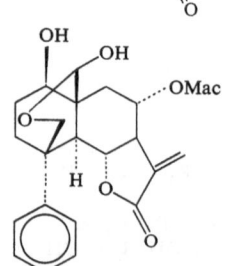

(486) Vernodesmin

(487) Eriolanin; R_1 = CH_2OH, R_2 = OMac, R_3 = OH
(488) Eriolangin; R_1 = CH_2OH, R_2 = OAng, R_3 = OH
(489) Ivangulin; R_1 = CO_2CH_3, R_2 = R_3 = H

(490) Isocritonilide

(491) Critonilide

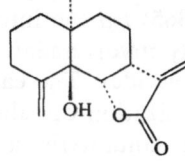

(492) (+) Frullanolide

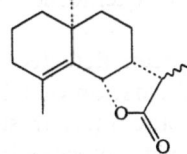

(493) (+)-Frullanolide, oxy

(494) (+) Frullanolide, dihydro

Chart IV-1 (continued)

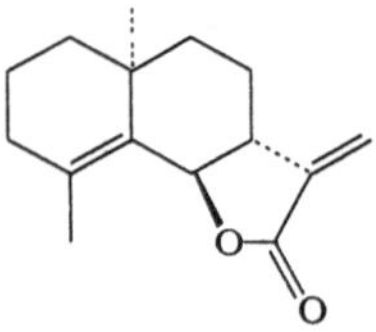

(**495**) (−)-Arbusculin B, ent

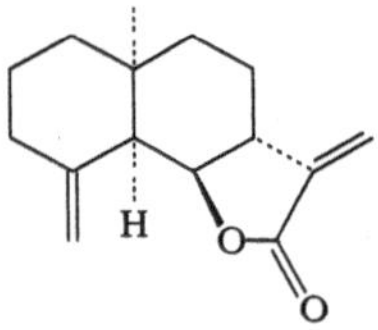

(**496**) (+) *cis*-β-Cyclocostunolide

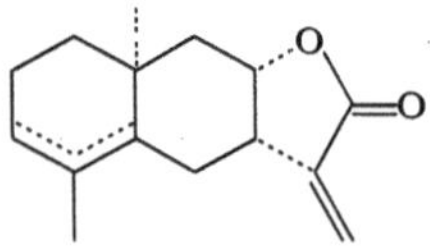

(**497**) Diplophyllolide A; Δ 3,4, H-5β
(**498**) Diplophyllin; Δ 4,5

Although the experimental information about the biosynthesis of eudesmanolides is scarce and limited to some work on santonin (**405**) by Barton and coworkers (*51*) a considerable number of stereospecific *in vitro* cyclization reactions of germacrolides and their epoxide derivatives has provided indirect evidence for their involvement in eudesmanolide biosynthesis (*1037*) (see also Chapter III-3). As outlined on top of Chart IV-2, cyclization of costunolide-1,10-epoxide (**327**) gives the eudesmanolides reynosin (**374**) and santamarine (**365**) *via* the intermediate cation (**499**) (*858*) thus exemplifying the likely involvement of 1,10-epoxygermacrolides in the biogenesis of eudesmanolides. The ease with which the acid-labile epoxide (**327**) undergoes cyclization on silica gel led Rabi and coworkers (*858*) to speculate that santamarin and reynosin might represent artifacts of (**327**), formed during chromatography over silica gel. Alternative biogenetic routes toward the eudesmanolides *via* melampolide intermediates (*393*) as well as artemorin type precursors (*288*) have been discussed.

Hydroperoxides of eudesmanolides have not yet been detected as natural products nor prepared in the laboratory. The formation of 1,10-*seco*-eudesmanolides is of considerable interest. Among other mechanisms (*393*) a biogenetic step that would involve a hydroperoxide intermediate

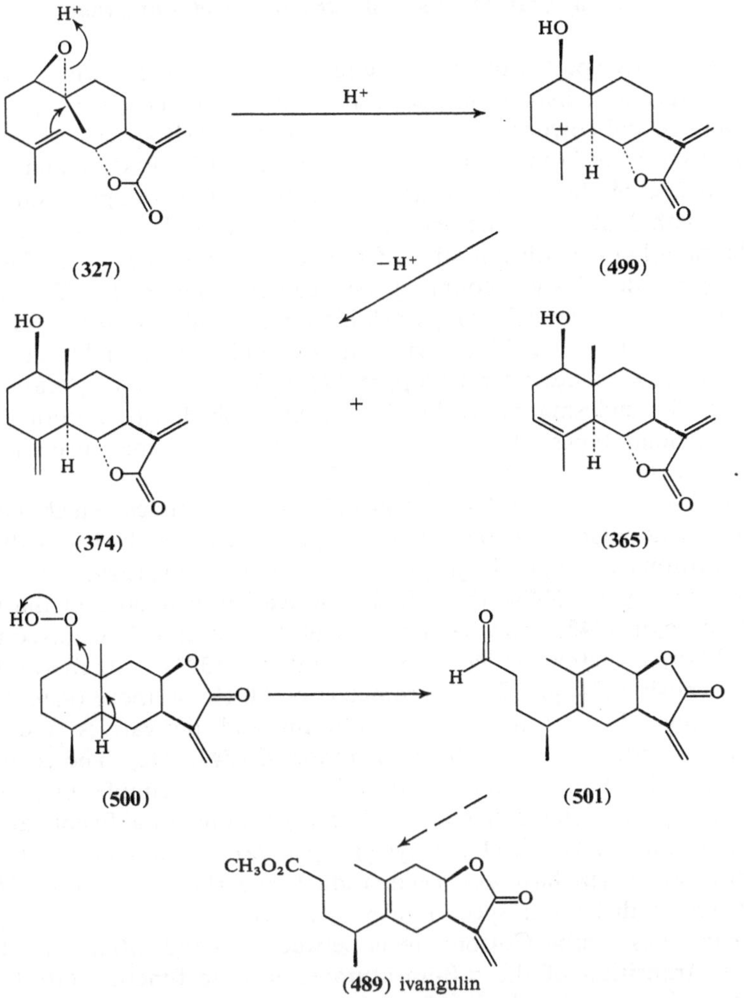

Chart IV-2. Biogenesis of eudesmanolides and 1,10-seco-Eudesmanolides

is outlined at the bottom of Chart IV-2. Ivangulin (489) could be derived from a eudesmanolide hydroperoxide (500) which by a fragmentation reaction, as shown by the arrows, would provide the aldehyde (501). Further oxidative biomodification followed by methyl ester formation would result in ivangulin (489). The other two known 1,10-*seco*-eudesmanolides can be obtained by similar reaction sequences.

The possible origin of the unusual eudesmanolides callitrisin (469) and dihydrocallitrisin (470) with "inverted" stereochemistry at the C-5, C-10 ring junction (*133*) is discussed in Chapter VI (Chart VI-3).

2. Physical Methods of Structure Determination

The NMR spectra of over 50 eudesmanolides with various patterns of unsaturation, hydroxylation and lactonization have been presented by MABRY and coworkers (*1172*). In addition, NMR spectra of santonin derivatives have been discussed in greater detail by PINHEY and STERN-HELL (*835*). Methods for the determination of the configuration of the C-11 methyl at the γ-lactone ring in C-6 as well as C-8-lactonized eudesmanolides involve studies of the solvent shift (*447, 753, 754*) and the magnitude of NMR couplings between H-7 and H-11 (*753*). Pseudo-equatorial C-11 methyl groups exhibit upfield shifts of about 0.23 ppm in benzene relative to chloroform solution, and pseudoaxial C-11 methyl groups show values near 0.05 ppm (*447*). This method appears to be reliable for eudesmanolides but has to be applied with caution to less rigid skeletal types of sesquiterpene lactones and may even lead to erroneous results in certain cases (*447*).

In C-6 as well as C-8-lactonized eudesmanolides which contain α-methylene-γ-lactones, the allylic couplings between H-7 and the two C-13 protons are in good agreement with SAMEK's rule (*929*). The 7,8-*cis*-lactones ivasperin (**448**), the dehydro derivative of microcephalin (**466**), and yomogin (**445**) show $J_{7,13}$ values of 1.5 Hz or below. In contrast, the 7,6-*trans*-lactonized compounds arglanin (**387**), tuberiferin (**386**), and reynosin (**374**) give $J_{7,13}$ parameters of 3.0 Hz or above (*929, 934*).

Carbon-13 spectral data of santonin and derivatives and other eudesmanolides have recently been reported (*741, 842*). The carbon-13 chemical shifts in α- and β-santonin provide a simple method for determining the stereochemistry of the γ-lactone ring fusion and the configuration of the C-11 methyl group (*842*). Compounds of the iso-alantolactone type have also been studied by carbon-13 methods (*1112*).

Circular dichroism spectra of α-methylene-γ-lactone-containing eudesmanolides exhibit Cotton effects between 250 and 270 nm due to the n → π* transition of the α,β-unsaturated lactone function (*1011*). The 7,6-*trans*-lactone douglanin (**363**) (*690*) as well as the 7,8-*cis*-lactone pinnatifidin (**443**) (*422*) exhibit negative bands at 250 and 269 nm, respectively, in agreement with the Waddel-Stöcklin-Geissman rule (*1124*). Also, the cyclized diacetate of chamissonin (**114**) (*657*), a 7,8-*trans*-lactone, shows a positive CE, as predicted by the rule (*1121, 1124*). Application of the HOREAU method (*493*) to the determination of the configuration of the hydroxyl groups at C-8 in the eudesmanolides, obtained by cyclization of the dihydroderivatives of tulipinolide (**38**) and epitulipinolide (**39**), gives results contrary to chemical and NMR spectral findings but data obtained by BREWSTER's benzoate method (*135 a*) were in accord with the NMR results (*228*). The stereochemistry of the hydroxyl group at

C-1 in asperilin (**447**) was established by chemical methods as C-1β-OH, a result which is in agreement with data obtained by the HOREAU method (*414a*).

X-ray diffraction data of several eudesmanolides are available and have been summarized in Table IV-1.

Table IV-1. *X-Ray Structures of Eudesmanolides and Derivatives*

Structure Number	Compound	References	Comments
(**363**)	Douglanin	(*289, 1104b*)	Bromoacetate of tetrahydro derivative
(**488**)	Eriolangin	(*140*)	Oxidation product mixture with Eriolanin
(**487**)	Eriolanin	(*140*)	Oxidation product mixture with Eriolangin
(**485**)	Lumisantonin	(*494*)	2-Bromo-derivative
(**452**)	Pulchellin C	(*200*)	*p*-Bromo-phenyl thiol adduct of diacetate
(**405**)	α-Santonin	(*38a*)	2-Bromo-derivative
(**406**)	β-Santonin	(*169a*)	2-Bromo-derivative
(**486**)	Vernodesmin	(*703*)	

3. Chemical Transformations of Eudesmanolides

Most of the classical chemistry of the eudesmanolides has been previously reviewed by SIMONSEN and BARTON (*991a*), COCKER and MCMURRY (*168a*), and BARTON and DE MAYO (*49a*). More recent examples of chemical transformations within the eudesmanolide series are described in the book by YOSHIOKA et al. (*1172*).

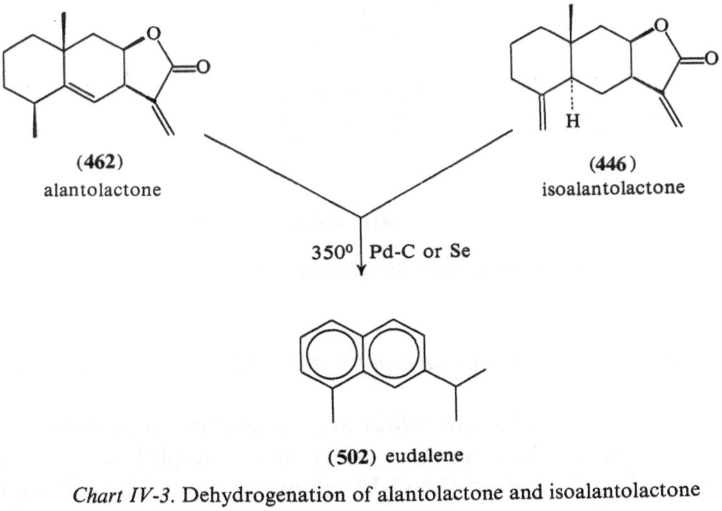

(**462**)
alantolactone

(**446**)
isoalantolactone

350° | Pd-C or Se

(**502**) eudalene

Chart IV-3. Dehydrogenation of alantolactone and isoalantolactone

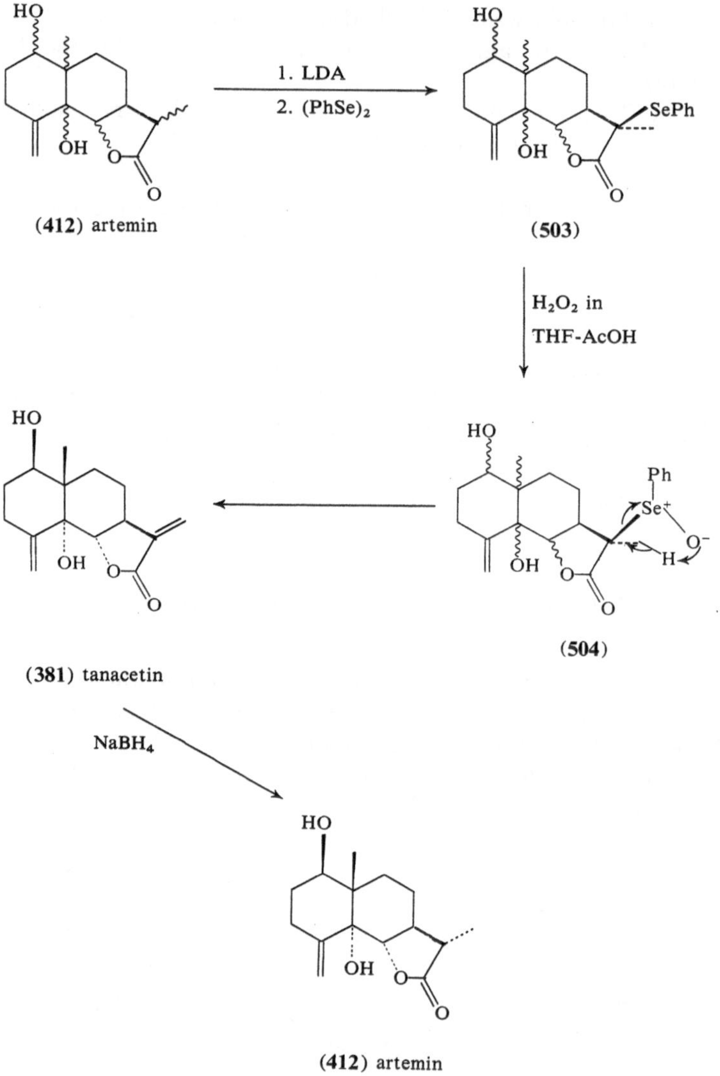

Chart IV-4. Conversion of artemin to tanacetin

3.1 Dehydrogenations, Hydrogenations, and Oxidations of Eudesmanolides

Pyrolysis of alantolactone (462) and isoalantolactone (446) at 350° in the presence of Pd-C or Se results in the naphthalene derivative eudalene (502) with loss of the C-10 methyl group (1102) (Chart IV-3).

More recently, GRIECO (*353*) described a novel and useful dehydrogenation process involving the conversion of α-methyl-γ-lactones into α-methylene-γ-lactones. This method was applied by GONZALEZ *et al.* (*328*) to the determination of the stereochemistry of artemin (**412**). As outlined in Chart IV-4, artemin after exposure to lithium diisopropylamide (LDA) and subsequent treatment of the enolate with diphenyldiselenide gave the phenylselenide (**503**). The selenide was oxidized with H_2O_2 to (**504**) which underwent spontaneous elimination to tanacetin (**381**), a compound of established stereochemistry (*934*). Reduction of compound (**381**) with NaBH$_4$ resulted in artemin which established its entire stereochemistry as (**412**) since all NaBH$_4$ reductions in this series provide the C-11α-methyl epimer (*175*).

Examples demonstrating catalytic hydrogenation of a eudesmanolide are the conversions of reynosin (**374**) to the C-4 epimeric mixture of the tetrahydroderivatives (**506**) and (**507**) with Pd-C and the formation of (**506**) and the isodihydroderivative (**505**) upon reaction with PtO$_2$ (*1174*) (see Chart IV-5). A number of other conversions involving hydrogenations and oxidations with various reagents have been summarized before (*1172*).

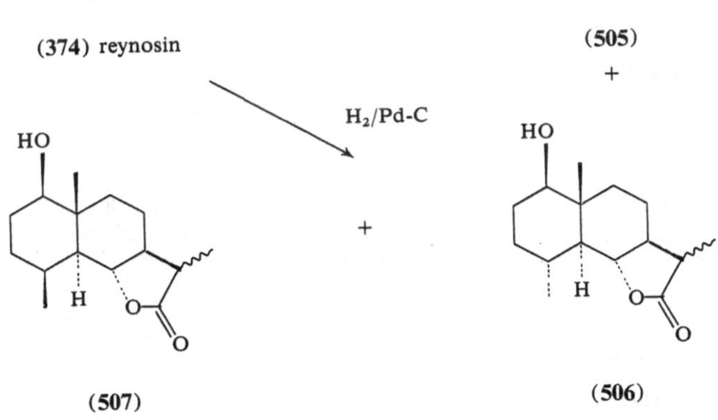

Chart IV-5. Catalytic hydrogenation of reynosin

3.2 Selected Chemical and Photochemical Modifications and Transformations of Eudesmanolides

Eudesmanolides have been frequently used as starting material for chemical and photochemical rearrangement processes which lead to other skeletal types of sesquiterpenes. In contrast to the reaction of dihydro-microcephalin (**508**) from whose mesylate (**509**) the expected elimination product (**510**) is formed with 2,4-lutidine (*413*), the reaction of dihydro-cyclopyrethrosin (**511**) takes a different course (Chart IV-6). Compound (**511**) upon tosylation provides the C-1 β-tosylate which in contrast to the previous example has no proton at C-2 which is *anti*-periplanar to

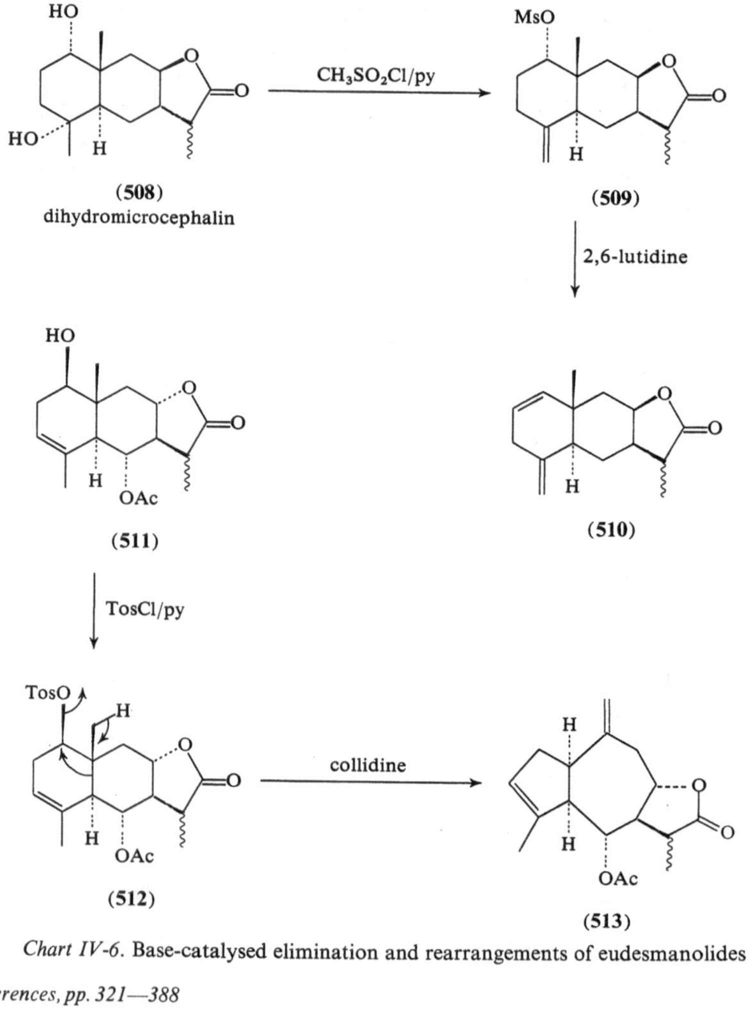

Chart IV-6. Base-catalysed elimination and rearrangements of eudesmanolides

the tosylate leaving group to permit *anti*-elimination. Instead, the stereo-chemistry is appropriate for the elimination-rearrangement reaction shown on the bottom of Chart IV-6 which occurs upon treatment of **(512)** with collidine to give a guaianolide **(513)** (*48*).

Starting with α-santonin-4,5-epoxide **(514)** HENDRICKSON and co-workers (*383*) constructed in a sequence of steps the *cis*-decalin type bromolactone **(515)** with a β-methyl group at C-5. Stereospecific re-arrangement of **(515)** resulted in the formation of the pseudoguaianolide **(516)**, as outlined in Chart IV-7.

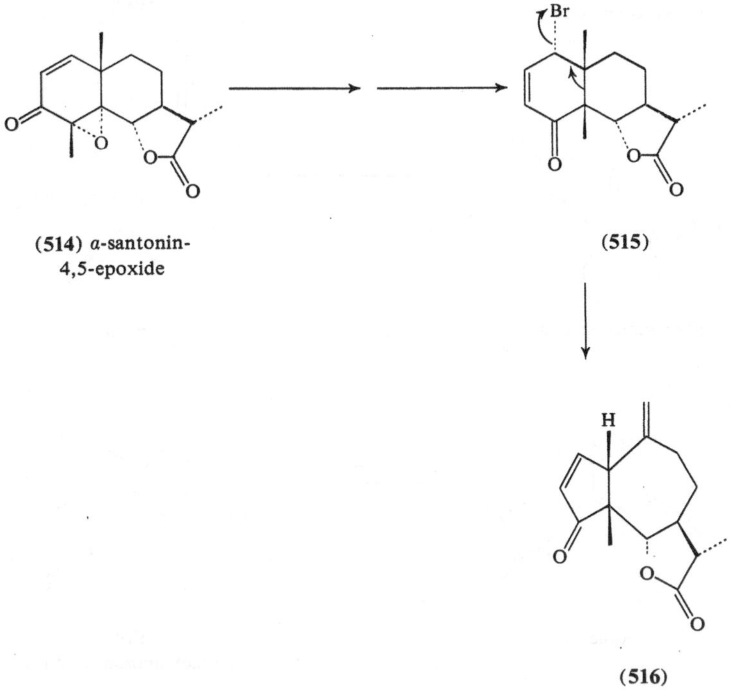

| (514) *a*-santonin- | (515) |
| 4,5-epoxide | |

(516)

Chart IV-7. Conversion of a eudesmanolide into a pseudoguaianolide

A highly interesting and now classical rearrangement reaction is the base-promoted conversion of α-santonin **(405)** into santonic acid **(518)** elucidated by WOODWARD and coworkers (*1144a*). The reaction can be interpreted as a base-mediated opening of the lactone ring providing the carbanion **(517)** which in an intramolecular Michael addition reaction, as shown for **(517a)** in Chart IV-8, forms santonic acid **(518)**. The acid-catalyzed rearrangement of santonin (bottom of Chart IV-8) yielding desmotroposantonin (J) **(519)** contributed strongly to the problems dogging the structure elucidation of santonin until CLEMO *et* *al.* (*167a*) pointed out that (**519**) could be formed by a methyl shift. This provided the impetus for all subsequent studies of the dienone-phenol rearrangement.

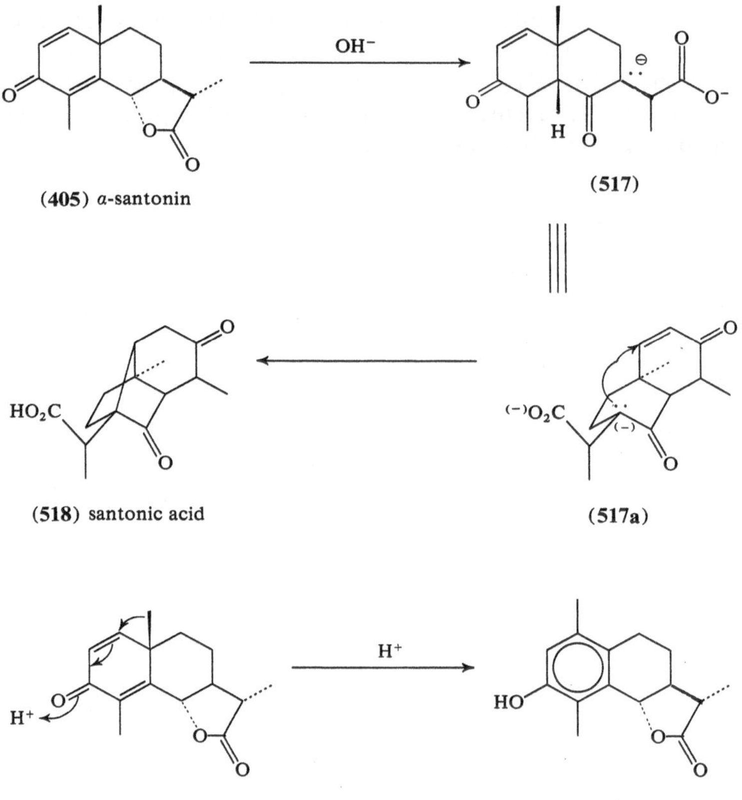

(**405**) α-santonin

(**517**)

(**518**) santonic acid

(**517a**)

(**405**)

(**519**)
desmotroposantonin (J)

Chart IV-8. Base-catalysed conversion of α-santonin to santonin acid and desmotroposantonin (J)

The pioneering work on the photochemistry of santonin and derivatives by BARTON, JEGER and collaborators has been reviewed (*157a, 945a*). The nature of products in the photochemistry of santonin strongly depends on the solvents. As shown on top of Chart IV-9, irradiation of santonin (**405**) in ethanol provides lumisantonin (**485**) (*47*) which upon further irradiation in aqueous acetic acid gives photosantonic acid (**520**). If irradiation of santonin is carried out in refluxing aqueous acetic acid the guaianolide isophotosantonic lactone (**521**) is formed which has been

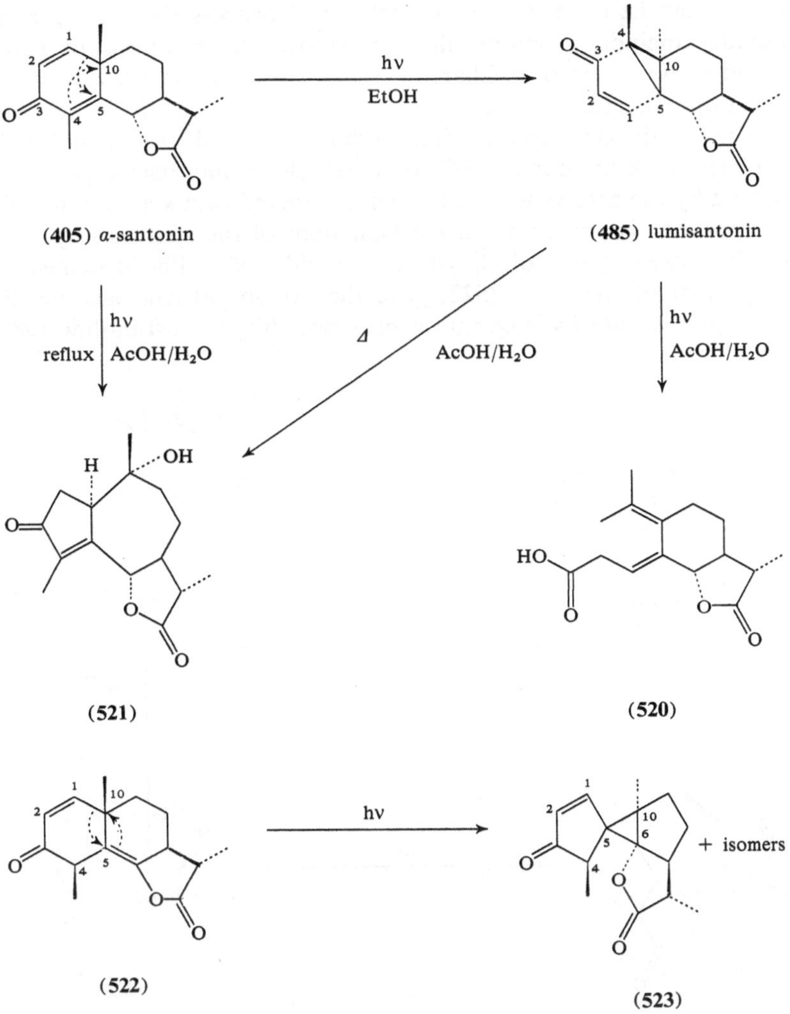

(405) a-santonin

(485) lumisantonin

(521)

(520)

(522)

(523)

+ isomers

Chart IV-9. Photochemistry of santonin and its isomer

used for numerous partial syntheses of other terpenoids. Isophotosantonic lactone may also be obtained from lumisantonin **(485)** in a thermal process (*30*). On the bottom of Chart IV-9, the photochemical conversion of the santonin isomer **(522)** to the spiro-derivative **(523)** is shown (*515*).

The conversion of α-santonin into lumisantonin **(485)** by a chemical exitation process has recently been reported (*529*). Further recent studies on the chemistry of santonin and its derivatives (*21, 85, 276, 277, 385, 559, 698, 700, 792, 941*) and their photochemistry (*40, 246, 515, 694,*

701, 827) can be cited. Other photochemical processes involving eudesmanolides include reactions directed toward the synthesis of germacranolides. As outlined in Chart IV-10, FUJIMOTO *et al.* (*278*) converted
the santonin derivative (**524**) to the chloroepoxide (**525**) which upon
treatment with zinc and sodium iodide provided the dienol (**526**).
Photolysis of compound (**526**) initiated electronic rearrangement as
indicated by the arrows to give the ten-membered ring ketolactone (**527**).
This approach is analogous to the final steps of the synthesis by COREY
and HORTMANN (*185*) of dihydrocostunolide (**98**). Photochemical rearrangement of the diene (**528**) gave the cyclodecatriene lactone (**529**)
which upon *in situ* hydrogenation provided dihydrocostunolide (**98**).

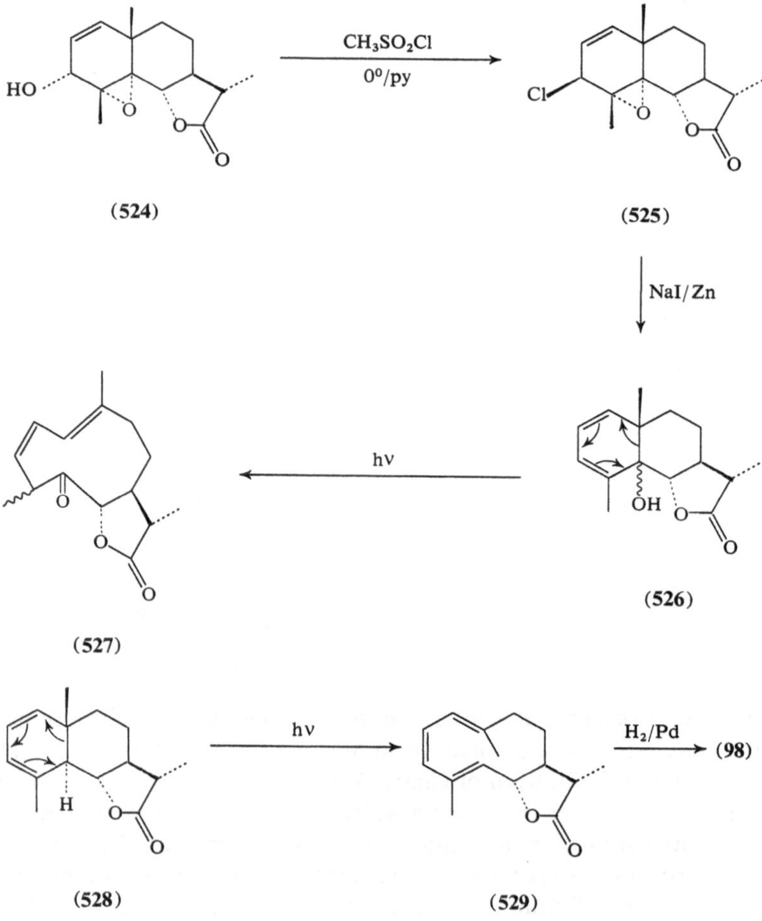

Chart IV-10. Photolysis of 1,2,3,4-unsaturated eudesmanolides

Table IV-2. Naturally Occurring Eudesmanolides and Biogenetic Derivatives

Structure Number	Name of Compound	Formula	m.p. °C	$[\alpha]_D$	Plant Source[c]	References[a,b]	Comments
(462)	Alantolactone	$C_{15}H_{20}O_2$	78–80	+175	Imula helenium I. grandis I. magnifica I. racemosa I. royleana	(913, 1102, 574, 571) (560, 661, 660) (769) (850a) (111)	
(473)	Alantolactone, dihydro	$C_{15}H_{22}O_2$	—	—	Telekia speciosa Inula helenium	(100) (574)	
(474)	Alantolactone, neo	$C_{15}H_{22}O_2$	118	—	Inula racemosa	(850a)	
(377)	Alantolactone, 1α,8α-dihydroxy	$C_{15}H_{20}O_4$	202–4	+206	Zexmenia phyllocephala	(110)	
(463)	Alantolactone, 1β-hydroxy	$C_{15}H_{20}O_3$	oil	—	Inula helenium	(111)	
(464)	Alantolactone, 2α-hydroxy	$C_{15}H_{20}O_3$	oil	+60	Inula royleana	(111)	
(379)	Alantolactone, 1α-hydroxy-8α-(2-hydroxymethylacryloxy	$C_{19}H_{24}O_6$	oil	+226	Zexmenia phyllocephala	(110)	
(378)	Alantolactone, 8α-hydroxy-1α-(2-hydroxymethylacryloxy	$C_{19}H_{24}O_6$	oil	+148	Zexmenia phyllocephala	(110)	
(446)	Alantolactone, iso	$C_{15}H_{20}O_2$	111–3	+172	Inula helenium I. grandis I. magnifica I. racemosa I. royleana Telekia speciosa Ambrosia camphorata	(551, 1102, 574, 719) (571, 560, 661) (769) (850a) (111) (63) (463)	
(475)	Alantolactone, iso, dihydro	$C_{15}H_{22}O_2$	174	+22	Inula helenium I. racemosa	(574, 571, 671) (850a)	
(457)	Alantolactone, iso, 3β-hydroxy-2α-senecioyloxy	$C_{20}H_{26}O_5$	222–3	—	Inula britannica	(115)	

[a] Reference reports structure revision;   [b] X-Ray data reported;   [c] When the plant sources are from a family other than the Compositae, the family name is given under Comments.

Table IV-2 (continued)

Structure Number	Name of Compound	Formula	m.p. °C	$[\alpha]_D$	Plant Source[c]	References	Comments
(465)	Alantolactone, 2-oxo	$C_{15}H_{18}O_3$	152–3	+301	Inula royleana	(111)	
(421)	Arabsin	$C_{15}H_{22}O_4$	188–9	+89	Artemisia absinthium	(17, 108)	
(391)	Arbusculin A	$C_{15}H_{22}O_3$	76–7	+25	Artemisia arbuscula	(508, 1154)	
					A. tridentata	(981)	
					Mochina vellutina		See footnote in ref. (523)
(392)	Arbusculin A, 4-epi	$C_{15}H_{22}O_3$	148–9	—	Frullania tamarisci	(37)	Hepaticae
(389)	Arbusculin, 1β-hydroxy	$C_{15}H_{22}O_4$	194–6	—	Tanacetum vulgare	(934, 26)	
(370)	Arbusculin B	$C_{15}H_{20}O_2$	86–8	+47	Artemisia arbuscula	(508, 350, 352, 1154)	
					A. tridentata	(981)	
					Frullania tamarisci	(173, 36)	Hepaticae
(495)	(−)-Arbusculin B, ent.	$C_{15}H_{20}O_2$	85–6	−35	Frullania dilatata	(36)	Hepaticae
(382)	Arbusculin C	$C_{15}H_{20}O_3$	150–1	+113	Artemisia arbuscula	(507, 1154, 508)	
					A. tridentata	(981)	
(427)	Arbusculin D	$C_{15}H_{22}O_4$	170–2	—	Artemisia arbuscula	(511)	
(394)	Arbusculin E	$C_{15}H_{24}O_4$	160–1	—	Artemisia arbuscula	(508)	
(387)	Arglanin	$C_{15}H_{18}O_4$	207	+111	Artemisia douglasiana	(689, 1159)	
					A. mexicana	(888, 905)	
(423)	Armexin diacetate	$C_{19}H_{24}O_6$	222–4	+4	Artemisia mexicana	(887)	
(373)	Armexifolin	$C_{15}H_{18}O_4$	207–8	—	Artemisia mexicana	(905)	
(420)	Arsanin	$C_{15}H_{22}O_4$	193–4	+26	Artemisia santolina	(12, 13, 14, 730, 1147, 1153)	
(419?)	Arsantin	$C_{15}H_{22}O_4$	168	+31	Artemisia santolina	(12, 730, 1147, 1153)	See ref. (1153) for comments on structure
(411)	Arsubin	$C_{15}H_{22}O_4$	233–4	—	Artemisia sublessingiana	(1067, 1068, 328[a])	
					A. hanseniana	(974)	
(393)	Artecalin	$C_{15}H_{20}O_4$	225–7	+45	Artemisia californica	(297, 1158)	
					Achillea biebersteinii	(1185)	

No.	Name	Formula	mp (°C)	$[\alpha]_D$	Species	Ref.	Remarks
(388)	Artemexifolin	$C_{17}H_{20}O_6$	260	+156	Artemisia mexicana	(905)	See Mibulactone
(412)	Artemin	$C_{15}H_{22}O_4$	238–40	+167	Artemisia maritima	(328)	
					A. tenuisecta	(31a, 561a)	
					A. halophila	(914, 1078)	
					A. hanseniana	(31)	
					A. kemrudica	(974)	
						(11)	
(407)	Artemisin	$C_{15}H_{18}O_4$	202–3	−84	Artemisia maritima	(1072, 1034[a], 751)	
					A. cina	(923)	
(400)	Artesin	$C_{15}H_{22}O_3$	172	+49	Artemisia santolina	(15)	
(476)	Ashurbin	$C_{15}H_{22}O_4$			Artemisia absinthium	(1191)	
					A. hanseniana	(974)	
(447)	Asperilin	$C_{15}H_{20}O_3$	151–2	+149	Iva asperifolia	(457)	
					I. texensis	(457)	
					Inula helenium	(111)	
(415)	Badkhysidin	$C_{20}H_{26}O_5$	117–8	—	Ferula oopoda	(966)	Umbelliferae
					F. badghysi	(967)	
(385)	Badkhysinin	$C_{20}H_{24}O_5$	104–5	−212	Ferula oopoda	(967, 965[a])	Umbelliferae
					F. badghysi		
(364)	Balchanin	$C_{15}H_{20}O_3$	142	+96	Artemisia balchanorum	(1012, 1154, 1002, 822)	See Santamarin ref. (1154)
(469)	Callitrisin	$C_{15}H_{20}O_2$	164	−41	Callitris columellaris	(133)	Cupressaceae
(470)	Callitrisin, dihydro	$C_{15}H_{22}O_2$	127–8	−39	Callitris columellaris	(133)	Cupressaceae
(467)	Carpesin	$C_{15}H_{20}O_3$	115–128	—	Carpesium eximium	(603)	
(439)	Chrysanin	$C_{20}H_{26}O_5$	202–4	+80	Chrysanthemum cinerariaefolium	(231, 1184)	Reported before Tachillin
(418)	Colartin	$C_{15}H_{24}O_3$	107–8	+11	Artemisia tripartita	(508, 1154)	
(484)	Commiferin	$C_{15}H_{20}O_3$	—	—	Commiphora myrra	(725)	Burseraceae
(491)	Critonilide	$C_{15}H_{20}O_2$	73	+140	Critonia morifolia	(105)	See also β-cyclo-costunolide
(490)	Critonilide, iso	$C_{15}H_{20}O_2$	57	—	Critonia morifolia	(105)	See also ref. (615)
(330)	α-Cyclocostunolide	$C_{15}H_{20}O_2$	82–3	+116	Frullania tamarisci	(173, 228, 687, 823)	Hepaticae
					Moquinea velutina	(1079)	

Table IV-2 (continued)

Structure Number	Name of Compound	Formula	m.p. °C	[α]$_D$	Plant Source[c]	References	Comments
(331)	β-Cyclocostunolide	$C_{15}H_{20}O_2$	68–9	+165	Frullania tamarisci	(37, 615, 105)	Hepaticae
(496)	β-Cyclocostunolide, (+)-cis	$C_{15}H_{20}O_2$	75–6	+38	Frullania dilatata	(36)	Hepaticae
(409)	β-Cyclocostunolide, dihydro	$C_{15}H_{22}O_2$	135–7	+165	Moquinea velutina	(1079, 524)	
(438)	β-Cyclopyrethrosin	$C_{17}H_{22}O_5$	166–7	+58	Chrysanthemum cinerariaefolium	(229, 1009)	
(441)	β-Cyclopyrethrosin, dihydro	$C_{17}H_{24}O_5$	212–3	+72	Chrysanthemum cinerariaefolium	(231, 229)	
(408a)	Decipienin A	$C_{20}H_{24}O_5$	185–6	+54.9	Melanoselinum decipiens	(332a)	Umbelliferae
(430)	Decipienin B	$C_{20}H_{26}O_6$	160–2	+44	Melanoselinum decipiens	(332a)	Umbelliferae
(398)	Decipienin G	$C_{20}H_{28}O_6$	gum	−22	Melanoselinum decipiens	(332)	Umbelliferae
(397)	Decipienin H	$C_{15}H_{22}O_5$	—	—	Melanoselinum decipiens	(332)	Umbelliferae
(380)	Dentatin A	$C_{15}H_{20}O_4$	—	—	Artemisia tridentata	(505)	
(498)	Diplophyllin	$C_{15}H_{20}O_2$	30–1	—	Diplophyllum albicans	(794)	Scrophulariaceae, or Hepataceae ent-eudesmanolide
(497)	Diplophyllolide A	$C_{15}H_{20}O_2$	60–2	—	D. taxifolium, Diplophyllum albicans	(794)	ent-eudesmanolide
(363)	Douglanin	$C_{15}H_{20}O_3$	115–7	+133	Artemisia douglasia, A. ludoviciana, A. mexicana	(690, 289b, 1104b), (642), (888)	
(461)	Encelin	$C_{15}H_{16}O_3$	195–6	−16	Encelia farinosa	(305)	Anhydrofarinosin
(488)	Eriolangin	$C_{20}H_{28}O_6$	94–6	−91	Eriophyllum lanatum	(621, 140b)	1,10-seco-eudesmanolide
(487)	Eriolanin	$C_{19}H_{26}O_6$	127–8	−93	Eriophyllum lanatum	(621, 140b)	1,10-seco-eudesmanolide
(413)	Erivanin	$C_{15}H_{22}O_4$	203–4	+98	Tanacetum balsamita, Artemisia fragrans	(930), (261)	

No.	Compound	Formula	M.p.	[α]	Source	Ref.	Remarks
(480)	Eudesma-4(15),7(11)-diene-8β,12-olide	$C_{15}H_{20}O_2$	129–30	+255	Inula helenium	(560)	
(481)	Eudesma-5,7(11)-diene-8β,12-olide	$C_{15}H_{20}O_2$	88–90	+470	Inula helenium	(560)	Umbelliferae
(482)	Eucesma-5,7(11)-diene-13-ol-8β,12-olide	$C_{15}H_{20}O_3$	gum	—	Inula helenium	(560)	Umbelliferae
(401)	Eudesm-4-en-6,12-olide, 1-hydroxy-6β,7α,11β-H	$C_{15}H_{22}O_3$	—	—	Artemisia granatensis	(335)	
					Hypochaenis setosus	(327, 335)	
(403)	Eudesm-4-en-6,12-olide, 1-oxo-6β,7α,11β-H	$C_{15}H_{22}O_3$	—	—	Artemisia granatensis	(335)	
(472)	Farinosin	$C_{15}H_{18}O_4$	200–1	−111	Encelia farinosa	(305, 449)	
					E. virginensis	(990)	
(399)	Feropodin	$C_{15}H_{20}O_2$	140–1	—	Ferula oopoda	(962, 961)	Umbelliferae
(433)	Ferula, hydroxylactone	$C_{15}H_{20}O_4$	—	—	Ferula oopoda	(971)	Umbelliferae
(424)	Finitin	$C_{15}H_{20}O_3$	153–5	−168	Artemisia finita	(565, 211)	
					A. ramosa	(336)	
(422)	(−)-Frullanolide	$C_{15}H_{20}O_2$	77	−113	Frullania tamarisci	(597, 173, 36, 350, 351)	Hepaticae; ent-eudesmanolide
(492)	(+)-Frullanolide	$C_{15}H_{20}O_2$	75–6	+94	Frullania dilatata	(36)	Hepaticae; ent-eudesmanolide
(494)	(+)-Frullanolide, dihydro	$C_{15}H_{22}O_2$	120–1	+55	Frullania dilatata	(36)	Hepaticae; ent-eudesmanolide
(493)	(+)-Frullanolide, oxy	$C_{15}H_{20}O_3$	179–80	+71	F. dilatata	(36)	Hepaticae; ent-eudesmanolide
(449)	Granilin	$C_{15}H_{20}O_4$	197–8	+157	Inula grandis	(770, 677)	
					Ambrosia polystachya	(1112)	
					Artemisia ashurbajevii	(1189)	
					Carpesium abrotanoides	(677)	
(450)	Grandulin	$C_{15}H_{20}O_3$	152–3	+158	Inula grandis	(771, 772)	See Ivalin
(468)	Graveolide	$C_{15}H_{22}O_4$	224–6	—	Inula graveolens	(1003)	
(477)	Hybrifarin	$C_{15}H_{20}O_4$	212–4	—	Encelia farinosa	(90)	
(434)	Ferula oopoda, hydrolactone				Ferula oopoda	(960)	Umbelliferae

Table IV-2 (continued)

Structure Number	Name of Compound	Formula	m.p. °C	[α]_D	Plant Source[a]	References	Comments
(450)	Ivalin	$C_{15}H_{20}O_3$	130-2	+142	Iva imbricata	(411, 457)	See also Grandulin
					I. microcephala	(411)	
					Carpesium eximium	(603)	
					C. abrotanoides	(676a)	
					Inula grandis	(771, 772)	
					I. royleana	(111)	
					Zaluzania triloba	(894)	
					Wedelia rugosa	(18a)	
(451)	Ivalin acetate	$C_{15}H_{22}O_4$	oil	—	Inula royleana	(111)	
(489)	Ivangulin	$C_{16}H_{22}O_4$	84-5	+109	Iva angustifolia	(454, 358)	Seco-eudesmanolide
(444)	Ivangustin	$C_{15}H_{20}O_3$	120-2	+85	Iva angustifolia	(454)	
(436)	Ivangustin, 1-desoxy, 8-epi	$C_{15}H_{20}O_2$	oil	+35	Inula royleana	(111)	
(437)	Ivangustin, 8-epi	$C_{15}H_{20}O_3$	oil	-25	Inula royleana	(111)	
(435)	Ivangustin, iso, 8-epi	$C_{15}H_{20}O_3$	135	-50	Inula helenium	(111)	
					I. royleana	(111)	
(448)	Ivasperin	$C_{15}H_{20}O_4$	157-9	+140	Iva asperifolia	(457)	
					I. texensis		
(416)	Judaicin	$C_{20}H_{28}O_5$	149-51		Ambrosia polystachya	(1112)	
					Artemisia judaica	(922a)	See Tauremisin
(429)	Lasolide	$C_{15}H_{18}O_2$	oil	+189	Laser trilobum	(475, 483)	Umbelliferae
(478)	Lindestrenolide	$C_{15}H_{16}O_2$	111-3	+180	Lindera strychnifolia	(1046)	Lauraceae
(483)	Lindestrenolide, dehydro	$C_{15}H_{18}O_3$	220-1	+209	Lindera strychnifolia	(1055)	
(479)	Lindestrenolide, hydroxy	$C_{17}H_{18}O_3$	179-8	-4	Lindera strychnifolia	(1046)	Lauraceae
(370a)	γ-Liriodenolide	$C_{17}H_{22}O_5$	169-71	+227	Liriodendron tulipifera	(235)	Magnoliaceae
(366)	Ludalbin	$C_{17}H_{22}O_5$			Artemisia ludoviciana	(306)	
(367)	Ludovicin A	$C_{15}H_{20}O_4$	215	128	Artemisia ludoviciana	(642)	
(375)	Ludovicin B	$C_{15}H_{20}O_4$	152	138	Artemisia ludoviciana	(642)	

(372)	Ludovicin C	$C_{15}H_{18}O_4$	193–5	+95	*Artemisia ludoviciana*	(642)	
(485)	Lumisantonin	$C_{15}H_{18}O_3$	153–5	−150	*Artemisia kurramensis*	(945, 30, 47, 494[b], 914)	
(431)	Mibulactone	$C_{15}H_{22}O_4$	228–9	+156	*Artemisia monogyna*	(280a, 553)	May be identical with Artemin
(466)	Microcephalin	$C_{15}H_{22}O_4$	206–8	+75	*Iva microcephala*	(413)	Umbelliferae
(414)	Oopodin	$C_{20}H_{26}O_4$	127–8	—	*Ferula oopoda*	(968)	Umbelliferae
(384)	Oopodin, dehydro	$C_{20}H_{24}O_4$	113–4	—	*Ferula oopoda*	(959, 968)	Umbelliferae
					F. badghysi	(967)	
(443)	Pinnatifidin	$C_{15}H_{18}O_3$	164–5	+302	*Helenium pinnatifidum*	(421, 422)	
(369)	*Pluchea*, lactone from	$C_{20}H_{26}O_6$	153	—	*Pluchea dioscorides*	(103)	
(454)	Pulchellin B	$C_{17}H_{22}O_5$	215–8	+92	*Gaillardia pulchella*	(414, 1168[a])	
(452)	Pulchellin C	$C_{15}H_{20}O_4$	199–202	+125	*Gaillardia pulchella*	(414, 1168[a], 200[b])	
					G. aristata	(725a)	
(453)	Pulchellin E	$C_{17}H_{22}O_5$	181–3	43	*Gaillardia pulchella*	(433, 1168[a])	
					G. aristata	(725a)	
(456)	Pulchellin F	$C_{20}H_{32}O_5$	144–6	+91	*Gaillardia pulchella*	(433, 1168[a])	
(374)	Reynosin	$C_{15}H_{20}O_3$	145–6	+180	*Ambrosia confertiflora*	(1174, 26, 821)	
					Chrysanthemum parthenium	(888)	
					Tanacetum vulgare	(934)	
(376)	Ridentin B	$C_{15}H_{20}O_4$	205–6		*Artemisia tripartita*	(513)	
					A. tripartita	(980)	
(371)	Rothin A	$C_{15}H_{20}O_3$	133–4	+122	*Artemisia rothrockii*	(507)	
(383)	Rothin B	$C_{15}H_{20}O_4$	254–6	+242	*Artemisia rothrockii*	(507)	
(364)	Santamarin	$C_{15}H_{20}O_3$	134–5	+97	*Chrysanthemum parthenium*	(900, 1159, 26, 1002, 821)	See Balchanin, ref. (1154)
					Artemisia spicata	(752a)	
					Ambrosia confertiflora	(1174)	
					Tanacetum tanacetiodes	(114)	
(395)	Santamarin, dihydro	$C_{15}H_{22}O_3$	134–6	+78	*Sonchus hierrensis*	(74)	Reported before Sant-3-en-6,12-olide C
(368)	Santamarin, epoxy	$C_{15}H_{20}O_4$	244–6	+87	*Ambrosia confertiflora*	(1174, 26, 900)	See Santamarin, dihydro
(395)	Sant-3-en-6,12-olide C, 1β-hydroxy				*Artemisia tridentata*	(981, 900, 690, 1025)	

11*

Table IV-2 (continued)

Structure Number	Name of Compound	Formula	m.p. °C	$[\alpha]_D$	Plant Source[c]	References	Comments
(410)	Sant-4(14)-en-6,12-olide C, 1β-hydroxy	$C_{15}H_{22}O_3$	130–1	+137	Artemisia tridentata	(981)	
(405)	α-Santonin	$C_{15}H_{18}O_3$	170–3	−172	Artemisia mexicana	(38a[b], 116, 1, 381, 741, 842)	For early reports see ref. (1138b)
					A. fragrans	(370)	
					A. halophila	(31)	
					A. herba-alba	(571a)	
					A. ramosa	(336)	
					A. schrenkiana	(837)	
					A. tenuisecta	(591a)	
					A. chasarica	(973)	
					A. hybrida	(280a)	
					A. kurramensis	(945)	
					A. incana	(277a)	
					A. mogoltavica	(339a)	
					A. transiliensis	(339a)	
					A. meyriana	(19a)	
					A. szovitsiana	(19a)	
(404)	Santonin, 1,2-dihydro	$C_{15}H_{20}O_3$	102–3	+79	Artemisia stelleriana	(51)	
(408)	Santonin, 11-oxy	$C_{15}H_{18}O_4$	295	−134	Artemisia maritima	(32, 835[a])	
(406)	β-Santonin	$C_{15}H_{18}O_3$	216–8	−137	Artemisia caerulescens	(169a[b], 565, 2, 166, 741, 842)	For early reports, see ref. (1138b)
					A. finita	(565)	
					A. hybrida	(564a)	
					A. kurramensis	(595)	
					A. monogyna	(564a)	

No.	Name	Formula	mp	[α]	Source (refs)	Notes
(426)	ψ-Santonin	$C_{15}H_{20}O_4$	183-4	-169	A. salina (564a); A. compacta (1107); A. schrenkiana (837); Artemisia maritima (167, 169); A. cina (210[a])	
(425)	ψ-Santonin, deoxy	$C_{15}H_{20}O_3$	101-2	-207	Artemisia sp. (211, 169)	
(432)	Semopodin	$C_{20}H_{24}O_5$	177-8	—	Ferula oopoda (964, 971[a])	Umbelliferae
(428)	Silerolide	$C_{22}H_{30}O_6$	—	—	Laserpitium siler (480, 487)	Umbelliferae
(417)	Tabarin	$C_{15}H_{20}O_5$	213-5	—	Artemisia canariensis (319)	
(439)	Tachillin				Tanacetum pseudoachillea (1184, 231)	See Chrysanin
(381)	Tanacetin	$C_{15}H_{20}O_4$	205	+175	Tanacetum vulgare (934)	
(442)	Tanapsin	$C_{20}H_{28}O_6$	191-2	-139	Tanacetum pseudoachillea (1181, 1183)	
(402)	Taurin	$C_{15}H_{20}O_3$	118-9	120	Artemisia hanseniana (566, 974)	
(416)	Tauremisin	$C_{15}H_{20}O_4$	176-7	+48	Artemisia taurica (812, 567, 917, 914, 491); A. chasarica (923); A. vulgaris (292); A. verlotorum (301); A. hanseniana (974); A. canariensis (319); A. granatensis (335)	
(460)	Telekin	$C_{15}H_{20}O_3$	159-60	+234	Telekia speciosa (65); Carpesium eximium (603)	
(458)	Telekin, iso	$C_{15}H_{20}O_3$	144-5	+100	Telekia speciosa (65, 717, 62); Ambrosia confertiflora (1174)	
(459)	Telekin, iso, 3-epi	$C_{15}H_{20}O_3$	176-7	+156	Gaillardia aristata (449); Inula britannica (115)	
(386)	Tuberiferin	$C_{15}H_{18}O_3$	160-2	+9	Sonchus tuberifer (72, 73, 609, 1158, 354)	
(390)	Vahlenin	$C_{19}H_{26}O_6$	200	+17	Centaurea hyssopifolia (321)	
(486)	Vernodesmin	$C_{25}H_{28}O_7$	190-4	+27	Vernonia pectoralis (703[b])	Modified eudesmanolide
(471)	Virginin	$C_{15}H_{18}O_3$	241-2	-128	Encelia virginensis (990)	
(416)	Vulgarin				Artemisia vulgaris (292)	See Tauremisin
(445)	Yomogin	$C_{15}H_{16}O_3$	201-2	-88	Artemisia princeps (286, 145, 1155)	

V. Guaianolides and Seco-Guaianolides (Xanthanolides)

1. Structural Types and Their Biogenesis

The guaianolides together with their *seco*-derivatives, the xanthano-lides, represent one of the largest groups of sesquiterpene lactones with over 200 known naturally occurring compounds. Structures of the guaianolides are summarized in Chart V-1 below, those of the xanthanolides in Chart V-2 on page 187. Table V-1 lists all guaianolides and xanthanolides whose structure has been established by X-ray crystallography (page 198). Names (in alphabetical order), physical properties (empirical formula, melting points, and optical rotations), plant sources and literature references are found in Table V-2 for the guaianolides (pages 199—211) and Table V-3 for the xanthanolides on page 212.

Chart V-1. Naturally occurring guaianolides and biogenetic derivatives

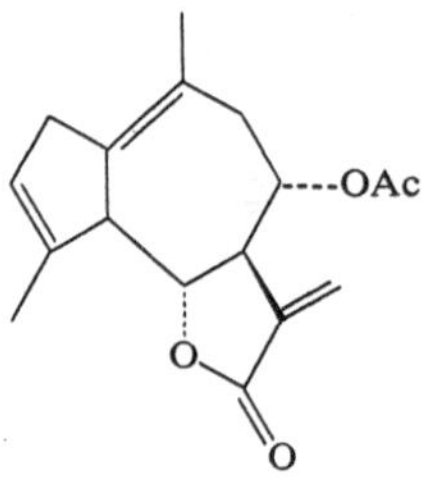

(530) Zuurbergenin

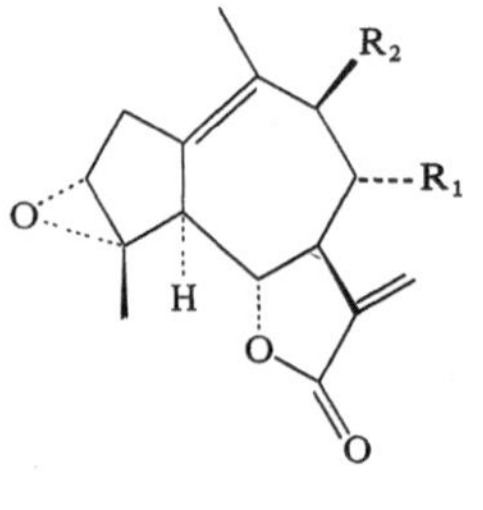

(531) Ludartin; $R_1 = R_2 = H$
(532) Arteglasin A; $R_1 = OAc, R_2 = H$
(533) Subacaulin; $R_1 = OH, R_2 = OAng$
(534) Berlandin; $R_1 = OAng, R_2 = OAc$

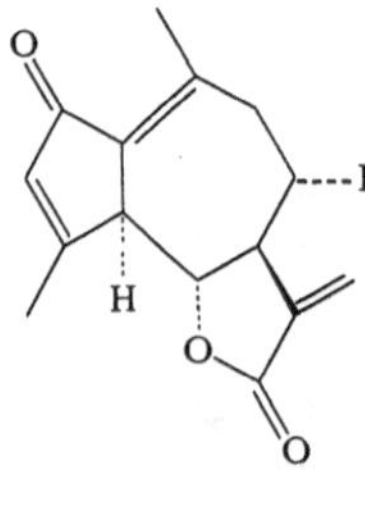

(535) Leucodin, dehydro; R = H
(536) Matricarin, 11,13-dehydro; R = OAc

(537) Estafietin, isoepoxy

(538?) Chrysartemin A

(539) Picridin

(540) Rupicolin A; R_1 = OH, R_2 = H
(541) Rupicolin A, 15-acetoxy-1 β-hydroxy-
8-(2 α-acetoxyethyl) acrylate;
R_1 = OA, R_2 = OAc

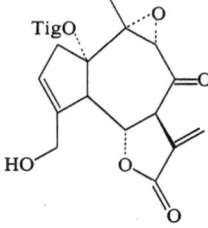

(542) Euperfolide

Chart V-1 (continued)

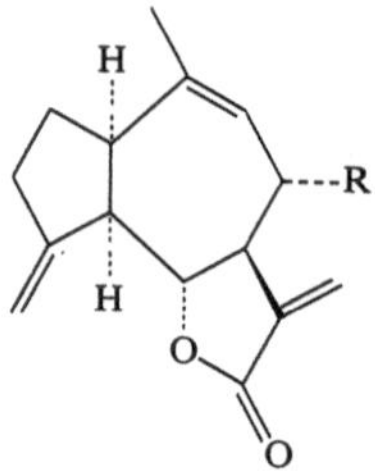

(**543**) Eremanthin; R = H
(**545**) Vanillosmin, 8 α-senecioyloxy;
        R = OSen

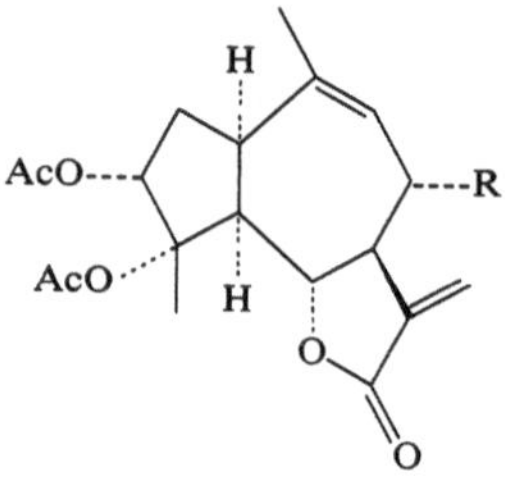

(**546**) Rupicolin A, 3 α,4 α-diacetoxy-3,4-
        dihydro-8-(2-α-acetoxyethyl)acrylate;
        R = OA

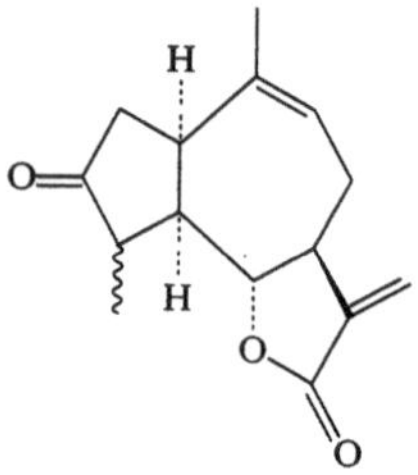

(**547**) Eregoyazin

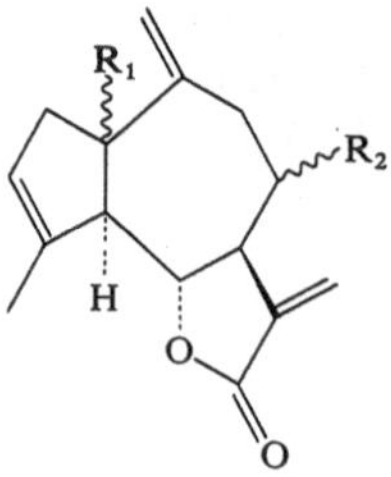

(**548**) Rupicolin B; $R_1 = R_2 = \alpha$-OH
(**549**) Ligustrin; $R_1 = H$, $R_2 = \beta$-OH

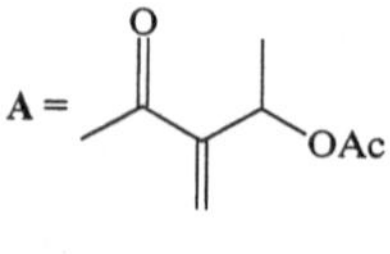

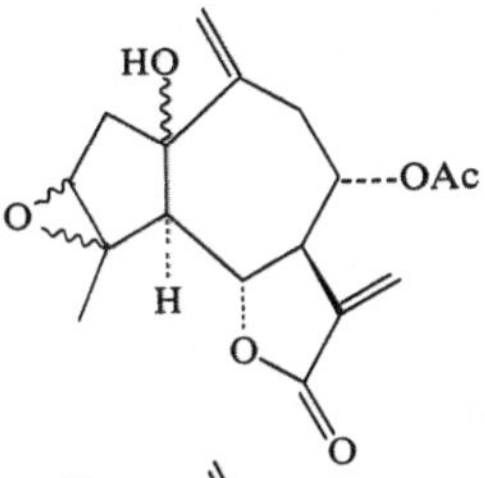

(550) Arteglasin B

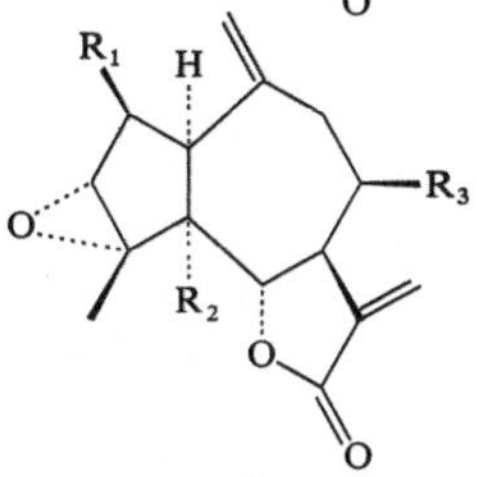

(551) Estafietin; $R_1 = R_2 = R_3 = H$
(552) Eupatundin; $R_1 = R_2 = OH$,
$R_3 = OAng$
·**(553)** Eupatundin acetate; $R_1 = OAc$,
$R_2 = OH$, $R_3 = OAng$

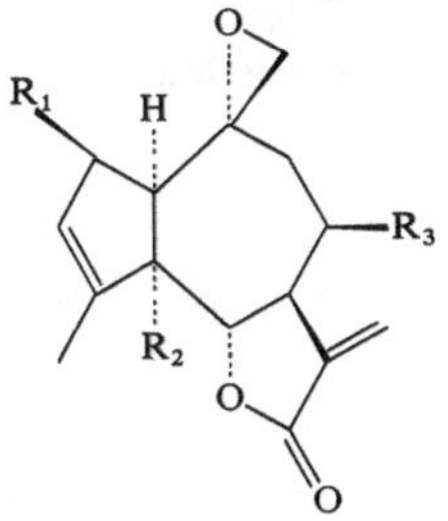

(554) Euparotin; $R_1 = R_2 = OH$, $R_3 = OAng$
(555) Preeupatundin, 8 β-tiglinoyloxy, 10,15-epoxide; $R_1 = OH$, $R_2 = H$, $R_3 = OTig$
(556) Graminiliatrin deoxy; $R_1 = OH$,
$R_2 = H$. $R_3 = OA$
(557) Euparotin acetate; $R_1 = OAc$,
$R_2 = OH$, $R_3 = OAng$
(558) Spicatin; $R_1 = OAc$, $R_2 = H$, $R_3 = OB$

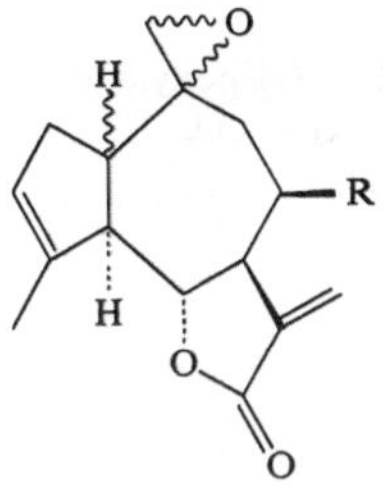

(559) Bahia I; $R = OH$
(560) Bahia II; $R = OC$
(561) Bahifolin; $R = OD$

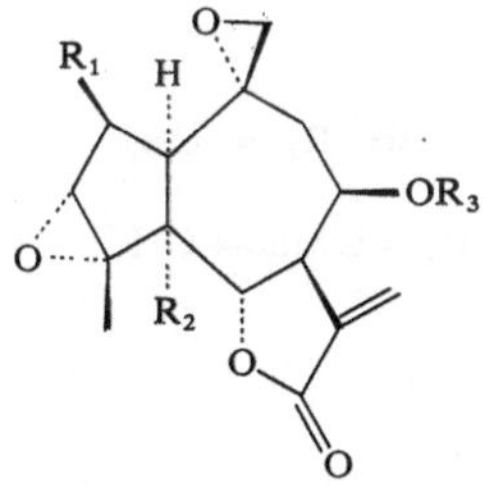

(562) Eupatoroxin; $R_1 = R_2 = OH$,
$R_3 = OAng$
(563) Graminiliatrin; $R_1 = OH$, $R_2 = H$,
$R_3 = OA$
(564) Spicatin, epoxy; $R_1 = OAc$, $R_2 = H$,
$R_3 = OB$

A =

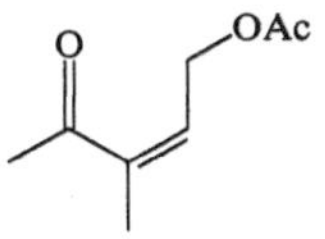

B =

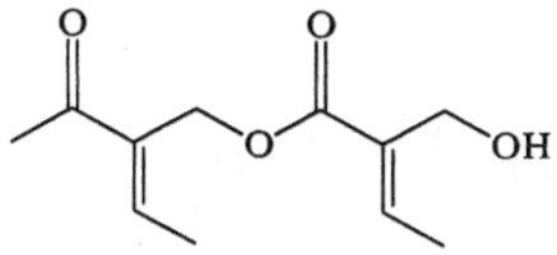

C =

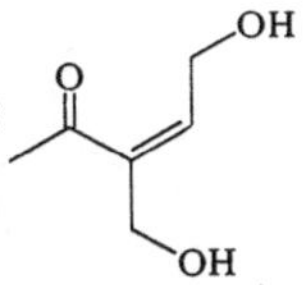

D =

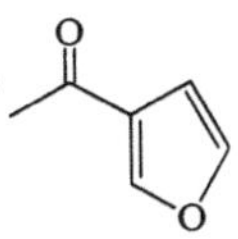

Chart V-1 (continued)

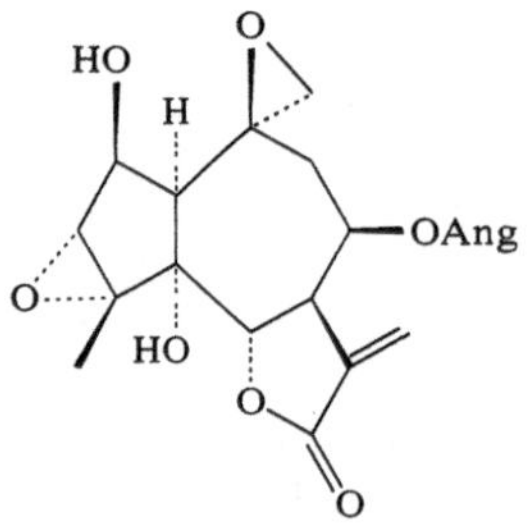

(565) Eupatoroxin, 10-epi

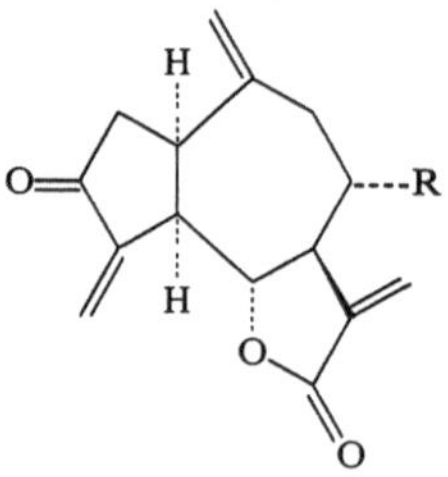

(566) Zaluzanin C, dehydro; R = H
(567) Cynaropicrin, dehydro; R = OMac-4-OH

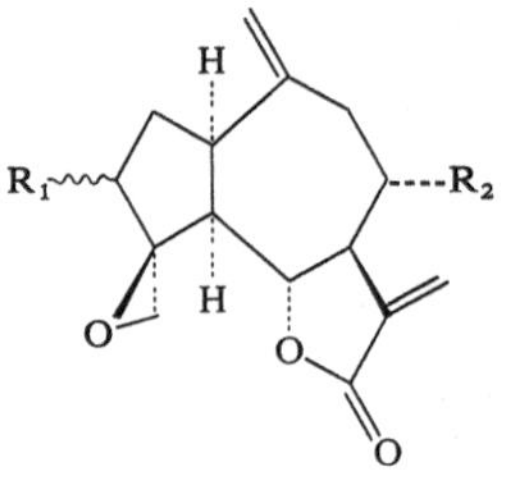

(568) Repin; R_1 = β-OH, R_2 = OEpoxymac
(569) Acroptilin; R_1 = OH, R_2 = OA
       (Chlorohyssopifolin C)

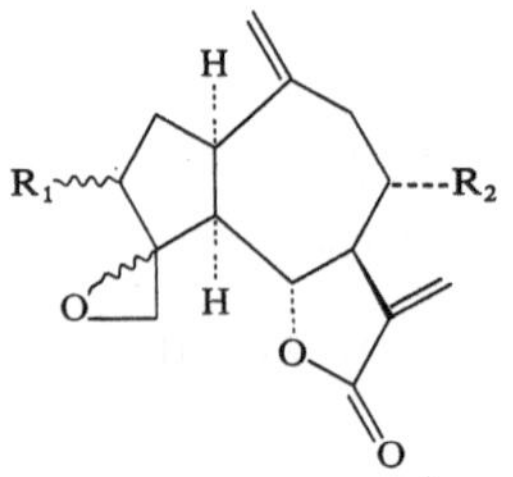

(571) Subluteolide; R_1 = OH, R_2 = OB, C-15β
(572) Janerin; R_1 = OH, R_2 = OMac-4-OH

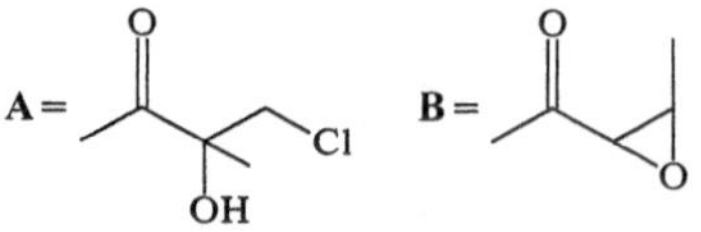

(573) Chlorohyssopifolin B; $R_1 = R_3 = \alpha$-OH,
$R_2 = \alpha$-OH, $R_4 = Cl$
(574) Chlorohyssopifolin E; $R_1 = OH$,
$R_2 = \alpha$-OH, $R_3 = OA$, $R_4 = Cl$
(575) Chlorohyssopifolin D; $R_1 = OH$,
$R_2 = \alpha$-OH, $R_3 = OB$, $R_4 = Cl$
(576) Chlorohyssopifolin A;
$R_1 = R_2 = \alpha$-OH, $R_3 = OC$, $R_4 = Cl$
(577) Muricatin; $R_1 = \beta$-OH, $R_2 = \beta$-H,
$R_3 = OMac$-4-OH, $R_4 = H$
(579) Janerin, chloro; $R_1 = R_4 = OH$,
$R_2 = Cl$, $R_3 = OMac$-4-OH

(580) Costus Lactone, dehydro;
$R_1 = R_2 = R_3 = H$
(581) Costus Lactone, dehydro, 8α-senecioyloxy; $R_1 = R_2 = H$, $R_3 = OSen$
(582) Zaluzanin C, 3-epi; $R_1 = \alpha$-OH,
$R_2 = R_3 = H$
(583) Zaluzanin C; $R_1 = \beta$-OH, $R_2 = R_3 = H$
(584) Zaluzanin D; $R_1 = \beta$-OAc, $R_2 = R_3 = H$
(585) Vernoflexin; $R_1 = \beta$-OSen, $R_2 = R_3 = H$
(586) Vernoflexuoside; $R_1 = \beta$-OD, $R_2 = R_3 = H$
(587) Cynaropicrin; $R_1 = \beta$-OH, $R_2 = H$,
$R_3 = OMac$-4-OH
(588) Zaluzanin C, 7α-hydroxy-3-desoxy;
$R_1 = R_3 = H$, $R_2 = OH$
(589) Zaluzanin C, senecioyl; $R_1 = OSen$,
$R_2 = R_3 = H$

(591) Zaluzanin C, 4β-14-dihydro-3-dehydro; $R = H$
(592) Grosshemin; $R = OH$

$A = $ $B = $ $C = $ $D = $

Chart V-1 (continued)

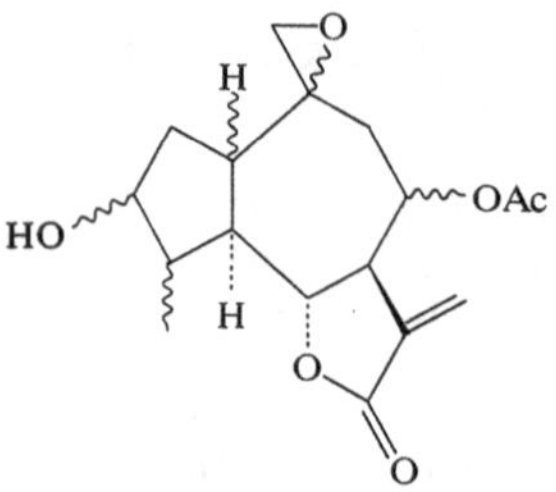

(593) Artefransin

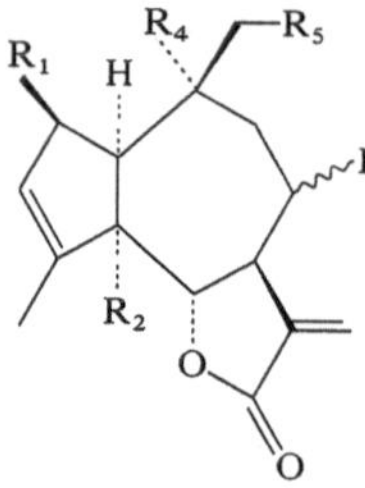

(594) Eupachlorin; $R_1 = R_2 = R_4 = OH$, $R_3 = \beta$-OAng, $R_5 = Cl$

(595) Eupachlorin acetate; $R_1 = OAc$, $R_2 = R_4 = OH$, $R_3 = \beta$-OAng, $R_5 = Cl$

(596) Spicatin hydrochloride; $R_1 = OAc$, $R_2 = H$, $R_3 = \beta$-OA, $R_4 = OH$, $R_5 = Cl$

(597) Cumambrin B; $R_1 = R_2 = R_5 = H$, $R_3 = \alpha$-OH, $R_4 = OH$

(599) Cumambrin A; $R_1 = R_2 = R_5 = H$, $R_3 = \alpha$-OAc, $R_4 = OH$

(600) Cumambrin B, 8-deoxy; $R_1 = R_2 = R_3 = R_5 = H$, $R_4 = OH$

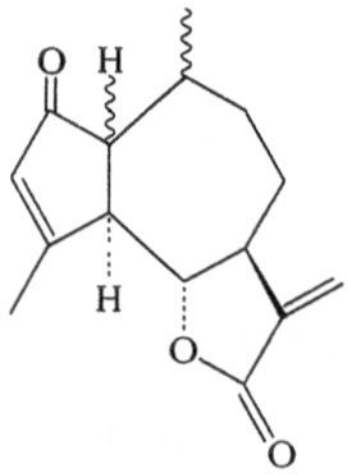

(601) Arbiglovin

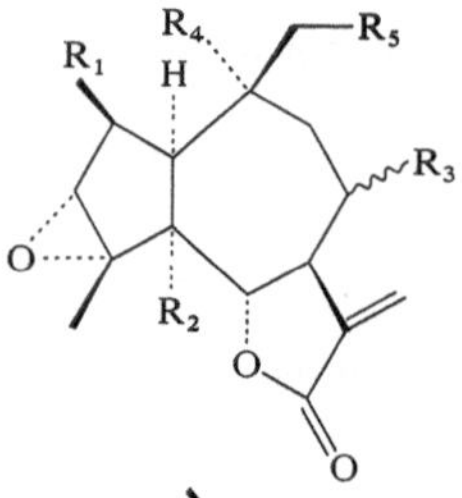

(602) Eupachloroxin; $R_1 = R_2 = R_4 = OH$, $R_3 = \beta$-OAng, $R_5 = Cl$

(603) Graminichlorin; $R_1 = R_4 = OH$, $R_2 = H$, $R_3 = \beta$-OB, $R_5 = Cl$

(604) Cumambrin B, 3,4-oxide; $R_1 = R_2 = R_5 = H$, $R_3 = \alpha$-OH, $R_4 = OH$

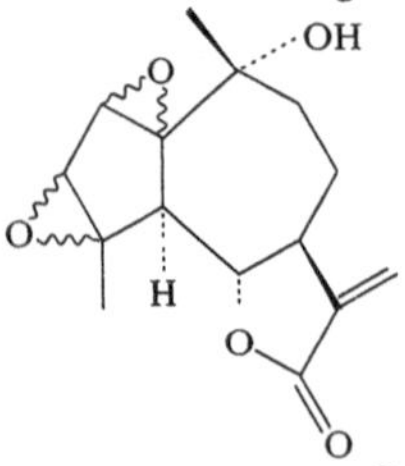

(605) Artecanin, both epoxides *cis*, α

(606) Canin; both epoxides *cis*, β

(607) Chrysartemin B; both epoxides *cis*, β

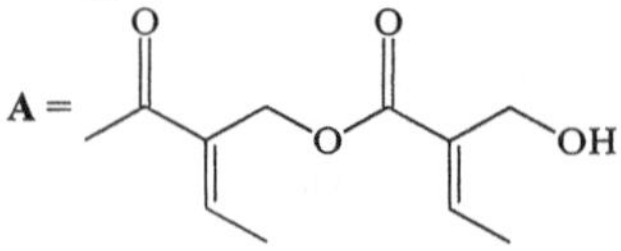

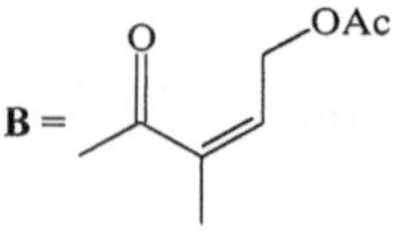

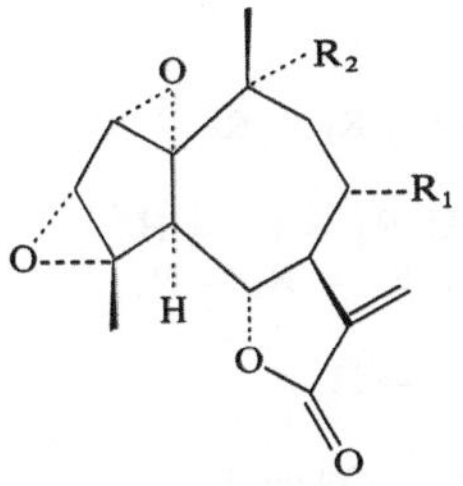

(608) Rupin A; $R_1 = R_2 = OH$
(609) Rupin B; $R_1 = OAc$, $R_2 = OH$

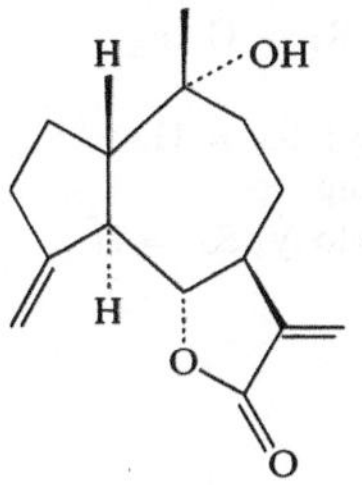

(610) Trifloculoside

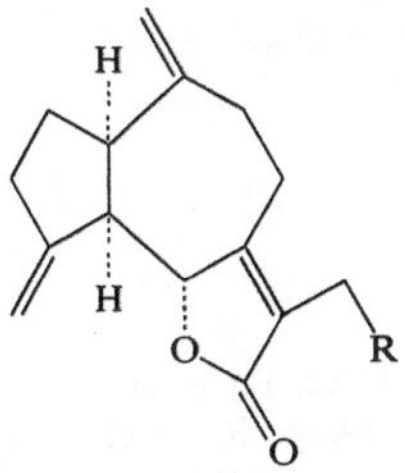

(611) Zaluzanin C, 11,13-dihydro-7,11-de-
hydro-3-desoxy; $R = H$
(612) Zaluzanin C, 13-acetoxy-11,13-di-
hydro-7,11-dehydro-3-desoxy;
$R = OH$

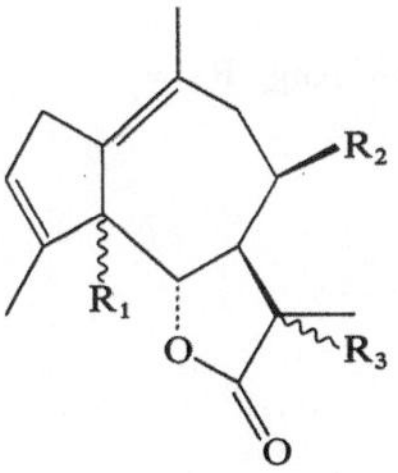

(613) Prutenin; $R_1 = \alpha$-H, $R_2 = H$, $R_3 = \beta$-OAng
(614) Petiolaride; $R_1 = H$, $R_2 = OA$, $R_3 = \alpha$-H

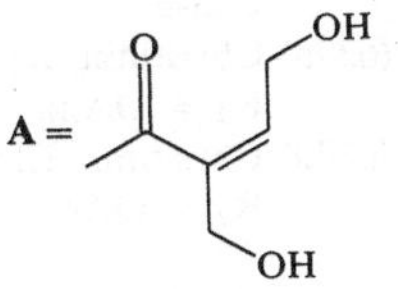

$A =$

Chart V-1 (continued)

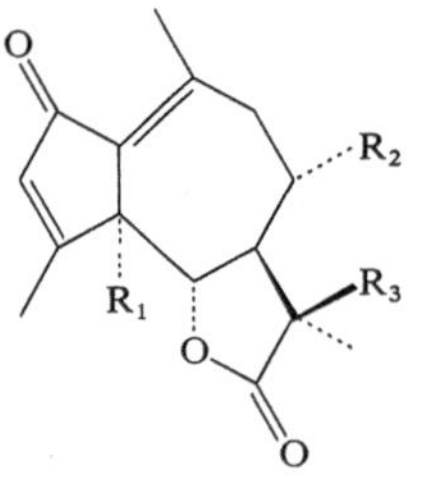

(616) Matricarin, desacetoxy; $R_1 = R_2 = R_3 = H$

(618) Parishin C; $R_1 = OH$, $R_2 = R_3 = H$

(619) Matricarin, desacetyl; $R_1 = R_3 = H$, $R_2 = OH$

(620) Matricarin; $R_1 = R_3 = H$, $R_2 = OAc$

(621) Laferin; $R_1 = H$, $R_2 = OAng$, $R_3 = OAc$

(623) Talassin B; $R_1 = H$, $R_2 = OAng$, $R_3 = O\text{-}i\text{-}But$

(624) Pruteninone, 8-acetoxy; $R_1 = H$, $R_2 = OAc$, $R_3 = OAng$

(625) Pruteninone, 8-angeloyloxy; $R_1 = H$, $R_2 = R_3 = OAng$

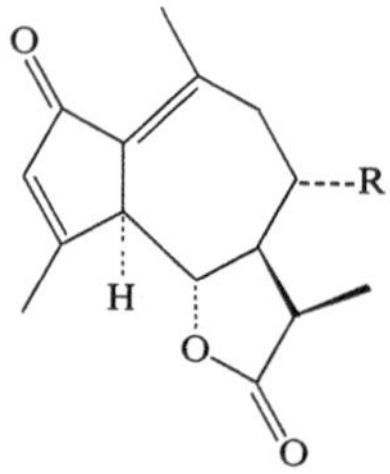

(626) Achillin; $R = H$

(627) Achillin, hydroxy; $R = OH$

(628) Grossmisin; $R = OH$

(630) Achillin, acetoxy; $R = OAc$

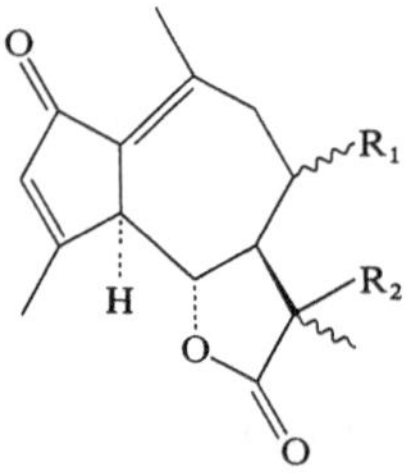

(632) Artilesin B; $R_1 = \beta\text{-}OAc$, $R_2 = H$

(633) Badkhysin; $R_1 = \alpha\text{-}OAng$, $R_2 = H$

(634) Olgin; $R_1 = \alpha\text{-}OAc$, $R_2 = OMac$

(635) Oferin; $R_1 = \alpha\text{-}OMac$, $R_2 = O\text{-}i\text{-}But$

(636) Olgoferin; $R_1 = \alpha\text{-}OMac$, $R_2 = OMac$

(637) Talassin A; $R_1 = \alpha\text{-}OAng$, $R_2 = OAng$

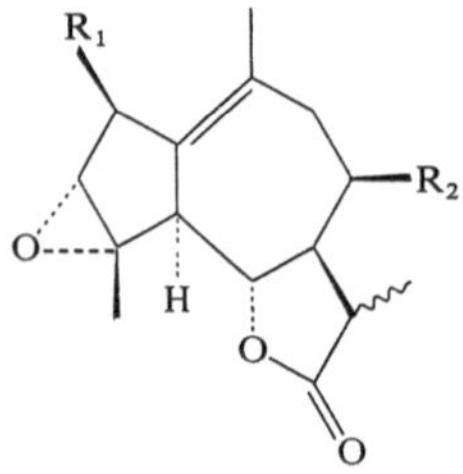

(638) Ludartin, dihydro; $R_1 = R_2 = H$

(639) Christinin; $R_1 = R_2 = OAc$, C13-β

(639a) Christinin II; $R_1 = O\text{-}i\text{-}But$, $R_2 = OAng$

(639b) Christinin III; $R_1 = O\text{-}i\text{-}But$; $R_2 = OAc$

(640) Arborescin; R = H,α-epoxide
(642) Globicin; R = OAc

(643) Achillin, 1,10-epoxy; R = H
(644) Achillin, 1,10-epoxy, 8α-hydroxy;
R = OH

(645) Picridin, dihydro

(646) Eufoliatorin

(647) Euperfolide, 11α-13-dihydro

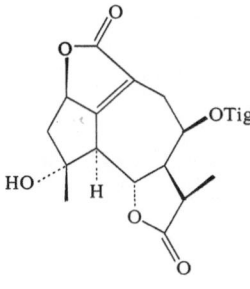

Chart V-1 (continued)

(648) Eregoyazidin

(649) Pruteninone, iso, 8-acetoxy

(650) Viscidulin C; R = OH
(651) Viscidulin B; R = OAc

(652) Viscidulin A

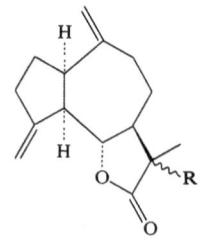

(653) Costus lactone, dehydrodihydro; R = β-H
(654) Mokko lactone; R = H

(655) Solstitialin A; $R_1 = R_4 = OH$,
      $R_2 = H$, $R_3 = \alpha\text{-OH}$
(656) Solstitialin acetate; $R_1 = OH$,
      $R_2 = H$, $R_3 = \alpha\text{-OH}$, $R_4 = OAc$
(657) Zaluzanin C, 7α-hydroxy-3-desoxy-
      11β,13-dihydro; $R_1 = R_4 = H$,
      $R_2 = OH$, $R_3 = \beta\text{-H}$

(658) Zaluzanin C, 8α-hydroxy-11β,13-
      dihydro-3-dehydro

(659) Lippidiol, iso; C-13α
(660) Lippidiol; C-13β

(661) Estafietone, dihydro; R = H, C-13α
      (Zaluzanin C, 4β,15,11β,13-tetrahydro-
      3-dehydro)
(664) Amberboin, iso; R = OH, C-13α
(666) Amberboin; R = OH, C-13β

(667) Archangelolide (see addendum)

Chart V-1 (continued)

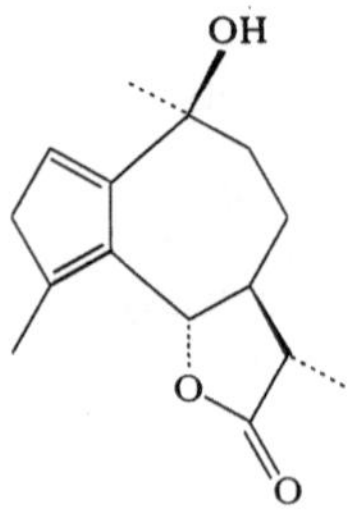

(668) Artabsin

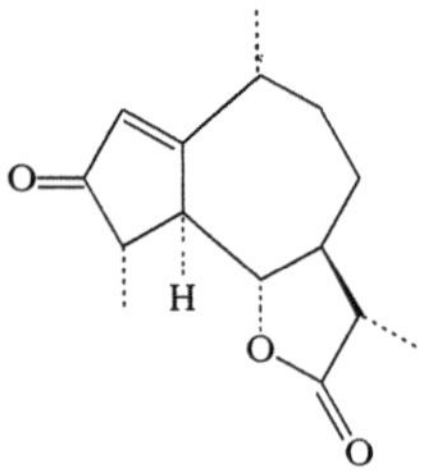

(669) Hypochaerin

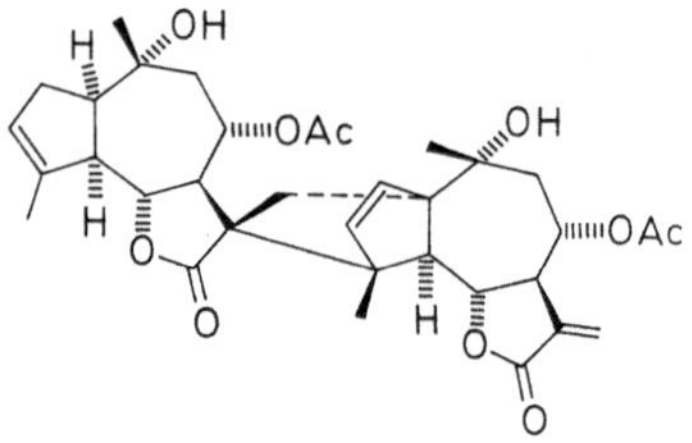

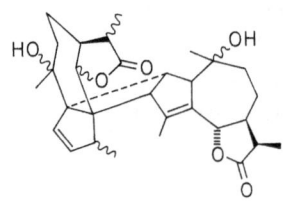

(670) Handelin

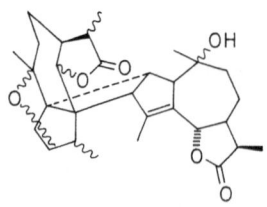

(671) Absinthin

(672) Anabsinthin

(673) Chlorochrymorin

(674?) Osmitopsin

(675?) Osmitopsin, 1,8-epoxy

(676?) Osmitopsin, 4,5-epoxy

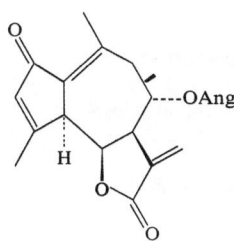

(677) Badkhysin, iso

Chart V-1 (continued)

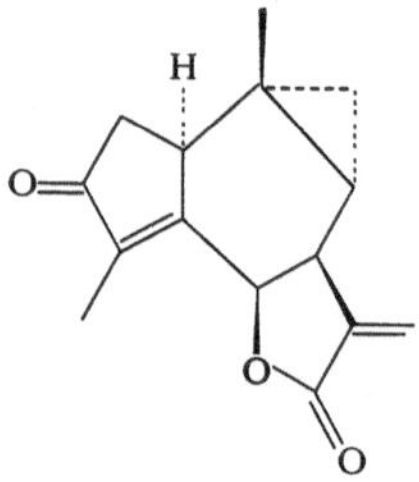

(678) Ivaxillarin, anhydro

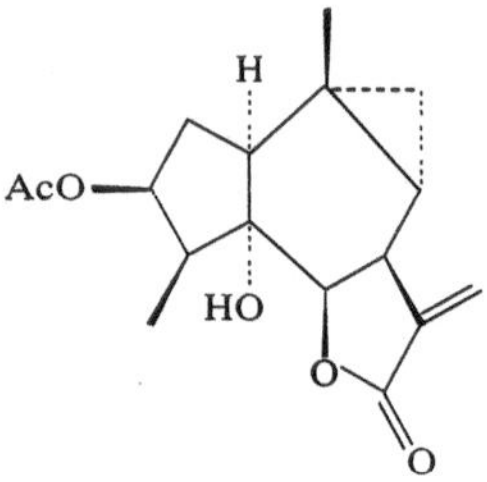

(679) Axivalin

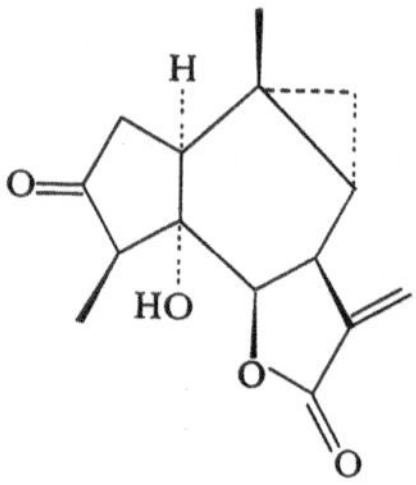

(680) Ivaxillarin

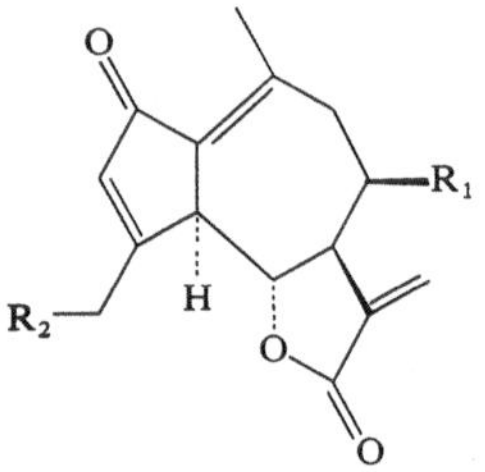

(681) Lactucin; $R_1 = R_2 = OH$
(682) Lactucopicrin; $R_1 = OA$, $R_2 = H$

$$A =$$

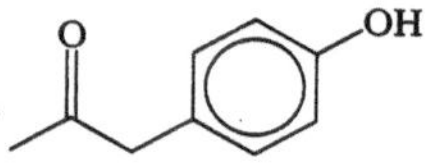

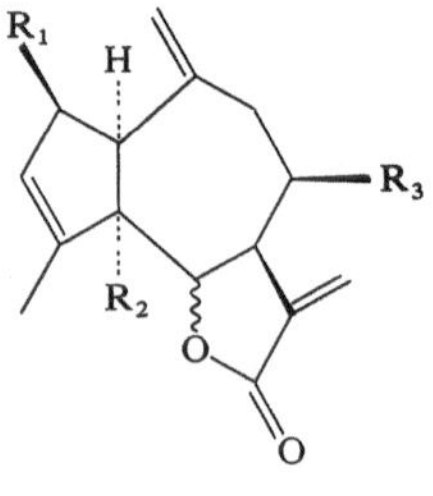

(683) Preeupatundin, 8β-angeloyloxy;
$R_1 = OH$, $R_2 = H$, $R_3 = OAng$
(684) Preeupatundin-2-acetate, 8β-angeloyl-
oxy-5α-hydroxy; $R_1 = OAc$, $R_2 = $
OH, $R_3 = OAng$
(685) Preeupatundin, 5α-hydroxy-8β-
tiglinoyloxy; $R_1 = R_2 = OH$, $R_3 = $
OTig

(686) Saurin; $R_1 = R_3 = OH$, $R_2 = H$
(687) Saupirin; $R_1 = H$, $R_2 = OH$, $R_3 =$ OMac-4-OH
(687a) Ferreyanthus lactone; H-1β, H-5α, H-6β; $R_1 = R_2 = H$, $R_3 = α$-OTig
(687b) Costus lactone, dehydro, 1-epi, 8α-senecioyloxy; H-1β; H-5α, H-6β; $R_1 = R_2 = H$, $R_3 = α$-OSen

(688) Cumambrin B, iso

-OH
-C bridge

(689) Yejuhua lactone

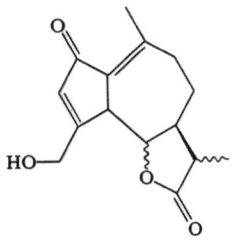

(691) Jacquinelin

Chart V-1 (continued)

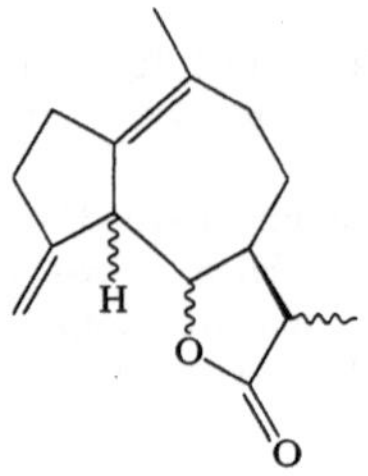

(**692**) Grilactone

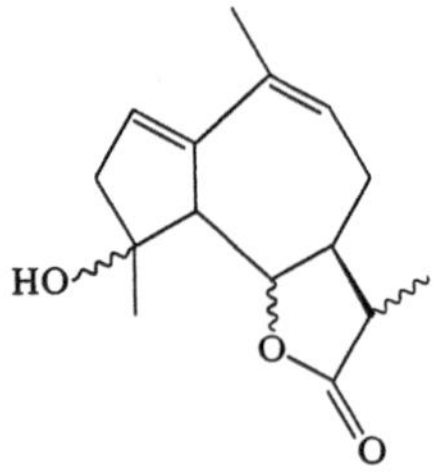

(**693**) Sieversin

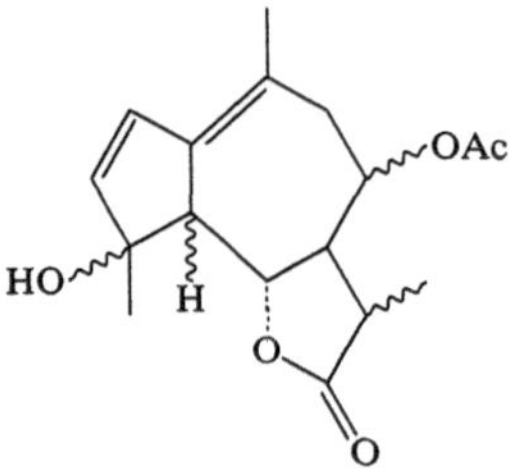

(**694**) Matricin

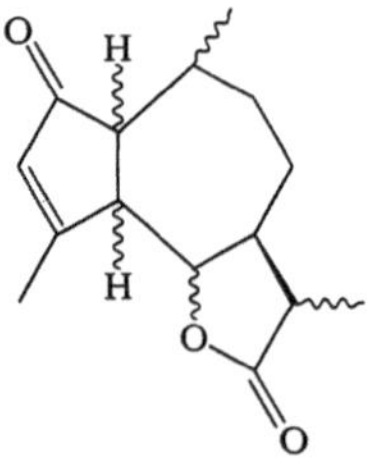

(**695**) Carpesia lactone

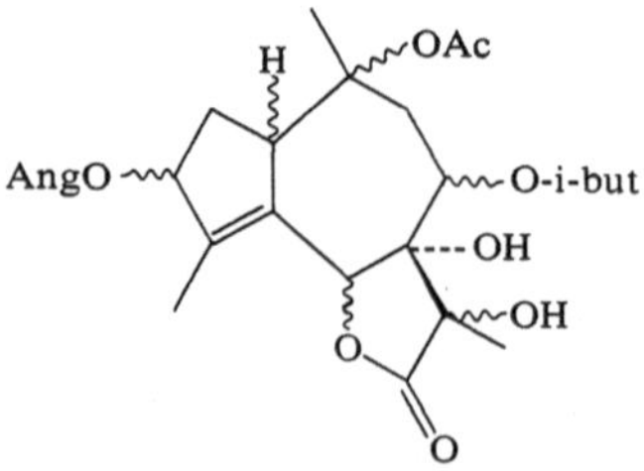

(**696**) Trilobolide

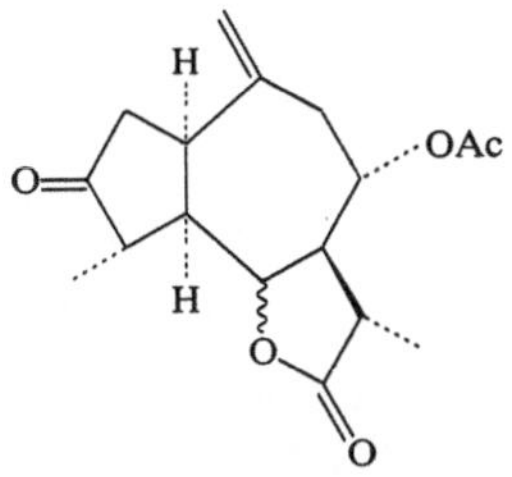

(697) Jurmolide

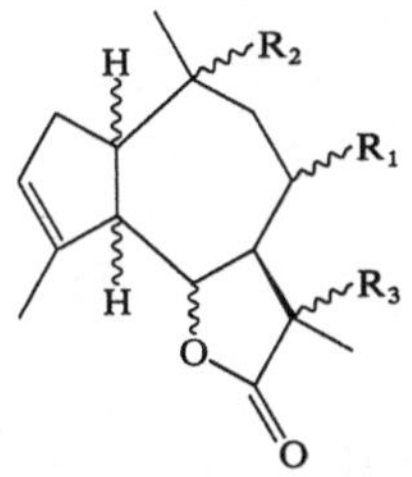

(698) Montanolide; R_1 = OSen, R_2 = OH, R_3 = OAc
(699) Montanolide, iso; R_1 = OAng, R_2 = OH, R_3 = OAc
(700) Montanolide, iso, acetyl; R_1 = OAng, R_2 = R_3 = OAc

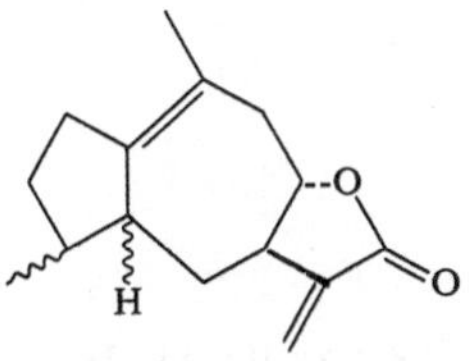

(701) Collumellarin

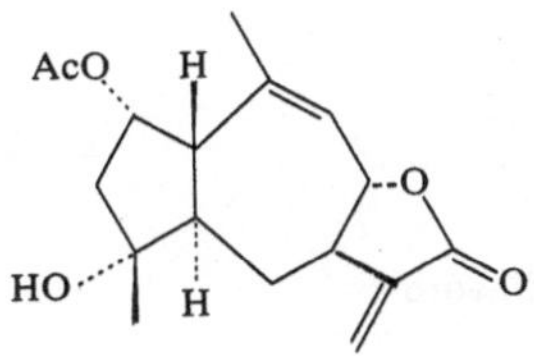

(702) Gaillardin

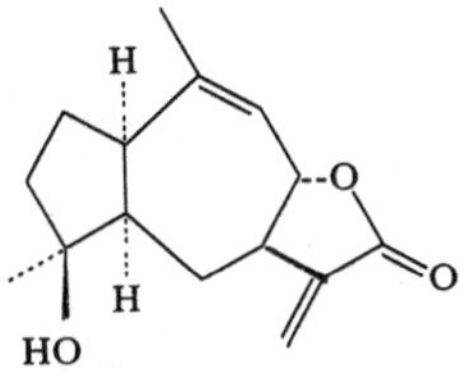

(703) Inuviscolide, iso, 4-epi

Chart V-1 (continued)

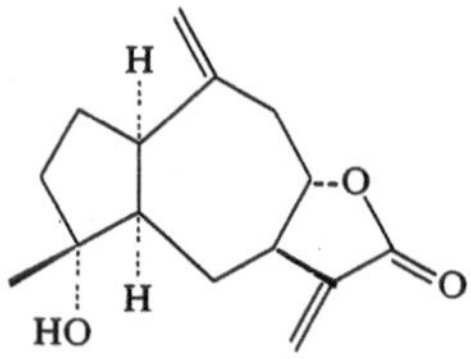

(704) Inuviscolide

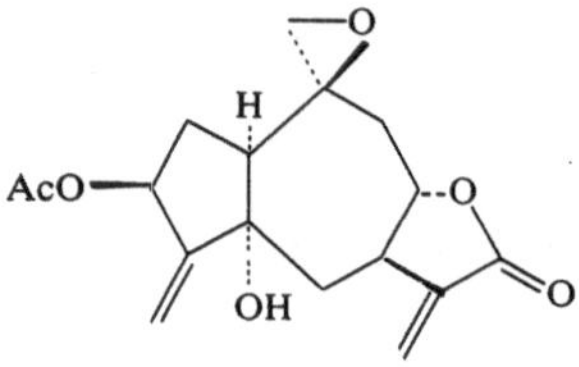

(705) Arctolide

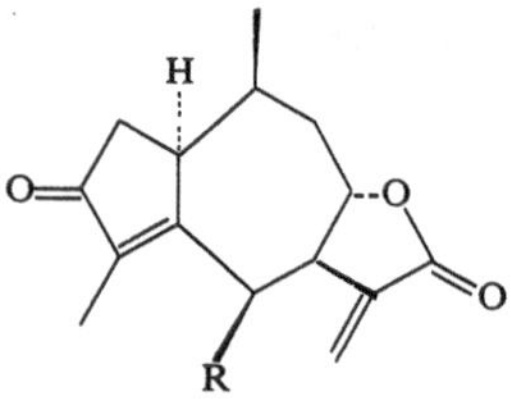

(706) Xerantholide; R = H
(707) Mikanocryptin; R = OH

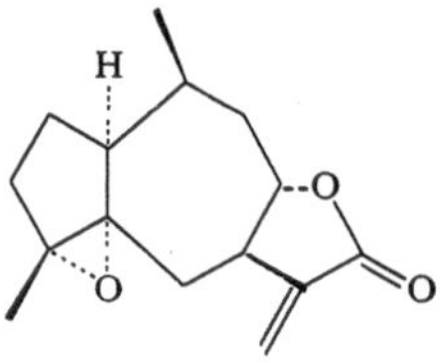

(708) Inuviscolide, 4α,5α-epoxy,10α,14α-H

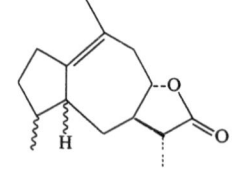

(709) Collumellarin, dihydro

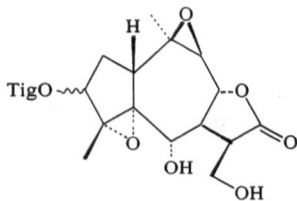

(710) Eufoliatin

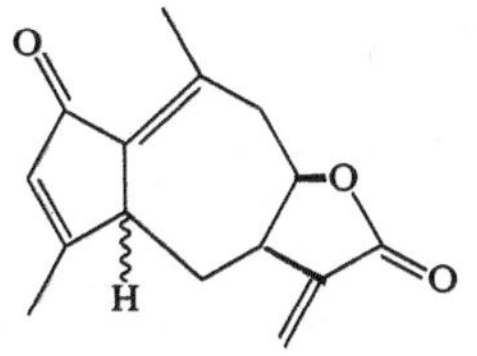

(711) Virginolide

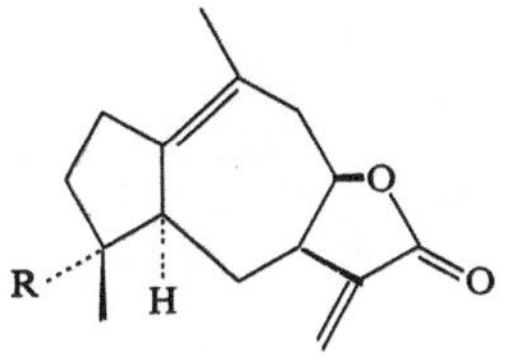

(712) Ivalin, pseudo; R = OH
(713) Ivalin, pseudo, acetate; R = OAc

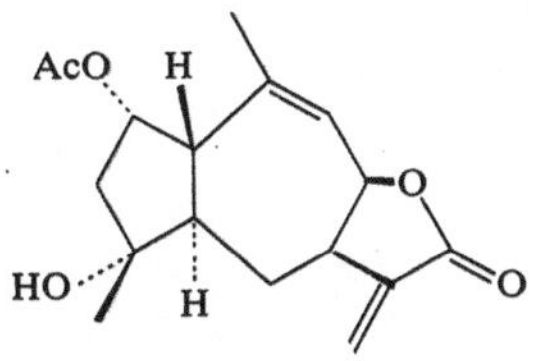

(714) Gaillardin, neo

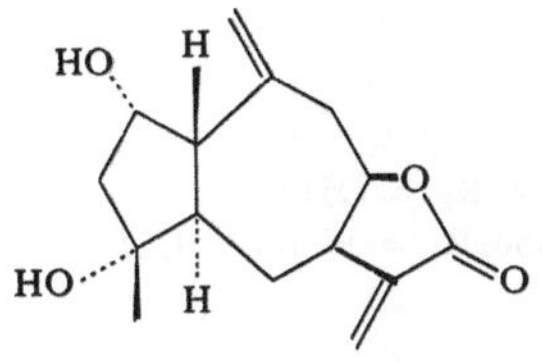

(715) Florilenalin

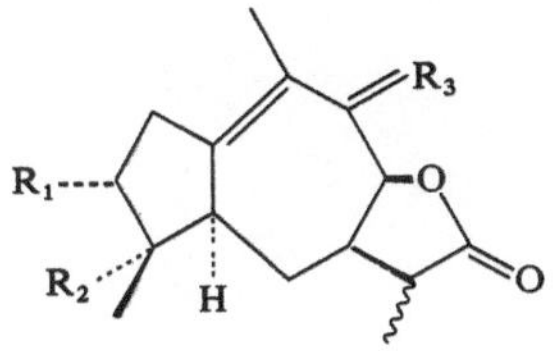

(716) Ivalin, pseudo, dihydro; R_1 = H,
R_2 = OH, R_3 = H_2
(717) Carolenalone; R_1 = R_2 = OH,
R_3 = O, C-13β

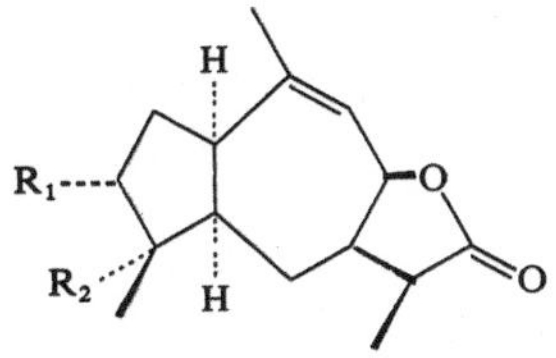

(718) Carolenalin; R_1 = R_2 = OH
(719) Carolenin; R_1 = OAng, R_2 = OH

Chart V-1 (continued)

(720) Florilenalin, dihydro

(721) Geigerin

(722) Pleniradin

(723) Halshalin; $R_1 = R_2 = OH$
(724) Helenium lactone, $R_1 = H$, $R_2 = OH$

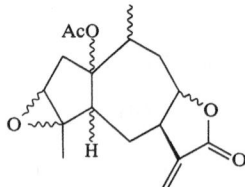

(725) Calocephalin

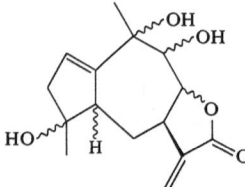

(726) Akihalin

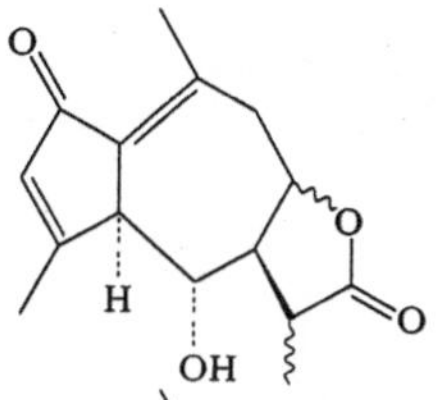

(728) Ferulidin

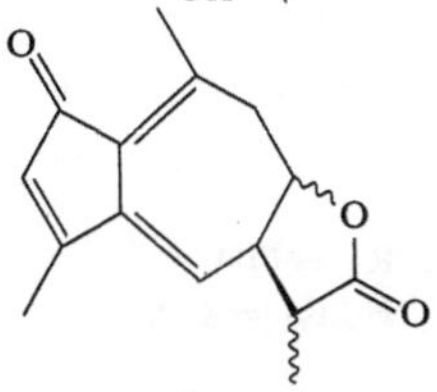

(729) Ferulin

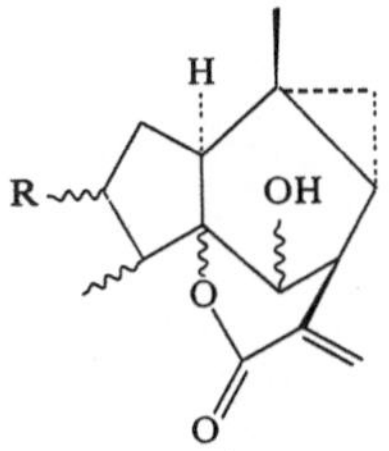

(730) Zaluzanin A; R = OH
(731) Zaluzanin B; R = OAc

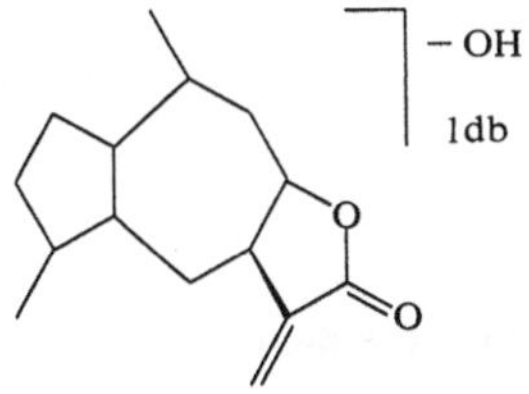

(732) Tanamyrin; position of lactone un-
certain.

Chart V-2. Naturally occurring seco-guaianolides (xanthanolides)

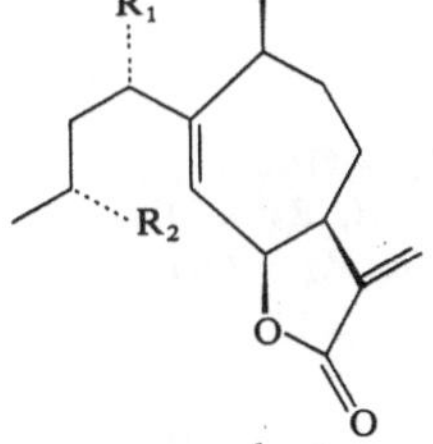

(733) Ivambrin; $R_1 = R_2 = OH$
(734) Apachin; $R_1 = OAc$, $R_2 = OH$

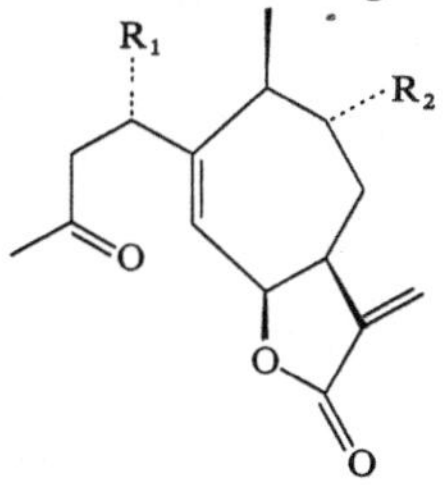

(735) Parthemollin; $R_1 = OH$, $R_2 = H$
(736) Parthemollin, acetyl; $R_1 = OAc$, $R_2 = H$
(737) Ivalbatin; $R_1 = H$, $R_2 = OH$

Chart V-2 (continued)

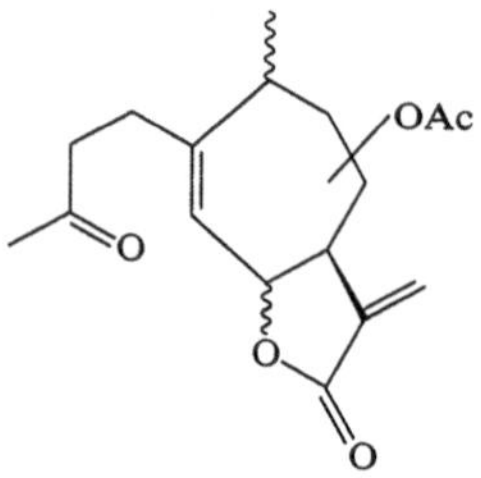

(738) Fruticosin;

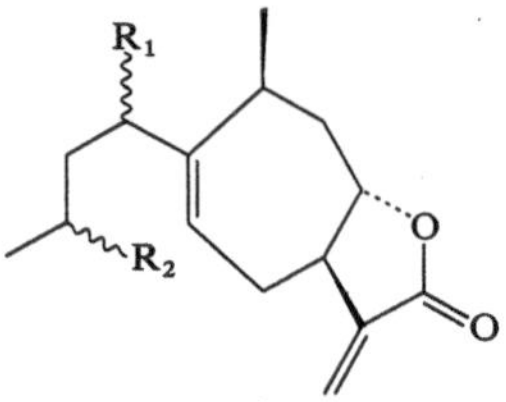

(739) Xanthanol; R_1 = OAc, R_2 = OH
(740) Xanthanol, iso; R_1 = OH, R_2 = OAc

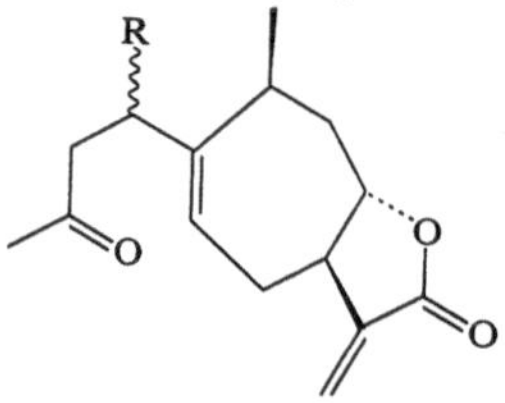

(741) Xanthinosin; R = H
(742) Xanthinin; R = OAc

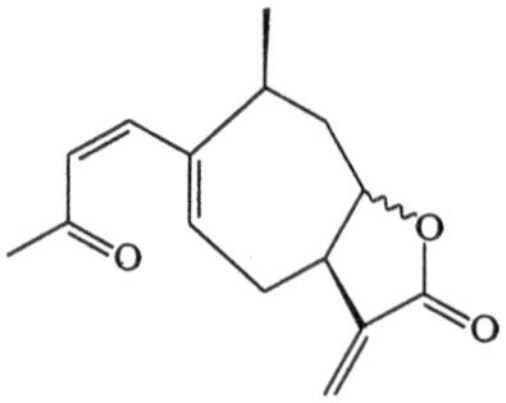

(743) Xanthatin; C-8 α-O
(744) Xanthumin, deacetoxy; C-8 β-O

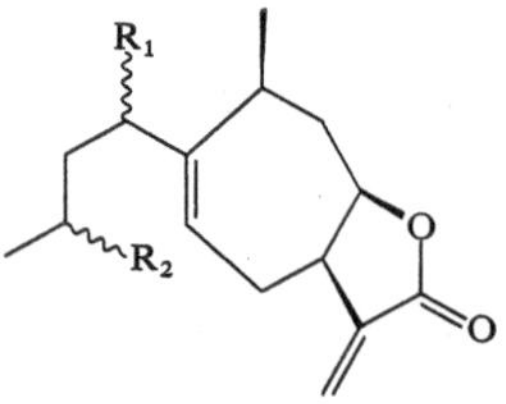

(745) Gafrinin; R_1 = OH, R_2 = OAc
(746) Tomentosin, 4-H; R_1 = H, R_2 = OH
(746a) Xanthuminol; R_1 = OAc, R_2 = OH

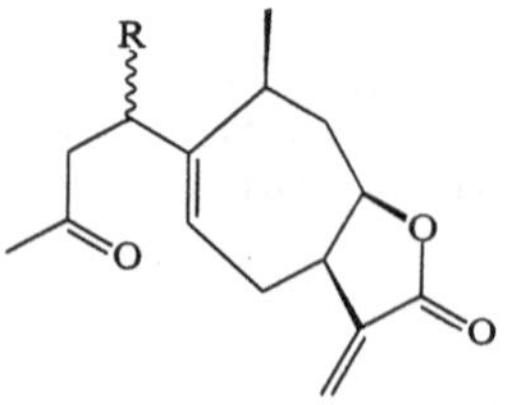

(747) Xanthumin; R = OAc

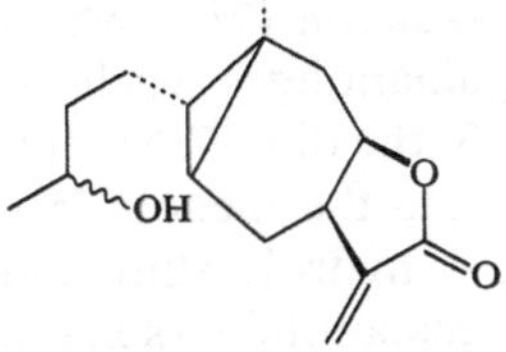

(748) Carabrone, 4-H;

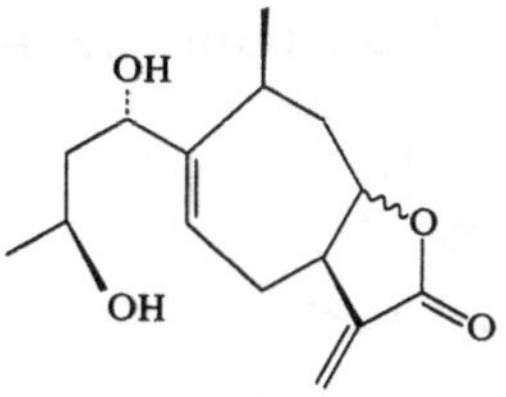

(751) Ivalbin;

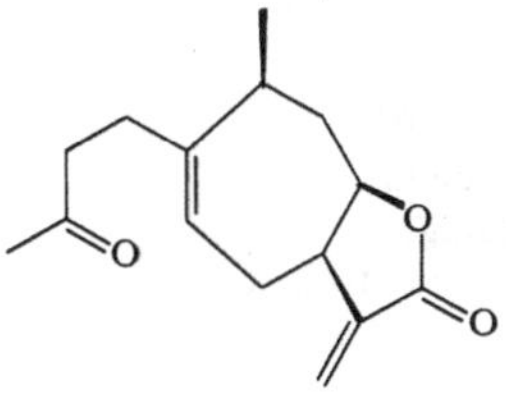

(752) Tomentosin;

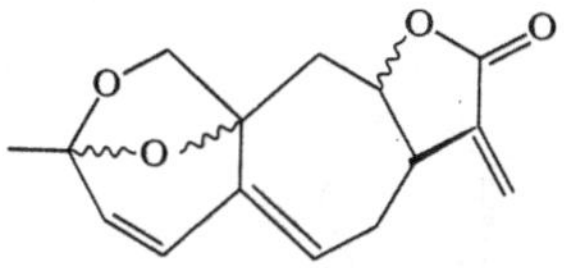

(753) Griesenin;

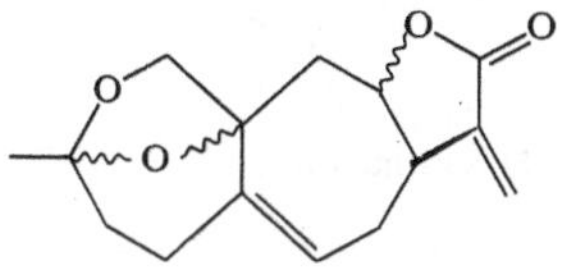

(754) Griesenin, dihydro;

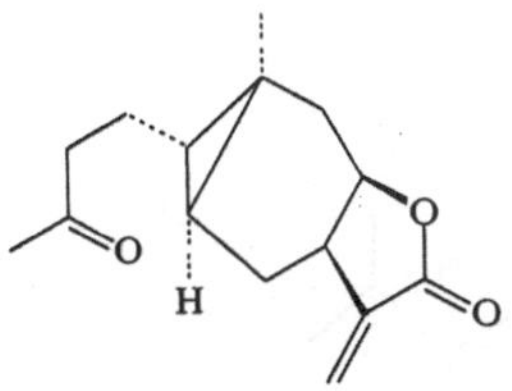

(749) Carabrone;
(Grandicin)

As outlined in Chart V-3, Markovnikov type cyclization of a germacrolide-4,5-epoxide (755) in a chair-like transition state would lead to the *cis*-fused guaianolide cation (756) from which the guaianolide skeleton (758) would be formed by an uptake of water. With few exceptions, this is the stereochemistry found in most guaianolides. Fragmentation of the 4,5-bond and H-1α to C-10α shift from cation (756), as indicated by arrows, would give the xanthanolide skeleton (757) which is typical for most seven-membered ring *seco*-guaianolides such as ivalbatin (737). As shown on the bottom of Chart V-3, the diol (758) after water elimination and oxidation at C-8 would provide the dienol (759) which upon intramolecular substitution, C-5 to C-1 hydride shift and uptake of water could provide the cyclopropane guaianolide skeleton (760) from which ivaxillarin (680) might be derived. A laboratory trans-

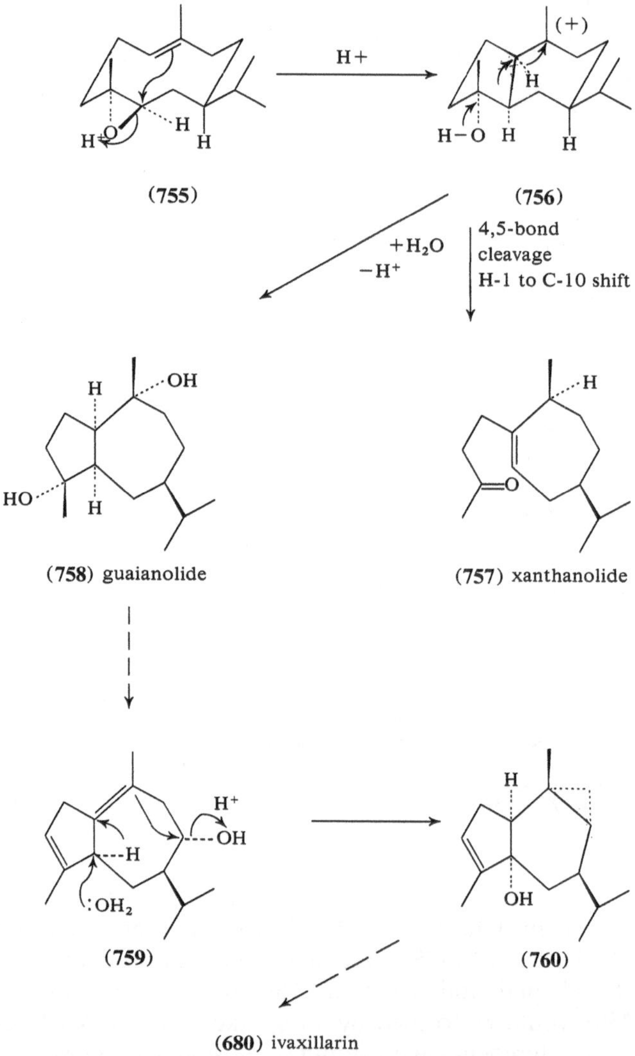

Chart V-3. Biogenesis of guaianolides and xanthanolides

formation of 8-hydroxyachillin (**627**) to a 8,10-cyclopropane type guaianolide has been reported in the literature (*1138*). It should be noted that ivaxillarin (**680**) belongs to a small group of guaianolides with 7,6-*cis*-lactone functions which are typical for the ambrosanolides.

A few *trans*-fused guaianolides, gaillardin (**702**) (*245*), neogaillardin (**714**) (*496*), florilenalin (**715**) (*705*) and its dihydroderivative (**720**) (*611*) have been found as natural products. Their biogenesis has been formulated to either proceed *via* the melampolide route (*393, 395*) or the germacrolide 4,5-epoxide pathway (*265, 273, 505*) which is outlined on top of Chart V-4. The germacrolide-4,5-epoxide (**761**), in its *anti*-conformation with the C-14 methyl group adopting an α-orientation,

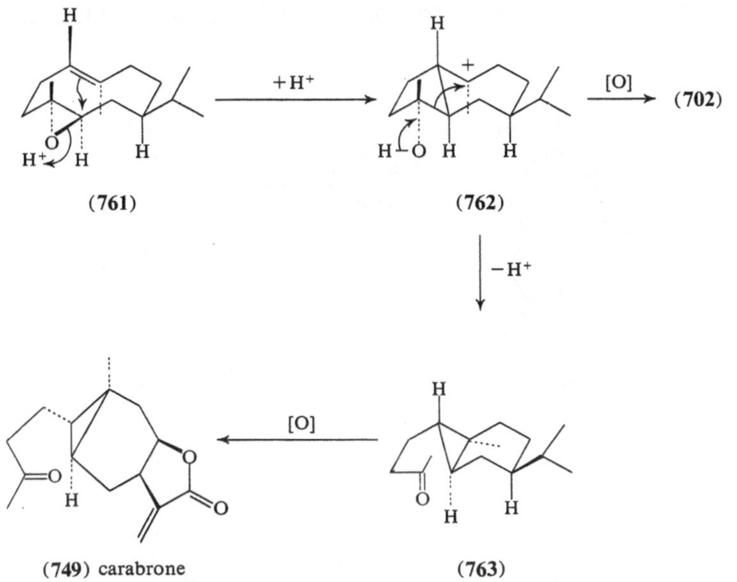

Chart V-4. Biogenesis of *trans*-fused guaianolides and cyclopropane-type xanthanolides

cyclizes in Markovnikov fashion to give the *trans*-fused guaianolide cation (**762**). Subsequent, 9,10-double bond formation, oxidative modification and lactonization would result in gaillardin (**702**). Re-arrangement of cation (**762**), as indicated by the arrows, would provide the cyclopropane derivative (**763**) which indicates the correct stereo-chemistry for the cyclopropane type xanthanolides such as carabrone (**749**).

2. Selected Chemical Transformations of Guaianolides and Xanthanolides

Correlations by simple chemical modifications have been frequently used in structure elucidation and making stereochemical assignment. The stereochemistry at C-10 of artabsin (**668**) was determined by a sequence of reactions which are outlined in Chart V-5 (*1119*). Isophoto-α-santonic lactone (**764**), a compound of established stereochemistry (*50*), was converted to the spiroepoxide (**765**) by a kinetically controlled SOCl₂/pyridine mediated elimination reaction followed by peracid epoxidation. Subsequent catalytic hydrogenation over Pd-SrCO₃ gave the tetrahydroartabsin (**766**) with a C-10-β-OH, a compound also obtained by saturation of the two double bonds with H₂/Pt in acetic acid.

Another example of chemical interconversion shown in Chart V-6 involves the unusual δ-lactone zaluzanin A (**730**) of unknown stereochemistry which by alkaline treatment and subsequent acidification gave

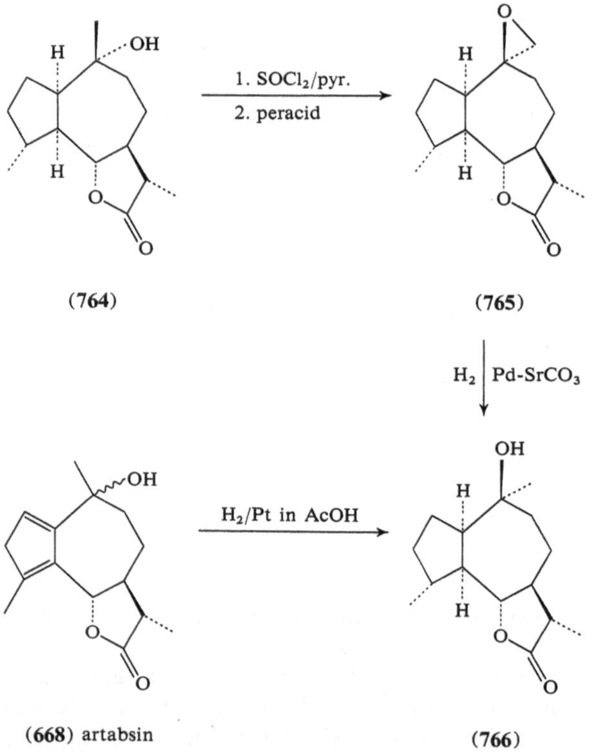

Chart V-5. Correlation of artabsin with isophoto-α-santonin lactone

the γ-lactone (**767**) (*884*). Oxidation of the C-3 hydroxyl function to the ketone and toluenesulfonic acid-catalyzed elimination of the angular C-5 hydroxyl group provided the α,β-unsaturated ketone (**768**) which was also obtained by dehydration of ivaxillarin (**680**). This established the stereochemistries of zaluzanin A at the carbon centers 1,6,7,8 and 10 (*884*). At the bottom of Chart V-6, the acid-mediated conversion of the C-3 anhydroderivative of zaluzanin A (**769**) into the trienone guaianolide (**770**) (*884*) is shown. The reaction involves an acid-catalyzed

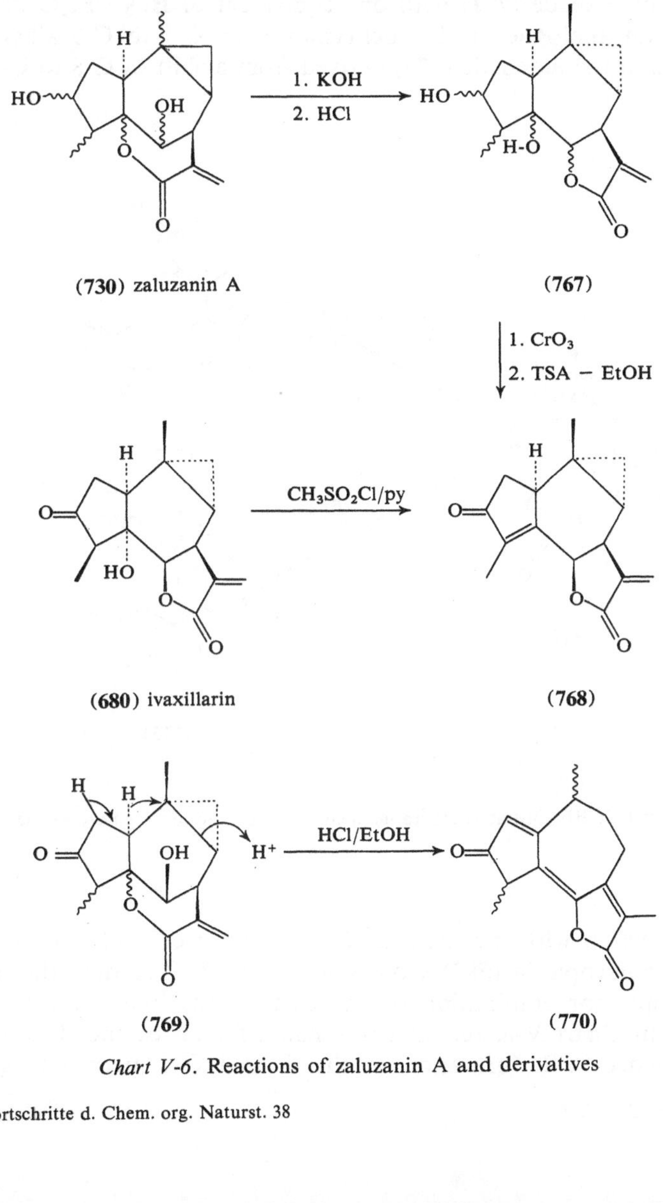

(**730**) zaluzanin A   (**767**)

(**680**) ivaxillarin   (**768**)

(**769**)   (**770**)

Chart V-6. Reactions of zaluzanin A and derivatives

lactone ring opening and double bond isomerizations, the opening of the cyclopropane ring possibly proceeding as indicating by the arrows.

The only naturally occurring guaianolide in which the seven-membered ring has undergone biogenetic ring contraction is chlorochrymorin (**673**), the biogenesis of which most likely involves a C-10 to C-9 alkyl shift of C-1 (*802*). As outlined in Chart V-7, Rabi and coworkers (*283, 667*) studied the Lewis acid catalyzed rearrangement of the 9,10-α-epoxide (**771**) which had been obtained by peracid oxidation of eremanthin (**543**). Treatment of epoxide (**771**) with one equivalent of BF$_3$ · Et$_2$O did not give (**772**), an analogue of chlorochrymorin, by C-1 to C-9 alkyl shift but resulted in the aldehyde (**773**) derived from a shift of C-8 to C-10.

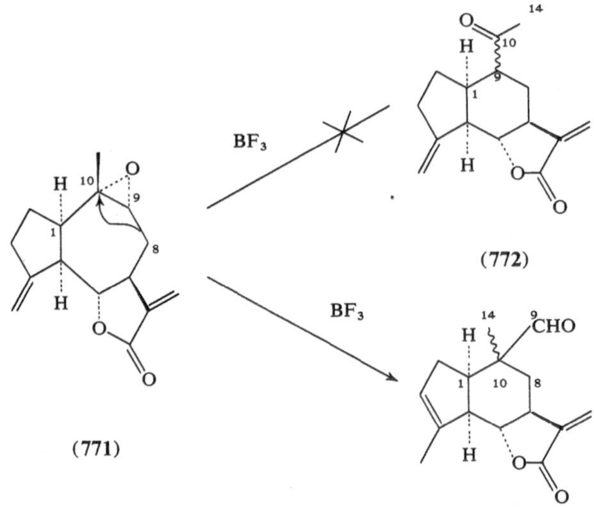

Chart *V-7.* BF$_3$-initiated rearrangement of a guaianolide 9,10-α-epoxide

In connection with the chemical correlation of chlorohyssopifolin A (**576**) and cynaropicrin (**587**) Gonzalez et al. (*320*) studied the use of Zn-Cu couple for elimination of the elements of hypochlorous acid. As shown in Chart V-8, reaction in ethanol for an extended period of time proceeded with formation of the desired C-4 methylene group.

This was accompanied by conversion of the ester side chain to an iso-
butyrate, most likely by way of the methacrylate which is further
reduced under the reaction conditions. Reduction of the α-methylene-γ-
lactone group also occurred stereospecifically giving (774) with the same
C-11 stereochemistry as in the NaBH₄ reduction product. Treatment of
spicatin (558) with Zn-Cu couple resulted in the expected conversion of
the C-10 epoxide to an exocyclic methylene function. In addition,
hydrogenolysis of the side chain and reduction of the lactonic methylene
group was observed (424, 556).

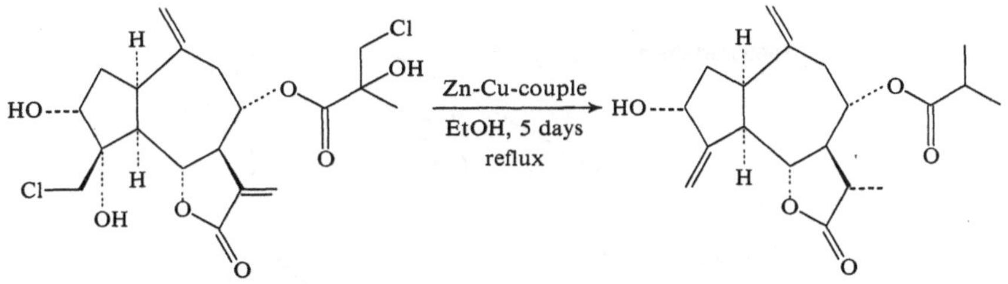

(576) chlorohyssopifolin A   (774)
(centaurepensin)

Chart V-8. Correlation of chlorohyssopifolin A with cynaropicrin

Chemical transformations as well as NMR studies of a number of
10-hydroxyguaianolides were recently reported by VOKÁC and SAMEK
(1117).

In an investigation of the constitutents of the "chamazulenogen"
mixture from wormwood (Artemisia absinthium) VOKÁC et al. (1118)
studied the chemistry of artabsin (668). As outlined in Chart V-9,
treatment of artabsin with 5% aqueous NaOH in the heat gave what is
presumed to be dianion (775) which upon treatment with acid and then
diazomethane yielded an orange ester. Structure (776) was ascribed to
the latter on the basis of NMR arguments. Acidification of the dianion
(775) with dilute H₂SO₄ followed by steam distillation provided a
mixture of colored hydrocarbons similar to those in wormwood oil.
Analysis of the mixture by NMR spectrometry (76) indicated that 3,6-
dihydrochamazulene (777) represents the major constituent (94%) and
its isomer (778) a minor component (6%) of the hydrocarbon mixture
which must be formed from (775) by decarboxylation during steam
distillation.

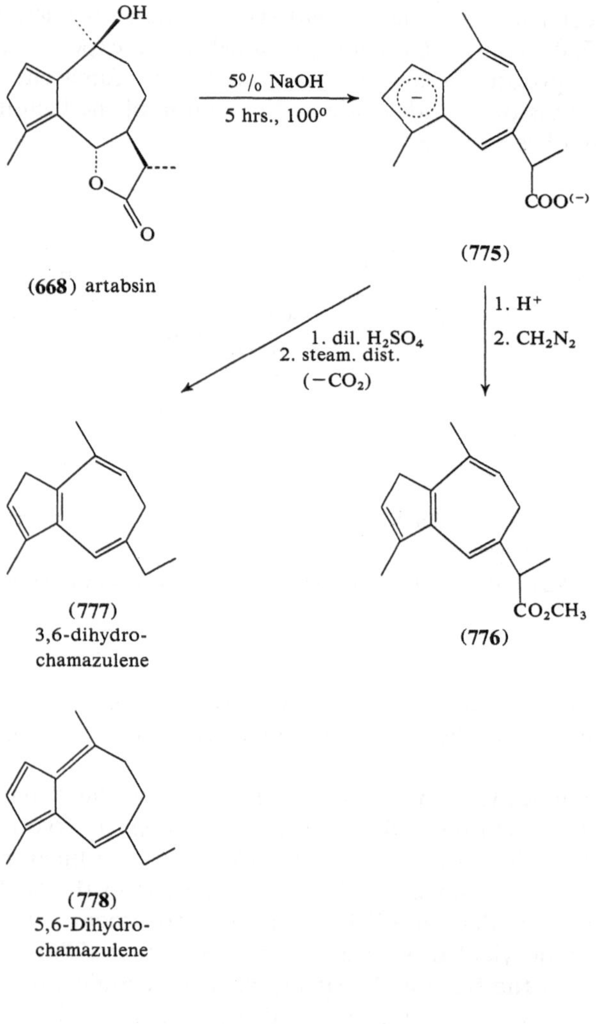

(668) artabsin

(775)

(777)
3,6-dihydro-
chamazulene

(776)

(778)
5,6-Dihydro-
chamazulene

Chart V-9. Major constituents of the "chamazulenogen" mixture

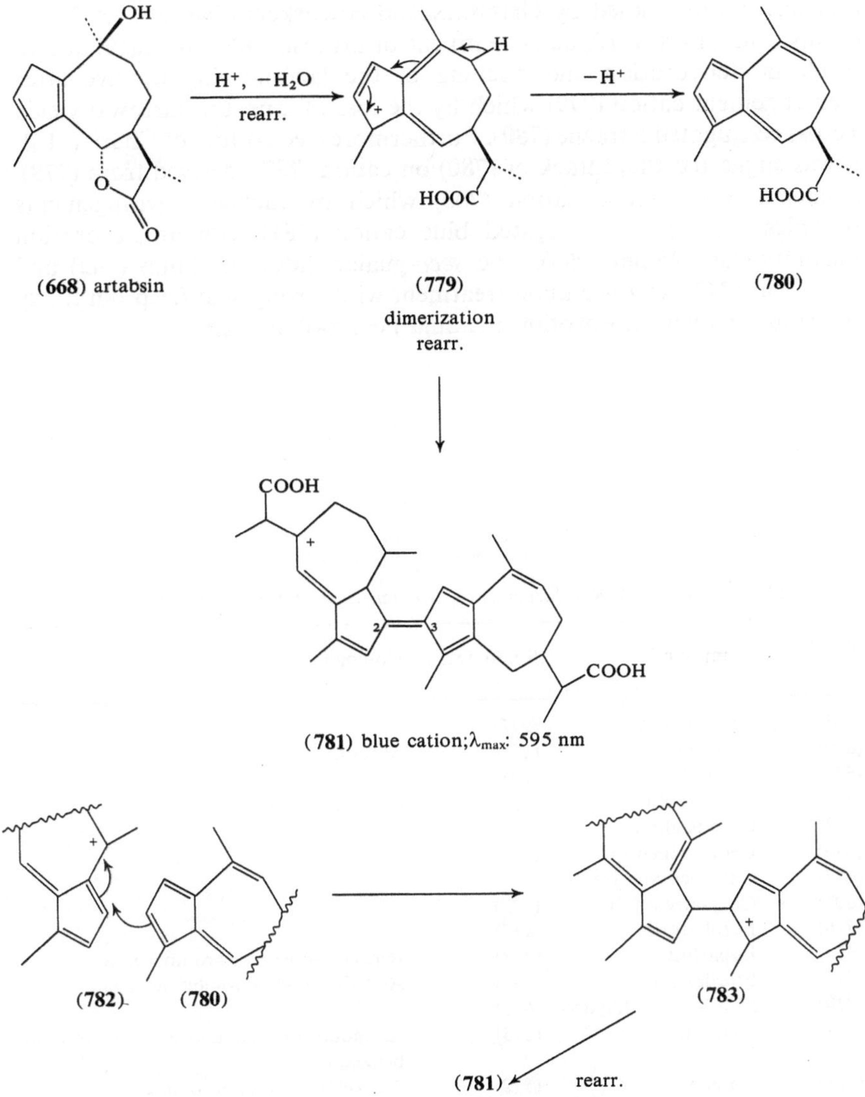
OH
H⁺, −H₂O
rearr.
H
−H⁺
HOOC
HOOC
(668) artabsin
(779)
dimerization
rearr.
(780)
COOH
2
3
COOH
(781) blue cation; λmax: 595 nm
(782)
(780)
(783)
(781)
rearr.

Chart V-10. Formation of a blue cation from artabsin

As in the case of the germacrolides, treatment of guaianolides and xanthanolides with strong mineral acids causes typical color reactions that have been studied by Geissman and coworkers (296, 361, 362). As outlined in Chart V-10, acid-treatment of artabsin (668) is postulated to result in dehydration and opening of the lactone ring to give after rearrangement cation (779) which by the loss of a proton (arrows) yields a cross conjugated tetraene (780). Furthermore (see bottom of Chart V-10), it was suggested that attack of (780) on cation (782), derived from (779), results in the dimeric cation (783) which by further rearrangements provides the highly conjugated blue cation (781) with an absorption maximum at 595 nm (296). The *seco*-guaianolides xanthinin (742) and xanthatin (743) produce upon treatment with strong acid deep burgundy red solutions with absorption maxima near 540 nm (296).

Table V-1. *Known X-Ray Structures of Guaianolides and Biogenetic Derivatives*

Structure Number	Compound	References	Comments
(532)	Arteglasin A	(947)	
(679)	Axivalin	(25)	Monohydrate
(534)	Berlandin	(193)	
(718)	Carolenalin	(702)	As monoacetate
(717)	Carolenalone	(707)	
(578)	Centaurepensin	(373)	
(673)	Chlorochrymorin	(802)	
(607)	Chrysartemin B	(803)	
(646)	Eufoliatorin	(417)	
(554)	Euparotin	(714)	Benzene solvate of bromoacetate
(715)	Florilenalin	(705)	As 4-O-acetyl-2-O-*p*-iodobenzoate
(720)	Florilenalin, dihydro	(611)	
(702)	Gaillardin	(245)	As mono-bromide and also as *p*-bromo-benzoate
(721)	Geigerin	(372)	As acetate of monobromide
(669)	Hypochaerin	(326)	Monohydrate
(712)	Ivalin, pseudo	(24)	As bromoacetate
(681)	Lactucin	(907a)	
(655)	Solstitialin A	(1073)	
(558)	Spicatin	(556)	As hydrobromide
(735)	Parthemollin	(1035)	Xanthanolide

Table V-2. *Naturally Occurring Guaianolides and Biogenetic Derivatives*

Structure Number	Name of Compound	Formula	m.p. °C	$[\alpha]_D$	Plant Source[c]	References[a, b]	Comments
(671)	Absinthin	$C_{30}H_{40}O_6$	180	+180	Artemisia absinthium A. siversiana	(385, 784, 785) (781, 1118)	Dimer
(626)	Achillin	$C_{15}H_{18}O_3$	144–5	+160	Achillea lamulosa A. millefolium Artemisia lanata A. lagocephala A. ludoviciana A. klotzchiana A. tripartita Hypochaeris setosus Matricaria suffructicosa	(1138, 1138a) (900a, 991b) (325) (837) (223) (223) (980) (327) (116)	
(630)	Achillin, acetoxy	$C_{17}H_{20}O_5$	193–4	+116	Achillea lamulosa	(1138, 116, 325)	Reported before Matricarin, 11-epi
(643)	Achillin, 1,10-epoxy	$C_{15}H_{18}O_4$	—	—	Artemisia lanata	(325)	
(644)	Achillin, 1,10-epoxy, 8α-hydroxy	$C_{15}H_{18}O_5$	—	—	Artemisia lanata	(325)	
(627)	Achillin, hydroxy	$C_{15}H_{18}O_4$	161–2	+110	Achillea lamulosa Artemisia lanata	(1138, 116, 325) (325)	Reported before Matricarin, 11-epi; desacetyl
(569)	Acroptilin	$C_{19}H_{23}O_7Cl$	196–8	+92	Acroptilon repens	(262, 320a)	Reported before Chlorohyssopifolin C
(726)	Akihalin	$C_{15}H_{20}O_5$	209–11	+75	Centaurea hyrcanica Helenium autumnale	(320) (600)	
(666)	Amberboin	$C_{15}H_{20}O_4$	145	+17	Amberboa lippii	(70, 1018, 174)	
(664)	Amberboin, iso	$C_{15}H_{20}O_4$	183	+138	Jurinea maxima	(174, 337, 1188, 1192, 1193)	Reported before Maximolide

[a] Reference containing structural revision; [b] Reference given X-ray data; [c] When plant sources are from a family other than Compositae, the family name is listed under Comments.

Table V-2 (continued)

Structure Number	Name of Compound	Formula	m.p. °C	$[\alpha]_D$	Plant Source[c]	References	Comments
(672)	Anabsinthin	$C_{30}H_{40}O_6$	267	+113	*Artemisia absinthium*	(385, 784, 785, 781)	
(601)	Arbiglovin	$C_{15}H_{18}O_3$	201–3	+199	*Artemisia bigelovii*	(435)	
(640)	Arborescin	$C_{15}H_{20}O_3$	145	+64	*Artemisia arborescens*	(695, 1026[a], 56[a])	Reported before Sieversinin
(667)	Archangelolide	$C_{19}H_{40}O_{10}$	69–71	+36	*A. jacutica* / *Matricaria globifera* / *Laserpitium archangelica*	(66) / (156) / (484, 1244)	Umbelliferae
(705)	Arctolide	$C_{17}H_{20}O_6$	144–5	+64	*Arctotis grandis*	(931)	Orange compound
(668)	Artabsin	$C_{15}H_{20}O_3$	133–5	−49	*Artemisia absinthium*	(387, 311[a], 388, 1118, 1119)	
(605)	Artecanin	$C_{15}H_{18}O_5$	244–5	+27	*A. sieversiana* / *Artemisia cana*	(781) / (81, 655, 803)	
(593)	Artefransin	$C_{17}H_{20}O_6$	197–8	+33	*Artemisia franserioides*	(641)	
(532)	Arteglasin A	$C_{17}H_{20}O_5$	207–8	+110	*Artemisia douglasiana* / *Chrysanthemum indicum*	(653, 947[b]) / (375)	
(550)	Arteglasin B	$C_{17}H_{20}O_6$	192–4	+150	*Artemisia douglasiana*	(653)	See Cumambrin B
(597)	Artenovin	$C_{17}H_{20}O_5$			*Artemisia nova*	(510)	
(632)	Artilesin	$C_{17}H_{22}O_5$	183–4	+19	*Artemisia tilesii*	(455, 1138)	
(679)	Axivalin	$C_{20}H_{24}O_5$	138–40	+132	*Iva axillaris*	(452, 25[b])	
(633)	Badkhysin	$C_{20}H_{24}O_5$	139–40	+68	*Ferula oopoda* / *F. grigoriewii*	(975, 580) / (579)	Umbelliferae
(677)	Badkhysin, iso	$C_{20}H_{24}O_5$	189–90	—	*Ferula oopoda*	(975, 580)	Umbelliferae / Artifact of Badkhysin
(559)	Bahia I	$C_{15}H_{18}O_4$	209–10	+4	*Bahia pringlei*	(901, 877)	
(560)	Bahia II	$C_{20}H_{24}O_7$	133–4	+48	*Bahia pringlei*	(901)	
(561)	Bahifolin	$C_{20}H_{20}O_6$	140–2	+14	*Bahia oppositifolia*	(405)	
(534)	Berlandin	$C_{22}H_{26}O_7$	183–5	+111	*Berlandiera subacaulis*	(407, 193[b])	
(725)	Calocephalin	$C_{17}H_{22}O_5$	146–7	+28	*Calocephalus brownii*	(58)	

No.	Name	Formula	m.p.	[α]	Species	References	Remarks
(606)	Canin	$C_{15}H_{18}O_5$	244-6	−35	Artemisia cana	(655, 506, 81[a], 803)	
					A. tripartita	(980)	
(718)	Carolenalin	$C_{15}H_{22}O_4$	oil	—	Helenium autumnale	(281, 702[a,b])	Monoacetate: m. p. 160-1°; [α]ᴅ −92°
(717)	Carolenalone	$C_{15}H_{20}O_5$	245-7	—	Helenium autumnale	(710, 707[b])	
(719)	Carolenin	$C_{20}H_{29}O_5$	oil	−102	Helenium autumnale	(281, 702[a,b])	
(695)	Carpesia lactone	$C_{15}H_{20}O_3$	oil	—	Carpesium abrotanoides	(555, 554)	
(576)	Centaurepensin	$C_{19}H_{24}Cl_2O_7$	214-5	+107	Centaurea repens	(373[b], 320[a])	See Chlorohyssopifolin A
(673)	Chlorochrymorin	$C_{15}H_{19}ClO_5$	—	−18	Chrysanthemum morifolium	(802[b])	Modified 7-Ring
(576)	Chlorohyssopifolin A	$C_{19}H_{24}Cl_2O_7$	218-9	+97	Centaurea hyssopifolia	(323, 321, 320)	See Centaurepensin
					C. repens	(324)	
					C. linifolia	(324)	
					C. nigra	(324)	
					C. solstitialis	(324)	
(573)	Chlorohyssopifolin B	$C_{15}H_{19}ClO_5$	192-4	+67	Centaurea hyssopifolia	(323, 321, 320)	See Acroptilin
(569)	Chlorohyssopifolin C			+89	Centaurea hyssopifolia	(321, 320[a])	
(575)	Chlorohyssopifolin D	$C_{21}H_{29}ClO_8$	186-8	+91	Centaurea hyssopifolia	(321, 320[a])	
(574)	Chlorohyssopifolin E	$C_{19}H_{25}ClO_8$	118-9	+20	Centaurea hyssopifolia	(321, 320[a])	
(639)	Christinin	$C_{19}H_{24}O_7$	164-5	+27.3	Stevia serrata	(927)	
(639a)	Christinin II	$C_{24}H_{32}O_7$	gum	—	Stevia serrata	(927a)	
(639b)	Christinin III	$C_{21}H_{28}O_7$	gum		Stevia serrata	(927a)	
(538)	Chrysartemin A	$C_{15}H_{18}O_5$	250	+51	Chrysanthemum parthenium	(888, 803)	Structure is doubtful; ref. (803)
					C. morifolium	(803)	
					Artemisia mexicana	(888)	
					A. klotzchiana	(888)	
(607)	Chrysartemin B	$C_{15}H_{18}O_3$	262-3	+37	Chrysanthemum parthenium	(888)	
					C. morifolium	(803[a])	
					Handelia trichophylla	(1062)	
(701)	Collumellarin	$C_{15}H_{20}O_2$	43-4	−45	Callitris columellaris	(133)	Cupressaceae
(709)	Collumellarin, dihydro	$C_{15}H_{22}O_2$	77.5	+19	Callitris columellaris	(133)	
(580)	Costus lactone, dehydro	$C_{15}H_{18}O_2$	60.5	−20	Saussurea lappa	(684, 681, 871, 465, 44, 539)	Reported before Zaluzanin C, desoxy
					Vernonia hirsuta	(94)	
					Podachaenium eminens	(109)	
					Zaluzania triloba	(895)	

Table V-2 (continued)

Struc-ture Number	Name of Compound	Formula	m. p. °C	[α]b	Plant Source[c]	References	Comments
(653)	Costus lactone, dehydro—dihydro	$C_{15}H_{22}O_3$	35–7	+18.2	Saussurea lappa	(539, 465, 44)	
(581)	Costus lactone, dehydro, 8α-senecioyloxy	$C_{20}H_{24}O_4$			Vernonia digocephala	(94)	
(687b)	Costus lactone, dehydro, 1-epi, 8α-senecioyloxy	$C_{20}H_{24}O_4$	81	+109	Vernonia nudiflora	(124)	
(599)	Cumambrin A	$C_{17}H_{22}O_5$	188–90	+103	Artemisia nova	(881, 510)	
					Handelia trichophylla	(1064, 1066)	
(597)	Cumambrin B	$C_{15}H_{20}O_4$	178–80	+93	Ambrosia cumanensis	(881, 510)	
					A. acanthicarpa	(298)	
					Artemisia nova	(510)	
					A. tripartita	(510)	See also Artenovin
(600)	Cumambrin B, 8-deoxy	$C_{15}H_{20}O_3$	117–9	—	Artemisia nova	(510)	
(688)	Cumambrin B, iso	$C_{15}H_{20}O_4$	130	−21	Croton divaricatum syn. Haplopappus rigidefolia	(225)	
(604)	Cumambrin B, 3,4-oxide	$C_{15}H_{20}O_5$	178–9.5	—	Artemisia tripartita	(506)	
(587)	Cynaropicrin	$C_{19}H_{22}O_6$	glass	+109	Cynara scolymus	(1018, 933, 174)	
					C. cardunculus	(1019, 948, 949)	
					Amberboa muricata	(397)	
					Centaurea americana	(793)	
(567)	Cynaropicrin, dehydro	$C_{19}H_{20}O_6$	126	+60	Cynara scolymus	(933, 174)	
(648)	Eregoyazidin	$C_{15}H_{20}O_3$	186–8	—	Eremanthus goyazensis	(1113)	
(547)	Eregoyazin	$C_{15}H_{18}O_3$	178–81	—	Eremanthus goyazensis	(1113)	
(543)	Eremanthin	$C_{15}H_{18}O_2$	73–4	—	Eremanthus elaeagnus	(1109, 175, 283, 667)	Reported before Vanillosmin
					Vanillosmopsis erythropappa	(175)	
					Vernonia hirsuta	(94)	

							Remarks
(551)	Estafietin	$C_{15}H_{18}O_3$	104-6	-10	Artemisia mexicana	(940)	
(537)	Estafietin, isoepoxy	$C_{15}H_{18}O_4$	168	+76	Pentzia elegans	(116)	
(661)	Estafietone, dihydro	$C_{15}H_{20}O_3$	82-3	+140	Centaurea webbiana	(331, 876)	Reported before Zaluzanin C, 4β,1, 11α, 13-tetrahydro-3-dehydro
(710)	Eufoliatin	$C_{20}H_{26}O_8$	227-9	-28	Eupatorium perfoliatum	(417)	
(646)	Eufoliatorin	$C_{20}H_{24}O_7$	224	—	Eupatorium perfoliatum	(417[b])	
(594)	Eupachlorin	$C_{20}H_{25}ClO_7$	219-21	-110	Eupatorium rotundifolium	(630)	
(595)	Eupachlorin acetate	$C_{22}H_{27}ClO_8$	161-4	-192	Eupatorium rotundifolium	(630)	
(602)	Eupachloroxin	$C_{20}H_{25}ClO_8$	—	—	Eupatorium rotundifolium	(630)	
(554)	Euparotin	$C_{20}H_{24}O_7$	199-200	-124	Eupatorium rotundifolium	(630, 714[b])	
(557)	Euparotin acetate	$C_{22}H_{26}O_8$	156-7	-191	Eupatorium rotundifolium	(630, 714[b])	
(562)	Eupatoroxin	$C_{29}H_{24}O_8$	197-200	-98	Eupatorium rotundifolium	(630)	
(565)	Eupatoroxin, 10-epi	$C_{20}H_{24}O_8$	230-2	-109	Eupatorium rotundifolium	(630)	
(552)	Eupatundin	$C_{20}H_{24}O_7$	188-9	-80	Eupatorium rotundifolium	(630)	
(553)	Eupatundin acetate	$C_{22}H_{26}O_8$	oil	—	Eupatorium rotundifolium	(112, 630)	
(542)	Euperfolide	$C_{20}H_{22}O_7$	oil	-48	Eupatorium perfoliatum	(112)	
(647)	Euperfolide, 11α,13-dihydro	$C_{29}H_{24}O_7$	oil	-110	Eupatorium perfoliatum	(112)	
					Vernonia natalensis	(94)	
(687 a)	Ferreyanthus, lactone	$C_{20}H_{24}O_4$	98.5	+113	Ferreyanthus verbascifolius	(104)	
(728)	Ferulidin	$C_{15}H_{18}O_4$	170-2	—	Ferula oopoda	(972, 970)	Umbelliferae
(729)	Ferulin	$C_{15}H_{16}O_3$	176-8	—	Ferula oopoda	(970, 969, 607)	Empirical formula not determined; Umbelliferae
(715)	Florilenalin	$C_{15}H_{20}O_4$	gum	—	Helenium autumnale	(646, 705[b])	
(720)	Florilenalin, dihydro	$C_{15}H_{22}O_4$	—	—	Helenium autumnale	(611[b])	
(702)	Gaillardin	$C_{17}H_{22}O_5$	199-200	-15	Gaillardia pulchella	(622, 245[b], 496)	
					Inula britannica	(1010)	
					I. oculus-christi	(583a)	
(714)	Gaillardin, neo	$C_{17}H_{22}O_5$	—	—	Gaillardia pulchella	(496)	Mixed with Gaillardin
(721)	Geigerin	$C_{15}H_{20}O_4$	191-2	-64	Geigeria africana ssp. syn. G. filifolia G. aspera	(54, 53, 372[b])	
(642)	Globicin	$C_{17}H_{22}O_5$	148-50	+66	Matricaria globifera	(156, 1026, 843)	
(603)	Graminichlorin	$C_{22}H_{27}ClO_9$	gum	-36	Liatris graminifolia	(424, 556[b])	

Table V-2 (continued)

Structure Number	Name of Compound	Formula	m. p. °C	$[\alpha]_D$	Plant Source[c]	References	Comments
(563)	Graminiliatrin	$C_{22}H_{26}O_9$	—	−49	Liatris graminifolia	(424, 556[b])	
(556)	Graminiliatrin, deoxy	$C_{22}H_{26}O_8$	gum	−49	Liatris graminifolia	(424, 556[b])	
(692)	Grilactone	$C_{15}H_{20}O_2$	79–81	−125	Ferula grigorjevii	(578, 579)	Umbelliferae
(592)	Grosshemin	$C_{15}H_{18}O_4$	205	+138	Grossheimia macrocephala	(915, 174, 933, 984, 135, 938, 322, 86)	
					Amberboa lippii	(337)	
					Chartolepis intermedia	(746)	
					Cynara scolymus	(938)	
					Venidium decurens	(347, 346)	
(628)	Grossmisin				Artemisia caucasica	(607, 1187)	See Achillin, hydroxy
(723)	Halshalin	$C_{15}H_{20}O_4$	180–1	−66(M)	Helenium autumnale	(600)	
(670)	Handelin	$C_{34}H_{44}O_{10}$	—		Handelia trichophylla	(1065)	Dimeric
(724)	Helenium lactone	$C_{15}H_{20}O_3$	112	−88	Helenium autumnale	(464)	
(669)	Hypochaerin	$C_{15}H_{20}O_3$	110–2	−64	Hypochaeris setosus	(327, 326[b])	
(704)	Inuviscolide	$C_{15}H_{20}O_3$	oil	−19	Inula viscosa	(97)	
(708)	Inuviscolide, 4α,5α-epoxy, 10α,14-α-H	$C_{15}H_{20}O_3$	oil	+40	Inula helenium	(111)	
					I. royleana	(111)	
(703)	Inuviscolide, iso, 4-epi	$C_{15}H_{20}O_3$	147	—	Inula helenium	(111)	
					I. royleana	(111)	
(712)	Ivalin, pseudo	$C_{15}H_{20}O_3$	122–3	−145	Iva microcephala	(430, 24[b], 312)	
					Calocephalus brownii	(58)	
(713)	Ivalin, pseudo, acetate	$C_{17}H_{22}O_4$	167–9	−165	Calocephalus brownii	(58)	
(716)	Ivalin, pseudo, dihydro	$C_{15}H_{22}O_3$	oil	—	Iva microcephala	(430)	
(680)	Ivaxillarin	$C_{15}H_{18}O_4$	186–8	−24	Iva axillaris	(452, 25[b])	
(678)	Ivaxillarin, anhydro	$C_{15}H_{16}O_3$	134	−18	Iva axillaris	(452, 884, 25[b])	
	Ivaxillin	$C_{15}H_{22}O_4$	173–6	−117	Iva axillaris	(452)	
(691)	Jacquinelin	$C_{15}H_{18}O_4$	165–7	+28	Sonchus jacquini	(71, 435)	Diepoxyguaianolide; no structure given
					S. pinnatus	(71)	

	Name	Formula	mp	[α]	Species	Ref.	Notes
(697)	Jurmolide	$C_{17}H_{22}O_5$	136–7	+102.7	Jurinea maxima	(1190)	
(572)	Janerin	$C_{19}H_{22}O_7$	—	—	Centaurea janeri	(323a)	
(579)	Janerin, chloro	$C_{19}H_{23}O_7$	184–9	+81	Centaurea janeri	(323a)	
(681)	Lactucin	$C_{15}H_{16}O_5$	224–8	+49	Lactuca virosa	(946, 221, 52, 42a, 907a[b])	
					L. serriola	(1010a)	
					Cichorium intybus	(221)	
(682)	Lactucopicrin	$C_{23}H_{22}O_7$	146–8	+67	Lactucosa virosa	(946, 221)	
					L. serriola	(1010a)	
					Cichorium intybus	(490)	
(621)	Laferin	$C_{22}H_{26}O_7$	142–4	−3	Ferula olga	(606)	See Pruteninone, 4-acetoxy; Umbelliferae
(616)	Leucodin				Artemisia leucodes	(477)	See Matricarin, des-acetoxy
(535)	Leucodin, dehydro	$C_{15}H_{16}O_3$	131	+77	Lidbeckia pectinata	(116)	
					Artemisia tridentata	(82)	
					Achillea cartilaginea	(766)	
(616)	Leucomisin						See Matricarin, des-acetoxy
(549)	Ligustrin	$C_{15}H_{18}O_3$	135–7	+56	Eupatorium ligustrinum	(877)	See Rupicolin B
(660)	Lippidiol	$C_{15}H_{22}O_4$	188–91	+100	Amberboa lippii	(337, 334)	
(659)	Lippidiol, iso	$C_{15}H_{22}O_4$	167–9	+39	Amberboa lippii	(337)	
(531)	Ludartin	$C_{15}H_{18}O_3$	100–2	—	Artemisia carruthii	(294)	
(638)	Ludartin, dihydro	$C_{15}H_{20}O_3$	—	—	Artemisia carruthii	(294)	Mixed with dihydro derivative
(620)	Matricarin	$C_{17}H_{20}O_5$	193–5	+23	Matricaria chamomilla	(155, 1135)	Mixed with Ludartin
					Artemisia tilesii	(455)	
					A. cana	(655)	
					A. tripartita	(980)	
					A. klotzchiana	(888)	
					Achillea lanulosa	(1138)	
(536)	Matricarin, 11,13-dehydro	$C_{17}H_{18}O_5$	146	+121	Athanasia coronopifolia	(117)	

Table V-2 (continued)

Structure Number	Name of Compound	Formula	m. p. °C	[α]D	Plant Source[c]	References	Comments
(616)	Matricarin, desacetoxy	$C_{15}H_{18}O_3$	204–6	+53	Artemisia tridentata	(477, 307, 1135)	See Leucodin and also Leucomisin
					A. leucodes	(477)	
					A. cana	(977)	
					Achillea cartilaginea syn. Ptarmica cartilaginea	(766)	
					A. millefolium	(900a)	
					A. eriophora	(908)	
					A. santolina	(657a)	
					Hypochaeris setosus	(327)	
					Matricaria suffructicosa	(116)	
(619)	Matricarin, desacetyl	$C_{15}H_{18}O_4$	149–50 (hydrate)	—	Artemisia tilesii	(455, 155, 1138, 1138a)	
					A. austriaca	(914a)	
					A. cana	(81, 655)	
					A. caucasica	(607)	
					A. klotzchiana	(223, 888)	
					A. incana	(852)	
					A. lercheana	(852)	
					A. juncea	(567a)	
					A. leucoides	(852)	
					A. ludoviciana	(223)	
					A. sibirica	(552)	
					A. tripartita	(980, 981)	
					Achillea lanulosa	(1138)	
					A. millefolium	(561, 1076)	
						(762)	

	Name	Formula	mp	$[\alpha]$	Species	Ref.	Notes
(630)	Matricarin, 11-epi				*Achillea lamulosa*	(1138, 552)	See Achillin, acetoxy
(627)	Matricarin, 11-epi, desacetyl				*Achillea lamulosa*	(1138)	See Achillin, hydroxy
					A. sibirica	(552)	
(694)	Matricin	$C_{17}H_{22}O_5$	158–60	—	*Matricaria chamomilla*	(153, 154)	
					Achillea asplenifolia		
					A. collina		
					A. millifolium		
					A. stricta		
					Artemisia caruthii	(1006)	
(664)	Maximolide				*Jurinea maxima*	(1188, 1192, 1193)	See Amberboin, iso
(707)	Mikanocryptin	$C_{15}H_{18}O_4$	248–50	+675	*Mikania*	(445)	
					Melampodium divaricatum	(415)	
(654)	Mokko lactone	$C_{15}H_{20}O_2$	35–7	+18	*Sen mokko*	(465)	See Costus lactone, dihydro
(698)	Montanolide	$C_{22}H_{30}O_7$	132–3	−72	*Laserpitium siler*	(480, 1244)	Umbelliferae
(699)	Montanolide, iso	$C_{22}H_{30}O_7$	176	−25	*Laserpitium siler*	(478, 1244)	Umbelliferae
(700)	Montanolide, iso, acetyl	$C_{24}H_{32}O_8$	134	−78	*Laserpitium siler*	(478, 1244)	Umbelliferae
(577)	Muricatin	$C_{19}H_{24}O_6$	—	—	*Amberboa muricata*	(329)	
(635)	Oferin	$C_{23}H_{28}O_7$	214–6	0	*Ferula olgae*	(606)	Umbelliferae
(634)	Olgin	$C_{21}H_{24}O_7$	176–8	+25	*Ferula olgae*	(606, 602)	Umbelliferae
(636)	Olgoferin	$C_{23}H_{26}O_7$	240–4	+47	*Ferula olgae*	(606, 602)	Umbelliferae
(674)	Osmitopsin	$C_{15}H_{20}O_2$	oil	—	*Osmitopsis asteriscoides*	(127)	Normal Gu-skeleton more likely; ref. (93)
(675)	Osmitopsin, 1,8-epoxy	$C_{15}H_{18}O_3$	132	−301	*Osmitopsis asteriscoides*	(127)	
(676)	Osmitopsin, 4,5-epoxy	$C_{15}H_{18}O_3$	119	−203	*Osmitopsis asteriscoides*	(127)	
(618)	Parishin C	$C_{15}H_{18}O_4$	241–3	—	*Artemisia tridentata*	(505)	
(614)	Petiolaride	$C_{20}H_{24}O_6$	oil	−53	*Ageratina petiolaris*	(105)	
(539)	Picridin	$C_{15}H_{18}O_4$	oil	—	*Picridium cristallinum*	(330)	
(645)	Picridin, dihydro	$C_{15}H_{20}O_4$	175–7	−26	*Picridium cristallinum*	(330)	
					P. ligulatum		
(722)	Pleniradin	$C_{15}H_{20}O_4$	94–7	+35	*Baileya pleniradiata*	(1165)	
(684)	Preeupatundin-2-acetate, 8β-angeloyloxy-5α-hydroxy	$C_{22}H_{26}O_7$	oil	−18	*Eupatorium rotundifolium*	(112)	
(683)	Preeupatundin, 8β-angeloyloxy	$C_{20}H_{24}O_5$	oil	−15	*Eupatorium rotundifolium*	(112)	

Table V-2 (continued)

Structure Number	Name of Compound	Formula	m. p. °C	$[\alpha]_D$	Plant Source	References	Comments
(685)	Preeupatundin, 5α-hydroxy-8β-tiglinoyloxy	$C_{20}H_{24}O_6$	oil	−20	Eupatorium rotundifolium	(112)	
(555)	Preeupatundin, 8β-tiglinoyl-oxy, 10,15-epoxide	$C_{20}H_{24}O_6$	oil	—	Eupatorium rotundifolium	(112)	Acetate: $[\alpha]_D$ −37°
(613)	Prutenin	$C_{20}H_{26}O_4$	oil	−61	Laserpitium prutenicum	(129)	See Laferin
(621)	Pruteninone, 4-acetoxy				Laserpitium prutenicum	(129)	Umbelliferae
(624)	Pruteninone, 8-acetoxy	$C_{22}H_{26}O_7$	142.5	−3	Laserpitium prutenicum	(129)	Umbelliferae
(625)	Pruteninone, 8-angeloyloxy	$C_{25}H_{30}O_7$	194	—	Laserpitium prutenicum	(129)	Umbelliferae
(649)	Pruteninone, iso, 8-acetoxy	$C_{22}H_{26}O_7$	oil	—	Laserpitium prutenicum	(129)	Umbelliferae
(568)	Repin	$C_{19}H_{22}O_7$	154–6	+101	Acroptilon repens / Centaurea hyrcanica	(257, 258, 320a)	
(540)	Rupicolin A	$C_{15}H_{18}O_4$	167–8.5	+98	Artemisia tripartita	(506)	$[\theta]_{245}$ −1412°
(541)	Rupicolin A, 15-acetoxy-1β-hydroxy-8-(2α-acetoxyethyl) acrylate	$C_{24}H_{28}O_9$	oil	—	Erlangea inyangana	(96)	
(546)	Rupicolin A, 3α,4α-diacetoxy-3,4-dihydro-8-(2α-acetoxy-ethyl) acrylate	$C_{26}O_{32}O_{10}$	oil	−175	Erlangea inyangana	(96)	
(548)	Rupicolin B	$C_{15}H_{18}O_4$	142–4	—	Artemisia tripartita	(506)	See Ligustrin
(608)	Rupin A	$C_{15}H_{18}O_4$	260–300	—	Artemisia tripartita / Achillea biebersteinii	(506) (1185)	$[\theta]_{250}$ −2660°
(609)	Rupin B	$C_{17}H_{20}O_7$	235–45	—	Artemisia tripartita	(506)	
(687)	Saupirin	$C_{19}H_{22}O_6$	75–84	+112	Sanssurea pulchella / S. neopulchella	(162)	
(686)	Saurin	$C_{15}H_{18}O_5$	87–8	+106	Saussurea pulchella	(10) (762)	
(693)	Sieversin	$C_{17}H_{22}O_5$	128–31	—	Artemisia siversiana / A. macrophylla	(762) (761a)	

No.	Name	Formula	mp	Rotation	Source	Ref.	Notes
(640)	Sieversinin				*Artemisia jacutica*	(66)	See Arborescin
					A. sieversiana	(762a)	
(655)	Solstitialin A	$C_{15}H_{20}O_5$	206-7	—	*Centaurea solstitialis*	(1074, 1073^b, 1194)	
(656)	Solstitialin acetate	$C_{17}H_{22}O_6$	—	—	*Centaurea solstitialis*	(1194)	
(558)	Spicatin	$C_{27}H_{32}O_{10}$	—	—	*Liatris spicata*	(424, 438, 556^b)	$[\theta]_{Hg} -146$
					L. pycnostachya		
					L. tenuifolia		
(564)	Spicatin, epoxy	$C_{27}H_{32}O_{11}$	gum	—	*Liatris pycnostachya*	(438)	
(596)	Spicatin, hydrochloride	$C_{27}H_{33}ClO_{10}$	122-3	—	*Liatris pycnostachya*	(424, 556^b)	$[\alpha]_{Hg} -80$
(533)	Subacaulin	$C_{20}H_{24}O_6$	160-2	+130	*Berlandiera subacaulis*	(438)	
(571)	Subluteolide	$C_{19}H_{22}O_7$	—	—	*Vernonia subluteus*	(407, 193^b)	$[\theta]_{249} +2226^c$
(637)	Talassin A	$C_{25}H_{30}O_7$	188-91	−30	*Talassia transiliensis*	(732)	
					Ferula olgae	(604, 581)	
						(606)	Umbelliferae
(623)	Talassin B	$C_{24}H_{30}O_7$	205-8	−72	*Talassia transiliensis*	(604, 581)	
					Ferula olgae	(606)	Umbelliferae
(732)	Tanamyrin	$C_{15}H_{20}O_3$	—	—	*Tanacetum myriophyllum*	(728, 319)	
(610)	Trifloculoside	$C_{15}H_{20}O_3$	oil	—	*Vernonia trifloculosa*	(128)	
					V. nudiflora	(128)	
(696)	Trilobolide	$C_{17}H_{24}O_6$	191-2	+73	*Laser trilobum*	(485, 475)	Umbelliferae
(543)	Vanillosmin				*Vanillosmopsis erythropappa*	(175, 283)	See Eremanthin
(545)	Vanillosmin, 8α-sencioyloxy	$C_{29}H_{24}O_4$	oil	—	*Vernonia oligocephala*	(94)	
(585)	Vernoflexin	$C_{20}H_{24}O_4$	73-5	—	*Vernonia flexuosa*	(581, 584)	
(586)	Vernoflexuoside	$C_{21}H_{28}O_8$	104-5	—	*Vernonia flexuosa*	(581, 584)	
(711)	Virginolide	$C_{15}H_{16}O_3$	133-5	+5	*Helenium virginicum*	(434)	
(652)	Viscidulin A	$C_{17}H_{22}O_5$	124	+77	*Artemisia cana*	(977)	
(651)	Viscidulin B	$C_{17}H_{22}O_5$	132-3	+60	*Artemisia cana*	(977)	
(650)	Viscidulin C	$C_{15}H_{20}O_4$	147	+49	*Artemisia cana*	(977)	
(706)	Xerantholide	$C_{15}H_{18}O_3$	175-7	+239	*Xeranthemum cylindraceum*	(932)	
(689)	Yejuhua lactone				*Chrysanthemum indicum*	(158)	
(730)	Zaluzanin A	$C_{15}H_{20}O_4$	265	−10	*Zaluzania augusta*	(884)	Modified Guaianolide
(731)	Zaluzanin B	$C_{17}H_{22}O_5$	223-5	−12	*Zaluzania augusta*	(884)	Modified Guaianolide

Table V-2 (continued)

Structure Number	Name of Compound	Formula	m. p. °C	[α]_D	Plant Source^c	References	Comments
(583)	Zaluzanin C	$C_{15}H_{18}O_3$	95–6	+37	Zaluzania augusta Z. robinsonii Z. triloba Podochaenium eminens Vernonia noveboracensis	(894) (528a) (894) (109) (94)	
(589)	Zaluzanin C, senecioyl	$C_{20}H_{24}O_4$			Vernonia hirsuta	(94)	
(612)	Zaluzanin C, 13-acetoxy-11,13-dihydro-7,11-dehydro-3-desoxy	$C_{17}H_{20}O_4$	oil	+26	Podochaenium eminens	(109)	
(566)	Zaluzanin C, dehydro	$C_{15}H_{16}O_3$	134–5	+160	Arctotis aspera A. revoluta	(107, 894)	
(581)	Zaluzanin C, desoxy				Vernonia noveboracensis Zaluzania robinsonii	(94) (684, 871, 886)	See Costus lactone, dehydro
(591)	Zaluzanin C, 4β-14-dihydro-3-dehydro	$C_{15}H_{18}O_3$	oil	—	Arctotis revoluta	(107)	
(611)	Zaluzanin C, 11,13-dihydro-7,11-dehydro-3-desoxy	$C_{15}H_{18}O_2$	108	+48	Podochaenium eminens	(109)	
(582)	Zaluzanin C, 3-epi	$C_{15}H_{18}O_3$	oil	−56	Vernonia anisochaetoides	(94)	
(588)	Zaluzanin C, 7α-hydroxy-3-desoxy	$C_{15}H_{18}O_3$	128	+69	Podochaenium eminens	(109)	
(657)	Zaluzanin C, 7α-hydroxy-3-desoxy-11β,13-dihydro	$C_{15}H_{20}O_3$	142–3	—	Podochaenium eminens	(109)	
(658)	Zaluzanin C, 8α-hydroxy-11β,13-dihydro-3-dehydro	$C_{15}H_{18}O_4$	168	+170	Vernonia noveboracensis	(94)	
(663)	Zaluzanin C, 4β,15,11α,13-tetrahydro-3-dehydro				Arctotis revoluta	(107)	See Estafietone, dihydro

(661)	Zaluzanin C, 4β-14,11β,13-tetrahydro-3-dehydro	$C_{15}H_{20}O_3$	oil	—	Arctotis revoluta	(107)
(584)	Zaluzanin D	$C_{17}H_{20}O_4$	103–4	0	Zaluzania augusta	(894)
					Z. triloba	(894)
					Podachaenium eminens	(109)
(530)	Zuurbergenin	$C_{17}H_{20}O_4$	176	+92	Matricaria zuurbergensis	(118)

Table V-3. *Naturally Occurring Seco-Guaianolides (Xanthanolides)*

Structure Number	Name of Compound	Formula	m.p. °C	$[\alpha]_D$	Plant Source	References	Comments
(734)	Apachin	$C_{17}H_{24}O_5$	80–1.5	−147	*Iva ambrosiaefolia*	(1166, 1036)	
(749)	Carabrone	$C_{15}H_{20}O_3$	90–1	+116.9	*Carpesium abrotanoides*	(724, 721, 1120, 676)	Reported before Grandicin; Modified Xanthanolide
					C. eximium	(603)	
					Arnica longifolia	(1140)	
					A. foliosa	(489)	
					Inula helenium	(111)	
					I. royleana	(111)	
					Helenium quadridentatum	(383a)	
(748)	Carabrone, 4-H	$C_{15}H_{22}O_3$	oil	+19.4	*Inula royleana*	(111)	Modified Xanthanolide
(738)	Fruticosin	$C_{17}H_{22}O_5$	127–9	—	*Parthenium fruticosum*	(1177, 866)	
(745)	Gafrinin	$C_{17}H_{24}O_5$	110–1	−16.1	*Geigeria africana*	(216a, 23, 1142)	
(749)	Grandicin				*Inula grandis*	(771, 772, 724)	See Carabrone; Modified Xanthanolide
(753)	Griesenin	$C_{15}H_{16}O_4$	196–7.5	+284	*Geigeria africana* ssp. Syn. *G. filifolia*	(214, 214b)	Modified Xanthanolide
(754)	Griesenin, dihydro	$C_{15}H_{18}O_4$	139.5–40.5	+92	*Geigeria africana*	(214, 214b)	Modified Xanthanolide
(737)	Ivalbatin	$C_{15}H_{20}O_4$	gum	−84	*Iva dealbata*	(159, 1036)	Acetate: m.p. 127–8°, $[\alpha]_D$ −136
(751)	Ivalbin	$C_{15}H_{22}O_4$	160–2	−44.7	*Iva dealbata*	(409, 159)	
(733)	Ivambrin	$C_{15}H_{22}O_4$	156–7	—	*Iva ambrosiaefolia*	(1166, 1036)	
(735)	Parthemollin	$C_{15}H_{20}O_4$	116–8	−130	*Parthenice mollis*	(406, 1035, 159)	
(736)	Parthemollin, acetyl	$C_{17}H_{22}O_5$	102–3	−135	*Iva ambrosiaefolia*	(1166, 406)	
(752)	Tomentosin	$C_{15}H_{20}O_3$	—	—	*Parthenium tomentosum*	(866, 97)	
					Inula helenium	(111)	
					I. royleana	(111)	

No.	Name	Formula	m.p.	$[\alpha]$	Source	Ref.	Notes
(741)	Tomentosin, 8-epi	$C_{15}H_{20}O_3$	oil	−38.9	Inula helenium	(111)	See Xanthinosin
(746)	Tomentosin, 4-H	$C_{15}H_{22}O_3$	oil	+37.0	Inula royleana	(111)	
(743)	Xanthatin	$C_{15}H_{18}O_3$	114–5	−20.0	Xanthium species	(291, 1142ᵃ, 216)	For other Xanthium sources see ref. (697a)
					X. riparium	(820a)	
					X. spinosum	(42)	
(739)	Xanthanol	$C_{17}H_{24}O_5$	78–9	−87.3	Xanthium strumarium	(1142)	For other Xanthium sources see ref. (697a)
					X. commune	(1142)	
(740)	Xanthanol, iso	$C_{17}H_{24}O_5$	101–2	+28	Xanthium strumarium	(1142)	
(742)	Xanthinin	$C_{17}H_{22}O_5$	126	−53	Xanthium pennsylvanicum	(291, 216, 1142, 287)	For other Xanthium sources see ref. (697a)
					X. italicum	(1092a)	
					X. orientale	(226)	
					X. commune	(1142)	
					X. riparium	(820a)	
					Angianthus tomentosus	(97)	
					Iva ambrosiaefolia	(1166)	
(741)	Xanthinin, 2-desacetoxy	$C_{15}H_{20}O_3$	oil	+14.1	Inula viscosa	(97, 866)	See Xanthinosin
(747)	Xanthumin	$C_{17}H_{22}O_5$	100.5–1.0	+48.2	Xanthium strumarium	(722, 1142)	For other Xanthium sources see refs. (697a) and (1142)
(741)	Xanthinosin	$C_{15}H_{20}O_3$	—	—	X. occidentale	(572)	See also Tomentosin, 8-epi, and Xanthinin, 2-desacetoxy For other Xanthium sources see ref. (697a)
					X. strumarium	(697a)	
(744)	Xanthumin, deacetoxy	$C_{15}H_{18}O_3$	—	—	X. strumarium	(697a)	
(746a)	Xanthuminol	$C_{17}H_{24}O_5$	—	—	X. strumarium	(697a)	

ᵃ Reference reports structural revision.

VI. Elemanolides

The biogenesis of elemanolides most likely involves Cope rearrangements of germacranolides which occur under laboratory reactions with great ease as discussed in Chapter III, 3.4. It has been suggested that elemanolides isolated from plants represent artifacts which are formed from the germacradienolides during the isolation procedures. There exists evidence for this assumption in the case of simple elemanolides, but in oxidatively modified divinylcyclohexane systems such as vernomenin (**795**), micordilin (**800**) and a number of other plant constituents shown in Chart VI-1 this can certainly not be true.

Chart VI-1. Naturally occurring elemanolides and biogenetic derivatives

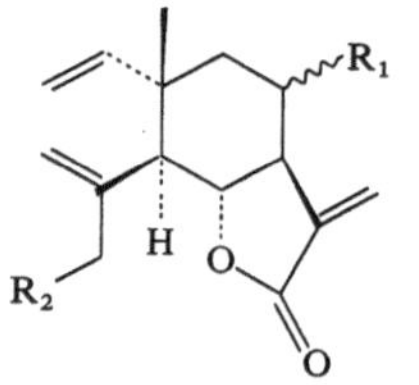

(**784**) Melitensin, 11(13)-dehydro; R_1 = α-OH, R_2 = OH
(**785**) Tulipdienolide, epi; R_1 = β-OAc, R_2 = H
(**786**) Melitensin, 11(13)-dehydro, β-hydroxyisobutyrate; R_1 = α-OA, R_2 = OH

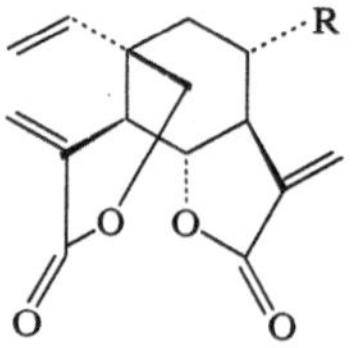

(**787**) Vernolepin; R = OH, 5 α-H
(**788**) Vernodalin; R = OMac-4-OH, 5 α-H

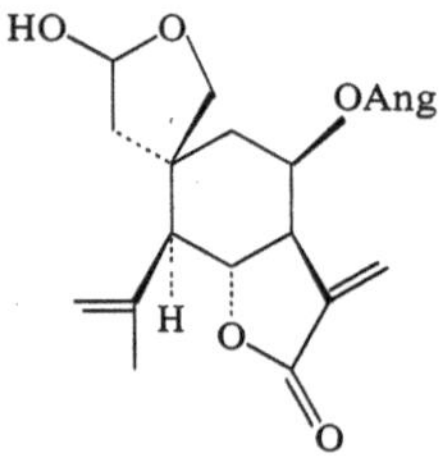

(**789**) Confertiphyllide

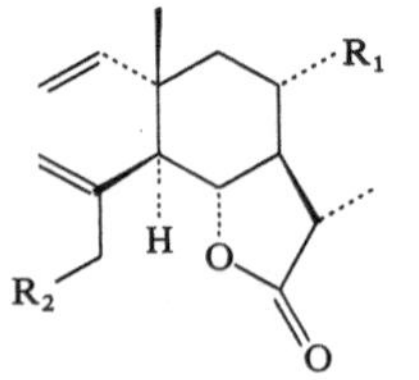

(**790**) Saussurea lactone; R_1 = R_2 = H
(**791**) Melitensin; R_1 = R_2 = OH
(**792**) Melitensin β-hydroxyisobutyrate; R_1 = OA, R_2 = OH

A =

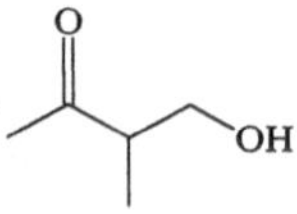

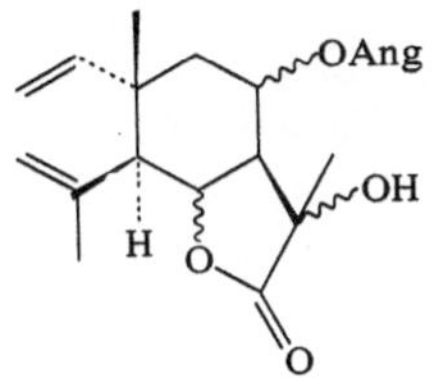

(**793**) Laserolide, iso

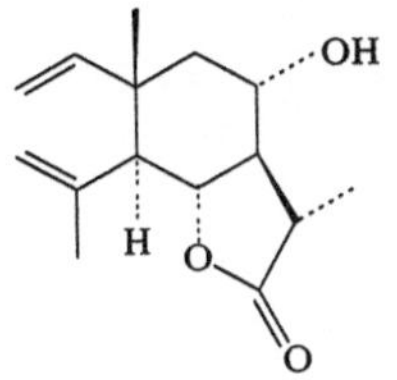

(**794**) Temisin

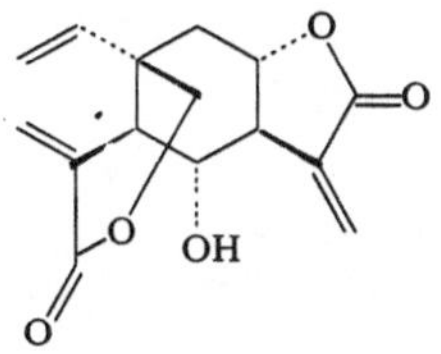

(**795**) Vernomenin

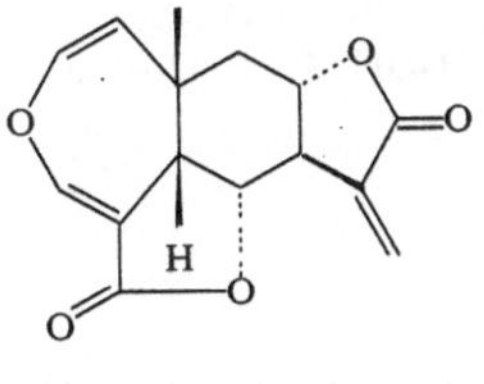

(**796**) Miscandenin

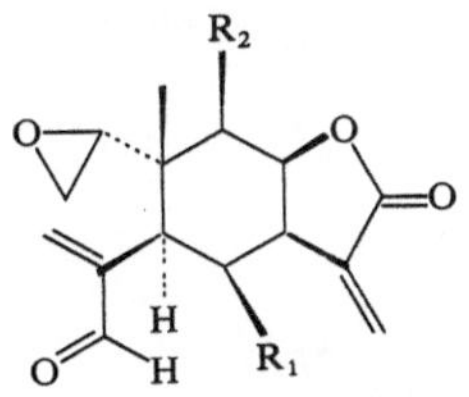

(**797**) Zinaflorin II; R_1 = OAng, R_2 = OH
(**798**) Zinaflorin III; R_1 = OMac, R_2 = OAc
(**799**) Zinaflorin I; R_1 = R_2 = OAng

Chart VI-1 (continued)

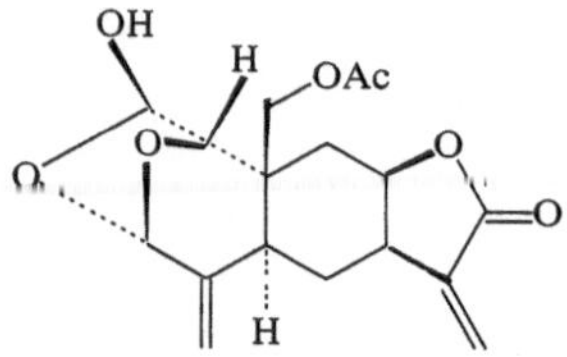

(800) Micordilin

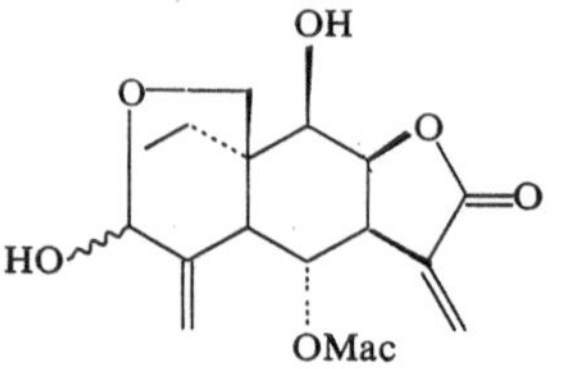

(801) Verafinin

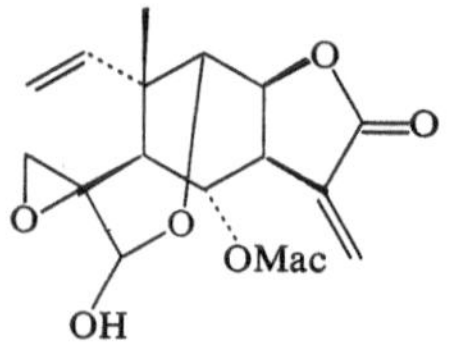

(802) Verafinin C

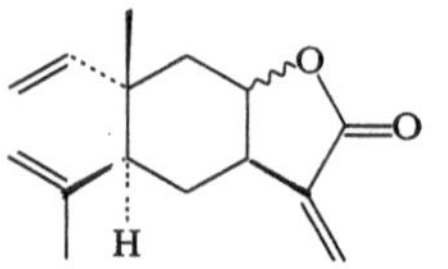

(803) Igalan

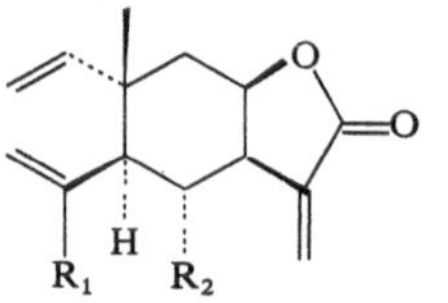

(803a) Zempoalin A; R_1 = CHO, R_2 = O-i-But
(803b) Zempoalin B; R_1 = CH$_2$OH, R_2 = O-i-But

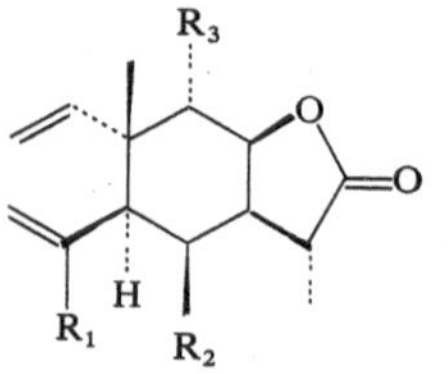

(804) Zinarosin; R_1 = CHO, R_2 = O-i-But, R_3 = OH
(805) Zinarosin, dihydro, diacetate; R_1 = CH$_2$OAc, R_2 = O-i-But, R_3 = OAc

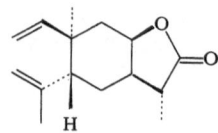

(806) Callitrin

(**807**) Germafurenolide, iso; R = H
(**808**) Germafurenolide, iso, hydroxy; R = OH

(**809**) Sericealactone, desoxy; R = H
(**810**) Sericealactone; R = OH

(**811**) Vernodalol

(**352**) Linderalactone, iso

The question whether the Cope rearrangement of the latter type of elemanolides is enzymatically controlled or occurs spontaneously remains open. The observation (*1043*) that the conversion of linderalactone (**233**) to isolinderalactone (**352**) is slow but does occur at ambient temperature might be an indication that germacradienes oxidized at C-14 and/or C-15 rearrange at lower temperatures than do the non-oxidized compounds. Possibly, slow but spontaneous Cope rearrangements may occur in the plant followed by enzyme-controlled oxidative modifications.

The structurally unique elemanolide miscandenin (**796**) (*190, 451*) contains a *cis*-ring fusion between the cyclohexane and the dihydro-oxepin ring. Its stereochemistry suggests that the biogenetic precursor is a 2,3-epoxidized germacradiene dilactone which undergoes Cope rearrangement *via* the boat-like transition state as (**812**) of Chart VI-2 which has the C-10 methyl group above the plane of the ring. Most likely, the 2,3-epoxide ring as well as the C-4, C-6-lactone group impose considerable constraint on the cyclodecadiene skeleton so as to favor in this case a boat-like transition state over the more common chair-like conformation which has the C-10 methyl below the plane of the ring.

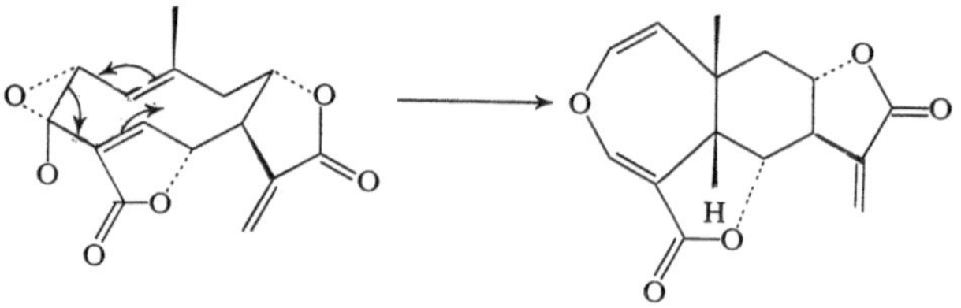

(812)   (796) miscandenin

Chart VI-2. Biogenesis of miscandenin

Even more unusual is callitrin from the heartwood of a member of the Cupressaceae which has been assigned (133) the stereochemistry shown in formula (806) (≡ 806 A) in Chart VI-3. The presumed precursor is a dihydroinunolide (135 a) which under the conditions of biogenesis, whether free or enzyme-bound, is postulated to assume conformation (135 a B) (which corresponds to conformation 116 C of laurenobiolide on p. 91) rather than the more usual conformation (135 a A). Cope rearrangement of (135 a B) would lead to callitrin whereas Cope rearrangement of (135 a A) would lead to an elemadienolide (806 a) belonging to the "normal" series that has so far not been isolated from a plant source. Indeed, callitrin is partially isomerized at elevated temperature to a substance whose physical properties are in accord with formula (806 a), presumably as the result of equilibration between (135 a A) and (135 a B). In this connection it is of interest that dihydroinunolide (135 b), C-11 epimeric with the postulated intermediate (135 a), has been prepared by reduction of inunolide (24) (847, 850 a). Unfortunately its high resolution NMR spectrum which might give a clue as to its conformation was not reported. Interestingly enough, the Cope rearrangement of (135 b) was said to proceed in poor yield only; no evidence was provided for the stereochemistry at C-5 and C-10 which the authors assigned to the product (850 a).

Cyclization of (135 a), possibly via a 1,10-epoxide (see Chapters III., 3.3, Chart III-26, and IV-1, Chart IV-2) would give rise to the eudesmanolide callitrisin (469), also of abnormal stereochemistry, and its dihydro derivative (470) which were isolated from the same source (133).

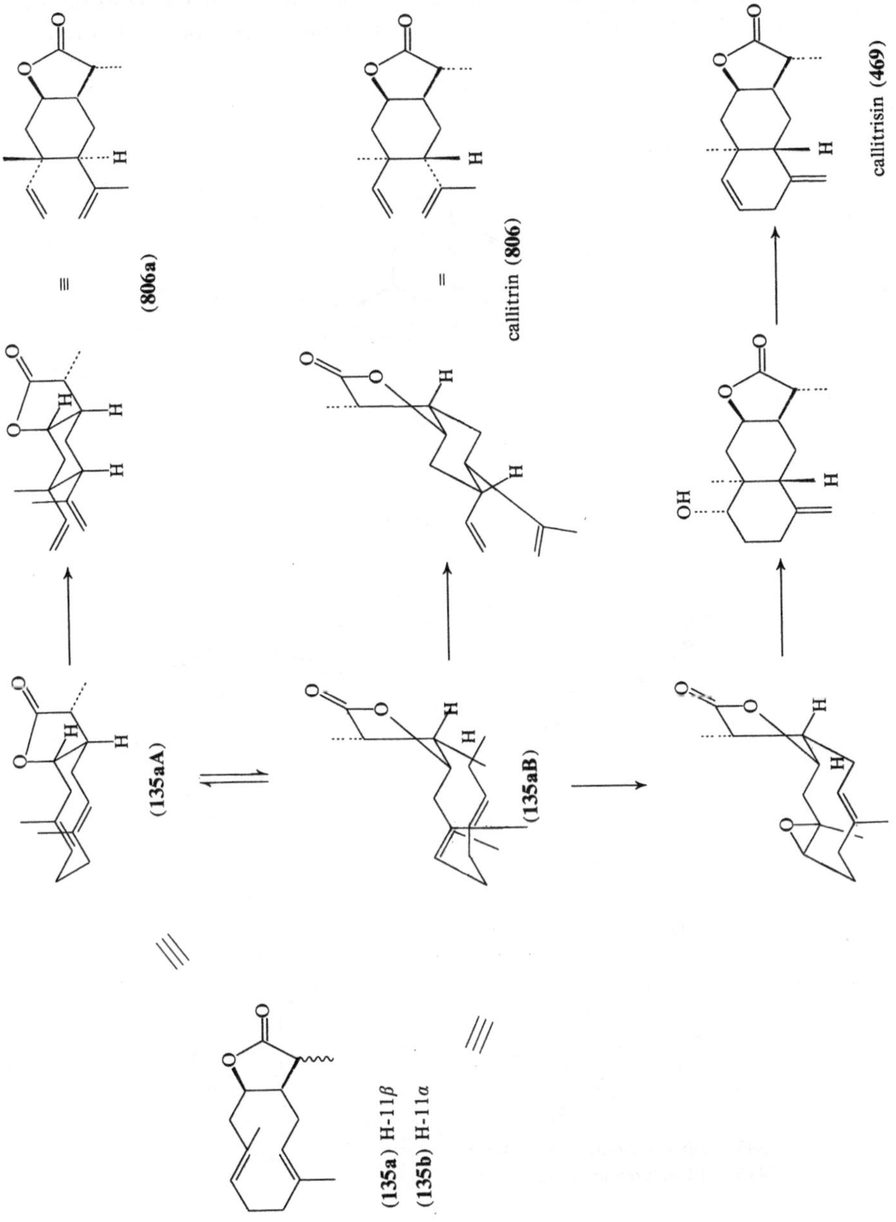

Chart VI-3. Biogenesis of callitrin and callitrisin

Recently, BANKAR and KULKARNI (45) described monoepoxidations of saussurea lactone (**790**) and the Lewis acid catalyzed cyclizations of the resulting monoepoxides. As outlined in Chart VI-4, peracid epoxid-

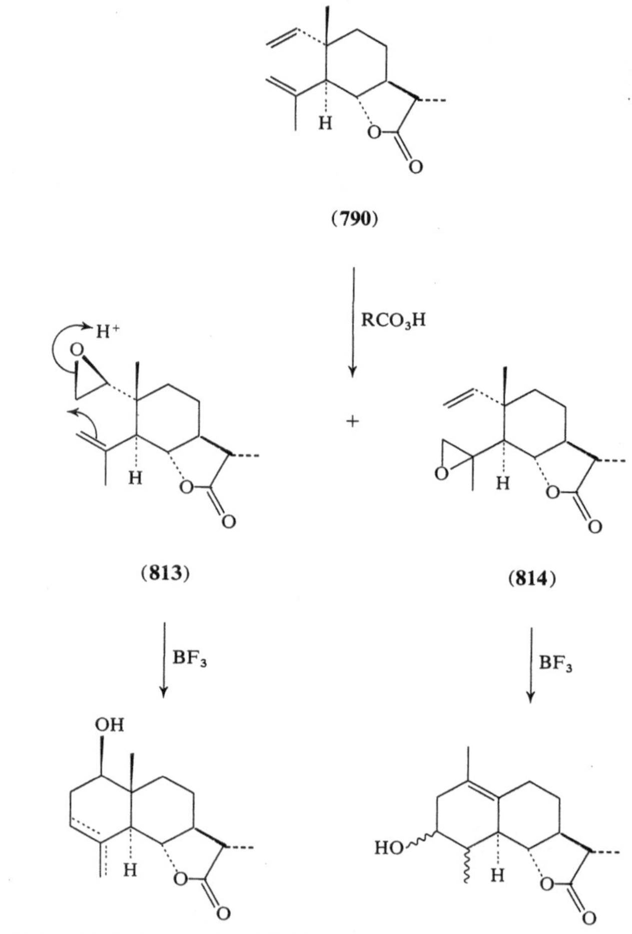

(**790**)

(**813**)   (**814**)

(**395**) dihydrosantamarin; 3,4-d.b.
(**410**) dihydroreynosin; 4,15-d.b.   (**816**)

Chart VI-4. Lewis acid catalysed cyclization of elemanolide monoepoxides

Table VI-1. Naturally Occurring Elemanolides and Biogenetic Derivatives

Structure Number	Name of Compound	Formula	m. p. °C	[α]D	Plant Source	References	Comments
(806)	Callitrin	$C_{15}H_{22}O_2$	82–3	+225	Callitris columellaris	(133)	
(789)	Confertiphyllide	$C_{20}H_{26}O_6$	169–71	−14.9	Eriophyllum confertiflorum	(924, 618)	
(807)	Germafurenolide, iso	$C_{15}H_{20}O_2$	85	+5	Lindera strychnifolia	(1046)	
(808)	Germafurenolide, iso, hydroxy	$C_{15}H_{20}O_3$	160–1	+4.7	Lindera strychnifolia	(1046)	
(803)	Igalan	$C_{15}H_{20}O_2$	79.0–.5	+105	Inula grandis	(773)	
(793)	Laserolide, iso	$C_{22}H_{30}O_6$	125–7	−81.1	Laser trilobum	(486, 475)	Family Umbelliferae
(352)	Linderalactone, iso	$C_{15}H_{16}O_3$	118–21	−247	Lindera strychnifolia	(1054, 1050, 1090, 1083)	
(791)	Melitensin	$C_{15}H_{22}O_4$	167–8	+85	Neolitsea aciculata Centaurea melitensis	(1048) (315, 317)	
(784)	Melitensin, 11(13)-dehydro	$C_{15}H_{20}O_4$	–		Centaurea pullata	(324)	
(786)	Melitensin, 11(13)-dehydro, β-hydroxyisobutyrate	$C_{19}H_{26}O_6$	115–7	+87	Centaurea melitensis C. pullata	(318) (33a)	
(792)	Melitensin, β-hydroxy-isobutyrate	$C_{19}H_{28}O_6$	107–8	+50	Centaurea melitensis	(318)	
(800)	Micordilin	$C_{17}H_{20}O_7$	176–8	–	Mikania cordifolia	(450)	
(796)	Miscandenin	$C_{15}H_{14}O_5$	232–5	−181.4	Mikania scandens	(451, 190)	
(790)	Saussurea lactone	$C_{15}H_{22}O_2$	148–9	+66	Saussurea lappa	(850, 991, 491)	
(810)	Sericealactone	$C_{16}H_{20}O_5$	150–1	–	Neolitsea sericea	(752, 376)	
(809)	Sericealactone, desoxy	$C_{16}H_{20}O_4$	137	–	Neolitsea sericea	(376)	
(794)	Temisin	$C_{15}H_{22}O_3$	228	+65.7	Artemisia cina	(33, 776a)	
(785)	Tulipdienolide, epi	$C_{17}H_{22}O_4$	134–5	+6	Liriodendron tulipifera	(235)	
(801)	Verafinin	$C_{19}H_{24}O_7$	145–6	−56	Verbesina coahuilensis	(366)	
(802)	Verafinin C	$C_{19}H_{22}O_7$	130–3	−37	Verbesina coahuilensis	(365)	
(788)	Vernodalin	$C_{19}H_{20}O_7$	gum	+124	Vernonia amygdalia V. guineensis	(627) (1095)	
(811)	Vernodalol	$C_{20}H_{24}O_8$	133–4	+36.5	Vernonia anthelmintica	(34)	

Table VI-1 (continued)

(787)	Vernolepin	$C_{15}H_{16}O_5$	179–80	+72	Vernonia hymenolepis	(628, 629, 716, 209, 357, 356)
					V. guineensis	(81)
(795)	Vernomenin	$C_{15}H_{16}O_5$	—	−62	Vernonia hymenolepis	(628, 357, 209)
(803a)	Zempoalin A	$C_{19}H_{24}O_5$	133–5	+75	Verbesina aff. stricta	(805a)
(803b)	Zempoalin B	$C_{19}H_{26}O_5$	141–3		Verbesina aff. stricta	(805a)
(799)	Zinaflorin I	$C_{25}H_{30}O_8$	178–80	+77	Zinnia pauciflora	(846)
(797)	Zinaflorin II	$C_{20}H_{24}O_7$	192–4	+115	Zinnia pauciflora	(846)
(798)	Zinaflorin III	$C_{21}H_{24}O_8$	133–5	—	Zinnia pauciflora	(846)
(804)	Zinarosin	$C_{19}H_{26}O_6$	151–3	+84	Zinnia acerosa	(886)
(805)	Zinarosin, dihydro, diacetate	$C_{23}H_{32}O_8$	100–1	+65	Zinnia acerosa	(886)

ation of lactone (790) provides a mixture of the two respective epoxides (813) and (814) in a 1 : 4 ratio. Treatment of the 1,2-epoxide (813) with BF₃ results in the formation of the eudesmanolide dihydrosantamarin (395) and its 4,15-double bond isomer, dihydroreynosin (410). From the β-configuration of the OH-group at C-1 in (395) it can be derived that epoxidation of the 1,2-vinyl group in (790) must have occurred specifically from the β-face of the double bond. The cyclization of the 3,4-epoxide (814) is more involved. Most likely, the epoxide ring is isomerized first with formation of an aldehyde at C-3 which is subsequently attacked by C-2 of the 1,2-double bond in a process resembling a Prins reaction. Methyl shift from C-10 to C-1 and 1,10-double bond formation finally provide alcohol (816). NBS-initiated cyclization of saussurea lactone was also reported (519).

Literature references and physical data of elemanolides are found in Table VI-1 on page 221.

VII. Pseudoguaianolides and Biogenetic Derivatives

Several reviews have appeared on biogenetic and chemical aspects of pseudoguaianolides. Chemical and chemotaxonomic consideration have been presented by ROMO and ROMO DE VIVAR (879, 891), HERZ (392, 394), and MABRY and coworkers (662, 663, 1172). The biogenesis of pseudoguaianolides has been discussed by GEISSMAN (288, 290), HERZ (393, 395), and FISCHER (265, 273). Furthermore, selected transformations and modifications as well as NMR spectra of pseudoguaianolides can be found in the book by YOSHIOKA, MABRY, and TIMMERMANN (1172).

1. Structural Types of Pseudoguaianolides and Their Biogenesis

The pseudoguaianolides are based upon the 5,7-ring skeleton which typically contain a methyl group at the C-5 ring junction. Formulae (953) and (954) illustrate the two major types of pseudoguaianolides both of

(953) ambrosanolide   (954) helenanolide

Chart VII-1. Naturally occurring pseudoguaianolides and biogenetic derivatives

(**817**) Damsin; $R_1 = R_2 = R_3 = H$
(**818**) Coronopilin; $R_1 = OH, R_2 = R_3 = H$
(**819**) Tetraneurin B; $R_1 = OH, R_2 = OAc, R_3 = H$
(**820**) Tetraneurin A; $R_1 = OH, R_2 = H, R_3 = OAc$
(**821**) Chiapin B; $R_1 = OH, R_2 = H, R_3 = O\text{-}i\text{-But}$
(**822**) Ligulatin B; $R_1 = R_3 = H, R_2 = OAc$
(**823**) Chiapin A; $R_1 = R_2 = H, R_3 = O\text{-}i\text{-But}$

(**824**) Ivoxanthin; $R_1 = \alpha\text{-}OH, R_2 = R_3 = H$
(**825**) Bipinnatin; $R_1 = \beta\text{-}OH, R_2 = R_3 = H$
(**826**) Damsin, 3-hydroxy; $R_1 = R_3 = H, R_2 = OH$
(**827**) Confertiflorin, desacetyl; $R_1 = R_2 = H, R_3 = OH$
(**828**) Confertiflorin; $R_1 = R_2 = H, R_3 = OAc$

(**829**) Conchosin A

(**830**) Ambrosin; $R_1 = \alpha\text{-}H, R_2 = R_3 = H$
(**831**) Parthenin; $R_1 = \alpha\text{-}OH, R_2 = R_3 = H$
(**832**) Hymenin; $R_1 = \beta\text{-}OH, R_2 = R_3 = H$
(**833**) Conchosin B; $R_1 = \alpha\text{-}OH, R_2 = H, R_3 = OAc$
(**834**) Oaxacin; $\alpha\text{-}H, R_2 = OAc, R_3 = H$

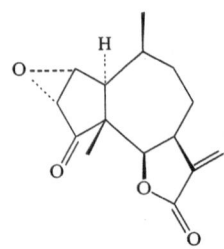

(**835**) Ambrosin, 2,3-H-2,3-epoxy;

(836) Ambrosin, neo

(838) Apoludin; $R_1 = R_3 = OH$, $R_2 = H$
(839) Salsolin; $R_1 = OAc$, $R_2 = H$, $R_3 = OH$
(840) Ambrosiol; $R_1 = H$, $R_2 = R_3 = OH$

(841) Tetraneurin D; $R_1 = R_2 = OH$, $R_3 = OAc$
(842) Tetraneurin C; $R_1 = OH$, $R_2 = R_3 = OAc$

(843) Tetraneurin E; $R_1 = R_3 = OH$, $R_2 = \beta\text{-OAc}$
(844) Tetraneurin F; $R_1 = OH$, $R_2 = \beta\text{-OAc}$,
       $R_3 = OAc$
(845) Hysterin; $R_1 = H$, $R_2 = \alpha\text{-OAc}$, $R_3 = OH$
(846) Hysterin acetate; $R_1 = H$, $R_2 = \alpha\text{-OAc}$,
       $R_3 = OAc$

(847) Ambrosin, tetrahydro; $R_1 = H$, $R_2 = \alpha\text{-H}$
(848) Hymenolin; $R_1 = OH$, $R_2 = \alpha\text{-H}$
(849) Franserin; $R_1 = H$, $R_2 = OH$

Chart VII-1 (continued)

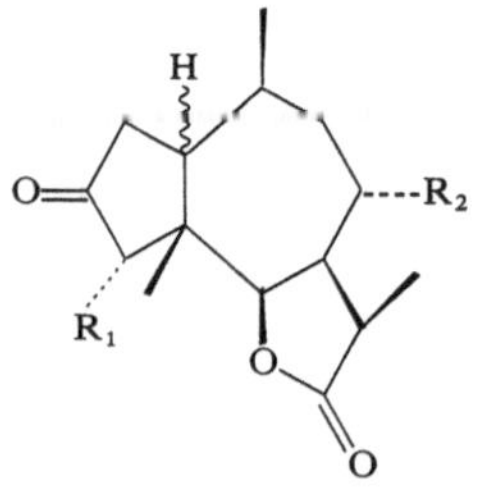

(850) Pseudoguaian-6,12-olide, 4-hydroxy-3-oxo;
R$_1$ = OH, R$_2$ = H
(851) Pseudoguaian-6,12-olide, 8-acetoxy-3-oxo;
R$_1$ = H, R$_2$ = OAc

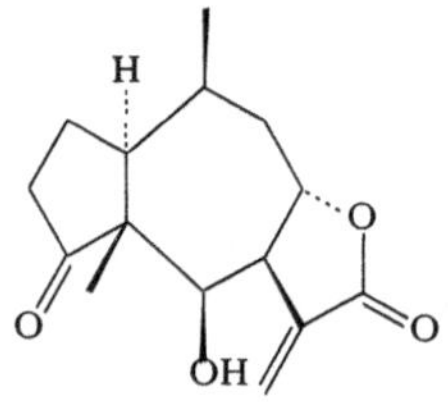

(852) Carpesiolin

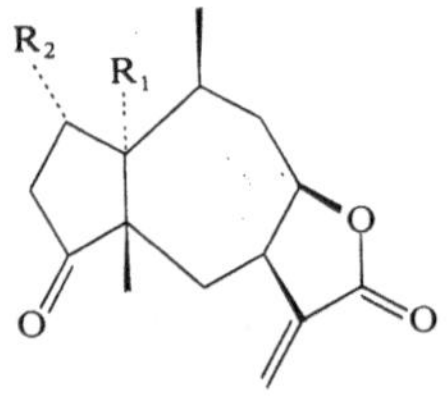

(853) Confertin; R$_1$ = R$_2$ = H
(854) Peruvin; R$_1$ = OH, R$_2$ = H
(855) Burrodin; R$_1$ = H, R$_2$ = OH

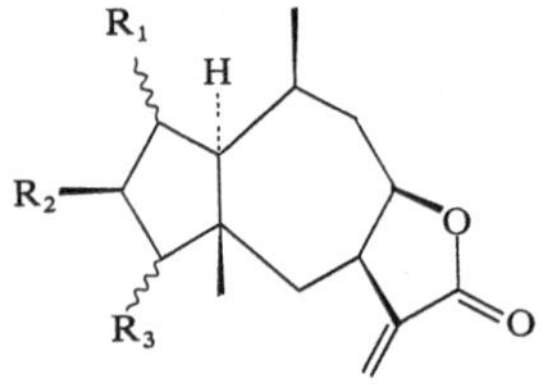

(856) Stevin; R$_1$ = OAc, R$_2$ = H, R$_3$ = OH
(857) Cumanin; R$_1$ = H, R$_2$ = OH, R$_3$ = β-OH
(858) Cumanin-3-acetate; R$_1$ = H, R$_2$ = OAc,
R$_3$ = β-OH
(859) Cumanin diacetate; R$_1$ = H, R$_2$ = OAc, R$_3$ =
β-OAc
(860) Confertin, 4-α-H; R$_1$ = R$_2$ = H, R$_3$ = α-OH

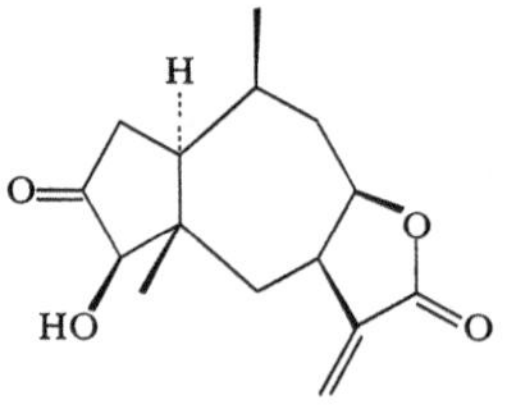

(861) Peruvinin

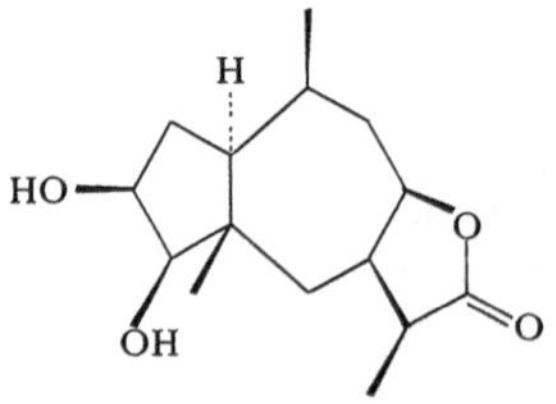

(862) Cumanin, dihydro

(863) Psilostachyin; R = α-OH
(864) Cordilin; R = β-OH

(865) Psilostachyin C

(866) Psilostachyin B

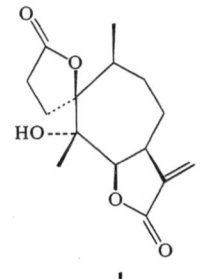

(867) Canambrin

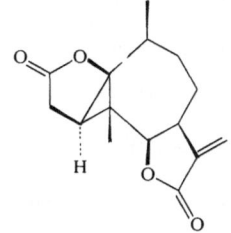

(868) Confertdiolide

Chart VII-1 (continued)

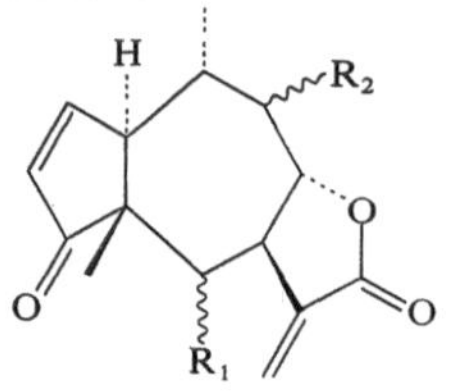

(**869**) Aromaticin; $R_1 = R_2 = H$
(**870**) Gaillardipinnatin, desacetyl; $R_1 = \alpha\text{-OH}$,
 $R_2 = \beta\text{-OAc}$
(**871**) Mexicanin I; $R_1 = \beta\text{-OH}$, $R_2 = H$
(**872**) Amblyodin; $R_1 = \beta\text{-OH}$, $R_2 = \beta\text{-OAc}$
(**873**) Gaillardipinnatin; $R_1 = \alpha\text{-OAc}$, $R_2 = \beta\text{-OAc}$
(**874**) Multigilin; $R_1 = \alpha\text{-OAng}$, $R_2 = \beta\text{-OH}$
(**875**) Fastigilin C; $R_1 = \alpha\text{-OSen}$, $R_2 = \beta\text{-OH}$
(**876**) Bigelovin; $R_1 = \alpha\text{-OAc}$, $R_2 = H$

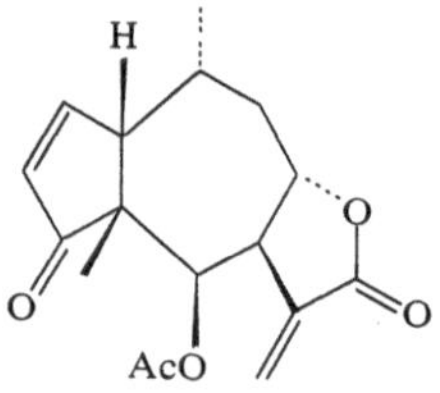

(**877**) Linifolin A

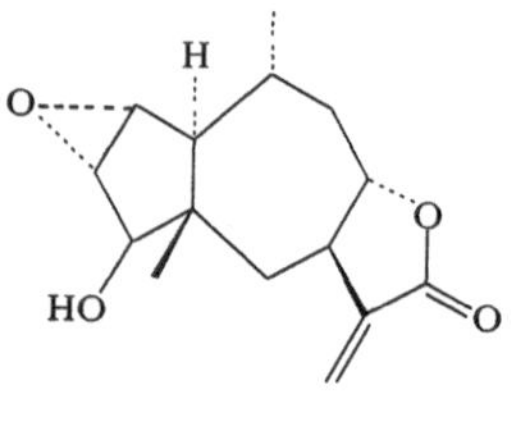

(**878**) Amarilin

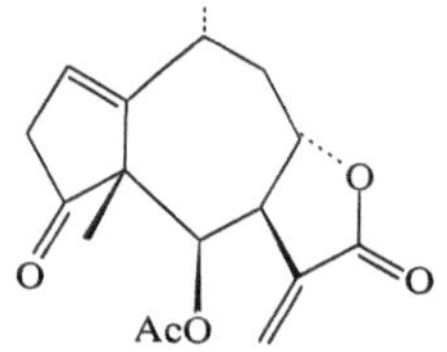

(**879**) Linifolin B

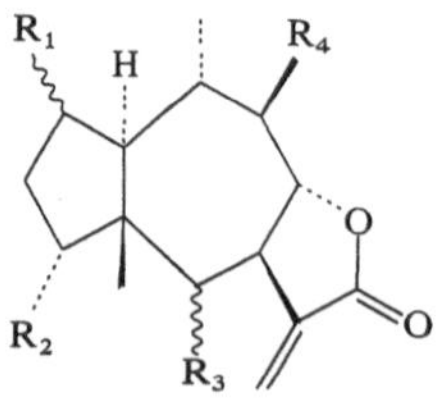

(**880**) Pulchellin; $R_1 = \alpha\text{-OH}$, $R_2 = OH$, $R_3 = H$,
 $R_4 = H$
(**881**) Flexuosin A; $R_1 = \beta\text{-OH}$, $R_2 = OH$,
 $R_3 = \beta\text{-OAc}$, $R_4 = H$
(**882**) Spathulin; $R_1 = \alpha\text{-OH}$, $R_2 = OH$,
 $R_3 = \beta\text{-OAc}$, $R_4 = \beta\text{-OAc}$
(**883**) Alternilin; $R_1 = \beta\text{-OAc}$, $R_2 = OH$,
 $R_3 = \alpha\text{-OH}$, $R_4 = H$
(**884**) Flexuosin A, 2-acetate; $R_1 = \beta\text{-OAc}$, $R_2 = OH$,
 $R_3 = \alpha\text{-OAc}$, $R_4 = H$

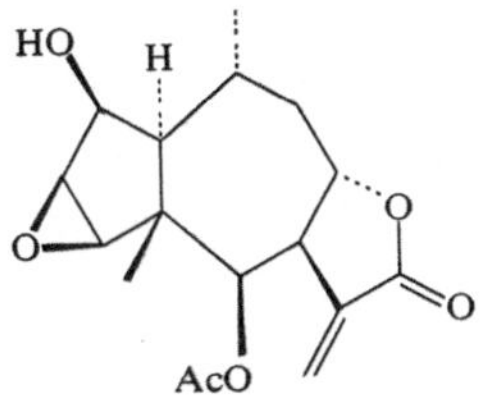

(885) Gaillardilin

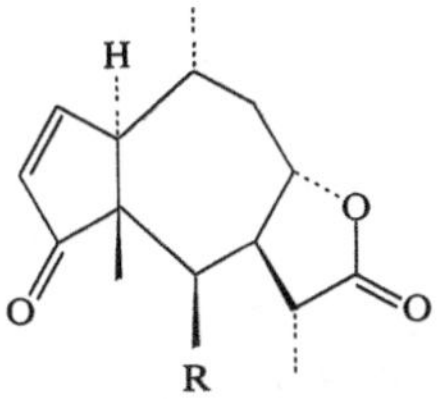

(886) Tenulin, iso, desacetyl; R = OH
(887) Tenulin, iso; R = OAc
(888) Thurberilin; R = OAng

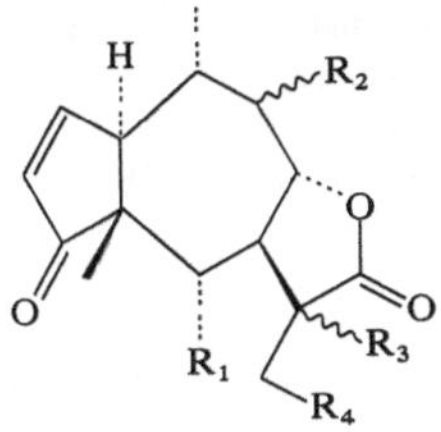

(889) Fastigilin A; R_1 = OAng, R_2 = β-OH,
R_3 = α-H, R_4 = H
(889a) Fastigilin B; R_1 = OSen, R_2 = OH, R_3 = α-H,
R_4 = H
(889b) Radiatin, R_1 = OMac, R_2 = β-OH, R_3 = α-H,
R_4 = H
(890) Amblyodiol; R_1 = H, R_2 = β-OAc,
R_3 = R_4 = OH

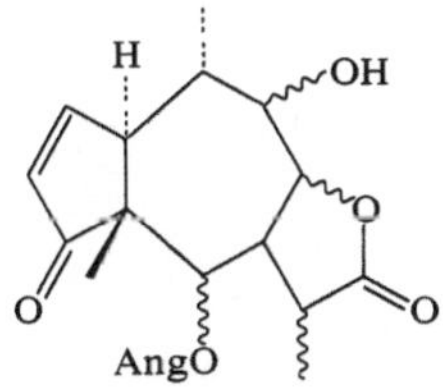

(891) Baileyolin

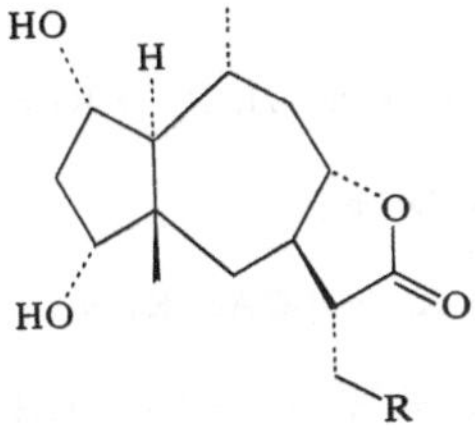

(892) Pulchellidine; R = **A** = —N⟨ ⟩

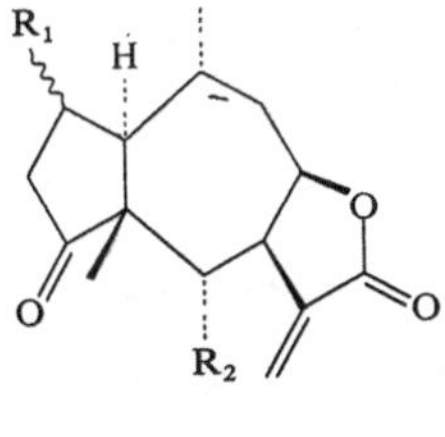

(893) Arnifolin; R_1 = α-OH, R_2 = OAng
(894) Helenalin, dihydro, 2-methoxy; R_1 = β-OCH$_3$,
R_2 = OH
(895) Paucin; R_1 = α-O**B**, R_2 = H

B =

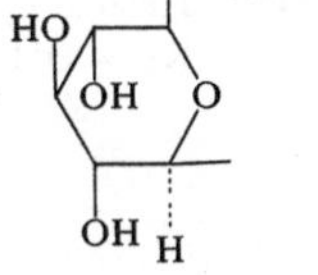

Chart VII-1 (continued)

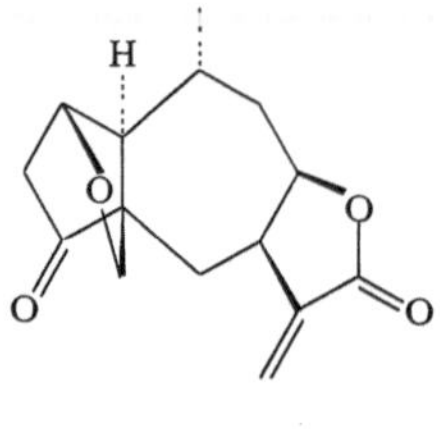

(**896**) Microhelenin A

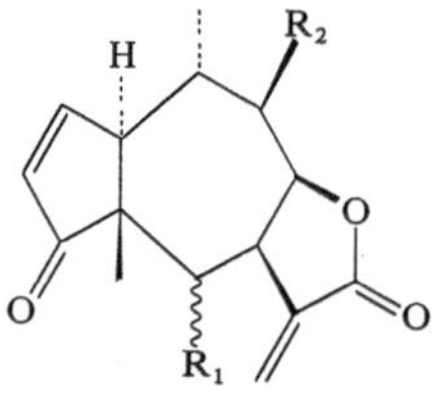

(**897**) Aromatin; $R_1 = R_2 = H$
(**898**) Helenalin; $R_1 = \alpha$-OH, $R_2 = H$
(**898a**) Helenalin, iso; 7,11 d.b. instead of 11,13 d.b.
(**899**) Angustibalin; $R_1 = \alpha$-OAc, $R_2 = H$
(**900**) Balduilin; $R_1 = \beta$-OAc, $R_2 = H$
(**901**) Linearifolin A; $R_1 = \alpha$-OTig, $R_2 = OH$

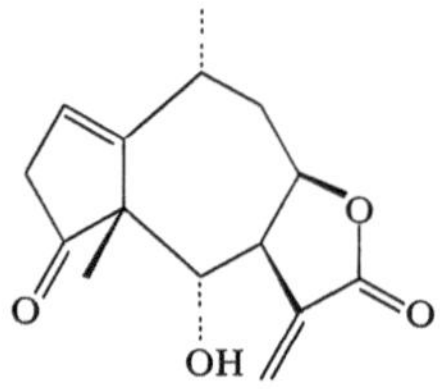

(**902**) Mexicanin A

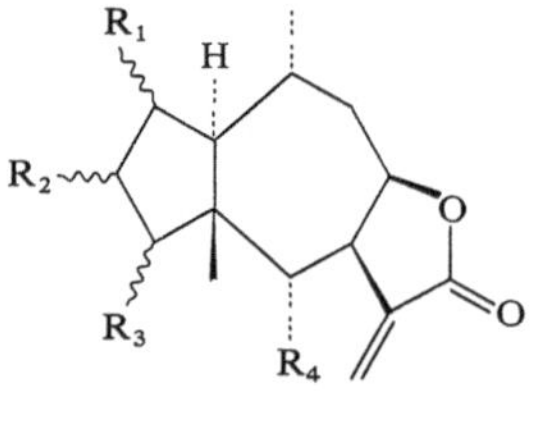

(**903**) Pulchellin, neo; $R_1 = R_3 = \alpha$-OH, $R_2 = R_4 = $ =H
(**904**) Picrohelenin; $R_1 = \alpha$-OAc, $R_2 = H$, $R_3 = \beta$-OH, $R_4 = OH$
(**905**) Hymenograndin; $R_1 = R_2 = \beta$-OAc, $R_3 = \alpha$-OH, $R_4 = H$
(**906**) Hymenoratin; $R_1 = R_4 = H$, $R_2 = R_3 = OH$,

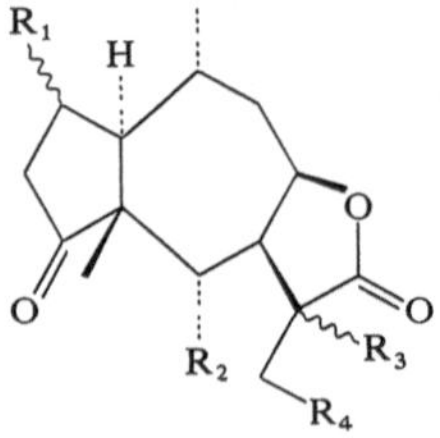

(**907**) Flexuosin B; $R_1 = OH$, $R_2 = OSen$, $R_3 = R_4 = H$
(**908**) Florigrandin; $R_1 = \alpha$-O-2-Mebut, $R_2 = H$, $R_3 = R_4 = OH$
(**909**) Helenalin, tetrahydro; $R_1 = R_4 = H$, $R_2 = OH$, $R_3 = \alpha$-H

(910) Plenolin; R = OH
(912) Arnicolide A; R = OAc
(913) Arnicolide D; R = OMac
(914) Arnicolide C; R = O-i-But
(915) Microhelenin B; R = O-2-Mebut
(916) Arnicolide B; R = O-i-Val
(917) Microhelenin C; R = OAng

(918) Mexicanin C; R = OH
(919) Brevilin A; R = OAng

(920) Hymenolane

(921) Pulchellidine, neo; R = —N⟨ ⟩

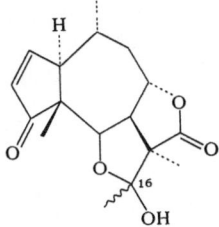

(922) Tenulin

Chart VII-1 (continued)

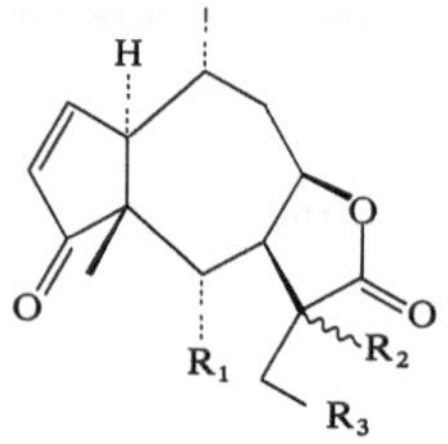

(923) Sulferalin; R_1 = OH, R_2 = H,, R_3 = SO_2CH_3
(924) Hymenoflorin; R_1 = H, R_2 = R_3 = OH

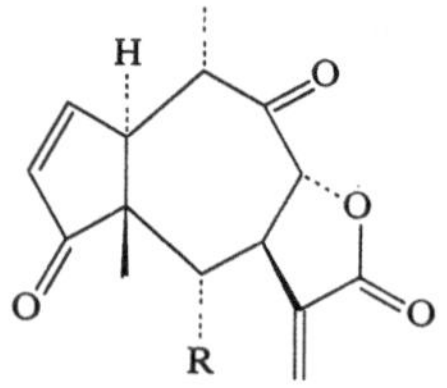

(925) Multiradiatin; R = OSen
(926) Multistatin; R = OAng

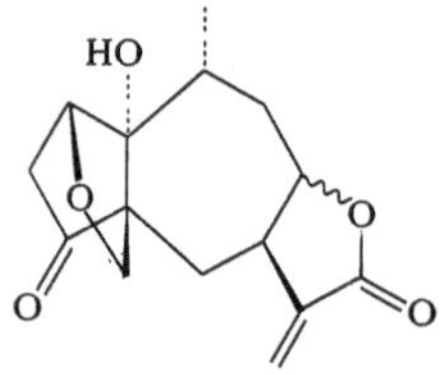

(927) Mexicanin H

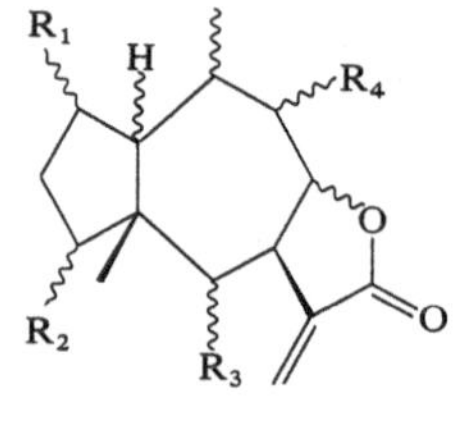

(928) Neoleonin; R_1 = R_2 = OH, R_3 = OAng,
R$_4$ = OAc
(929) Brittanin; R_1 = R_3 = OAc, R_2 = OH, R_4 = H

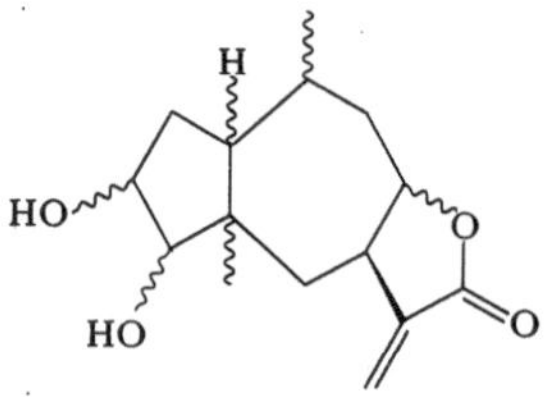

(930) Geigerinin

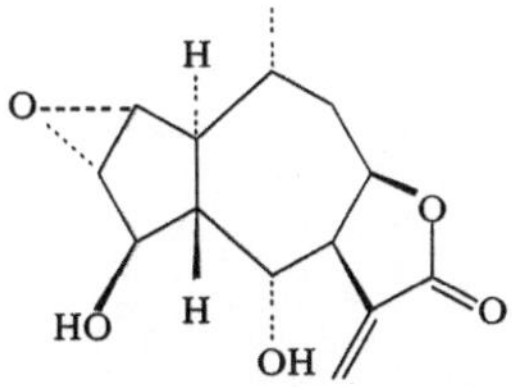

(931) Autumnolide

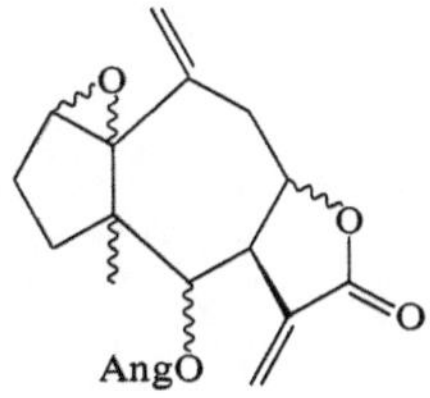

(932) Badkhysinin

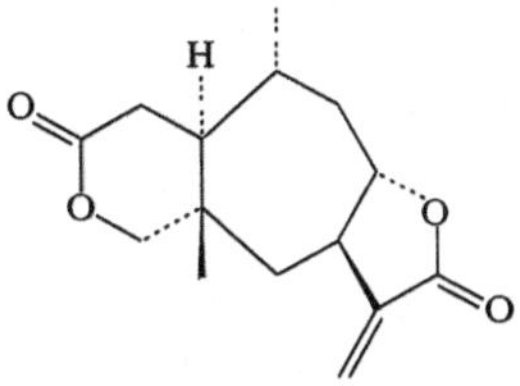

(933) Vermeerin

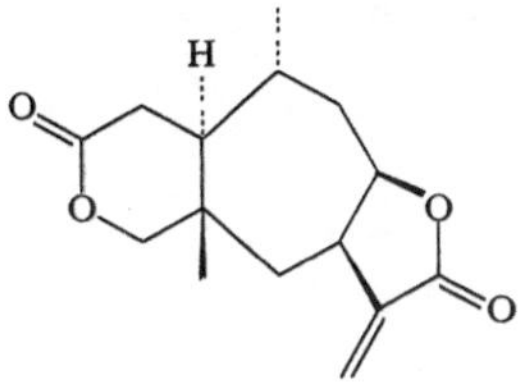

(934) Psilotropin

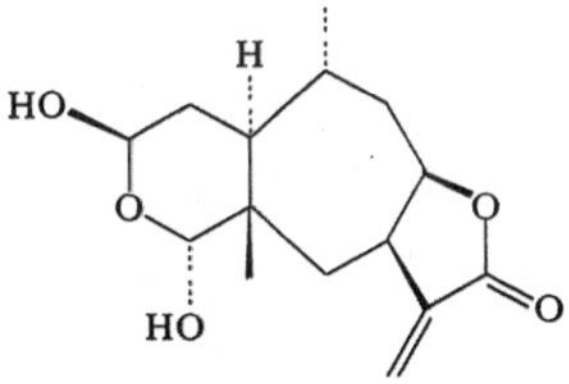

(936) Hymenoxon
(937) Hymenovin

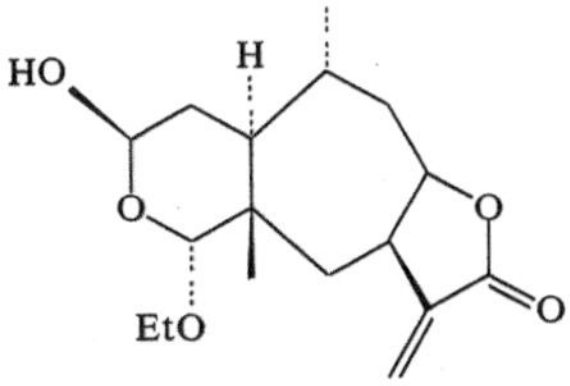

(938) Hymenolide

Chart VII-1 (continued)

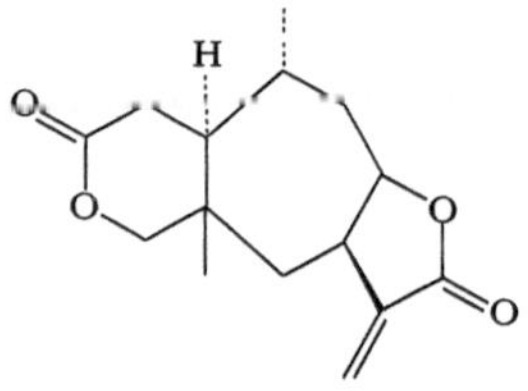

(939) Greenein

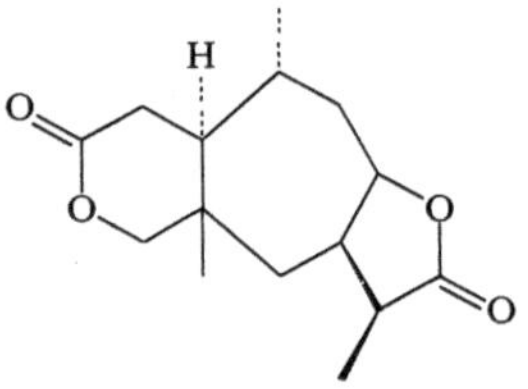

(940) Themoidin

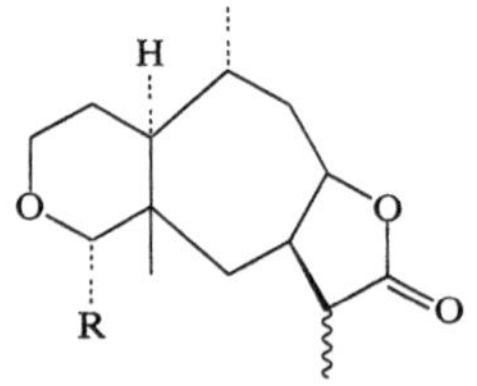

(942) Hymenoxynin; R = OA

A =

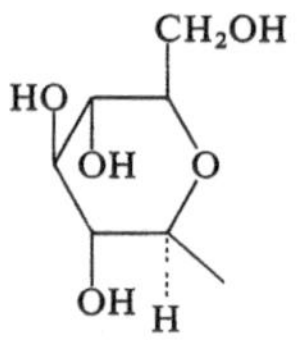

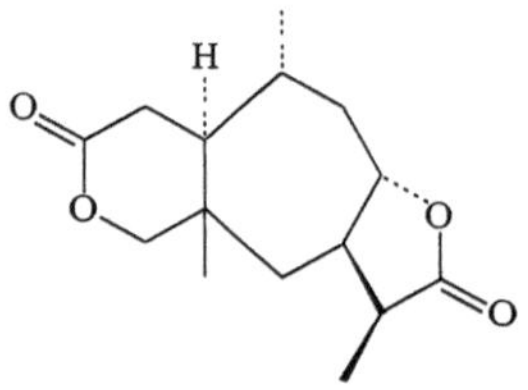

(943) Anthemoidin

(945) Microlenin

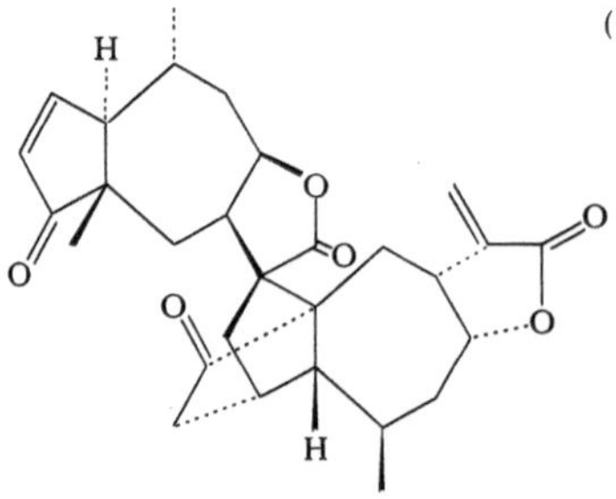

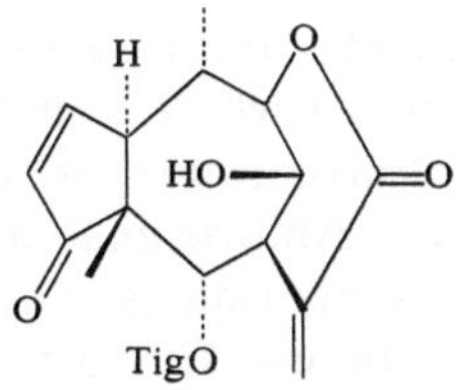

(946) Linearifolin B

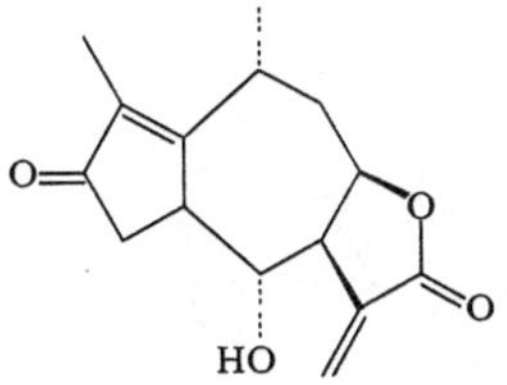

(947) Helenalin, neo

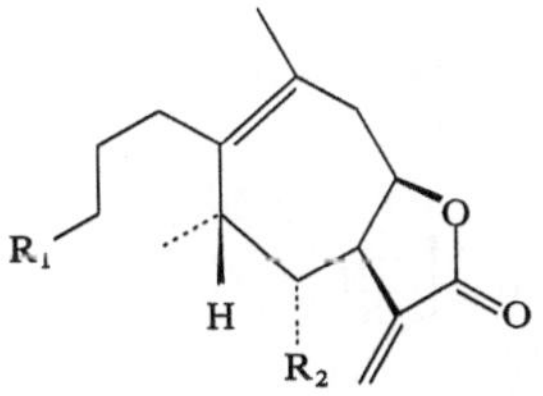

(949) Inulicin, desacetyl; $R_1 = R_2 = OH$
(950) Inulicin; $R_1 = OAc$, $R_2 = OH$

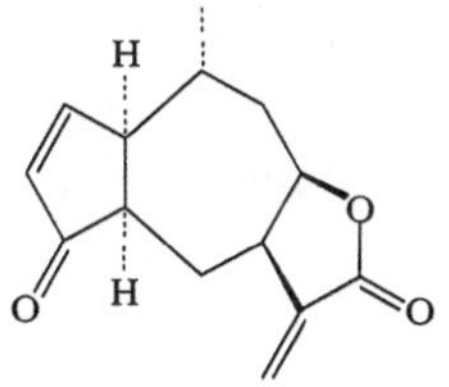

(951) Mexicanin E

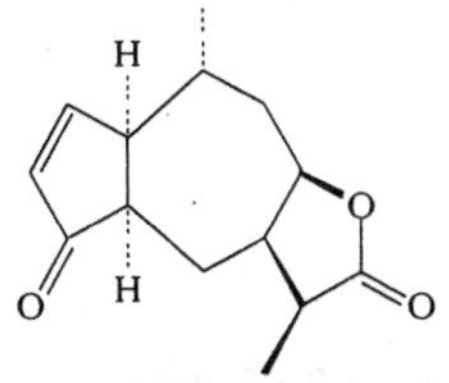

(952) Mexicanin E, dihydro

which possess a *trans*-fusion of the 5,7-ring system. The ambrosanolide skeleton (953) typically contains a C-10β methyl group whereas in the helenanolides (954) the C-10 methyl group is α. Furthermore, in the ambrosanolides, which usually occur in the subtribe Ambrosiinae and the genus *Parthenium* of the Compositae, the lactone ring is closed predominantly toward C-6 and then exclusively with the C-6 oxygen function β. In the helenanolides, which commonly occur in the tribe Heleniae, the lactone ring is invariably closed toward C-8 with the C-8 oxygen bond oriented either α or β.

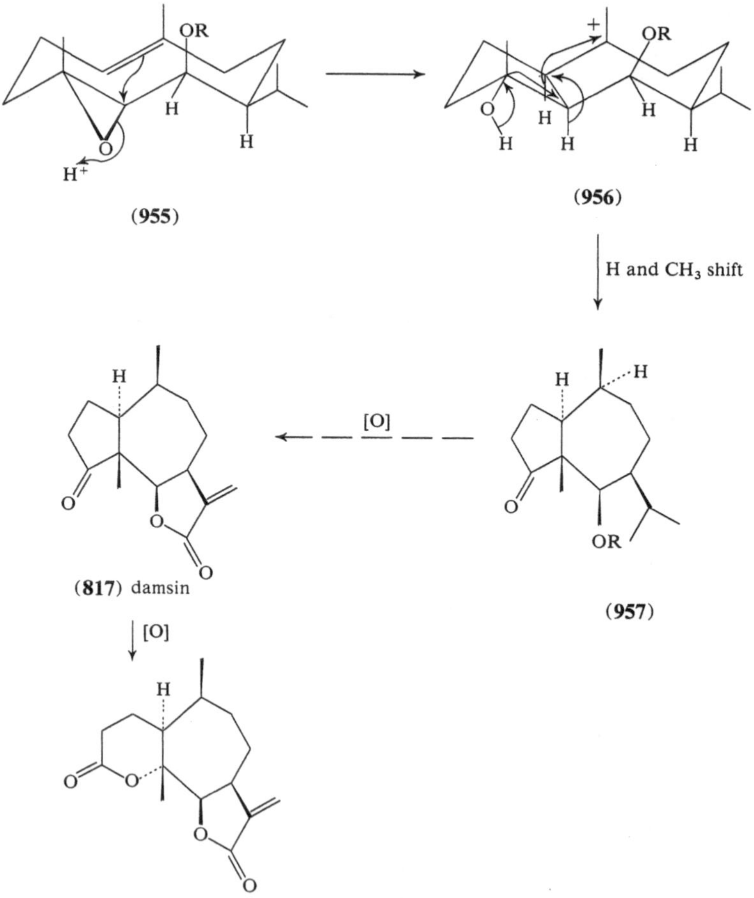

(955)   (956)

H and CH₃ shift

(817) damsin   [O]   (957)

[O]

(865) psilostachyin C

Chart VII-2. Biogenesis of ambrosanolides and psilostachyanolides

The biogenesis of the ambrosanolides and their *seco*-derivatives, the psilostachyanolides, can be envisioned as shown in Chart VII-2 (*265, 288, 393, 395*). The germacrolide-4,5-epoxide (**955**) will initially undergo Markovnikov cyclization to the guaianolide type cation (**956**) which upon double hydride and methyl shift, as indicated by the arrows, gives the ambrosanolide skeleton (**957**) from which the 7,6-*cis*-lactone damsin (**817**) is formed. Subsequent biological Baeyer-Villiger oxidation, most likely *via* a hydroperoxide intermediate, gives the dilactone psilostachyin C (**865**). Peracetic acid oxidation of damsin which proceeds highly stereospecific to give psilostachyin was used as a key step in the structure elucidation of the naturally occurring dilactone (*549*). The interesting fact that all ambrosanolide 7,6-lactones represent *cis*-lactones has led to the suggestion that possibly a C-6-β-oxygen is assisting the rearrangement step (**956**) to (**957**) (*265*). Intramolecular frontside stabilization of the cationic center at C-10 by the C-6 oxygen would allow a C-1 to C-10 hydride shift to occur before competing elimination or nucleophilic attack at C-10 might occur to form a guaianolide.

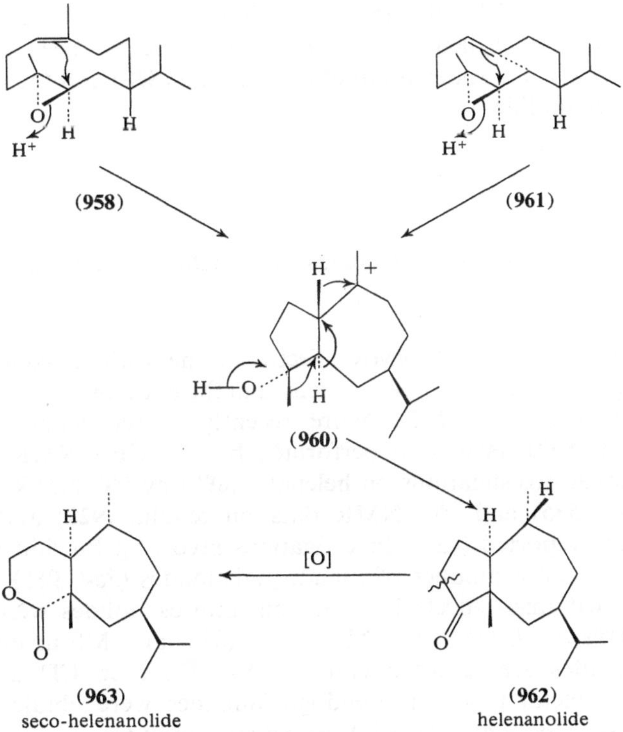

Chart VII-3. Biogenesis of helenanolides

A plausible suggestion for the biogenesis of helenanolides, shown in Chart VII 3, was presented by Herz (*393, 395*). Acid-induced cyclization of the melampolide-4,5-epoxide (**958**) would give cation (**960**) from which by the indicated shifts the skeleton (**962**) results, showing a stereochemical arrangement typical for most naturally occurring helenanolides. Oxidative modification between C-3 and C-4 of the cyclopentanone ring of (**962**) gives the skeleton typical of the *seco*-helenanolides (**963**) typical representatives being vermeerin (**933**) and greenein (**939**). An alternative route to the helenanolide intermediate (**960**) could involve a germacrolide 4,5-epoxide precursor which from conformation (**961**) would provide the same cyclized skeleton (**960**) which is formed *via* the melampolide route (*265, 273*). Helenalin (**898**) as well as many other helenanolides contain a cyclopentenone ring for which C-2 hydroxylated germacranolides would be logical precursors.

The biogenesis of the norsesquiterpene lactones mexicanin E (**951**) and derivatives proceeds most likely *via* helenanolide precursors since they cooccur with C-15 oxygenated helenanolides which could be transformed to the nor-compounds by decarboxylation and/or by a retroaldol process of the C-15 alcohol (*393*). The isolation of the first dimeric helenanolide, microlenin (**945**) (*649*), has recently been described. It represents a Diels-Alder product of helenalin (**898**) and the enol form of a norhelenanolide.

2. Physical Methods of Structure Elucidation

A large number of NMR spectra of the various skeletal types of pseudoguaianolides has been presented in the book by Yoshioka, Mabry, and Timmermann (*1172*). More recently, stereochemical studies on spathulin (**882**) have been performed by 270 MHz NMR (*444*). Conformational investigations on helenalin (**898**) by ^{13}C NMR were carried out (*535, 536*) and ^{13}C NMR data on tenulin (**922**) and derivatives have been reported (*437*). Investigations involving Eu-shift studies were carried out on a number of pseudoguaianolides (*364, 985*).

Detailed mass spectral studies on ambrosanolides were performed by Cortes *et al.* (*186*) and Matsueda (*688*) and MS-investigations on helenanolides are reported (*1101, 1186*). Data on CD and ORD of various skeletal types of pseudoguaianolides were obtained by Herz *et al.* (*396, 399, 420, 428, 432*). Pseudoguaianolides whose structure has been established by X-ray crystallography are summarized in Table VII-1.

Table VII-1. *Pseudoguaianolides of Known X-Ray Structure*

Structure Number	Compound	Structure Type	References	Comments
(830)	Ambrosin	PA	(251)	As monobromide
(931)	Autumnolide	PH	(235a)	
(898)	Helenalin	PH	(706, 711, 801)	As monobromide and as oxide
(920)	Hymenolane	PH	(830)	
(936)	Hymenoxon	PH	(830)	
(951)	Mexicanin E	PH	(1104)	As monobromide
(945)	Microlenin	PH	(649)	
(895)	Paucin	PH	(191)	Monohydrate
(910)	Plenolin	PH	(647, 709)	As p-iodobenzoate
(880)	Pulchellin	PH	(27, 956)	As dibromide and also as 3-bromo-anhydrodehydrodihydro-derivative
(889b)	Radiatin	PH	(248a)	
(882)	Spathulin	PH	(497)	As diacetate
(922)	Tenulin	PH	(869, 1105)	Bromoisotenulin examined

3. Selected Chemical Transformations

Chart VII-4 lists a number of elimination reactions which are presented as examples showing the use of simple laboratory modifications in structure elucidation of closely related ambrosanolides. Treatment of the 3-tosyl derivative of ambrosiol with dilute sulfuric acid resulted in damsin, a compound with known stereochemistry, thus establishing five chiral centers in ambrosiol (840) (666). The β-hydroxyl group of bipinnatin (825) was eliminated under mild acylation conditions to provide ambrosin (830) (866). Acid-mediated eliminations of the epimeric angular C-1 hydroxyls in parthenin (831) and hymenin (832) provided the same doubly conjugated ketone (965) thus allowing correlation between hymenin and the known parthenin (1092).

Treatment of helenalin (898) or ambrosin (830) with mineral acids resulted in epimerization of the C-1 position (432, 903). As shown at the top of Chart VII-5, helenalin (898), upon treatment with HCl in methanol, gave 1-epiallohelenalin (966) which is the product of C-1 epimerization and relactonization to a 7,6-*trans*-lactone (903). Acid treatment of helenalin under different conditions afforded different products, mexicanin A (902) (432) and the ether (968) (639), a compound whose structure was established by X-ray crystallography (711). The authors interpreted the sequence of steps as follows: Initial formation

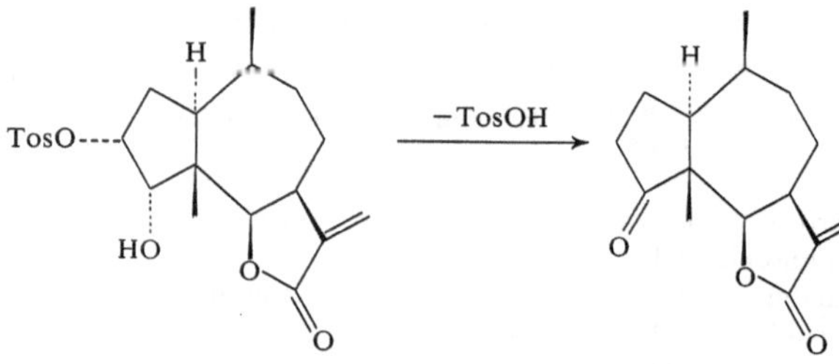

(964)   (817) damsin

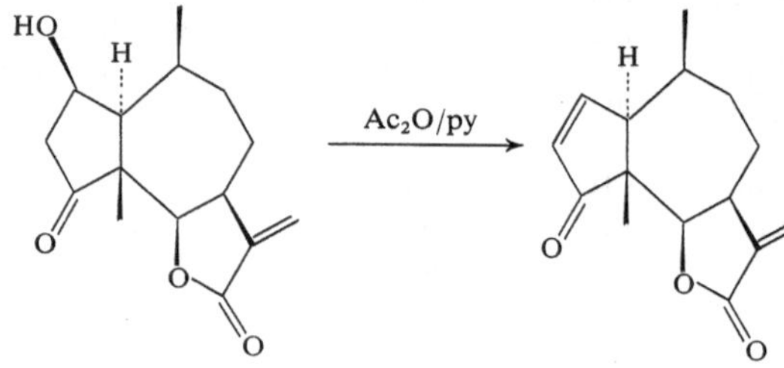

(825) bipinnatin   (830) ambrosin

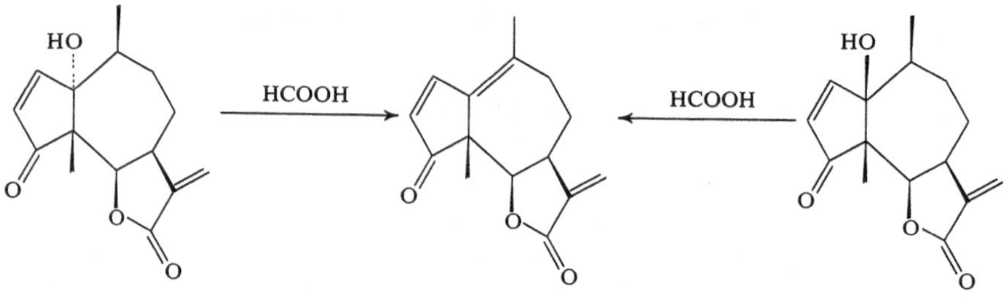

(831) parthenin   (965)   (832) hymenin

Chart VII-4. Structural correlations of ambrosanolides *via* elimination reactions

of the dienol (966a) and subsequent C-3 protonation of (966a) gives the nonconjugated cyclopentenone, mexicanin A (902). Protonation at C-1 results in the helenalin epimer (967) in which due to the spatial arrangement of the newly formed *cis*-fused ring system intramolecular Michael addition can occur. An abnormal reaction of mexicanin E (951) with N-bromosuccinimide has been reported by Romo and coworkers (763). Bromination at C-1 of (951) gives (969) which is postulated to undergo bromide attack at C-13 and subsequent intramolecular nucleophilic displacement (arrows) of C-11 at C-1 in (969) to give the tetracyclic bromo lactone (970).

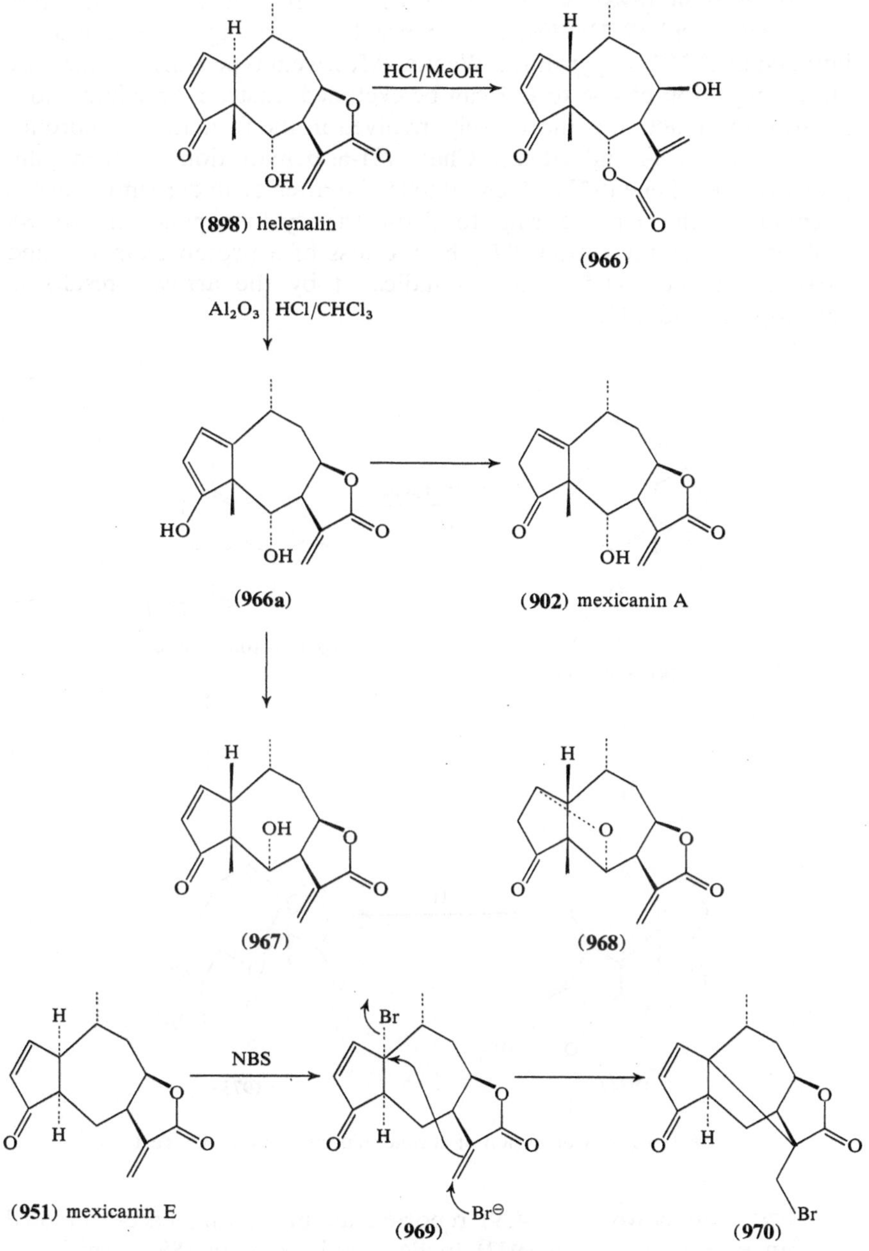

Chart VII-5. Reactions of helenalin and derivatives

Coronopilin (**818**), when heated with H_2SO_4 in acetic acid, gives coronopilic acid (**971**) (*309, 548*). Since this rearrangement cannot be initiated by $SOCl_2$ in pyridine a Wagner-Meerwein type rearrangement by direct migration of C-6 to C-1 can be excluded. Instead, an elimination-substitution reaction is more likely involved in the formation of corono-pilic acid (**971**). As indicated in Chart VII-6, dehydration of coronopilin provides the alkene (**972**) which attacks C-6 after or under simultaneous opening of the lactone ring to form the cyclopropane ring (**973**). Stabilization of the cation (**973**) by the loss of a proton from C-6 and cleavage of the C-1,5 bond, as indicated by the arrows, results in coronopilic acid (**971**).

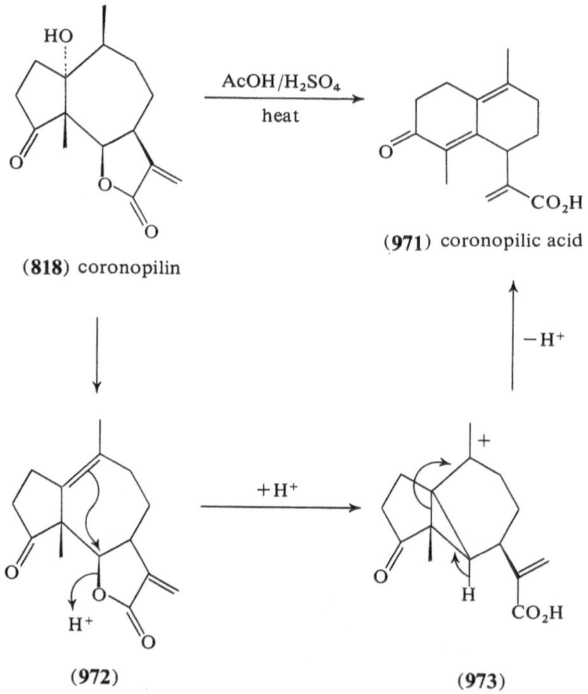

Chart VII-6. Conversion of coronopilin into coronopilic acid

Herz and coworkers (*429*) reported an interesting base-promoted rearrangement of tenulin (**922**) to desacetylisotenulin (**886**) which the authors interpreted as a reaction involving a retroaldol process. Chart VII-7 outlines the key steps of the analog retroaldol reaction leading

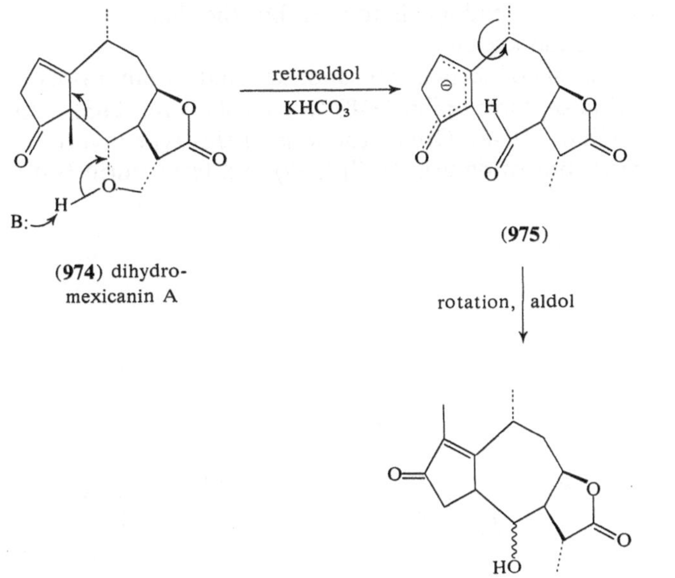

(974) dihydro-
mexicanin A

(975)

rotation, aldol

(976) dihydroneohelenalin

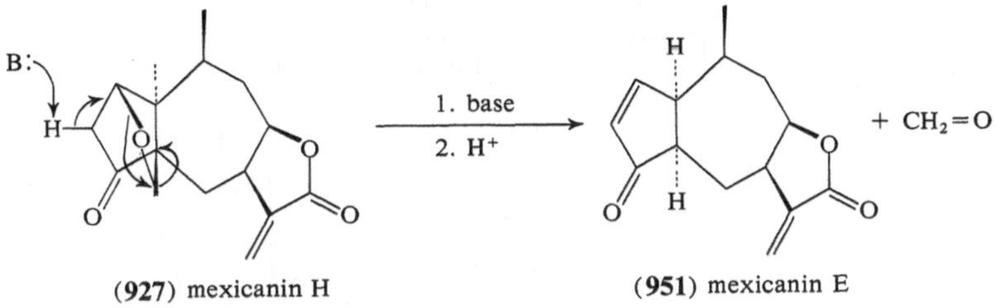

(927) mexicanin H

(951) mexicanin E

+ CH₂=O

Chart VII-7. Base-promoted reactions of pseudoguaianolides

from dihydromexicanin A (974) to dihydroneohelenalin (976) (432). Base-initiated opening of the seven-membered ring of (974) gives the resonance stabilized carbanion (975) which after rotation of the 5-membered ring aldolizes again to provide the rearranged 5,7-ring skeleton (976). An example for another extended retroaldol process of a β-hydroxy ketone is shown on the bottom of Chart VII-7. Treatment of mexicanin H (927) with base and subsequent acidification gives the norpseudoguaianolide mexicanin E (951) and formaldehyde, a fragmentation that also occurs upon electron impact in the mass spectrometer (883). The above reaction may be a model for the possible

16*

biogenesis of norpseudoguaianolides by the loss of C-15 from the pseudoguaianolide skeleton.

The biogenesis of *seco*-ambrosanolides finds a laboratory analogy in the conversion of coronopilin (**818**) to psilostachyin (**863**) with peracetic acid (*665*) (Chart VII-8). Other reactions of this type involve the Baeyer-Villiger oxidations of damsin (**817**) (*549*) and tetraneurin B (**819**) (*1177*).

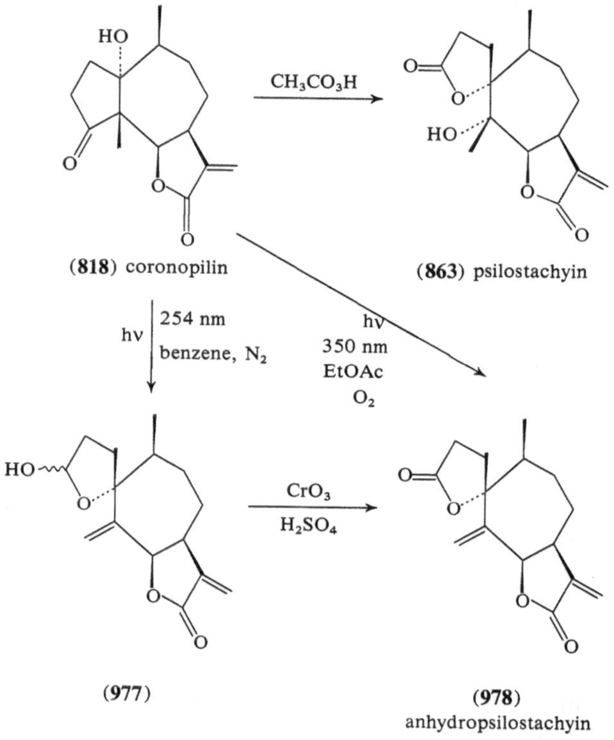

(**818**) coronopilin   (**863**) psilostachyin

(**977**)

(**978**)
anhydropsilostachyin

Chart VII-8. Chemical and photochemical conversion of coronopilin into *seco*-derivatives

Also shown in Chart VII-8 are photochemical conversions of coronopilin into the *seco*-ambrosanolides (**977**) and (**978**). Irradiation at 254 nm of compound (**818**) under nitrogen in benzene results in the spirolactol (977) (*1173*). Oxidation of compound (977) with CrO_3 provides anhydropsilostachyin (978). The latter compound is directly obtained by photolysis of coronopilin (818) at 350 nm in ethyl acetate in the presence of oxygen (*550*). The photochemistry of parthenin (831) is outlined in

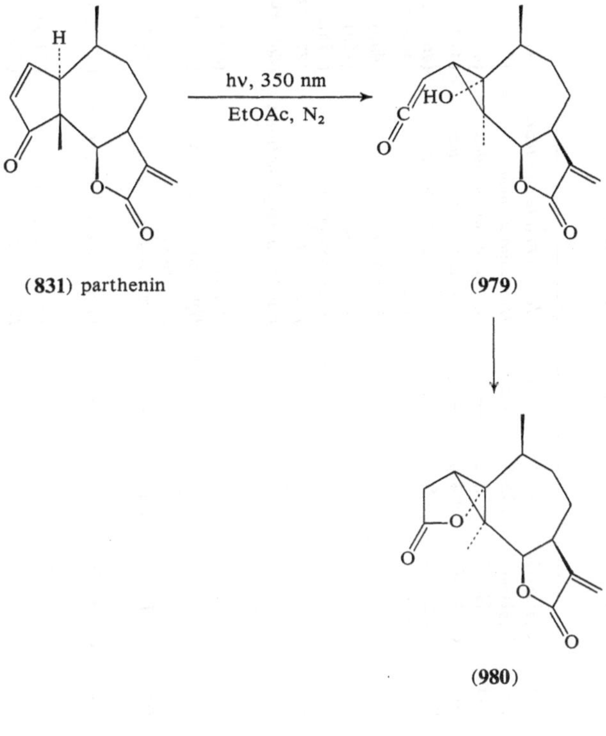

(831) parthenin   (979)

(980)

Chart VII-9. Photochemistry of parthenin

Chart VII-9. As indicated by the arrows, the photolytic rearrangement involves the intermediate formation of a cyclopropane ketene (979) which reacts with the C-1 hydroxyl to give the dilactone (980) (*550*). A similar reaction is reported for the conversion of the C-1 epimer of parthenin, hymenin (832), to the dilactone confertdiolide (868) (*902*).

Table VII-2. *Naturally Occurring Pseudoguaianolides and Seco-Derivatives*

Structure Number	Name of Compound	Type[c]	Formula	m.p. °C	$[\alpha]_D$	Plant Source[d]	References	Comments
(883)	Alternilin	PH	$C_{17}H_{24}O_6$	193–5	+11	*Helenium alternifolium*	(410, 448)	
(878)	Amarilin	PH	$C_{15}H_{20}O_4$	195–8	+5	*Helenium amarum*	(658)	
(872)	Amblyodin	PH	$C_{17}H_{20}O_6$	220–2	—	*Gaillardia amblyodon*	(443, 442)	
(890)	Amblyodiol	PH	$C_{17}H_{22}O_7$	204–6	−50	*Gaillardia amblyodon*	(442)	
(830)	Ambrosin	PA	$C_{15}H_{18}O_3$	146	−154	*Ambrosia maritima*	(461, 423[a], 359)	
						A. cumanensis	(398, 130)	
						A. hispida	(453)	
						A. "jamaicensis"	(463)	
						A. psilostachya	(718a)	
						Iva xanthifolia	(777, 778)	
						Parthenium bipinnatifidum	(860, 893)	Source misidentified in ref. (893)
						P. incanum	(896)	
						Hymenoclea monogyra	(1091)	
						H. salsola	(308)	
(835)	Ambrosin, 2,3-H-2,3-epoxy	PA	$C_{15}H_{18}O_4$	oil	—	*Ambrosia cumanensis*	(130)	
(836)	Ambrosin, neo	PA	$C_{15}H_{18}O_3$	126–7	−66	*Parthenium bipinnatifidum*	(903, 308, 866, 860)	Reported before Coronopilin, anhydro
						Hymenoclea monogyra	(1091)	
						H. salsola	(308)	
						Ambrosia cumanensis	(130)	
(847)	Ambrosin, tetrahydro	PA	$C_{15}H_{22}O_3$	—	—	*Ambrosia peruviana*	(394, 423)	
(840)	Ambrosiol	PA	$C_{15}H_{22}O_4$	116–7	−111	*Ambrosia psilostachya*	(666, 463)	
						A. dumosa	(304)	
(899)	Angustibalin	PH	$C_{17}H_{20}O_5$	181	−54	*Balduina angustifolia*	(636)	Helenalin acetate

Ref	Name		Formula	m.p.	[α]	Source	Ref	Comments
(943)	Anthemoidin	PH	$C_{15}H_{22}O_4$	220-1	−115	Hymenoxys anthemoides	(401)	3,4-Seco-dilactone; see Vermeerin B, dihydro
(838)	Apoludin	PA	$C_{15}H_{22}O_4$	133-5	−42	Ambrosia dumosa	(1092, 304)	
(912)	Arnicolide A	PH	$C_{17}H_{22}O_5$	177-8	−112	Arnica montana	(836, 9)	
(916)	Arnicolide B	PH	$C_{20}H_{28}O_5$	116-7	−95	Arnica montana	(836)	
(914)	Arnicolide C	PH	$C_{19}H_{26}O_5$	137-8	−91	Arnica montana	(836)	
(913)	Arnicolide D	PH	$C_{19}H_{24}O_5$	100-2	+7	Arnica montana	(836)	
(893)	Arnifolin	PH	$C_{20}H_{26}O_6$	128-137	+52	Arnica foliosa	(922, 260, 985, 1186, 254)	NMR: H-3' (6.8 δ) indicates tiglate side chain
						A. montana		
(869)	Aromaticin	PH	$C_{15}H_{18}O_3$	232-4	+18	Helenium aromaticum	(873)	
						H. amarum	(658)	
(897)	Aromatin	PH	$C_{15}H_{18}O_3$	159-6	−6	Helenium aromaticum	(873)	
(931)	Autumnolide	PH	$C_{15}H_{20}O_5$	188-90	+20	Helenium autumnale	(448, 235a[b])	
(932)	Badkhysinin	P	$C_{19}H_{21}O_5$	—	—	Ferula grigoriewii	(579)	
(891)	Baileyolin	PH	$C_{20}H_{26}O_6$	251-3	—	Baileya pleniradiata	(1123)	Possibly a guaianolide
						B. multiradiata	(1123, 93)	
(900)	Balduilin	PH	$C_{17}H_{20}O_5$	231-2	+57	Balduina uniflora	(431)	
(876)	Bigelovin	PH	$C_{17}H_{20}O_5$	190-1	+46	Helenium bigelovii	(817, 419[a])	
						Gaillardia pinnatifida	(427)	
(825)	Bipinnatin	PA	$C_{15}H_{20}O_4$	196-8	−9	Parthenium bipinnatifidum	(866, 860)	
(919)	Brevilin A	PH	$C_{20}H_{26}O_5$	116-127	—	Helenium brevifolium	(421, 410, 447)	
						H. alternifolium		
(929)	Brittannin	P	$C_{19}H_{26}O_7$	192-4	−26	Inula britannica	(163, 921)	
(855)	Burrodin	PA	$C_{15}H_{20}O_4$	167-8	+108	Ambrosia dumosa	(304)	
(867)	Canambrin	PA	$C_{15}H_{20}O_5$	209-210	−134	Ambrosia canescens	(878)	
(852)	Carpesiolin	PA	$C_{15}H_{20}O_4$	122-3	+115	Carpesium abrotanoides	(676)	
(823)	Chiapin A	PA	$C_{19}H_{26}O_5$	120-1	—	Parthenium fruticosum	(866, 865)	
(821)	Chiapin B	PA	$C_{19}H_{26}O_6$	156-7	—	Parthenium fruticosum	(866, 865)	
(829)	Conchosin A	PA	$C_{15}H_{18}O_5$	150-2	−29	Parthenium confertum	(892)	Seco-dilactone
(833)	Conchosin B	PA	$C_{17}H_{20}O_4$	143-4	—	Parthenium confertum	(892)	

[a] Reference giving structural revision; [b] Reference presenting X-ray data; [c] P = pseudoguaianolides of undetermined or special structure, PA = ambrosanolide, PH = helenanolide; [d] When plant sources are from a family other than Compositae, the family name is listed under Comments.

Table VII-2 (continued)

Structure Number	Name of Compound	Type[c]	Formula	m.p. °C	$[\alpha]_D$	Plant Source[d]	References	Comments
(868)	Confertdiolide	PA	$C_{15}H_{18}O_4$	156–8	–95	Parthenium confertum	(902)	Modified ambrosanolide
(828)	Confertiflorin	PA	$C_{17}H_{22}O_5$	144–5	+25	Ambrosia confertiflora	(267, 463)	
						A. acanthicarpa	(298)	
(827)	Confertiflorin, desacetyl	PA	$C_{15}H_{20}O_4$	202–4	+17	Ambrosia confertiflora	(267)	
						A. acanthicarpa	(298)	
						A. pumila	(463)	
(853)	Confertin	PA	$C_{15}H_{20}O_3$	145–6	—	Ambrosia confertiflora	(889, 672)	Source misidentified as A. tenuifolia in ref. (889)
(860)	Confertin, 4α-H	PA	$C_{15}H_{22}O_3$	147	+83	Parthenium schottii	(865, 860)	4 α-H or 4 β-H?
(864)	Cordilin	PA	$C_{15}H_{20}O_5$	210–1	–100	Inula helenium	(111)	Seco-dilactone
(818)	Coronopilin	PA	$C_{15}H_{20}O_4$	177–8	–30	Ambrosia cordifolia	(428)	
						Ambrosia psilostachya	(461, 412, 550, 1173, 548)	
						A. dumosa Gray	(304)	
						A. arborescens	(398)	
						A. artemisiifolia	(412)	
						Parthenium schottii	(865, 860)	
						P. incanum	(860, 896)	
						Hymenoclea salsola	(1092)	
						Iva acerosa	(264a)	
						I. nevadensis	(264a)	
						I. xanthifolia syn.	(778)	
						Cyclochaena xanthifolia		
(836)	Coronopilin, anhydro	PA				Cyclachaena xanthifolia	(777, 778)	See Ambrosin, neo
						Parthenium incanum	(896)	

No.	Compound	Type	Formula	mp	$[\alpha]$	Species	Refs.	Notes
(848)	Coronopilin, dihydro	PA				Hymenoclea salsola	(1092)	See Hymenolin
(857)	Cumanin	PA	$C_{15}H_{22}O_4$	120	+16	Ambrosia artemisiifolia	(838)	
						A. psilostachya	(875, 298)	Source misidentified as A. cumanensis in ref. (875)
(858)	Cumanin-3-acetate	PA	$C_{17}H_{24}O_5$	140–3	+147	Ambrosia psilostachya	(298)	
(859)	Cumanin diacetate	PA	$C_{19}H_{26}O_6$	109	+64	Ambrosia psilostachya	(298, 875)	
(862)	Cumanin, dihydro	PA	$C_{15}H_{24}O_4$	179–81	+77	Ambrosia artemisiifolia	(838, 875)	
(817)	Damsin	PA	$C_{15}H_{20}O_3$	111	–72	Ambrosia maritima	(6, 1022, 213, 359, 612)	
						A. cumanensis	(398, 130)	
						A. hispida	(453)	
						A. ambrosioides	(889, 232)	
						A. arborescens	(398)	
						A. chenopodiifolia	(398, 463)	
						A. deltoidea	(549, 463)	
						A. "jamaicensis"	(463)	
						Parthenium bipinnatifidum	(866, 860)	
(826)	Damsin, 3-hydroxy	PA	$C_{15}H_{20}O_4$	142–5	+27	Ambrosia psilostachya	(718)	
(889)	Fastigilin A	PH	$C_{20}H_{26}O_6$	175–7	–81	Gaillardia fastigiata	(427)	
						Baileya multiradiata	(831b)	
(889 a)	Fastigilin B	PH	$C_{20}H_{26}O_6$	259–61	—	Gaillardia fastigiata	(427)	
						Baileya multiradiata	(831a, 831b)	
(875)	Fastigilin C	PH	$C_{20}H_{24}O_6$	197–9	–85	Gaillardia fastigiata	(427)	
						Baileya multiradiata	(1123, 831a)	
						Hymenoxys acaulis	(399)	
(881)	Flexuosin A	PH	$C_{17}H_{24}O_6$	220–1	+12	Helenium flexuosum	(414b, 418, 448)	
						H. autumnale	(446)	
						Gaillardia parryi	(425)	
(884)	Flexuosin A, 2-acetate	PH	$C_{19}H_{26}O_7$	124–6	—	Helenium autumnale	(448)	
(907)	Flexuosin B	PH	$C_{20}H_{28}O_6$ + H_2O	132–7	+44	Helenium flexuosum	(414b, 418)	
(934)	Floribundin	PH					(401)	See Psilotropin 3,4-Seco-dilactone;
(940)	Floribundin A, dihydro	PH					(401)	see Themoidin

Table VII-2 (continued)

Structure Number	Name of Compound	Type[c]	Formula	m.p. °C	[α]D	Plant Source[d]	References	Comments
(908)	Florigrandin	PH	$C_{20}H_{30}O_7$	173–5	+187	Hymenoxys grandiflora	(400)	
(849)	Franserin	PA	$C_{15}H_{22}O_4$	225–7	±0	Ambrosia ambrosioides	(889)	
(885)	Gaillardilin	PH	$C_{17}H_{22}O_6$	197–9	−2	Gaillardia pinnatifida	(426)	
(873)	Gaillardipinnatin	PH	$C_{19}H_{22}O_9$	270	—	Gaillardia pinnatifida	(427, 442)	
						G. amblyodon	(443)	
(870)	Gaillardipinnatin, desacetyl	PH	$C_{17}H_{20}O_6$	167–70	−523	Gaillardia amblyodon	(442)	
(930)	Geigerinin	P	$C_{15}H_{22}O_4$	202–3	−10	Geigeria aspera	(246b)	3,4-Seco-dilactone
						G. africana syn G. filifolia	(22)	
(939)	Greenein	PH	$C_{15}H_{20}O_4$	175–6	+114	Hymenoxys greenei	(401)	
(898)	Helenalin	PH	$C_{15}H_{18}O_4$	169–72	−102	Helenium autumnale	(8, 432, 801[b], 706[b], 91, 157, 535, 1144, 831, 654, 371[a])	
						H. microcephalum	(1144)	
						H. tenuifolium	(1080)	
						H. aromaticum	(873, 91)	
						H. campestre	(414b)	
						H. bloomquistii	(396)	
						H. laciniatum	(396)	
						H. mexicanum		
						H. ooclinium	(396)	
						H. quadridentatum	(88)	
						H. scorzoneraefolium	(880)	
						H. vernale	(421)	
						Gaillardia megapotamica	(414)	
						G. multiceps	(414)	
						G. pinnatifida	(427)	

No.	Compound	PH/PA	Formula	m.p.	[α]	Species	Refs.	Remarks
(910)	Helenalin, dihydro	PH	$C_{15}H_{20}O_4$	230–5	—	*Arnica longifolia*	(1140)	
(894)	Helenalin, dihydro, 2-methoxy	PH	$C_{16}H_{22}O_5$	—	—	*Anaphalis morrisonicola* (643); *Balduina angustifolia* (420a, 636); *Arnica montana* (836, 9); *Helenium autumnale* (598)		See Plenolin
(898a)	Helenalin, iso	PH	$C_{15}H_{18}O_4$	260–2	−183	*Helenium microcephalum*	(144)	Structure deduced from literature data
(947)	Helenalin, neo	PH	$C_{15}H_{18}O_4$	252–3	+107	*Helenium mexicanum*	(421, 432, 431, 880)	Reported before Mexicanin D
(909)	Helenalin, tetrahydro	PH	$C_{15}H_{22}O_4$	177–9		*Arnica montana*	(836, 9, 8, 481)	
(832)	Hymenin	PA	$C_{15}H_{18}O_4$	173–4	−88	*Hymenoclea salsola* (1092); *Parthenium confertum* (866, 902)		
(924)	Hymenoflorin	PH	$C_{15}H_{20}O_5$	197–9	−54	*Hymenoxys grandiflora*	(400)	
(905)	Hymenograndin	PH	$C_{19}H_{26}O_7$	153–4	+80	*Hymenoxys grandiflora*	(400)	
(920)	Hymenolane	PH	$C_{21}H_{30}O_8$	216	—	*Hymenoxys odorata*	(575, 830b)	3,4-Seco-derivative
(938)	Hymenolide	PH	$C_{17}H_{26}O_5$	136–8	−48	*Hymenoxys odorata*	(401)	
(848)	Hymenolin	PA	$C_{15}H_{20}O_4$	186–8	+47	*Hymenoclea salsola*	(1092)	
(906)	Hymenoratin	PH	$C_{15}H_{22}O_4$	165–7	+71	*Hymenoxys odorata*	(807)	Hydroxyl groups *trans*. Name changed from odoratin
(937)	Hymenovin	PH	$C_{15}H_{22}O_5$			*Hymenoxys odorata*	(518a)	3,4-Seco-derivative; mixture of two epimers
(936)	Hymenoxon	PH	$C_{15}H_{22}O_5$	136	—	*H. richardsonii* (518b); *Dugaldia hoopesi* (518b); *Hymenoxys odorata* (575, 830)		3,4-Seco-derivative; one of two epimers in Hymenovin mixture
(942)	Hymenoxynin	PH	$C_{21}H_{34}O_9$	125–8	−37	*Baileya multiradiata* (466); *Dugaldia hoopesii* (466); *Hymenoxys odorata* (401)		3,4-Seco-derivative; glucoside

Table VII-2 (continued)

Structure Number	Name of Compound	Type[e]	Formula	m.p. °C	$[\alpha]_D$	Plant Source[d]	References	Comments
(845)	Hysterin	PA	$C_{17}H_{24}O_5$	168	−80	Parthenium bipinnatifidum	(893, 866, 860)	Source misidentified in ref. (893)
(846)	Hysterin acetate	PA	$C_{19}H_{26}O_6$	—	—	P. confertum	(902)	
						Parthenium bipinnatifidum	(893, 866, 860)	
(950)	Inulicin	P	$C_{17}H_{24}O_5$	125–6	+90	Inula japonica	(583, 259)	4,5-Seco-derivative
(949)	Inulicin, desacetyl	P	$C_{15}H_{22}O_4$	174–6	+90	Inula japonica	(259)	4,5-Seco-derivative
(824)	Ivoxanthin	PA	$C_{15}H_{20}O_3$	164–6	+47	Iva xanthifolia syn. Cyclachaena xanthifolia	(778, 937)	
(822)	Ligulatin B	PA	$C_{17}H_{22}O_5$	175	−52	Parthenium tomentosum	(866, 896, 860)	Name changed from incanin
						P. incanum	(896, 860)	
						P. ligulatum	(860)	
(901)	Linearifolin A	PH	$C_{20}H_{24}O_6$	187–8	−90	Hymenoxys linearifolium	(399)	
(946)	Linearifolin B	PH	$C_{20}H_{24}O_6$	214–5	−103	Hymenoxys linearifolium	(399)	
(877)	Linifolin A	PH	$C_{17}H_{20}O_5$	195–8	+33	Helenium linifolium	(396, 410)	
						H. aromaticum	(91)	
						H. plantagineum	(987a)	
						H. scorzoneraefolium	(880)	
(879)	Linifolin B	PH	$C_{17}H_{20}O_5$	149–51	—	Helenium linifolium	(396, 410)	
(902)	Mexicanin A	PH	$C_{15}H_{18}O_4$	138–40	−27	Helenium mexicanum	(904, 432)	
(918)	Mexicanin C	PH	$C_{15}H_{20}O_4$	251–2	−80	Helenium mexicanum	(431, 904)	
(947)	Mexicanin D	PH				Helenium mexicanum	(432)	See Helenalin, neo
(951)	Mexicanin E	PH	$C_{14}H_{16}O_3$	100–1	−47	Helenium mexicanum	(659, 885, 1104[b], 650[b])	
(952)	Mexicanin E, dihydro	PH	$C_{14}H_{18}O_3$	133–5	−188	H. microcephalum	(650)	
						Helenium autumnale	(659)	
(927)	Mexicanin H	PH	$C_{15}H_{18}O_4$	150–1	+44	Helenium mexicanum	(883)	

No.	Compound		Formula	m.p.	[α]	Source	Ref.	Notes
(871)	Mexicanin I	PH	$C_{15}H_{18}O_4$	257–60	+42	Helenium mexicanum	(227)	
						H. aromaticum	(873)	
						H. autumnale	(446)	
						H. plantagineum	(987a)	
						H. tenuifolium	(1080)	
						Gaillardia pinnatifida	(427)	
						Hymenoxys linearis	(880)	
(896)	Microhelenin A	PH	$C_{15}H_{18}O_4$	111–3	−84	Helenium microcephalum	(648)	
(915)	Microhelenin B	PH	$C_{20}H_{28}O_5$	gum	−85	Helenium microcephalum	(650, 647)	
(917)	Microhelenin C	PH	$C_{20}H_{26}O_5$	280	+10	Helenium microcephalum	(650, 647)	
(945)	Microlenin	PH	$C_{29}H_{34}O_7$	—	—	Helenium microcephalum	(649)	Dimeric helenanolide
(874)	Multigilin	PH	$C_{20}H_{24}O_6$	257–60	—	Baileya multiradiata	(831b)	
(926)	Multistatin	PH	$C_{20}H_{22}O_6$	226–30	—	Baileya multiradiata	(831b)	
(925)	Multiradiatin	PH	$C_{20}H_{22}O_6$	218–20	+36	Baileya multiradiata	(831a)	
(928)	Neoleonin	P	$C_{22}H_{30}O_8$			Gaillardia mexicana	(222)	
(834)	Oaxacin	PA	$C_{17}H_{20}O_5$			Parthenium tomentosum	(866, 860)	
(831)	Parthenin	PA	$C_{15}H_{18}O_4$	163–6	+7	Parthenium hysterophorus	(461, 423a, 550)	
						Iva nevadensis	(264a)	
						Ambrosia psilostachya	(298)	
(895)	Paucin	PH	$C_{23}H_{32}O_{10}$	178–9	+51	Baileya pauciradiata	(1123, 1122, 399)	Glycoside
						B. pleniradiata	(1123)	
						Hymenoxys grandiflora	(400)	
						H. rusbyi	(399)	
(854)	Peruvin	PA	$C_{15}H_{20}O_3$	191–3	+155	Ambrosia peruviana	(537)	
						A. confertiflora	(1174)	
						A. artemisiifolia	(838)	
(861)	Peruvinin	PA	$C_{15}H_{20}O_4$	169–71	+34	Ambrosia peruviana	(874)	
(904)	Picrohelenin	PH	$C_{17}H_{24}O_6$	—	—	Helenium autumnale	(599)	
(910)	Plenolin	PH	$C_{15}H_{20}O_4$	223–6	—	Baileya pleniradiata	(1123, 647b, 709b)	Isolated before Helenalin, dihydro
(851)	Pseudoguaian-6,12-olide, 8-acetoxy, 3-oxo	PA	$C_{15}H_{24}O_5$	—	—	Helenium autumnale	(647)	
						Ambrosia artemisiifolia	(1007)	

Table VII-2 (continued)

Structure Number	Name of Compound	Type[c]	Formula	m. p. °C	$[\alpha]_D$	Plant Source[d]	References	Comments
(850)	Pseudoguaian-6,12-olide, 4-hydroxy, 3-oxo	PA	$C_{15}H_{22}O_4$	—	—	Ambrosia artemisiifolia	(1007)	
(863)	Psilostachyin	PA	$C_{15}H_{20}O_5$	215	−125	Ambrosia psilostachya	(665)	4,5-Seco-dilactone
						A. arborescens	(398, 428)	
						A. artemisiifolia	(428)	
						A. confertiflora	(398, 428, 1176)	
						A. cumanensis	(428)	
						A. dumosa	(304)	
						A. tenuifolia	(398)	
						A. "jamaicensis"	(463)	
						A. pumila	(398)	
(866)	Psilostachyin B	PA	$C_{15}H_{18}O_4$	123	−5	Ambrosia psilostachya	(664)	4,5-Seco-dilactone
						A. artemisiifolia	(850b)	
						A. cumanensis	(428)	
						A. confertiflora	(398, 1174)	
						A. cordifolia	(428)	
(865)	Psilostachyin C	PA	$C_{15}H_{20}O_4$	223–5	−82	Ambrosia psilostachya	(549, 359)	4,5-Seco-dilactone
						A. acanthicarpa	(298)	
						A. arborescens	(398)	
						A. confertiflora	(1174)	
						A. cordifolia	(428)	
						A. dumosa	(304)	
						A. cumanensis	(428, 398)	
						A. deltoidea	(549)	
						A. peruviana	(398)	
						A. pumila	(398)	

No.	Compound	Type	Formula	mp	[α]	Source	Refs	Notes
(934)	Psilotropin	PH	$C_{15}H_{20}O_4$	144–5	+84	Psilostrophe cooperi	(215[a], 401)	3,4-Seco-dilactone; see Floribundin
						Hymenoxys anthemoides	(401)	
						H. rusbyi	(401)	
						H. richardsonii	(401)	
						H. subintegra	(401)	
(892)	Pulchellidine	PH	$C_{20}H_{33}O_4N$	185–6	−22	Gaillardia pulchella	(1164, 1162, 956[b])	
(921)	Pulchellidine, neo	PH	$C_{20}H_{33}O_4N$	131–4	—	Gaillardia pulchella	(1162, 956[b])	
(880)	Pulchellin	PH	$C_{15}H_{22}O_4$	165–8	−43	Gaillardia pulchella	(456, 1162, 956[b], 496, 27[b])	
(903)	Pulchellin, neo	PH	$C_{15}H_{22}O_4$	166–7	+43	Gaillardia pulchella	(1162, 956[b], 496)	
(889b)	Radiatin	P	$C_{19}H_{24}O_6$	202–4	−84	Baileya pleniradiata	(1165, 248a[b])	
						B. multiradiata	(831a)	
(839)	Salsolin	PA	$C_{17}H_{24}O_5$	148–9	−72	Hymenoclea salsola	(1092)	
(882)	Spathulin	PH	$C_{19}H_{26}O_8$	259–61	+17	Gaillardia spathulata	(425, 444, 497[b])	
						G. aristata	(425, 449, 725a)	
						G. grandiflora	(425)	
						G. mexicana	(425)	
						G. pulchella	(426)	
(856)	Stevin	PA	$C_{17}H_{24}O_5$	184–6	+161	Stevia rhombifolia	(853)	
(923)	Sulferalin	PH	$C_{16}H_{22}O_6S$	255–6	−167	Helenium autumnale	(600)	
(922)	Tenulin	PH	$C_{17}H_{22}O_5$	195–7	−20	Helenium amarum	(429, 869[b], 437, 1105[b], 371[a])	
						H. autumnale	(446)	
						H. bigelovii	(817)	
						H. tenuifolium	(1080)	
						H. elegans	(88)	
						H. thurberi	(419)	
(887)	Tenulin, iso	PH	$C_{17}H_{22}O_5$	162	+4	Helenium bigelovii	(817, 396)	
						H. arizonicum	(396)	
(886)	Tenulin, iso, desacetyl	PH	$C_{15}H_{20}O_4$	245–8	−15	Helenium bigelovii	(817, 431, 396)	
(820)	Tetraneurin A	PA	$C_{17}H_{22}O_6$	186–8	+3	Parthenium alpinum	(910, 1177)	Mixture of C-16 epimers
						P. cineraceum	(860)	
						P. confertum	(1176)	
						P. fruticosum	(866, 865)	

N. H. FISCHER, E. J. OLIVIER, and H. D. FISCHER:

Table VII-2 (continued)

Structure Number	Name of Compound	Type[c]	Formula	m.p. °C	[α]_D	Plant Source[d]	References	Comments
(819)	Tetraneurin B	PA	$C_{17}H_{22}O_6$	194–5	−44	Parthenium alpinum	(910, 1177, 860)	
						P. fruticosum	(1177)	
						P. ligulatum	(860)	
						P. lozanianum	(1177, 860)	
						P. schottii	(865, 860)	
(842)	Tetraneurin C	PA	$C_{19}H_{26}O_7$	145	−109	Parthenium alpinum	(910, 1177, 860)	
						P. fruticosum	(1177)	
						P. integrifolium	(1176, 860)	
						P. lozanianum	(1177, 860)	
						P. schottii	(860)	
(841)	Tetraneurin D	PA	$C_{17}H_{24}O_6$	203–5	−72	Parthenium lozanianum	(1177, 860)	
						P. fruticosum		
						P. schottii	(865)	
(843)	Tetraneurin E	PA	$C_{17}H_{24}O_6$	197–9	−70	Parthenium confertum	(1176, 902)	
						P. integrifolium	(1176, 860)	
(844)	Tetraneurin F	PA	$C_{19}H_{26}O_7$	135–6	−47	Parthenium confertum	(1176)	
(940)	Themoidin	PH	$C_{15}H_{22}O_4$	214–5	+61.8	Hymenoxys anthemoides	(401)	See Floribundin, A dihydro
(888)	Thurberilin	PH	$C_{20}H_{26}O_5$	162	+20	Helenium thurberi	(419)	NMR suggests angelate side chain
(933)	Vermeerin	PH	$C_{15}H_{20}O_4$	147	−58	Geigeria africana ssp. syn. Geigeria filifolia	(22)	3,4-Seco-dilactone
						Hymenoxys anthemoides	(401)	
						H. richardsonii	(401)	
(943)	Vermeerin B, dihydro	PH				Hymenoxys anthemoides	(401)	3,4-Seco-dilactone; see Anthemoidin

VIII. Eremophilanolides and Bakkenolides

The chemistry of eremophilanes and related sesquiterpenes has recently been reviewed by PINDER (*834*). Therefore, this class of sesquiterpene lactones will only be discussed in relation to their biogenesis. For some time, reactions attempting to mimick the postulated methyl migration from C-10 to C-5 of a eudesmanolide leading to an eremophilanolide were studied with little success (*834*). More recently, in a biogenetic-type conversion, dihydroalantolactone epoxide (**981**) was successfully transformed to the eremophilanolide (**982**) upon treatment with formic acid in acetone (*588, 590*). Further transformations of this type have been reported more recently (*587, 589*).

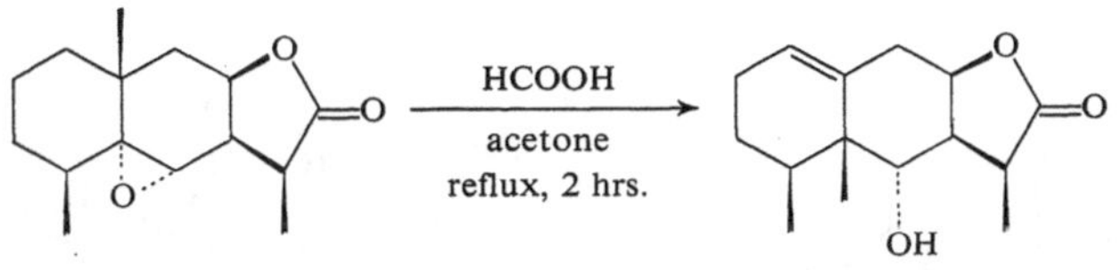

(981)   **(982)**

Chart VIII-1. Transformation of a eudesmanolide to an eremophilanolide

Evidence for the formation of the eremophilanolide lactone ring from furan precursors has been provided by chemical (*790*) as well as photochemical transformations (*739, 759*). As shown on top of Chart VIII-2, oxidative biogenetic conversion of the furanoeremophilane (**983**) would provide the naturally occurring furan derivative (**984**) which can be transformed *in vitro* to eremophilenolide (**986**) *via* the intermediate (**985**) (*790*).

On the bottom of Chart VIII-2, the photosensitized autoxidation of the furanolactone (**987**) to the dilactone (**988**) is used as an example that naturally occurring lactones of type (**988**) might possibly represent artifacts (*739*). Most recently, two dimeric eremophilanolides (**1011**) and (**1012**) have been reported (*109 a*).

Biogenetically, bakkenolides (fukinanolides) are being considered as derivatives of the eremophilanolides which result from ring contraction

of ring B of the eremophilane skeleton followed by biomodifications (*834*).
A successful laboratory synthesis of bakkenolide A (**992**) has recently
been reported (*377 a*). In Chart VIII-3, epoxidation of fukinone (**989**)
gave the α,β-epoxiketone (**990**) which upon treatment with base underwent a Favorskii-type rearrangement forming after methylation the ring
contraction product (**991**). Subsequent elimination, SeO_2-oxidation and
spontaneous lactonization yielded bakkenolide A (**992**) (*377 a*).

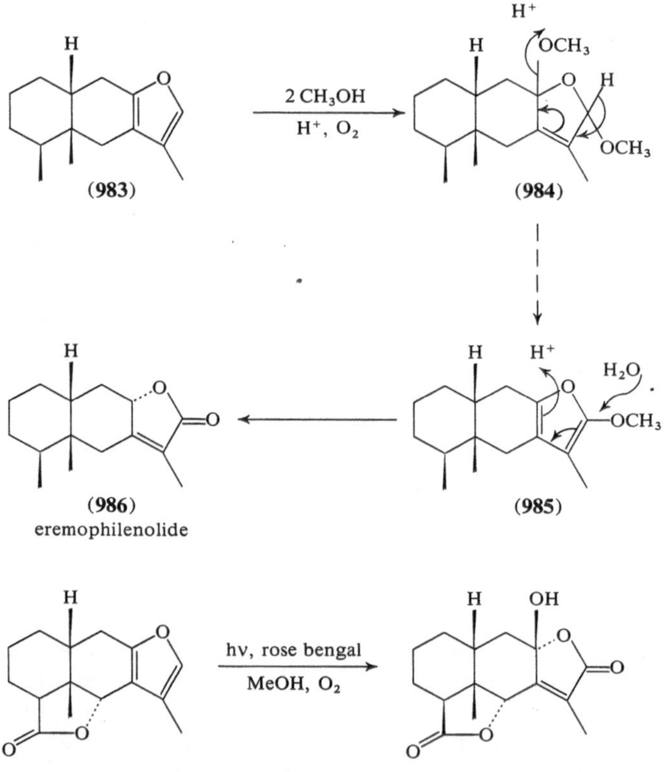

Chart VIII-2. Conversion of eremophilanofurans to eremophilanolides

The structures of naturally occurring eremophilanolides and bakkenolides are given in Chart VIII-4 and Chart VIII-5, respectively. Physical
data as well as plant sources and literature references of the two skeletal
types are found in Table VIII-1 and Table VIII-2.

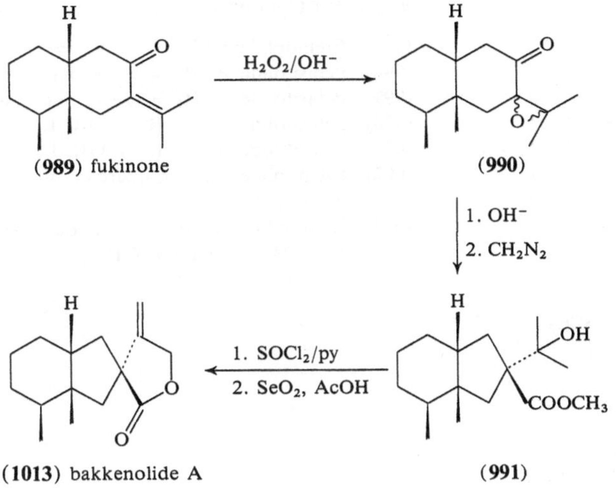

(989) fukinone   H₂O₂/OH⁻   (990)

1. OH⁻
2. CH₂N₂

(1013) bakkenolide A   1. SOCl₂/py   2. SeO₂, AcOH   (991)

Chart VIII-3. Synthesis of bakkenolide A from fukinone

Chart VIII-4. Naturally occurring eremophilanolides and biogenetic derivatives

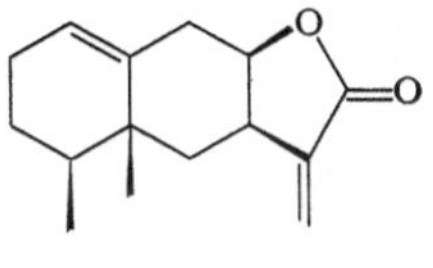

(993) Xanthanodiene

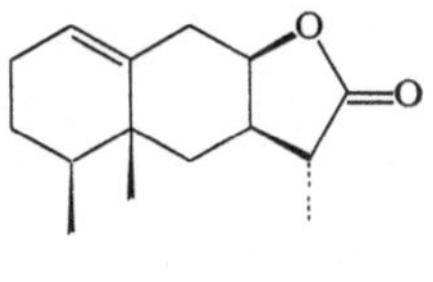

(994) Xanthanene

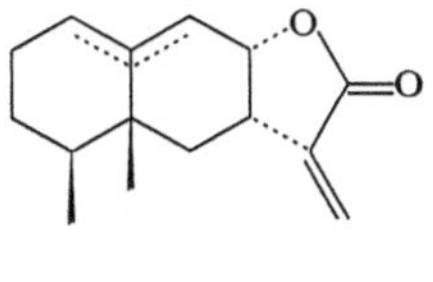

(995?) Dugesialactone; Δ1,10
(996) (+)-Eremophofrullanolide; Δ9,10

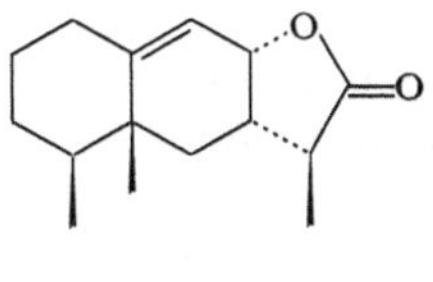

(997) (+)-Eremophofrullanolide, dihydro

17*

Chart VIII-4 (continued)

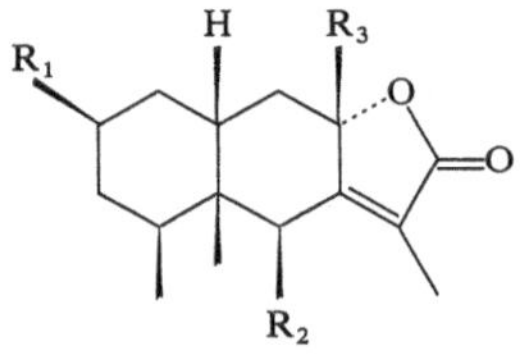

(**986**) Eremophilenolide; $R_1 = R_2 = R_3 = H$
(**998**) Petasitolide B; $R_1 = OTig$, $R_2 = R_3 = H$
(**999**) Petasitolide A; $R_1 = OAng$, $R_2 = R_3 = H$
(**1000**) Petasitolide A, –S; $R_1 = OA$, $R_2 = R_3 = H$
(**1001**) Petasitolide B, –S; $R_1 = OB$, $R_2 = R_3 = H$
(**1002**) Eremophilenolide, 6-hydroxy; $R_1 = R_3 = H$, $R_2 = OH$
(**1003**) Eremophil-7(11)-en-12,8α-olide,6β,8β-dihydroxy; $R_1 = H$, $R_2 = R_3 = OH$

$A =$

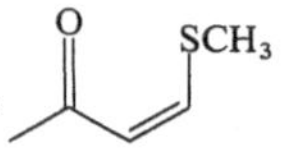

$B =$

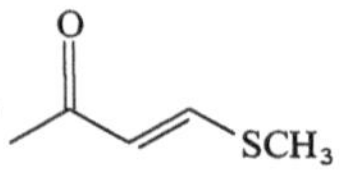

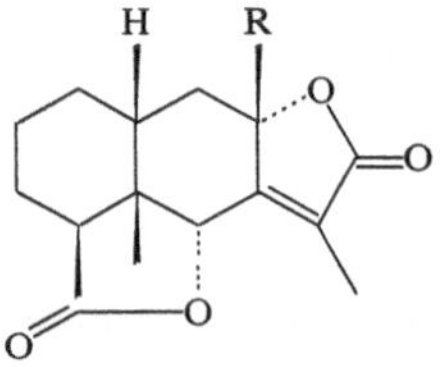

(**1004**) Eremophil-7(11)-ene-12,8α, 14β,6α-diolide; $R = H$
(**988**) Eremophil-7(11)-ene-12,8α, 14β,6α-diolide; 8β-hydroxy; $R = OH$

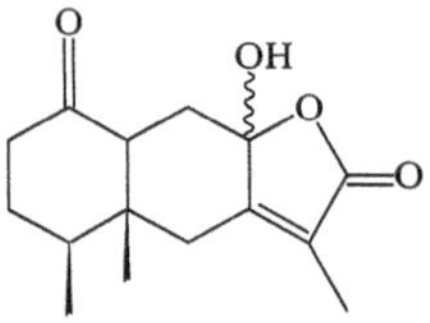

(**1005**) Istanbulin A

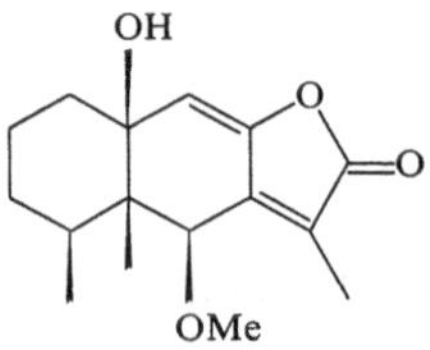

(**1006**) Ligolide

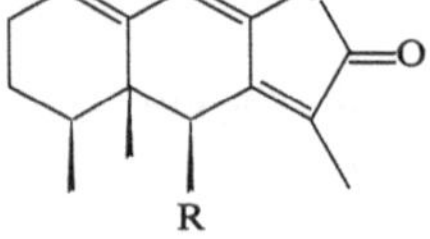

(**1007**) Ligularenolide; $R = H$
(**1008**) Ligularenolide, 6β-hydroxy; $R = OH$
(**1009**) Ligularenolide, 6β-acetoxy; $R = OAc$

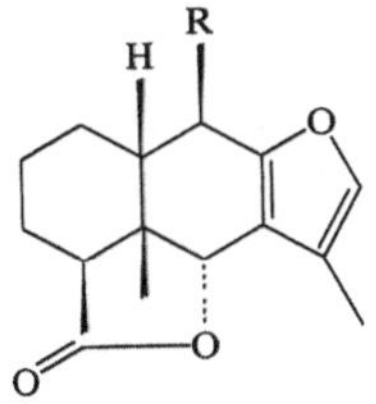

(**987**) Furanoeremophilan-14β,6α-olide; R = H
(**1010**) Ligucalthaefolin; R = OTig

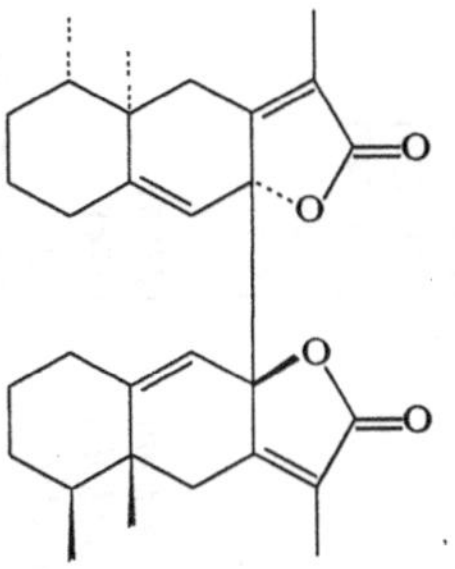

(**1011**) 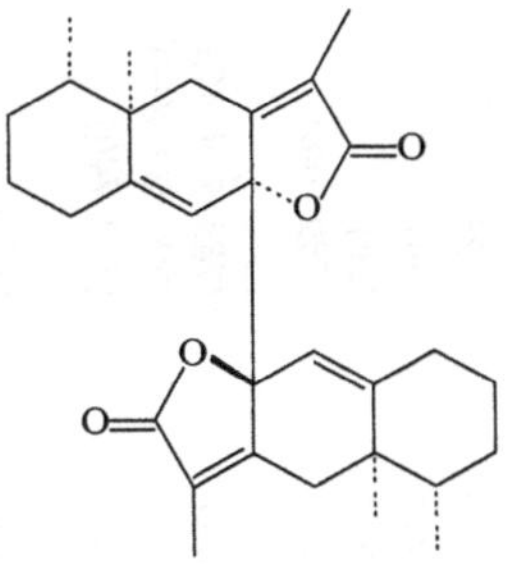   (**1012**)

Chart VIII-5. Naturally occurring bakkenolides (fukinanolides)

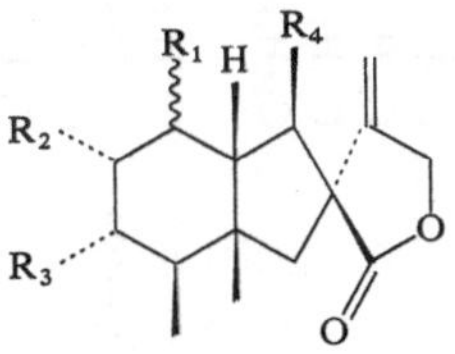

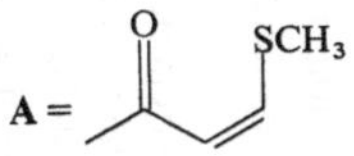

(**1013**) Fukinanolide; $R_1 = R_2 = R_3 = R_4 = H$
(Bakkenolide A)
(**1014**) Bakkenolide C; $R_1 = \alpha$-OAng, $R_2 = R_3 = H$,
$R_4 = OH$
(**1015**) Fukinolide; $R_1 = \alpha$-OAng, $R_2 = R_3 = H$,
$R_4 = OAc$
(Bakkenolide B)
(**1016**) Fukinolide S; $R_1 = \alpha$-OA, $R_2 = R_3 = H$,
$R_4 = OAc$
(Bakkenolide D)
(**1017**) Bakkenolide E; $R_1 = \beta$-OAng, $R_2 = R_3 = H$,
$R_4 = OAc$
(**1018**) Homofukinolide; $R_1 = \beta$-OAng, $R_2 = R_3 = H$,
$R_4 = OAng$
(**1019**) Fukinolide, dihydro; $R_1 = \beta$-O-2-Mebut,
$R_2 = R_3 = H$, $R_4 = OAc$
(**1020**) Bakkenolide A, 2-hydroxy, angeloyl;
$R_1 = R_3 = R_4 = H$, $R_2 = OAng$
(**1021**) Bakkenolide A, 3-α-hydroxy, tigloyl;
$R_1 = R_2 = R_4 = H$, $R_3 = OTig$
(**1022**) Fukinanolide, 9-acetoxy; $R_1 = R_2 = R_3 = H$,
$R_4 = OAc$

Table VIII-1. *Naturally Occurring Eremophilanolides and Biogenetic Derivatives*

Structure Number	Name of Compound	Formula	m.p. °C	$[\alpha]_D$	Plant Source	References	Comments
(995)	Dugesialactone	$C_{15}H_{20}O_2$	oil	−13	Dugesia mexicana	(123)	See ref. (393)
(996)	(+)-Eremofrullanolide	$C_{15}H_{20}O_2$	82–82.5	+9	Frullania dilatata	(36)	Hepaticae
(997)	(+)-Eremofrullanolide, dihydro	$C_{15}H_{22}O_2$	70–1	+108	Frullania dilatata	(36)	Hepaticae
(1004)	Eremophil-7(11)-ene-12,8α; 14β,6α-diolide	$C_{15}H_{18}O_4$	186	+93	Ligularia fauriei	(739, 740)	
(986)	Eremophilenolide	$C_{15}H_{22}O_2$	125	+16	Petasites hybridus	(783, 788, 473, 546[b], 748, 832, 833)	
(988)	Eremophil-7(11)-ene-12,8α, 14β,6α-diolide, 8β-hydroxy	$C_{15}H_{18}O_5$	254	+94	Ligularia fauriei	(739, 740)	
(1003)	Eremophil-7(11)-en-12, 8α-olide, 6β,8β-dihydroxy	$C_{15}H_{22}O_4$	217–8	+82	Petasites japonicus	(756)	
(1002)	Eremophilenolide, 6-hydroxy	$C_{15}H_{22}O_3$	208	−35	Ligularia fauriei Petasites albus P. japonicus Ligularia fauriei L. fischeri	(739, 740) (782, 546[b]) (779, 756) (739, 740) (514)	
(987)	Furanoeremophilan-14β,6α-olide	$C_{15}H_{18}O_3$	136–8	−47	Ligularia hodgsoni	(517)	
(1005)	Istanbulin A	$C_{15}H_{20}O_4$	246	+81	Smyrnium olusantrum	(1106)	
(1006)	Ligolide	$C_{16}H_{22}O_4$	—	—	Ligularia macrophylla	(774)	
(1010)	Ligucalthaefolin	$C_{20}H_{24}O_5$	oil	—	Ligularia calthaefolia	(99)	
(1007)	Ligularenolide	$C_{15}H_{18}O_2$	134–5	—	Aster tataricus Ligularia species	(518, 1070, 832) (1069, 1060)	
(1009)	Ligularenolide, 6β-acetoxy	$C_{17}H_{20}O_4$	oil	—	Euryops brevipapposus	(93)	
(1008)	Ligularenolide, 6β-hydroxy	$C_{15}H_{18}O_3$	oil	—	Senecio vellereus	(106)	
(999)	Petasitolide A	$C_{20}H_{28}O_4$	147	+48	Petasites officinalis	(783, 787, 786, 592[a])	
(1000)	Petasitolide A, -S	$C_{19}H_{26}O_4S$	201–3	−15	Petasites officinalis	(783, 787, 786, 592[a])	

(998)	Petasitolide B	$C_{20}H_{28}O_4$	146	+31	*Petasites officinalis*	*(783, 787, 786, 592[a])*
(1001)	Petasitolide B, -S	$C_{19}H_{26}O_4S$	199–200	−32	*Petasites officinalis*	*(787, 786, 592[a])*
(994)	Xanthanene	$C_{15}H_{22}O_2$	—	—	*Xanthium canadense*	*(1061)*
(993)	Xanthanodiene	$C_{15}H_{20}O_2$	—	—	*Xanthium canadense*	*(1061)*

Dimeric Lactones

(1011)	Symmetric lactone	$C_{30}H_{38}O_4$	186	+88	*Bedfordia salicina*	*(109 a)*
(1012)	Unsymmetric lactone	$C_{30}H_{38}O_4$	203–4	+94	*Bedfordia salicina*	*(109 a)*

[a] Reference contains structural revision;   [b] Reference giving X-ray data.

Table VIII-2. *Naturally Occurring Bakkenolides (Fukinanolides)*

Struc-ture Number	Name of Compound	Formula	m. p. °C	[α]$_D$	Plant Source	References	Comments
(1013)	Bakkenolide A						See Fukinanolide
(1020)	Bakkenolide A, 2-hydroxy, angeloyl	$C_{20}H_{28}O_4$	62–5	—	*Homogyne alpina*	(374)	
(1021)	Bakkenolide A, 3-α-hydroxy, tigloyl	$C_{20}H_{28}O_4$	gum	—	*Homogyne alpina*	(374)	
(1015)	Bakkenolide B						See Fukinolide
(1014)	Bakkenolide C	$C_{20}H_{28}O_5$	167	—	*Petasites japonicus*	(3)	
(1016)	Bakkenolide D						See Fukinolide S
(1017)	Bakkenolide E	$C_{22}H_{30}O_6$			*Petasites japonicus*	(564)	Reported before Bakkenolide A
(1013)	Fukinanolide	$C_{15}H_{22}O_2$	80	+20	*Petasites japonicus*	(760, 986, 21, 4, 755, 252, 253, 526)	
					P. albus	(789)	
					P. fragrans	(526)	
					P. hybridus	(526)	
					Cacalia hastata	(377)	
					C. hastata	(758)	
					Homogyne alpine	(374)	
					Ligularia hodgsonii	(516)	
(1022)	Fukinanolide, 9-acetoxy	$C_{17}H_{24}O_4$	96–7	–28	*Petasites japonicus*	(757)	Reported before Bakkenolide B
(1015)	Fukinolide	$C_{22}H_{30}O_6$	101–2	–126	*Petasites japonicus*	(760[a], 986, 3, 562[a], 755)	
(1019)	Fukinolide, dihydro	$C_{22}H_{32}O_6$	125–6	–105	*Petasites japonicus*	(755)	Side chain has +–(S) configuration
(1016)	Fukinolide S	$C_{21}H_{28}O_6S$	200–1	–161	*Petasites japonicus*	(760[a], 986, 3, 562[a], 755)	Reported before Bakkenolide D
(1018)	Homofukinolide	$C_{25}H_{34}O_6$	184–6	–127	*Petasites japonicus*	(755)	

[a] Reference gives X-ray data.

IX. Special Structural Types and Minor Classes of Sesquiterpene Lactones

1. Drimanolides

The biogenesis of the drimanolide skeleton most probably involves the sequence of steps outlined in Chart IX-1 (*195*). A likely precursor is farnesol epoxide (**1023**) which can be directly cyclized to the drimane cation (**1024**). Loss of a proton and oxidative modifications of ring (**1024**) would give iresin (**1025**), a member of the antipodal series (*907*)

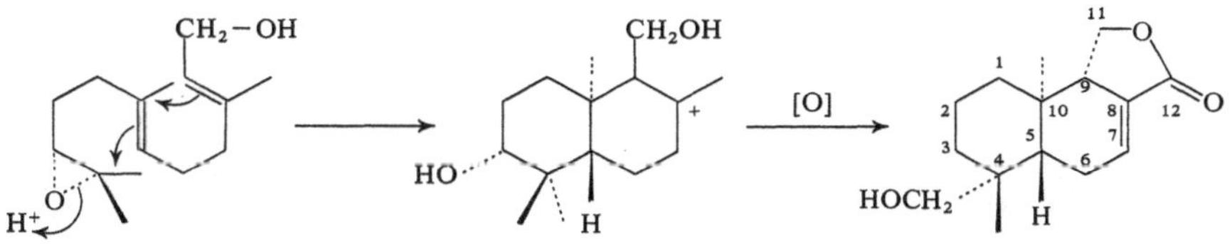

(**1023**)   (**1024**)   (**1025**) iresin

Chart IX-1. Biogenesis of drimanolides

representatives of which seem to be common in the genus *Iresine* of the family Amarantacea. Drimanolides of the more common structure with C-10β-methyl groups occur in the genera *Cinnamosma* and *Porella* of the family Canellaceae and other taxa. In addition, analogues of cinnamolide (**1041**) have recently been isolated from liquid cultures of the birds nest fungus *Mycocalia reticulata* Petch of the Nidulariaceae (*41*) and from *Penicillum purpurogenum* Stoll. (*576*). Chart IX-2 lists the structures of all known naturally occurring drimanolides and Table IX-1 gives the compounds in alphabetical order with physical data and literature references (Chart IX-2 on page 266; Table IX-1 on page 269).

Chart IX-2. Naturally occurring drimanolides

(1026) Drimenin

(1027) Drimenin, iso; $R_1 = R_2 = H$
(1028) Ugandensolide; $R_1 = OAc$, $R_2 = OH$
(1029) Futronolide; $R_1 = H$, $R_2 = OH$

(1030) Colorata-4(13),8-dienolide

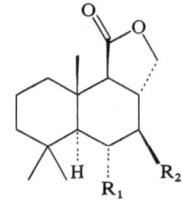

(1031) Drimenin, 6α,7β-dihydroxy dihydro; $R_1 = R_2 = OH$
(1032) Drimenin, 7β-hydroxydihydro; $R_1 = H$, $R_2 = OH$

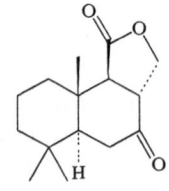

(1033) Drimenin, 7-ketodihydro

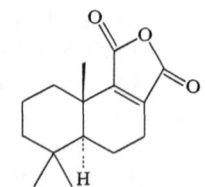

(1034) Winterin

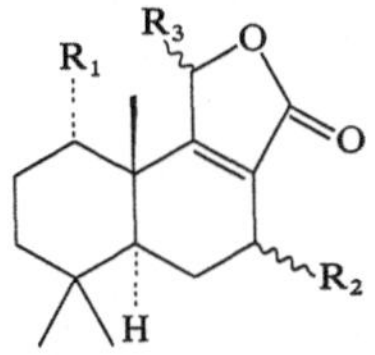

(1035) Confertifolin; $R_1 = R_2 = R_3 = H$
(1036) Purpuride; $R_1 = OA$, $R_2 = R_3 = H$
(1037) Fuegin; $R_1 = H$, $R_2 = R_3 = OH$
(1038) Valdiviolide; $R_1 = R_2 = H$, $R_3 = OH$

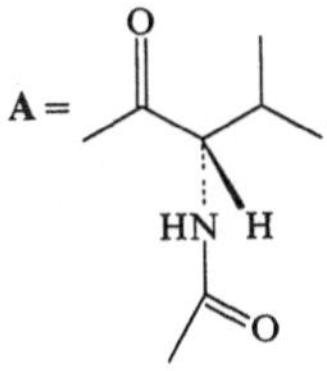

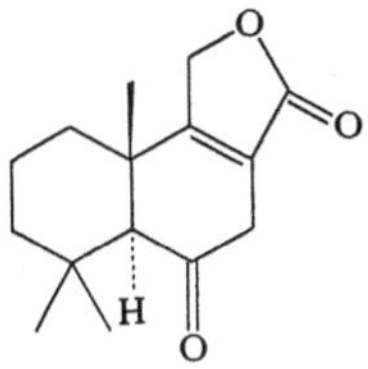

$A =$

(1039) Fragrolide

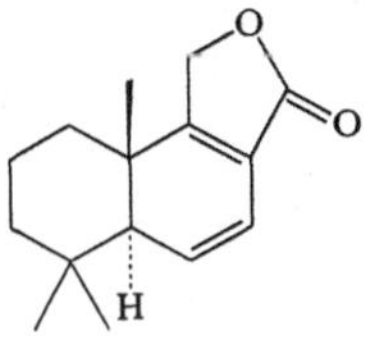

(1040) Bemadienolide

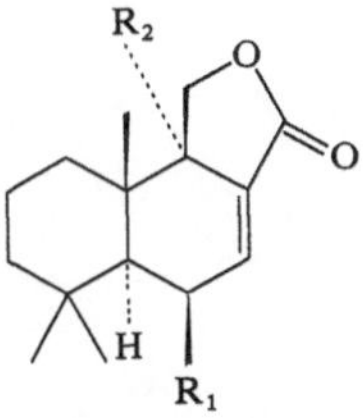

(1041) Cinnamolide; $R_1 = R_2 = H$
(1042) Bemarivolide; $R_1 = OAc$, $R_2 = H$
(1043) Cinnamosmolide; $R_1 = OAc$, $R_2 = OH$

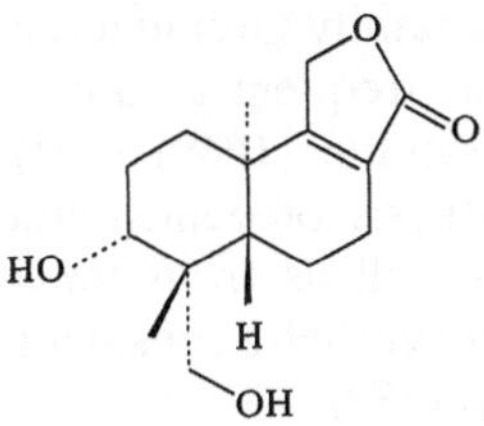

(1044) Iresin, iso

Chart IX-2 (continued)

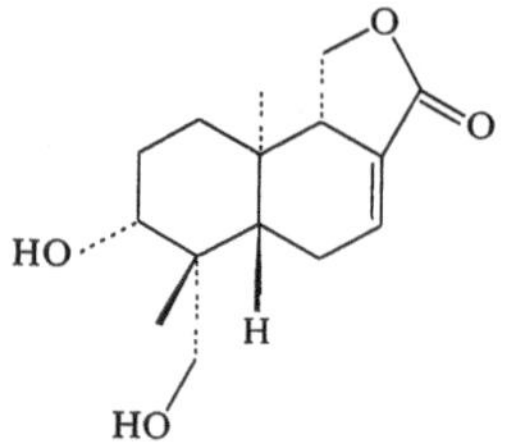

(1025) Iresin

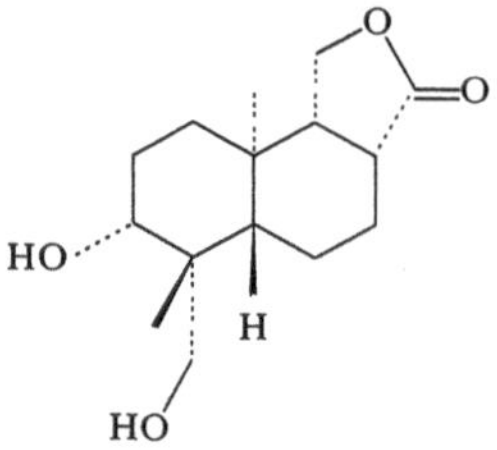

(1045) Iresin, dihydro

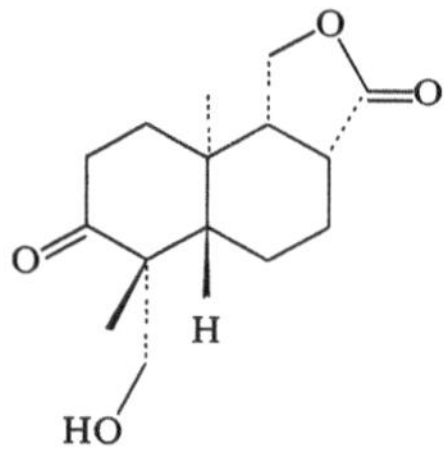

(1046) Iresone, dihydro

2. Tutinanolides (Picrotoxins)

The tutinanolides represent bitter, toxic lactones isolated from the genera *Coriaria, Dendrobium* and *Hyaenanche* of the family Orchidaceae. These compounds are generally highly oxidized and frequently contain nitrogen such as dendrobine (1048) (*1160*), dendramine (1049) (*501*), and dendrine (1051) (*348*). A study of the biosynthesis of dendrobine (1048) has demonstrated that mevalonate (*248*) as well as *trans,trans*-farnesol (*177*) represent precursors. More detailed biosynthetic considerations can be found in the recent review by Cordell (*182*).

Table IX-1. *Naturally Occurring Drimanolides*

Structure Number	Name of Compound	Formula	m. p. °C	$[\alpha]_D$	Plant Source	References	Comments
(1040)	Bemadienolide	$C_{15}H_{20}O_2$	124–5	+22	*Cinnamosma fragrans*	(148)	Canellaceae
(1042)	Bemarivolide	$C_{17}H_{24}O_4$	137–8	−253	*Cinnamosma fragrans*	(148)	
(1041)	Cinnamolide	$C_{15}H_{22}O_2$	125–6	−29	*Porella species* *Cinnamosma fragrans*	(147, 149, 1038, 38)	
(1043)	Cinnamosmolide	$C_{17}H_{24}O_5$	204	−332	*Cinnamosma fragrans*	(147, 149, 530, 1161)	
(1030)	Colorata-4(13), 8-dienolide	$C_{15}H_{20}O_2$	132–3	+292	*Pseudowintera colorata*	(181)	Modified Drimanolide
(1035)	Confertifolin	$C_{15}H_{22}O_2$	152–3	+72	*Drimys confertifolia* *D. winteri* *Polygonum hydropiper* *Cinnamosma fragrans*	(29, 1134) (530) (29, 1134, 591, 1161)	
(1026)	Drimenin	$C_{15}H_{22}O_2$	133	−42	*Drimys winteri* *Porella species*	(38)	
(1031)	Drimenin, 6α,7β-dihydroxy-dihydro	$C_{15}H_{24}O_4$	43–5	—	*Mycocalia reticulata*	(41)	Nidulariaceae
(1032)	Drimenin, 7β-hydroxydihydro	$C_{15}H_{24}O_3$	157–8	−65	*Mycocalia reticulata*	(41)	
(1027)	Drimenin, iso	$C_{15}H_{22}O_2$	131–2	+87	*Drimys winteri*	(29, 1134)	
(1033)	Drimenin, 7-keto-dihydro	$C_{15}H_{22}O_3$	120–2	−118	*Mycocalia reticulata*	(41)	
(1039)	Fragrolide	$C_{15}H_{20}O_3$	165–6	+149	*Cinnamosma fragrans*	(148)	
(1037)	Fuegin	$C_{15}H_{22}O_4$	170–2	+76	*Drimys winteri*	(28)	
(1029)	Futronolide	$C_{15}H_{22}O_3$	215–7	—	*Drimys winteri*	(28, 563[a])	
(1025)	Iresin	$C_{15}H_{22}O_4$	140–2	+21	*Iresine celosioides*	(219, 218, 825, 907[b])	Amarantacea
(1045)	Iresin, dihydro	$C_{15}H_{24}O_4$	—	—	*Iresine celosioides*	(195, 825)	Identified as di-acetate, m. p. 212–3°
(1044)	Iresin, iso	$C_{15}H_{22}O_4$	—	—	*Iresine celosioides*	(195, 825)	Identified as di-acetate, m. p. 166–8°
(1046)	Iresone, dihydro	$C_{15}H_{22}O_4$	215–9	+62	*Iresine celosioides*	(195)	
(1036)	Purpuride	$C_{11}H_{33}NO_5$	200–1	—	*Penicillium purpurogenum*	(576)	
(1028)	Ugandensolide	$C_{17}H_{24}O_5$	218	+23	*Warburgia ugandensis*	(685)	
(1038)	Valdiviolide	$C_{15}H_{22}O_3$	177–8	+111	*Drimys winteri*	(28)	
(1034)	Winterin	$C_{15}H_{20}O_3$	158	+109	*Drimys winteri*	(28, 136)	

[a] Reference gives structural revision; [b] Reference reports X-ray data.

The naturally occurring tutinanolides are listed in Chart IX-3 and their physical properties, plant sources and literature references are tabulated in Table IX-2 on page 272.

Chart IX-3. Naturally occurring tutinanolides

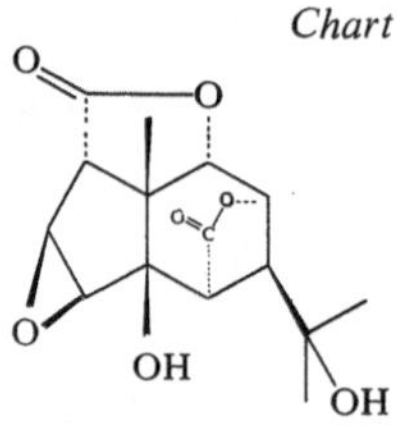

(**1047**)   Picrotin

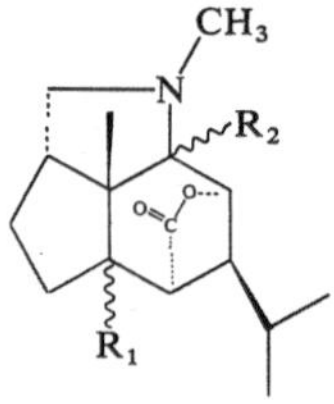

(**1048**)   Dendrobine; R_1 = H, R_2 = β-H
(**1049**)   Dendramine; R_1 = β-OH, R_2 = β-H
(**1050**)   Dendrobine, 2-hydroxy; R_1 = H, R_2 = OH

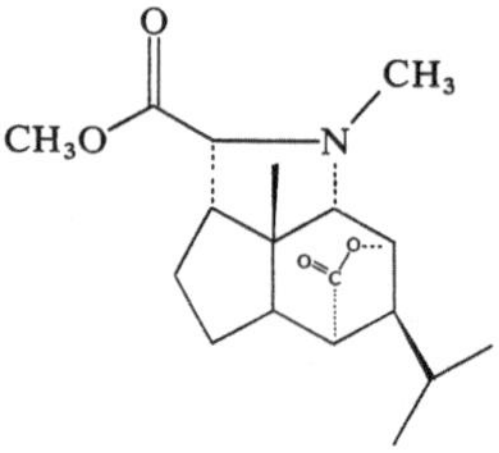

(**1051**)   Dendrine

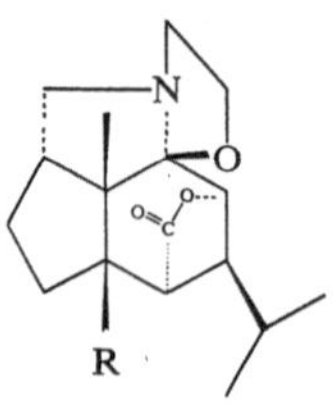

(**1052**)   Dendroxine; R = H
(**1053**)   Dendroxine, 6-hydroxy; R = OH

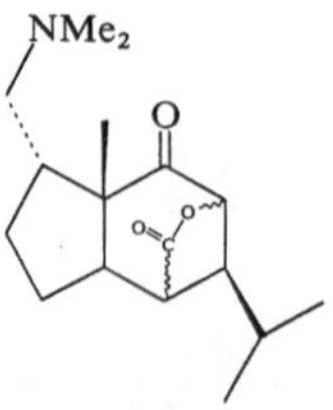

(**1054**)   Nobiline

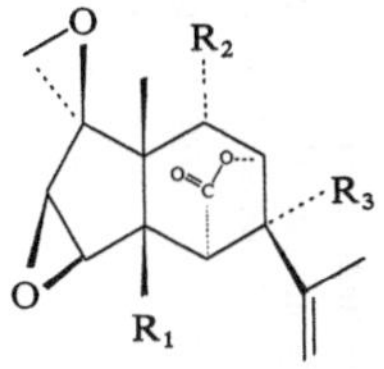

(1055) Coriamyrtin; R_1 = OH, R_2 = R_3 = H
(1056) Tutin; R_1 = R_2 = OH, R_3 = H
(1057) Hyenancin; R_1 = R_2 = R_3 = OH
(1060) Substance C; R_1 = R_3 = OH, R_2 = OCH_3

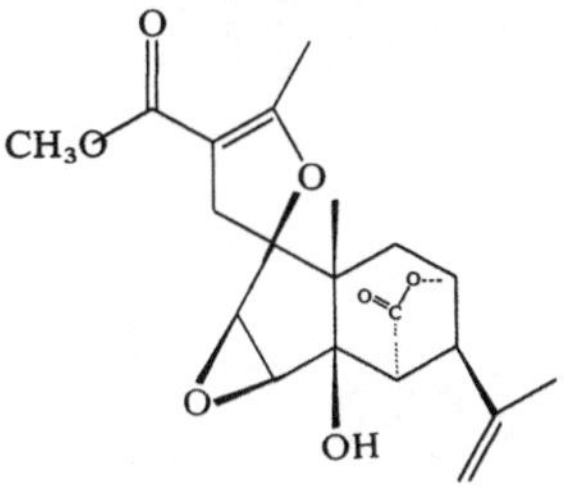

(1061) Capenicin

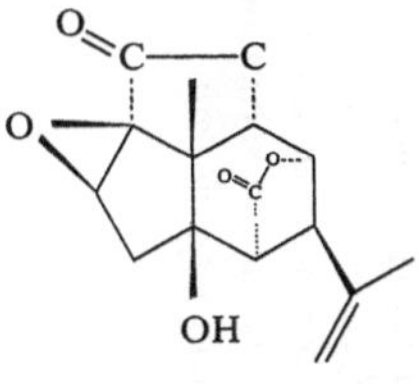

(1062) Picrotoxinin

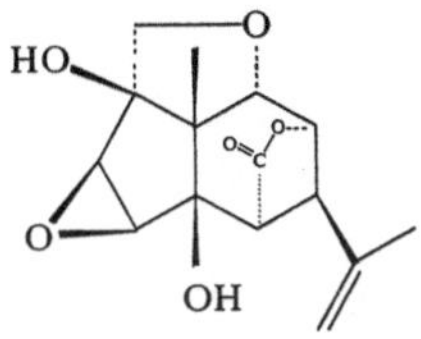

(1063) Corianin

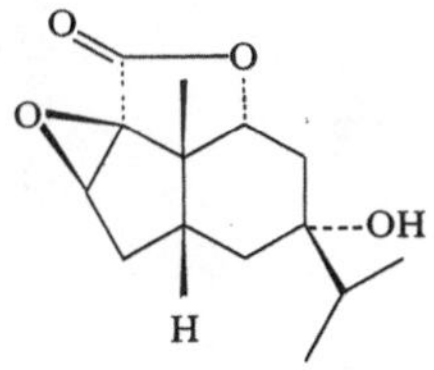

(1064) Aduncin

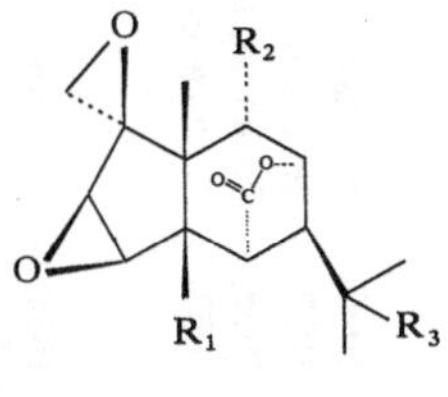

(1065) Hyenancin, iso; R_1 = R_2 = R_3 = OH
(1067) Substance D; R_1 = R_3 = OH, R_2 = OMe

Table IX-2. *Naturally Occurring Tutinanolides (Picrotoxins)*

Structure Number	Name of Compound	Formula	m.p. °C	$[\alpha]_D$	Plant Source	References	Comments
(1064)	Aduncin	$C_{15}H_{18}O_6$	—	—	*Dendrobium aduncum*	(284)	
(1061)	Capenicin	$C_{20}H_{27}O_7$	243-4	+77	*Hyaenanche globosa* *Toxicodendrum capense*	(532) (530)	
(1055)	Coriamyrtin	$C_{15}H_{18}O_5$	229-30	—	*Coriaria japonica* *C. myrtifolia*	(799) (799)	
(1063)	Corianin	$C_{15}H_{18}O_6$	215-16	-27	*Coriaria japonica*	(798)	
(1049)	Dendramine	$C_{16}H_{25}O_3N$	187	-114	*Dendrobium nobile*	(501)	
(1051)	Dendrine	$C_{19}H_{29}NO_4$	191	—	*Dendrobium nobile*	(348)	
(1048)	Dendrobine	$C_{15}H_{25}O_2N$	135-6	-45	*Dendrobium nobile*	(1160, 177, 248)	
(1050)	Dendrobine, 2-hydroxy	$C_{16}H_{25}NO_3$	104	-30	*Dendrobium findlayanum*	(349)	
(1052)	Dendroxine	$C_{17}H_{25}O_3N$	115	—	*Dendrobium nobile*	(795)	
(1053)	Dendroxine, 6-hydroxy	$C_{17}H_{25}O_4N$	—	—	*Dendrobium nobile*	(796)	
(1057)	Hyenancin	$C_{15}H_{18}O_7$	225-40	—	*Coriaria arborea*	(531, 474, 533)	Reported before Mellitoxin; see also Ienancin
(1065)	Hyenancin, iso	$C_{15}H_{18}O_7$	298	—	*Hyenanche globosa* *Miel empoisonne*	(531) (474)	See Ienancin, iso
(1057)	Ienancin				*Hyenanche globosa*	(531)	See Hyenancin
(1065)	Ienancin, iso				*Toxicodendrum capense*	(530)	See Hyenancin, iso
(1057)	Mellitoxin				*Toxicodendrum capense* *Scolypopa australis*	(530) (474)	See Hyenancin; the source, an insect, feeds on the leaves and stems of *Coriaria arborea*
(1054)	Nobiline	$C_{17}H_{27}O_3N$	88	—	*Dendrobium nobile*	(1160)	
(1047)	Picrotin	$C_{15}H_{18}O_7$	252	—	*Anamirta cocculus* *A. paniculata* *Cocculus indicus*	(165, 198)	

(1062)	Picrotoxinin	$C_{15}H_{16}O_6$	210	—		(165, 171, 172, 198[a], 170, 179)
(1060)	Substance C	$C_{16}H_{20}O_7$	—		Hyenanche globosa	(179)
(1067)	Substance D	$C_{16}H_{22}O_7$	213	—	Hyenanche globosa	(534, 179)
(1056)	Tutin	$C_{15}H_{18}O_6$	212–3	−59	Coriaria angustissima	(247, 577, 197[a], 176, 1103, 179)
					C. japonica	
					C. lurida	
					C. ruscifolia	
					C. thymifolia	
					Toxicodendrum capense	(530)

[a] Reference presents X-ray data.

3. Special Structural Types

Chart IX-4 lists a number of compounds with various skeletal arrangements most of them biogenetically derived from farnesyl pyrophosphate by processes other than the germacradiene route. Compounds (1068) to (1076) represent oxidatively modified derivatives of the farnesol skeleton in which oxidations at both ends of the carbon chain have

Chart IX-4. Special structural types of sesquiterpene lactones

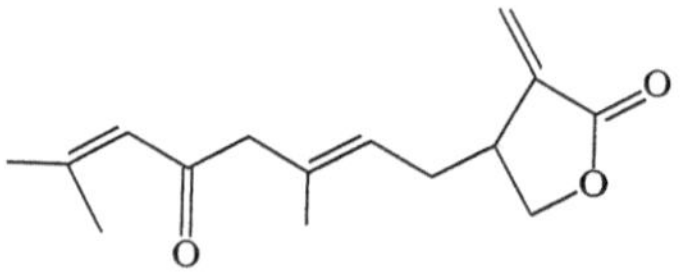

(1068) Anthemis, lactone from

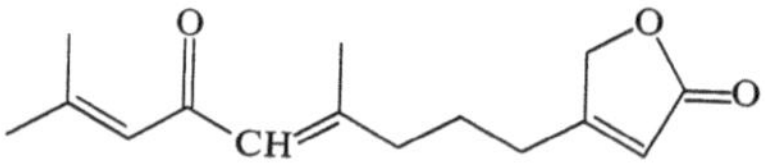

(1069) *cis*-Linifolon; *cis*-double bond
(1070) *trans*-Linifolon; *trans*-double bond

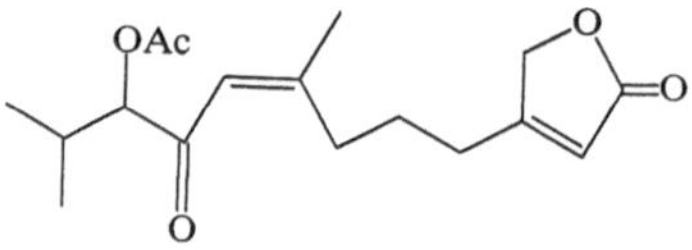

(1071) Linifolon, Z-2,3-dihydro-3-acetoxy

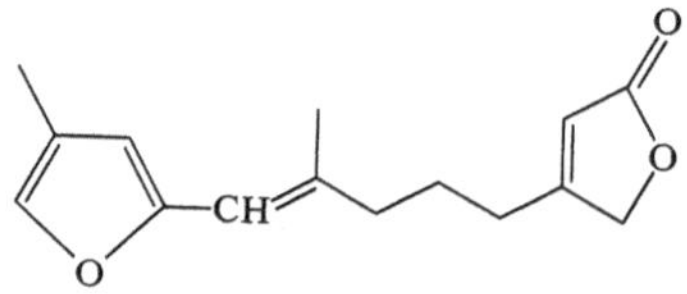

(1072) Lasiosperman, Z-7,12,13H-5-dehydro-12-oxo; *cis*-double bond
(1073) Lasiosperman E-7,12,13H-5-dehydro-12-oxo; *trans*-double bond

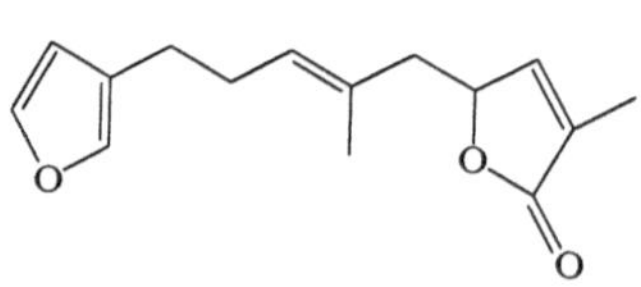

(1074) Freelingnite

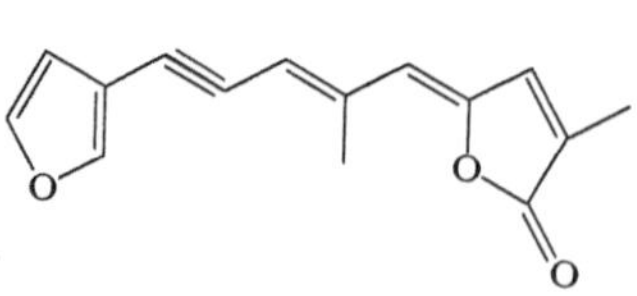

(1075) Freelingyne

(1076) Freelingyne, dihydro

(1077) Pallescensin 3

(1078) Paniculide A; R_1 = OH, R_2 = R_3 = H
(1079) Paniculide B; R_1 = R_3 = OH, R_2 = H
(1080) Paniculide C; R_1 = R_2 = O, R_3 = OH

(1081) Lemnalactone; R = H
(1082) 12-Oxolemnacarnol; R = OH

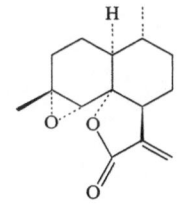

(1083) Arteannuin B

(1084) Quing Hau Sau

Chart IX-4 (continued)

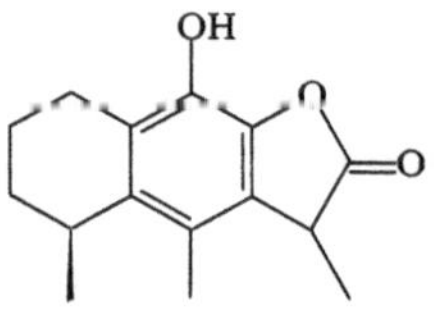

(1085) Cacalolide

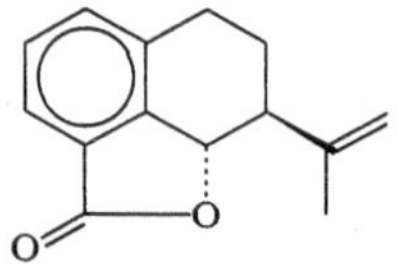

(1086) Platyphyllide

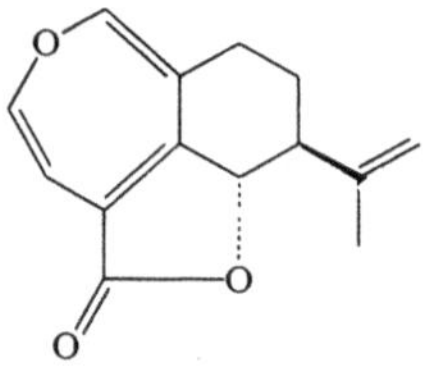

(1087) Senoxepin

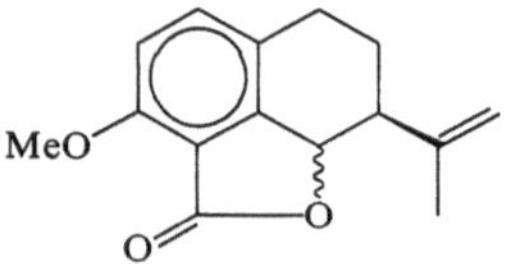

(1088) Othonna, lactone from

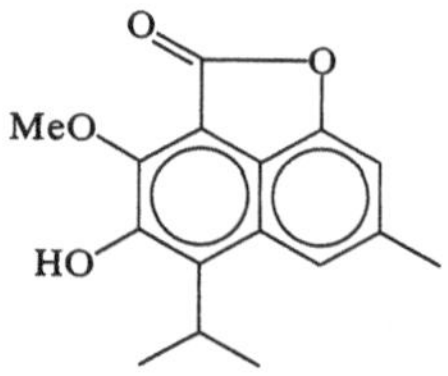

(1089) Bombax, lactone from

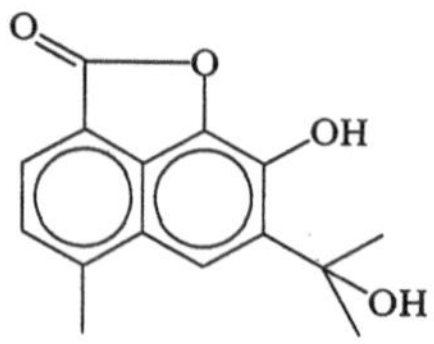

(1090) Emmotin D

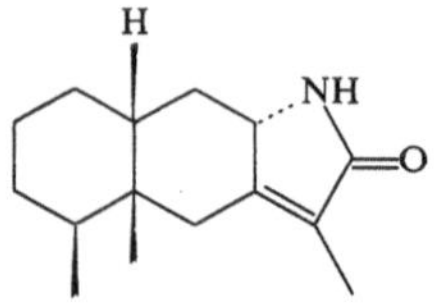

(1091) Eremophilene lactam

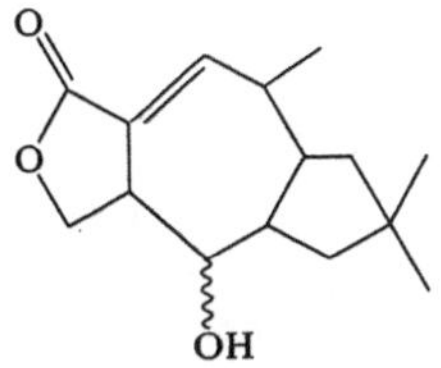

(1092) Blennin A

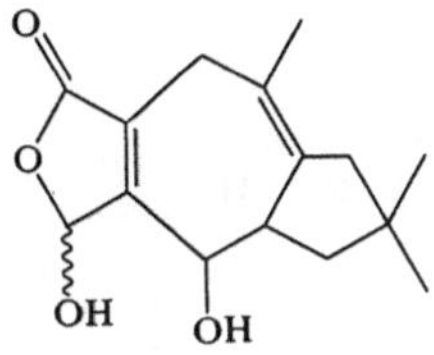

(1093) Blennin B

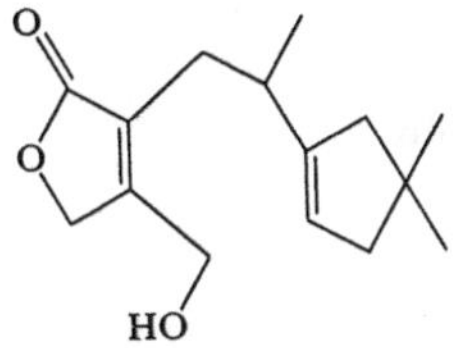

(1094) Blennin C
(Lactaronecatorin)

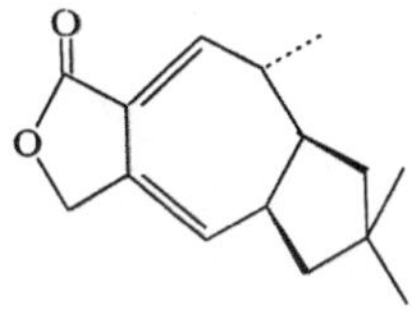

(1095) Lactarius lactone I
(Vallerolactone)

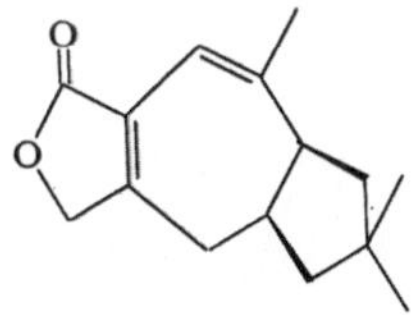

(1096) Lactarius lactone II
(Pyrovallerolactone)

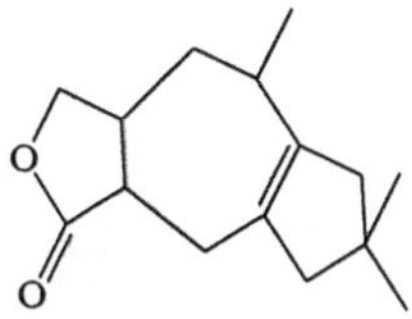

(1097) Lactarius lactone III

Chart IX-4 (continued)

(1098) Lactarius epoxylactone

(1099) Lactarorufin A; R = OH
(1100) Lactarorufin A, 3-deoxy; R = H

(1101) Lactarorufin A, anhydro

(1102) Lactarorufin B

(1103) Lactarorufin, iso

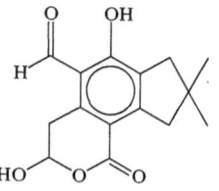

(1104) Illudalic acid

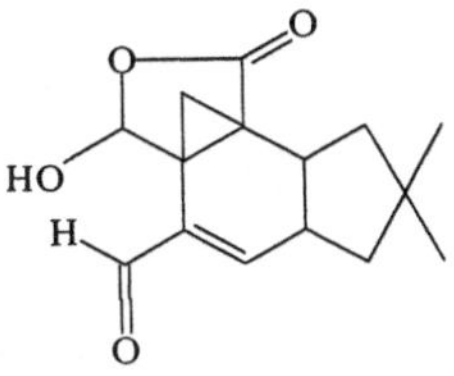

(1105) Marasmic acid

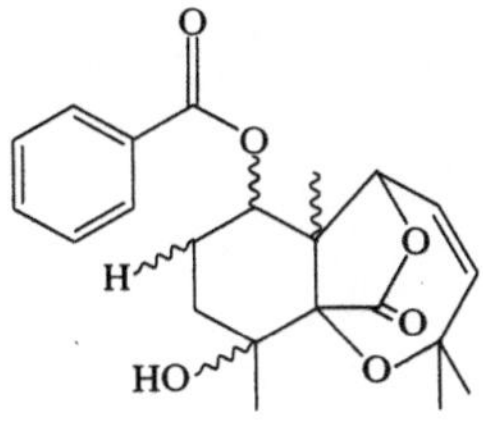

(1106) Mortonin A

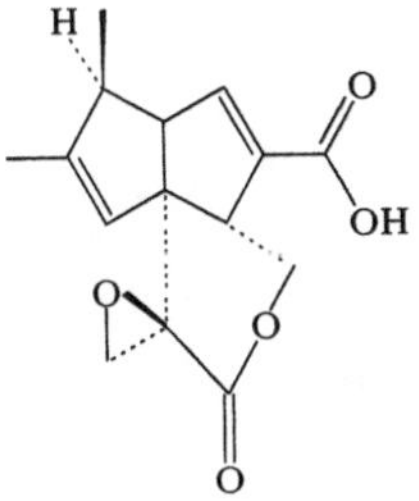

(1107) Pentalenolactone

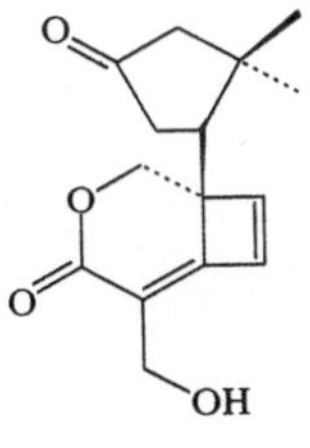

(1108) Fomannosin

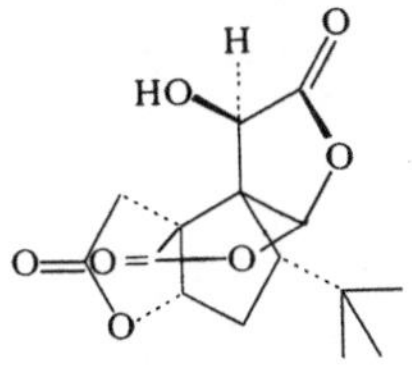

(1109) Bilobalide

Chart IX-4 (continued)

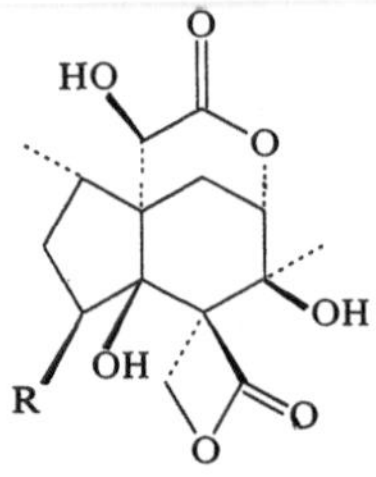

(1110)   Anisatin, neo; R = H
(1111)   Anisatin; R = OH

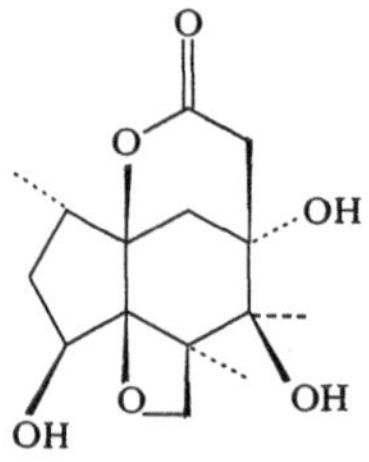

(1112)   Anisatin, pseudo

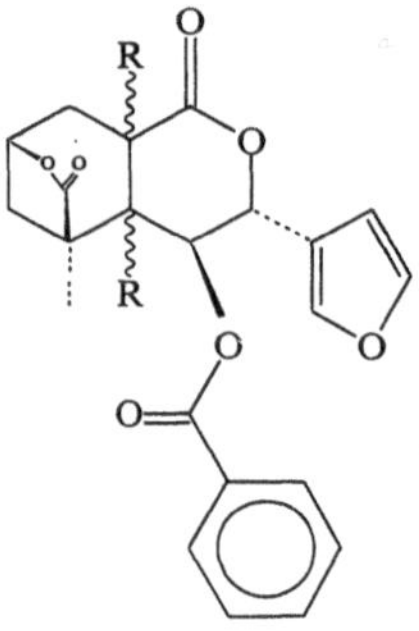

(1113)   Collybolide; R = H
(1114)   Collybolide, iso; R = α-H

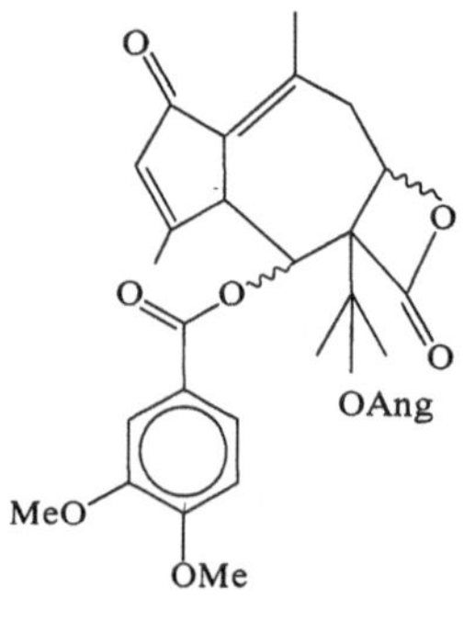

(1115)   Diversolide (misprint?)

occurred. Pallescensin 3 (**1077**) (*164*) and the paniculides A to C (**1078**) to (**1080**) (*20*) are monocyclization-oxidation products in which the head of the farnesol unit has been cyclized in pallescensin 3 whereas the paniculides represent oxidized bisabolenolides, their biogenesis being sketched in Chart IX-5. Arteannuin (**1083**) (*1106a*) and lactone (**1089**) (*976*)

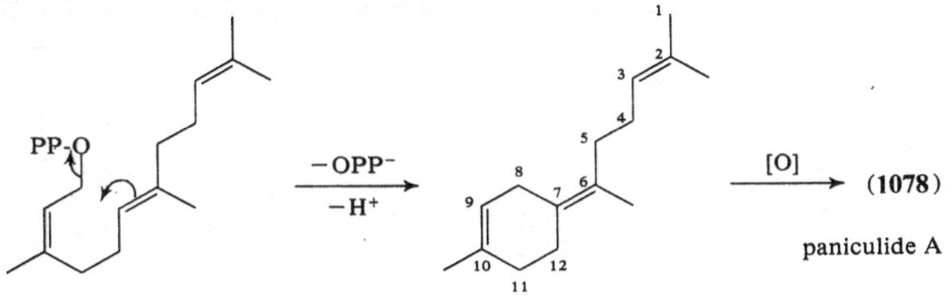

Chart IX-5. Biogenesis of bisabolenolides

represent the only sesquiterpene lactones with a cadinane skeleton the biogenesis being shown in Chart IX-6 (*393*). Cyclization of *cis,trans*-farnesol pyrophosphate gives cation (**1118**) which after 1,3-hydride shift provides the allylic cation (**1119**). Upon cyclization the cation (**1120**) is formed which by oxidative modifications gives arteannuin (**1083**).

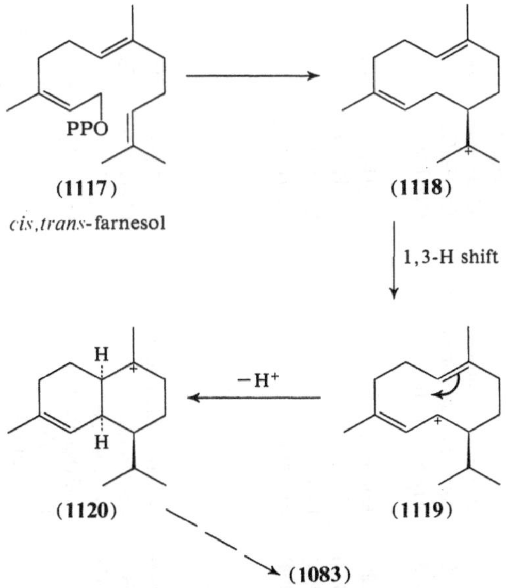

Chart IX-6. Biogenesis of cadinanolides

Table IX-3. *Special Structural Types of Sesquiterpene Lactones*

Structure Number	Name of Compound	Formula	m.p. °C	[α]D	Plant Source[b]	References	Comments
(1111)	Anisatin	$C_{15}H_{20}O_8$	227–8	−28	*Illicium anisatum*	(635, 1149, 925, 1145, 1146, 1147)	Illiciaceae
(1110)	Anisatin, neo	$C_{15}H_{20}O_7$	237–8	−25	*Illicium anisatum*	(1147, 1145, 1146)	Illiciaceae
(1112)	Anisatin, pseudo	$C_{15}H_{22}O_6$	207–8	—	*Illicium anisatum*	(797)	Illiciaceae
(1068)	Anthemis, lactone from	$C_{15}H_{20}O_3$	oil	+116	*Anthemis cotula*	(129 a)	
(1083)	Arteannuin B	$C_{15}H_{20}O_3$	152	−6	*Artemisia annua*	(527, 656[a], 1106 a[a])	Cadinane skeleton; $[\theta]_{256} = -4545$
(1109)	Bilobalide	$C_{15}H_{18}O_8$	300	−64	*Ginkgo biloba*	(750, 1133)	Ginkoaceae
(1092)	Blennin A	$C_{15}H_{22}O_3$	—	—	*Lactarius blennius*	(1114)	Russulaceae
(1093)	Blennin B	$C_{15}H_{20}O_4$	—	—	*Lactarius blennius*	(1114)	Russulaceae
(1094)	Blennin C	$C_{15}H_{20}O_3$	—	—	*Lactarius blennius*	(1114, 212)	Russulaceae; reported before Lactarionecatorin
					L. necator	(204)	Russulaceae
					L. scrobiculatus	(212)	Russulaceae
(1089)	Bombax, lactone from	$C_{16}H_{16}O_4$	210	—	*Bombax malabaricum*	(976)	Bombacaceae
(1085)	Cacalolide	$C_{15}H_{18}O_3$	158–9	+18	*Cacalia delphinijfolia*	(758)	
(1113)	Collybolide	$C_{22}H_{20}O_7$	—	—	*Collybia maculata*	(819[a])	Tricholomataceae
(1114)	Collybolide, iso	$C_{22}H_{20}O_7$	—	—	*Collybia maculata*	(819[a])	Tricholomataceae
(1115)	Diversolide	$C_{30}H_{34}O_9$	—	—	*Ferula diversivittata*	(582)	Umbelliferae
(1090)	Emmotin D	$C_{15}H_{14}O_4$	209–10	—	*Emmotum nitens*	(215 b, 215 c)	Icacinaceae
(1091)	Eremophilene lactam	$C_{15}H_{23}NO$	215–8	+199	*Petasites hybridus*	(528)	
(1108)	Fomannosin	$C_{15}H_{18}O_4$	oil	—	*Fomes annosus*	(570, 146, 726, 608)	Polyporaceae
(1074)	Freelingnite	$C_{15}H_{18}O_3$	162–3	+11	*Eremophila freelingii*	(596)	Myoporaceae
(1075)	Freelingyne	$C_{15}H_{12}O_3$	181–3	—	*Eremophila freelingii*	(680, 500, 499, 595)	Myoporaceae
(1076)	Freelingyne, dihydro	$C_{15}H_{14}O_3$	200	±0	*Eremophila freelingii*	(500)	Myoporaceae
(1104)	Illudalic acid	$C_{15}H_{16}O_5$	—	—	*Clitocybe illudens*	(749)	Trichomataceae

Ref	Name	Formula	mp (°C)/oil	[α]	Source	References	Family / Comments
(1095)	Lactarius Lactone I	$C_{15}H_{20}O_2$	oil	+364	Lactarius vellereus	(669)	Russulaceae
					L. pergamenus	(669)	Russulaceae
(1096)	Lactarius Lactone II	$C_{15}H_{20}O_2$	41–4	−73	Lactarius vellereus	(668)	Russulaceae
					L. pergamenus	(668)	Russulaceae
(1097)	Lactarius Lactone III	$C_{15}H_{20}O_2$	—	—	Lactarius scrobiculatus	(212)	Russulaceae
(1098)	Lactarius epoxylactone	$C_{15}H_{20}O_3$	oil	+102	Lactarius scrobiculatus	(1115)	Russulaceae
(1094)	Lactaronecatorin				Lactarius necator	(204)	See Blennin C; Russulaceae
(1099)	Lactarorufin A	$C_{15}H_{22}O_4$	156–8	+7	Lactarius rufus	(203, 202, 46)	Russulaceae
(1101)	Lactarorufin A, anhydro	$C_{15}H_{20}O_3$	—	—	Lactarius necator	(204)	Russulaceae
(1100)	Lactarorufin A, 3-deoxy	$C_{15}H_{22}O_3$	—	—	Lactarius necator	(205)	Russulaceae
(1102)	Lactarorufin B	$C_{15}H_{20}O_4$	—	—	Lactarius rufus	(208, 92)	Russulaceae
(1103)	Lactarorufin, iso	$C_{15}H_{22}O_4$	—	—	Lactarius rufus	(202, 601)	Russulaceae
(1073)	Lasiosperman, E-7,12,13H-5-dehydro-12-oxo	$C_{15}H_{18}O_3$	oil	—	Athanasia linifolia	(102)	
(1072)	Lasiosperman, Z-7,12,13H-5-dehydro-12-oxo	$C_{15}H_{18}O_3$	oil	—	Athanasia linifolia	(102)	
(1081)	Lemnalactone	$C_{15}H_{22}O_2$	—	—	Paralemmalia digitoformis	(201)	
(1069)	cis-Linifolon	$C_{15}H_{20}O_3$	68	—	Athanasia linifolia	(102)	
(1070)	trans-Linifolon	$C_{15}H_{20}O_3$	48	—	Athanasia linifolia	(102)	
(1071)	Linifolon, Z-2,3-dihydro-3-acetoxy	$C_{17}H_{24}O_5$	oil	—	Athanasia linifolia	(102)	
(1105)	Marasmic Acid	$C_{15}H_{18}O_4$	173–4	+182	Marasmius conigenus	(244)	
(1106)	Mortonin	$C_{22}H_{26}O_6$	199–200	+80	Mortonia gregii	(899)	
(1088)	Othonna, lactone from	$C_{15}H_{16}O_5$	107	+32	Othonna cylindrica	(101)	
(1082)	12-Oxolemnacarnol	$C_{15}H_{22}O_3$	161–2	−20	Paralemmalia thyrsoides	(201)	
(1077)	Pallescensin 3	$C_{15}H_{20}O_3$	—	—	Disidea pallescens	(164)	
(1078)	Paniculide A	$C_{15}H_{20}O_4$	120–1	+4	Andrographis paniculata	(20)	Bisabolene skeleton; source is a tissue culture
(1079)	Paniculide B	$C_{15}H_{20}O_5$	145–6	+37	Andrographis paniculata	(20)	Bisabolene skeleton; source is a tissue culture
(1080)	Paniculide C	$C_{15}H_{18}O_5$	oil	—	Andrographis paniculata	(20)	Bisabolene skeleton; source is a tissue culture
(1107)	Pentalenolactone	$C_{15}H_{16}O_5$	61–2	−172	Streptomyces 8403-MC$_1$	(1057, 674)	
(1086)	Platyphyllide	$C_{14}H_{14}O_2$	oil	−17	Senecio platyphylloides	(106)	
(1084)	Quing Hau Sau	$C_{15}H_{22}O_5$	—	—	Artemisia annua	(844)	
(1087)	Senoxepin	$C_{14}H_{14}O_3$	oil	−126	Senecio platyphylloides	(106)	

a Reference gives X-ray data;   b Family name is given under Comments.

Quing Hau Sau (**1084**) (*846*) appears to be a biomodified cadinanolide. The lactone (**1081**) and (**1082**) (*201*) show features of the cadinanolides as well as the eremophilanolides. The latter skeletal type must be a precursor for cacalolide (**1085**) (*758*) since this skeletal system has been obtained in a laboratory conversion from an eremophilanolide (*109 a*). The nor-sesquiterpenoids (**1086**) to (**1088**) (*101, 106*) seem to be biogenetically derived from either the eudesmane or the eremophilane skeleton most likely by an oxidation decarboxylation process. The biogenesis of senoxepin (**1087**) can be rationalized as outlined in Chart IX-7 (*106*). Enzymatic oxidation of platyphyllide (**1086**) gives the epoxide (**1121**) which rearranges to the oxepin (**1087**). Emmotin D (**1090**) (*215b*)

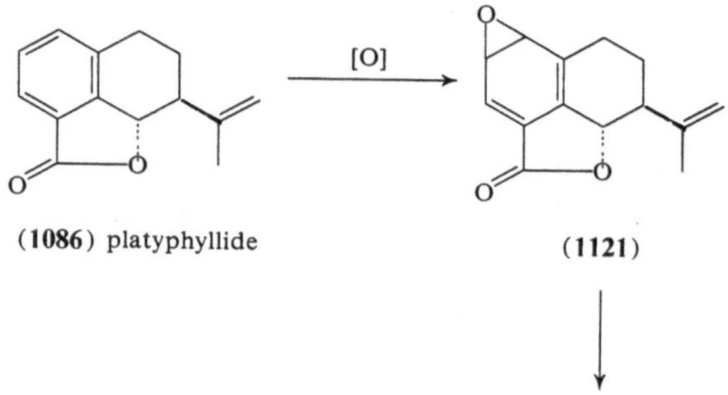

(**1086**) platyphyllide   (**1121**)

(**1087**) senoxepin

Chart *IX-7*. Biogenesis of senoxepin

appears to be biogenetically derived from the eudesmanolide skeleton in which a C-10 to C-1 methyl shift has occurred thus allowing formation of the aromatic ring system. Although it is not a lactone, it should be mentioned that the first sesquiterpene lactam, eremophilene lactam (**1091**), has recently been reported (*528*).

An increasing number of lactones (**1092**) to (**1105**) from higher mushrooms, most of them belonging to the genus *Lactarius* of the Russulaceae, is being isolated. Their biogenesis seems to involve a number of rearrangements which are outlined in Chart IX-8 (*244*). Cyclization of *cis,trans*-farnesol pyrophosphate (**1117**) provides the medium ring cation (**1122**) which by a 1,2-hydride shift rearranges to (**1123**). Subsequent cyclization gives cation (**1124**) which upon further rearrangement

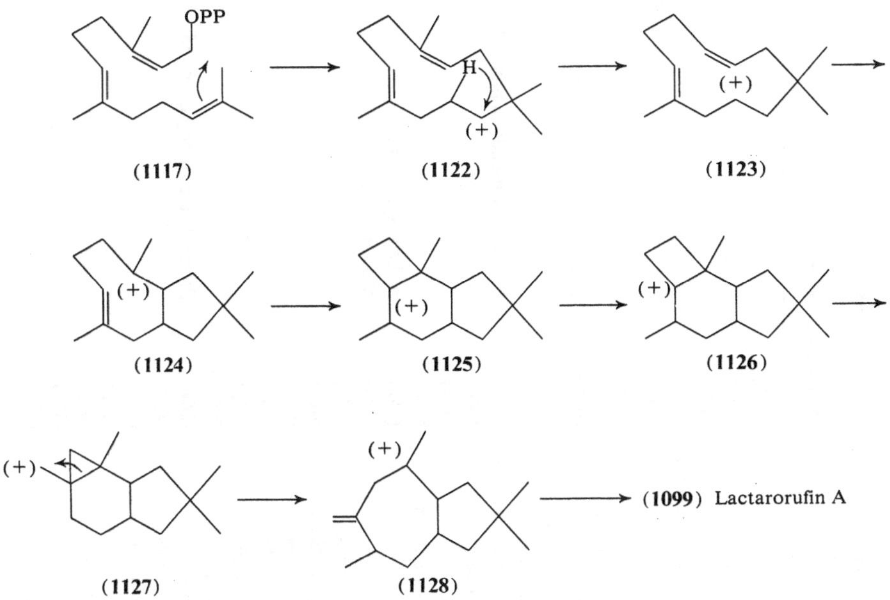

Chart IX-8. Biogenesis of lactarorufin A and analogs

via (1125) and (1126) produced the tricyclic cyclopropane derivative (1127). Ring opening of (1127) gives the basic 5,7-ring skeleton (1128) from which lactarorufin A (1099) could be derived by oxidative modifications. Evidence for the cyclopropane intermediate (1127) is provided by the natural occurrance of isolactarorufin (1103) (601) and marasmic acid (1105) (244).

More complex lactones are shown at the end of Chart IX-4. Among those the trilactone bilobalide (1109) (750) from Ginkgo biloba of the Ginkgoaceae represents a rare naturally occurring compound with a tert-butyl group.

X. Sesquiterpene Lactones of Unknown Structure

A considerable number of sesquiterpene lactones of unestablished structure has been reported in the literature. These compounds, tabulated in alphabetical order with available physical data and plant sources, are given in Table X-1.

Table X-1. *Naturally Occurring Sesquiterpene Lactones of Undetermined Structure*

Name of Compound	Formula	m. p. °C	[α]$_D$	Plant Source	References	Comments
Achillea lactone	C$_{15}$H$_{22}$O$_3$	138	−66	*Achillea millefolium*	(472)	
Arnicolide E	C$_{20}$H$_{30}$O$_5$	151–3	—	*Arnica montana*	(836)	
Aristalin	C$_{20}$H$_{28}$O$_7$	204–6	—	*Gaillardia grandiflora*	(425)	
Baldvernin				*Vernonia baldwini* and other *V. species*	(5)	Forms diacetate, m. p. 194–5°
Calendin	C$_{15}$H$_{22}$O$_4$	153	—	*Calendula officinalis*	(1014)	
Diplophyllolide B	C$_{15}$H$_{22}$O$_2$	92–100	—	*Diplophyllum albicans*	(68)	
Elegin	C$_{19}$H$_{23}$O$_6$	158–9	—	*Saussurea elegans*	(982)	
Eremantholide B	C$_{20}$H$_{26}$O$_6$	—	—	*Eremanthus elaeagus*	(848)	
Grossheimia lactone	C$_{15}$H$_{18}$O$_4$	201–3	—	*Grossheimia ossica*	(837)	Probably a pseudoguaianolide
Helenalin isomer	C$_{15}$H$_{18}$O$_4$	149–50	—	*Helenium scorzoneraefolium*	(396)	
Hirsutolide	C$_{16}$H$_{20}$O$_5$	110–2	−149	*Venidium hirsutum*	(344)	
Igalin	C$_{15}$H$_{20}$O$_3$	134–5	—	*Inula grandis*	(775)	
Istanbulin B	C$_{15}$H$_{20}$O$_3$	167	—	*Smyrnium olusantrum*	(1106)	A diepoxyguaianolide
Ivaxillin	C$_{15}$H$_{22}$O$_4$	173–6	−117	*Iva axillaris*	(452)	
Jurinea lactone	C$_{19}$H$_{26}$O$_6$	59–61	+64	*Jurinea olata*	(243)	Monohydrate
Libanotis lactone	C$_{15}$H$_{20}$O$_3$	140–1	—	*Libanotis intermedia*	(742)	
Ludovicin D	C$_{15}$H$_{18}$O$_4$	230–2	—	*Artemisia ludoviciana*	(642)	
Mexicanin B	C$_{17}$H$_{24}$O$_5$	212–4	+39	*Helenium mexicanum*	(904)	
Micranthin	C$_{19}$H$_{26}$O$_7$	284–6	+123	*Achillea micrantha*	(767)	
Millefolide	C$_{15}$H$_{22}$O$_3$	138–40	+107	*Achillea millefolium*	(472)	
Monogynin	C$_{15}$H$_{20}$O$_3$	138	—	*Artemisia monogyna*	(553 a)	
Nevadivalin	C$_{15}$H$_{22}$O$_3$	129–31	+173.5	*Iva nevadensis*	(264 a)	
Pulchellin D	C$_{17}$H$_{24}$O$_6$	182–5	—	*Gaillardia pulchella*	(414)	
Rutifolin	C$_{15}$H$_{18}$O$_5$	234–6	±0	*Artemisia caucasica*	(607)	Probably a pseudoguaianolide
				A. rutafolia	(255)	
Saussurea elegans, lactone from	C$_{16}$H$_{22}$O$_3$	194–5	—	*Saussurea elegans*	(982)	
Senecio aegyptius, ketolactone from	C$_{15}$H$_{20}$O$_3$	235–8	−94	*Senecio aegyptius*	(371)	
Stenophyllolide	C$_{15}$H$_{20}$O$_4$	186–8	+71	*Centaurea aspera*	(939)	

Tachillin	$C_{20}H_{26}O_5$	208–12	—	*Tanacetum pseudoachillea*	(1182)	Monoacetate, m. p. 100–1°; tetrahydro der., m. p. 133–5°
Tanacetum lactone I	$C_{15}H_{22}O_4$	194–6	—	*Tanacetum vulgare*	(345)	
Tanacetum lactone II	$C_{15}H_{22}O_4$	137–40	—	*Tanacetum vulgare*	(345)	
Tanacetum lactone III	$C_{15}H_{22}O_4$	133–7	—	*Tanacetum vulgare*	(345)	
Temisin, dihydroiso	$C_{15}H_{24}O_3$	70–3	−24.4	*Artemisia cina*	(749, 1138b)	
Trilobolide, iso	$C_{27}H_{38}O_{10}$	235–40	—	*Laser trilobum*	(475)	
Tutin, pseudo		184	—	*Coriaria japonica*	(797a)	
Ursiniolide A	$C_{22}H_{28}O_7$	140–1	—	*Ursinia anthemoides*	(343)	Monohydrate
Ursiniolide B	$C_{24}H_{32}O_9$	153–5	—	*Ursinia anthemoides*	(343)	
Ursiniolide C	$C_{20}H_{26}O_6$	173–5	—	*Ursinia anthemoides*	(343)	
Venidolide	$C_{20}H_{26}O_7$	58	+38	*Venidium hirsutum*	(344)	
Verafinin B	$C_{19}H_{20}O_6$	137–40	—	*Verbesina aff. coahuilensis*	(365)	

XI. Addendum

Since completion of the main body of this review a large number of publications related to sesquiterpene lactones has appeared. Most data in this Addendum which covers reports published for the most part in 1978 and early 1979 will be presented in Charts and Tables as in the main part of the review. The addenda to the Tables will also contain information about new plant sources of known sesquiterpene lactones as well as references to significant new chemical data concerning known compounds. Notes for the Tables are the same as in the main Tables.

Reports dealing with various aspects of biological activities of sesquiterpene lactones (*1218, 1255, 1276a, 1280, 1284a, 1302*) and of biochemical systematic interest (*1219, 1232, 1239, 1268*) have appeared.

1. Germacranolides

The revision of the structure of baileyin from a germacrolide (**129**) to a melampolide (**1193**) is of considerable significance to hypotheses dealing with the biogenesis of helenanolides. [Herz *et al.*, (*1241*).] The biogenetic aspects of the new findings will be discussed in detail in the pseudoguaianolide section.

Samek and Harmatha (*1290*) discuss stereochemical assignments of α-methylene-γ-lactones in germacra-1(10),4-dienolides on the basis of allylic and vicinal couplings of the lactonic bridgehead protons. They also present a convenient symbolism to describe conformations of germacra-1(10),4-dienolides as follows:

"The symbols ^xD_y and $_xD^y$, $_xD_y$ and $^xD^y$ express the absolute configuration of the side chains of the endocyclic double bond, two *trans* and two *cis*. The letters x and y symbolize the numbers of the atoms X and Y that are directly bound to the double bond, in arbitrary numbering. These letters, given in the position of the index, mean α-configuration and in the position of the exponent β-configuration $\binom{x}{y} = \binom{\beta}{\alpha}$. The symbols do not require additional definitions, as the C(crossed) and T(parallel) symbols do (*1037*). Crossed and parallel conformations follow immediately from the symbols."

For instance, the conformations of the germacradienolide skeleton (**28**) presented in Chart III-13 (page 89) could be described as follows: $[_1D^{14}, {}^{15}D_5]$ for (**305**), $[^1D_{14}, {}^{15}D_5]$ for (**306**), $[^1D_{14}, {}_{15}D^5]$ for (**308**) and $[_1D^{14}, {}^{15}D^5]$ for (**307**). A conformational description of costunolide (**17**) would be $[_1D^{14}, {}^{15}D_5]$-costunolide.

Modifications involving protection of the lactonic exocyclic me-thylene function were used in catalytic hydrogenations of endocyclic germacranolides (*1290*) as well as in Cope rearrangements and acid-catalyzed cyclizations of germacranolides (*1195a*). Cyclizations of germa-

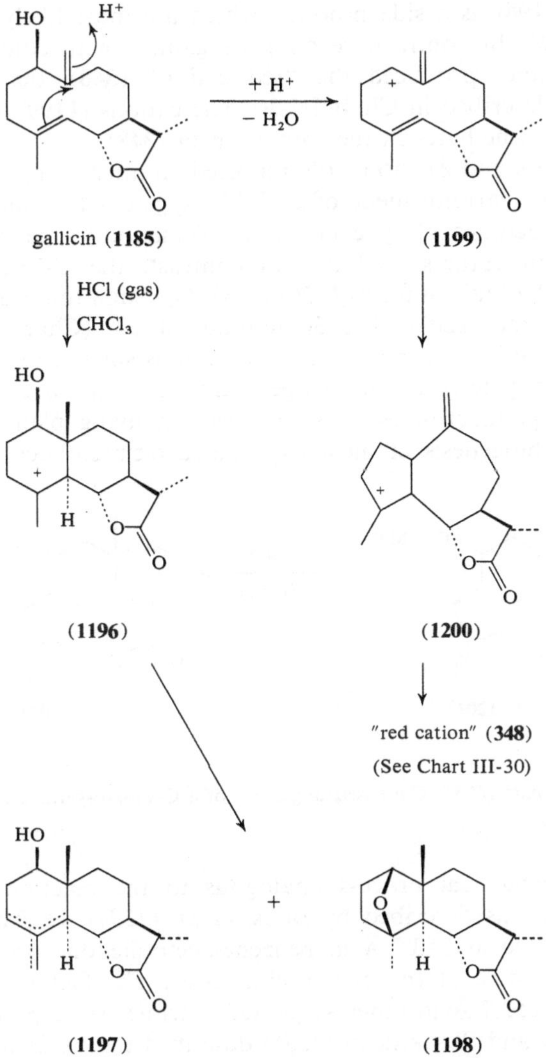

Chart III-37. Acid-catalyzed cyclization of gallicin

cranolides to eudesmanolides without protection of the lactonic methylene function was facilitated by oxymercuration-demercuration (1235).

Gonzalez and coworkers (1231) provided experimental evidence for Geissman's hypothesis (288) that C-1β-hydroxylated eudesmanolides might be formed via C-1(14)-unsaturated germacranolides. As shown in Chart III-37, treatment of gallicin (1185) with dry HCl in CHCl$_3$ provided a mixture of C-1β-hydroxy eudesmanolides (1197), as well as the ether (1198) as a side product, which are most likely derived from cation (1196). In contrast, reaction of gallicin with cold concentrated HCl in ethanol produced the "red cation" (348), the formation of which was described in Chart III-30. The cations (1199) and (1200) are probable intermediates in the formation of (348).

Bohlmann and Zdero (1209) reported an interesting reaction related to the Cope rearrangement of a C-15-oxygenated germacrolide. Conversion of lactones of type (1136) to elemanolides generally required elevated temperatures (> 140°). In contrast, the C-15-aldehyde derivative (1201), obtained from (1136) by MnO$_2$-oxidation, was converted to the elemanolide (1202) after 30 minutes at 60° (Chart III-38). These considerably milder rearrangement conditions suggest that spontaneous, instead of enzyme-controlled Cope rearrangements of C-14- and C-15-oxygenated germacrolides possibly occur in living plants. This would explain the biogenesis of most oxygenated elemanolides.

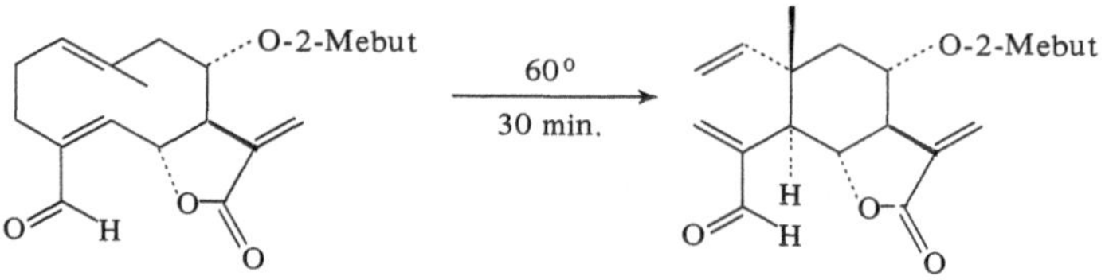

Chart III-38. Cope rearrangement of a C-15-oxo-germacrolide

A photochemical process analogous to the reaction presented in Chart III-35 was described by Blum et al. (1204). Herbolide A (106), when irradiated at 253.7 Å in benzene-acetophenone provided the 9β-acetoxy derivative of the guaianolide (358). The fact that no reaction occurred at 253.7 Å in ether suggested a triplet state process.

Ruecker and Schikarski (1288) described the oxidation of a furan-germacradiene with 2,3-dichloro-5,6-dicyanobenzoquinone (DCC) to a germacrolide. This gentle oxidative modification of the furan ring resembles the reactions shown in Chart VIII-4. Another furan modification involves peracid oxidation (*1212*).

A 1,6-Michael addition with attack of a thiol at C-5 of eremantholide A (**222**) was described by LeQuesne et al. (*1258*).

2. Eudesmanolides

In addition to reports on new eudesmanolides several publications have been concerned with chemical modifications and syntheses of this group of lactones.

An intramolecular Diels-Alder approach toward the synthesis of eudesmanolides was used by WILSON and MAO (*1305*). The synthesis of yomogin (**445**) and pinnatifidin (**443**) from α-santonin (**405**) has been described by YAMAKAWA et al. (*1308*) and further transformations of α-santonin (*1201, 1222, 1309*) and badkhysinin (**385**) (*1294*) have been reported.

GOVINDAN and BHATTACHARYYA (*1236*) transformed isoalantolactone (**446**) to isotelekin (**458**) by SeO₂-oxidation which resulted in the introduction of an OH-group at C-3 of (**446**). Oxidation of α-cyclocostunolide (**330**) and its 11,13-dihydroderivative by Rose-Bengal-sensitized photo-

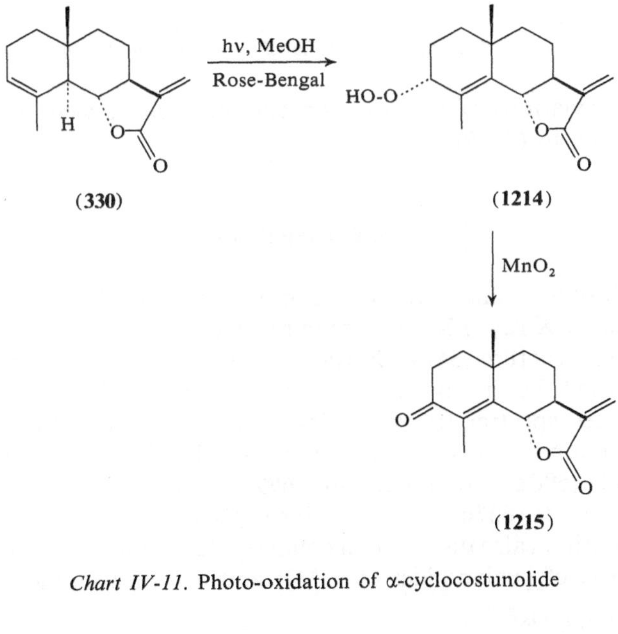

Chart IV-11. Photo-oxidation of α-cyclocostunolide

19*

reaction provided the hydroperoxide (1214) which upon MnO_2-treat-ment gave the 1,2 dihydro 11,13-dehydrosantonin (1215). (Chart IV-11.) The authors suggested that this photoreaction might be a model for the biosynthesis of santonin.

Corbet and Benezra (1215) described a degradation-synthesis pro-cedure for the specific introduction of labeled carbon at C-13 in iso-alantolactone (446). The reaction sequence involved a Pummerer re-arrangement of sulfoxide intermediates with an overall transformation of isoalantolactone (446) to the lactone intermediate (1216) back to the labeled isoalantolactone (446) (Chart IV-12).

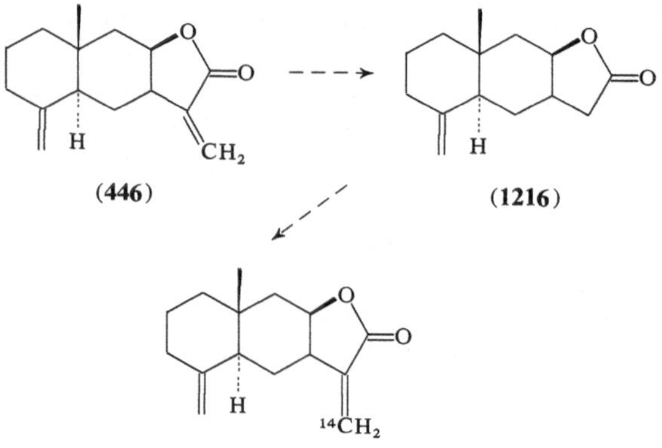

Chart IV-12. Specific C-13-labeling of isoalantolactone

C-11 hydroxylations of 11,13-dihydroeudesmanolides were described by Gonzalez et al. (1229).

3. Guaianolides

The absolute configuration of compound (341) in Chart III-28 was determined by X-ray (1248) and required revision of configuration at C-5 as shown in structure (341a). X-Ray analysis of pleniradin [old structure (722)] revealed that it represents a trans-guaianolide (722a) (1241). More detailed mass spectral fragmentations on grossmisin (628) (1312) and other guaianolides (1216) were reported. The full stereochemistry of montanolide (698a), isomontanolide (699a), isoacetylmontanolide (700a) and archangolide (667a) was determined (1244) (see Addendum to Chart V-1). The authors also discussed in detail the determination of the absolute configuration of guaianolides by NMR and CD measurements.

4. Pseudoguaianolides

Of the theories related to the biogenesis of helenanolides, the melampolide route proposed by HERZ (*395*) and the germacrolide pathway suggested by FISCHER (*265*), (Chart VII-3), the melampolide route has received strong support recently (*1241*). The structural revision of baileyin from a germacrolide (**129**) to a melampolide skeleton (**1193**) together with the X-ray finding that pleniradin represents a trans-guaiananolide (**722a**) strongly suggests that these two co-occurring lactones are biogenetically related as outlined in Chart VII-3 (*1241*).

Oxidations with $KMnO_4$ of the lactonic α-methylene groups in pseudoguaiananolides to provide 11,13-dihydroderivatives were reported by ROMO *et al.* (*1286*). Finally, the finding of the first 11,13-epoxylactone, stramonin B (**1247**) is of interest (*1237*).

Addendum to Charts III-2 to III-8

(**1129**) Mollisorin A; R = OH, R₁ = H, R₂ = OAng
(**1130**) Mollisorin B; R = OH, R₁ = H, R₂ = OEpoxyang
(**1130a**) Costunolide, 3β-isovaleryloxy; R = R₂ = H, R₁ = O-*i*-Val

(**1131**) Costunolide, 3β,9β-dihydroxy-8β-[2-methylbutanoyloxy];

(**1132**) Salonitenolide, 8-desoxy,15-(3-hydroxy-2-methylacryloxy); R₁ = H, R₂ = O-Mac-4-OH
(**1133**) Salonitenolide, 8-desoxy,15-(2,3-epoxyisobutyryloxy); R₁ = H, R₂ = O-Epoxymac
(**1134**) Salonitenolide, 8-desoxy,15-(3-hydroxyisobutyryloxy); R₁ = H, R₂ = OA
(**1135**) Salonitenolide, 8-desoxy,15-(2,3-dihydroxyisobutyryloxy); R₁ = H, R₂ = OB
(**1136**) Salonitenolide-8(O)-[2-methylbutyrate]; R₁ = O-2-Mebut, R₂ = OH
(**1136a**) Costunolide, 15-hydroxy-8α-[α-methylacryl]; R₁ = OMac, R₂ = OH
(**1136b**) Costunolide, 15-hydroxy-8α-isobutyryl; R₁ = O-*i*-Bu, R₂ = OH

Addendum to Charts III-2 to III-8 *(continued)*

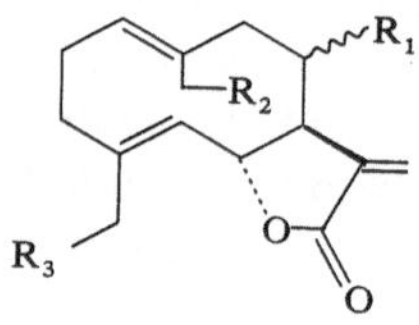

(1137) Linearilobin A; $R_1 = \beta$-OH, $R_2 =$ O-*i*-But, $R_3 =$ OAc

(1138) Linearilobin B; $R_1 = \beta$-OH, $R_2 =$ O-2-Mebut, $R_3 =$ OAc

(1139) Linearilobin C; $R_1 = \beta$-OA, $R_2 =$ OH, $R_3 =$ OAc

(1140) Linearilobin D; $R_1 = \beta$-OMac, $R_2 =$ OH, $R_3 =$ OAc

(1141) Linearilobin E; $R_1 = \beta$-OH, $R_2 =$ OMac, $R_3 =$ OAc

(1142) Linearilobin F; $R_1 = \beta$-OH, $R_2 =$ O-2-Mebut, $R_3 =$ OH

(1143) Albicolide, 8α-[2-methylbutyryloxy]; $R_1 = \alpha$-O-2-Mebut, $R_2 = R_3 =$ OH

A =

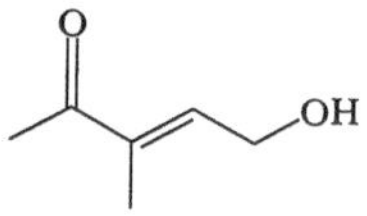

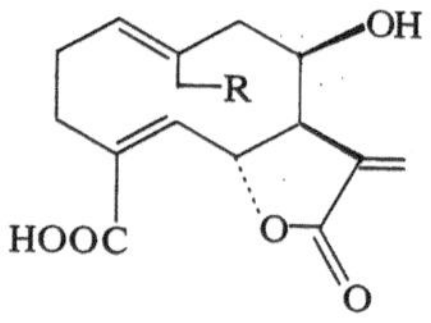

(1144) Linearilobin G; R = O-2-Mebut

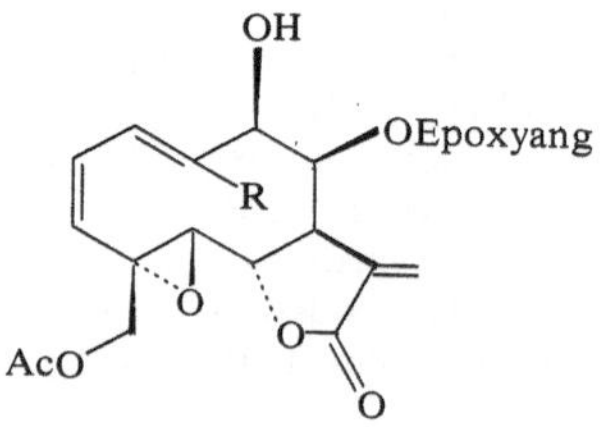

(1145) Melfusin; R = CO_2CH_3

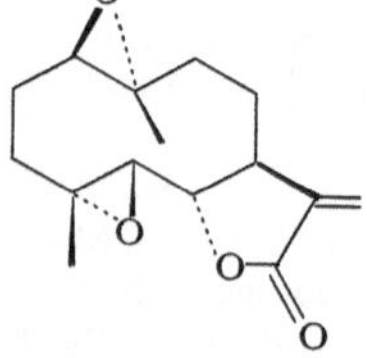

(1146) Costunolide diepoxide;

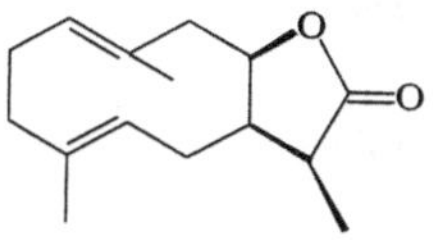

(1147) Inunolide, dihydro;

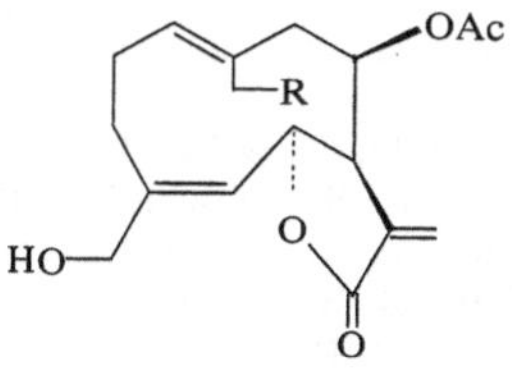

(1148) Linearilobin H; R = OMac
(1149) Linearilobin I; R = OTig

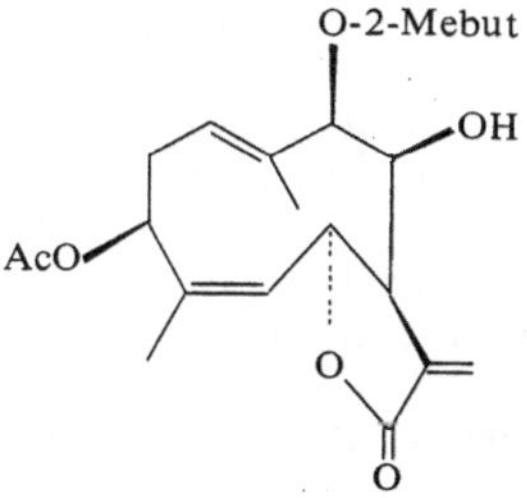

(1150) Heliangolide from *Eupatorium re-curvans*;

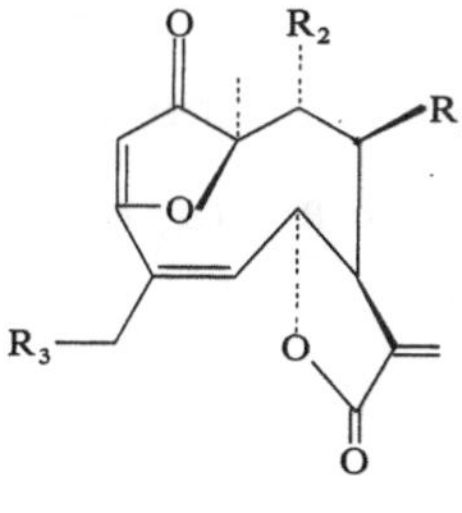

(1150a) Viguiepinin; R_1 = O-*i*-But, R_2 = H, R_3 = OH
(1151) Atripliciolide, 9α-hydroxy, 8-O-[2-methylacrylate]; R_1 = OMac, R_2 = OH, R_3 = H
(1152) Atripliciolide, 9α-[angeloyloxy]-15-hydroxy-8-O-[2-methylacrylate]; R_1 = OMac, R_2 = OAng, R_3 = OH
(1153) Atripliciolide, 9α-[isovaleryloxy]-15-hydroxy-8-O-[2-methylacrylate]; R_1 = OMac, R_2 = O-*i*-Val, R_3 = OH
(1154) Atripliciolide, 9α-[senecioyloxy]-15-hydroxy-8-O-[2-methylacrylate]; R_1 = OMac, R_2 = OSen, R_3 = OH

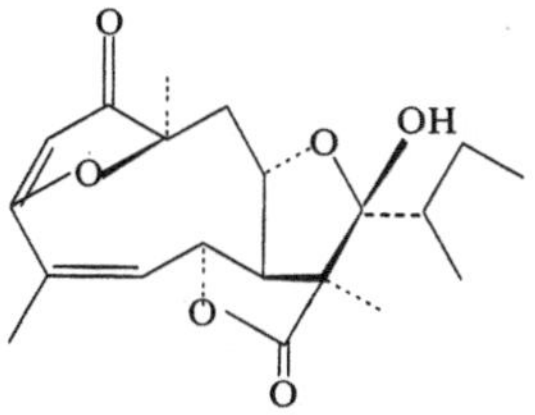

(1154a) Eremantholide B;

Addendum to Charts III-2 to III-8 *(continued)*

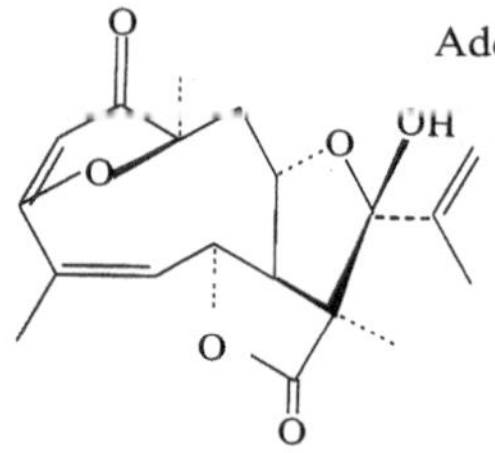

(1154b) Eremantholide C;

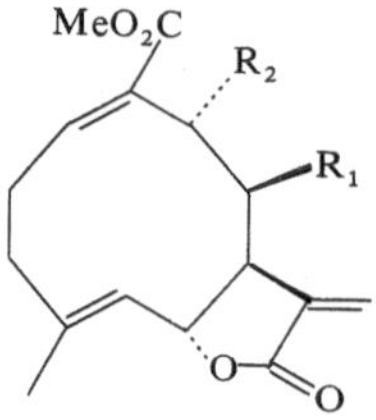

(1155) Polymatin A; R_1 = OAng, R_2 = OH
(1156) Polymatin B; R_1 = OAng, R_2 = OAc
(1157) Polymatin C; R_1 = OAc, R_2 = OEpoxyang
(1158) Tetrahelin A; R_1 = OA, R_2 = OAc
(1159) Tetrahelin C; R_1 = OB, R_2 = O-2-Mebut
(1160) Tetrahelin D; R_1 = OB, R_2 = OAc
(1161) Tetrahelin E; R_1 = OB, R_2 = OC
(1162) Tetraludin A; R_1 = OD, R_2 = OAc
(1163) Tetraludin B; R_1 = OD, R_2 = OE
(1164) Tetraludin C; R_1 = OE, R_2 = OD
(1165) Tetraludin D; R_1 = OF, R_2 = O-2-Mebut
(1166) Tetraludin E; R_1 = OF, R_2 = O-2-Mebut
(1167) Tetraludin F; R_1 = F, R_2 = O-*i*-But
(1168) Tetraludin G; R_1 = OF, R_2 = O-*i*-But
(1169) Tetraludin H; R_1 = OF, R_2 = OAc
(1170) Tetraludin I; R_1 = O-Epoxyang; R_2 = OE
(1171) Tetraludin J; R_1 = OD, R_2 = O-2-Mebut
(1172) Tetraludin K; R_1 = OD, R_2 = O-2-Mebut
(1173) Tetraludin L; R_1 = OD, R_2 = O-*i*-But
(1174) Tetraludin M; R_1 = OD, R_2 = O-*i*-But
(1175) Tetraludin N; R_1 = OF, R_2 = OE

A =   B =

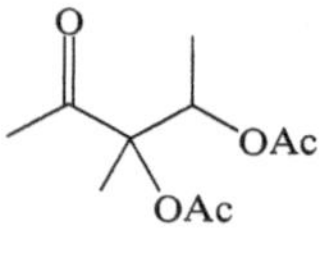

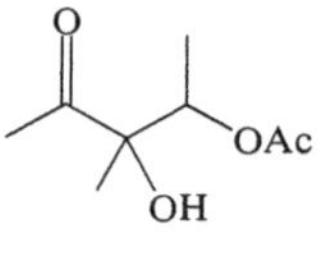

C 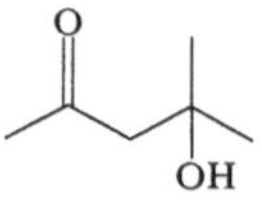 =   D =

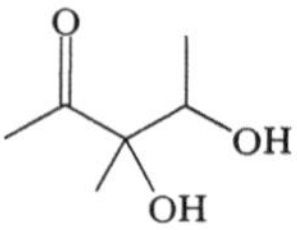

E 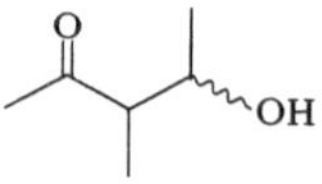 =   F =

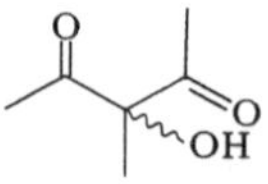

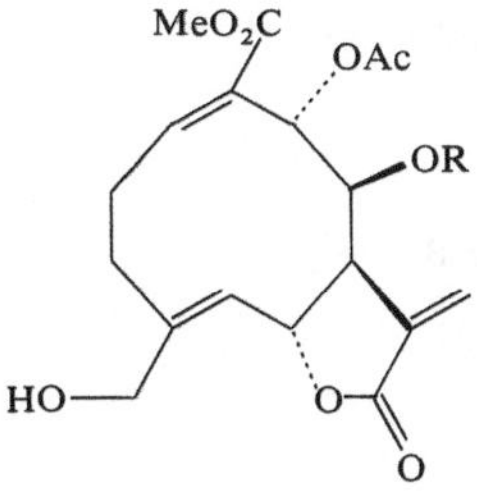

(1176) Tetrahelin B; R = OA
(1177) Tetrahelin F; R = OB

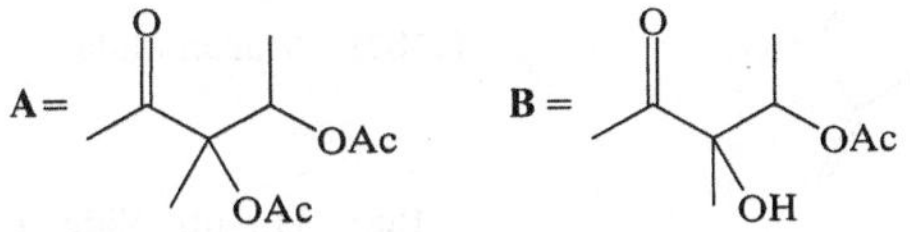

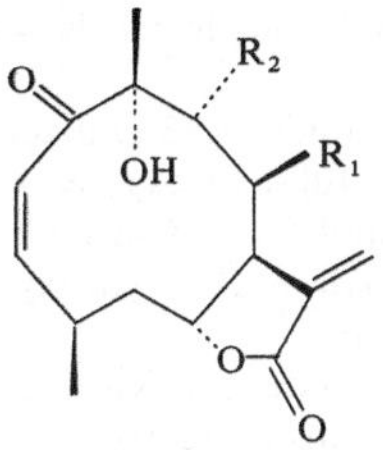

(1178) Calein A; R_1 = OAc, R_2 = OAng
(1179) Calein B; R_1 = OAng, R_2 = OAc
(1180) Neurolenin A; R_1 = O-i-Val, R_2 = H
(1181) Neurolenin B; R_1 = O-i-Val, R_2 = OAc

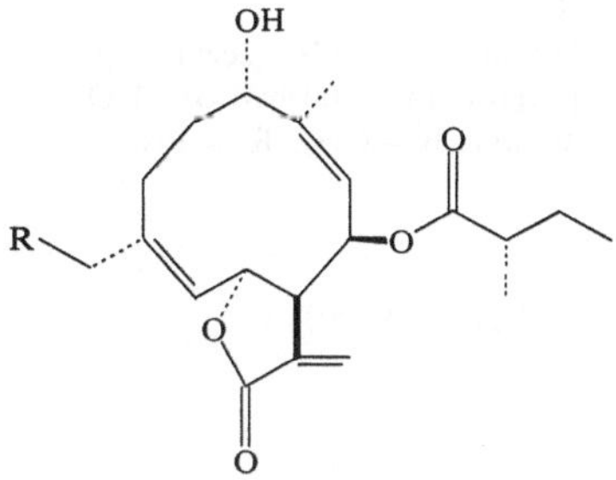

(1182) Eurecurvin, 15-deshydroxy; R = H
(1183) Eurecurvin; R = OH

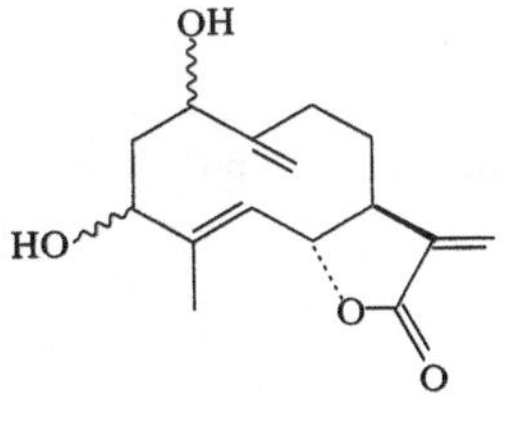

(1184) Ridentin, iso;

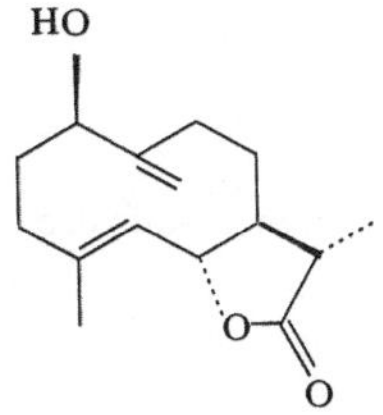

(1185) Gallicin

Addendum to Charts III-2 to III-8 *(continued)*

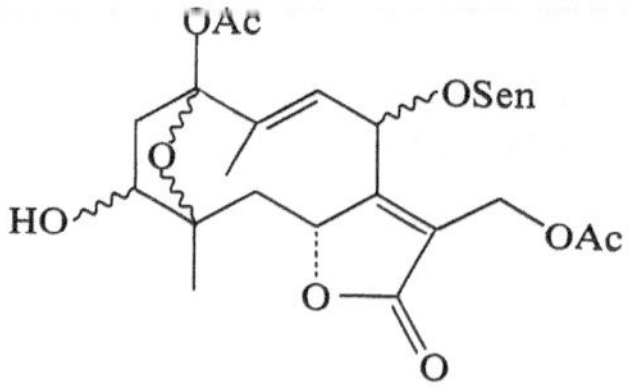

(1186) Fasciculide B

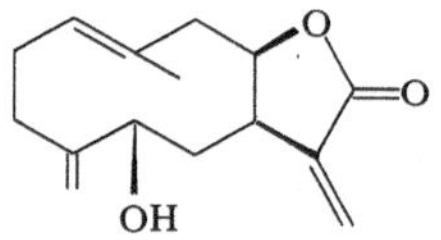

(1187) Maroniolide

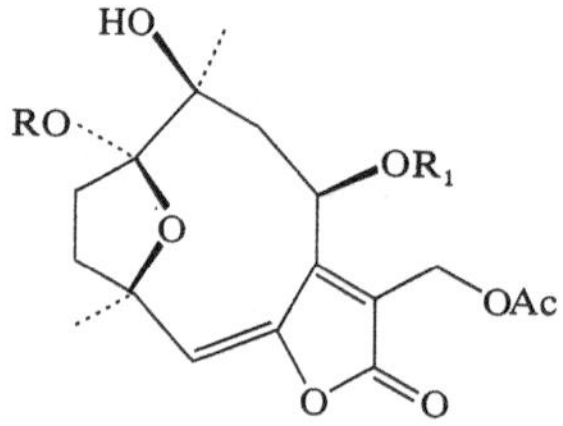

(1188) Hirsutinolide, 13-O-acetate-8β-acetoxy-10β-hydroxy; R = H, R$_1$ = Ac

(1189) Hirsutinolide, 1,13-O-diacetate-8β-acetoxy-10β-hydroxy; R = R$_1$ = Ac

(1190) Hirsutinolide, 13-O-acetate;-8β-propionyloxy-10β-hydroxy; R = H, R$_1$ = Pro

(1191) Hirsutinolide, 1,13-O-diacetate-8β-propionyloxy-10β-hydroxy; R = Ac, R$_1$ = Pro

(1192) Hirsutinolide, 13-O-acetate-8β-propionyloxy-10β-hydroxy-1-O-methyl; R = CH$_3$, R$_1$ = Pro

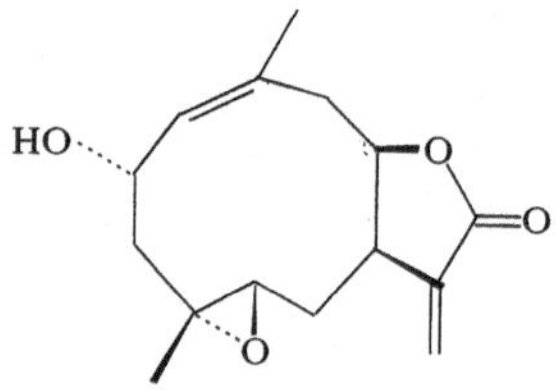

(1193) Baileyin (revised)

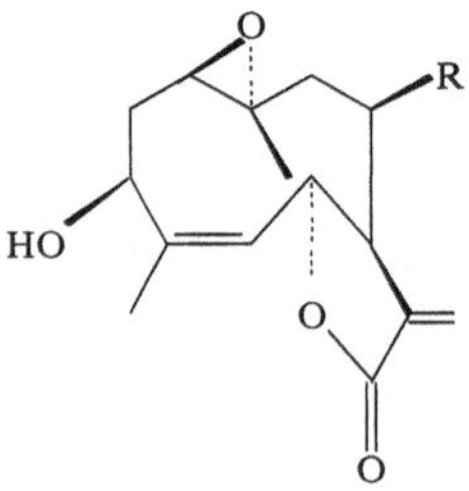

(1194) Tagitinin E; R = β-O-*i*-But

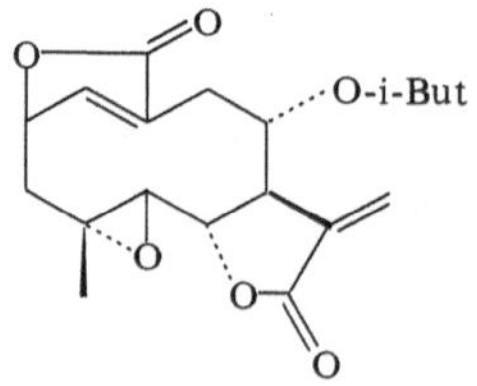

(1195) Elephantopin, 3′-dihydro;

Addendum to Chart IV-1

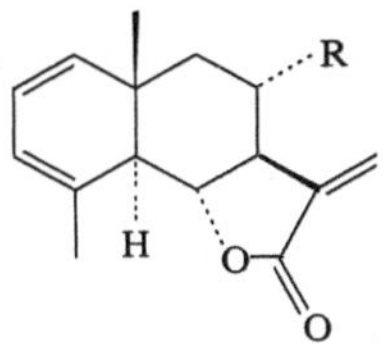

(**1203**) Gazaniolide; R = H
(**1204**) Gazaniolide, 8α-isovaleroyloxy;
R = O-*i*-Val

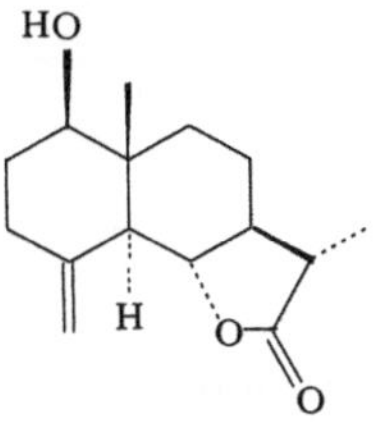

(**1205**) Reynosin, dihydro;

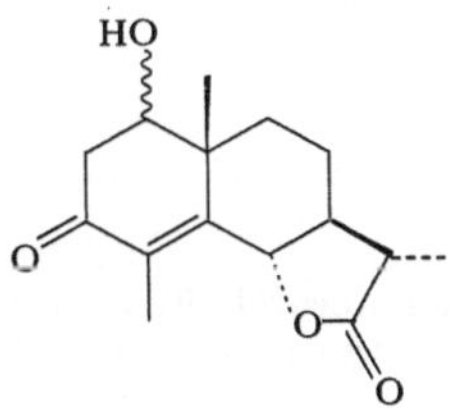

(**1206**) Alkhanin;

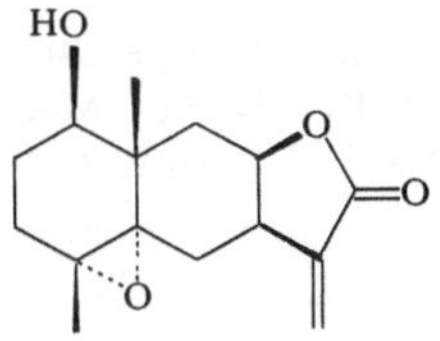

(**1207**) Meridianone;

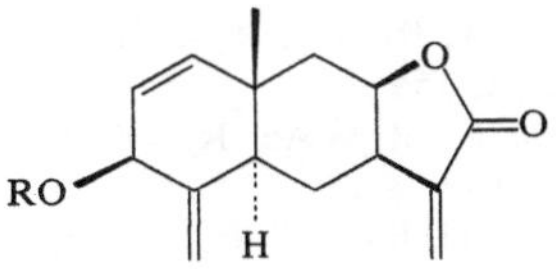

(**1208**) Telekin, 3-epiiso-1,2-dehydro;
R = H
(**1209**) Telekin, 3-epiiso-1,2-dehydro,
acetate; R = Ac

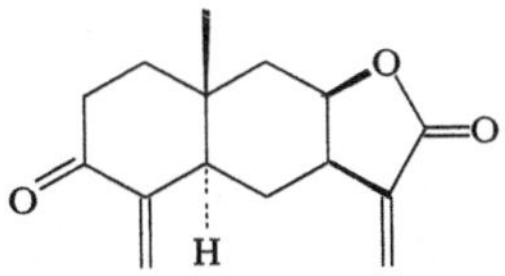

(**1210**) Telekin, iso, dehydro;

Addendum to Chart IV-1 *(continued)*

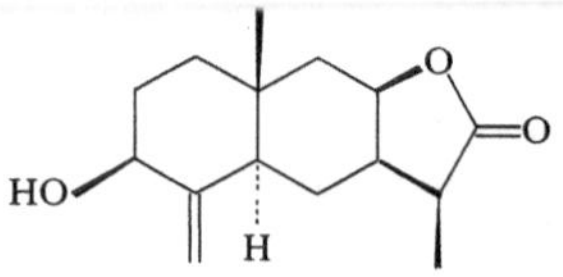

(1211) Telekin, 3-epiiso,11,13-dihydro;

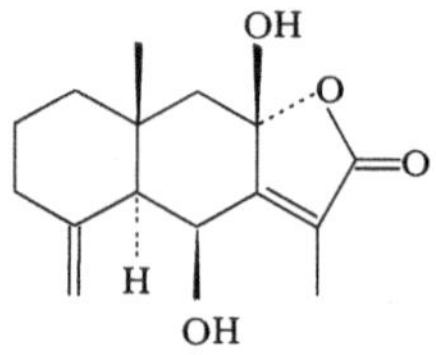

(1212) Ocotealactol;

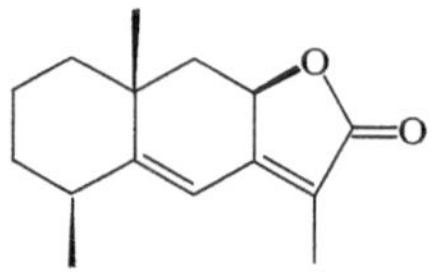

(1213) Alantolactone, neo;

Addendum to Chart V-1

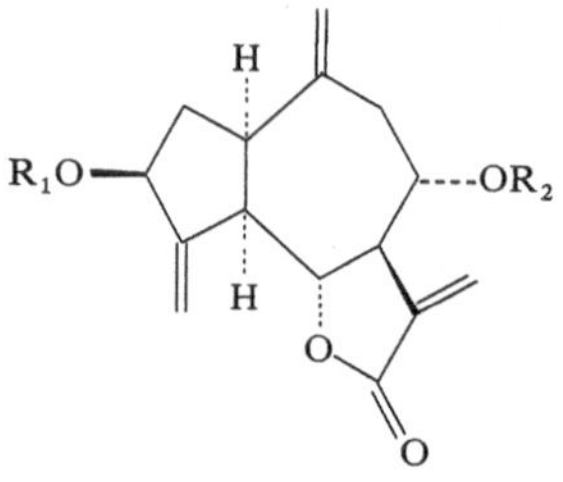

(1217) Cynaropicrin, deacyl; $R_1 = R_2 = H$
(1218) Zaluzanin C, 8α-acetoxy; $R_1 = H$, $R_2 = Ac$
(1219) Zaluzanin D, 8α-acetoxy; $R_1 = R_2 = Ac$
(1220) Aguerin A; $R_1 = H$, $R_2 = i$-But
(1221) Aguerin B; $R_1 = H$, $R_2 = Mac$
(1222) Linichlorin B; $R_1 = H$, $R_2 = A$

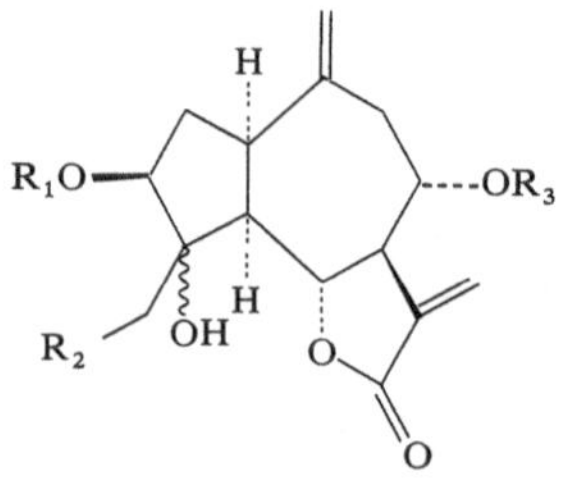

(1223) Linichlorin A; $R_1 = H$, $R_2 = Cl$, $R_3 = Mac$, C-15α
(1224) Linichlorin C; $R_1 = Ac$, $R_2 = OH$, $R_3 = A$, C-15α
(1225) Elegin, $R_1 = H$, $R_2 = Cl$, $R_3 = Mac$, C-15β

$A=$

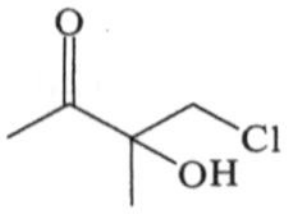

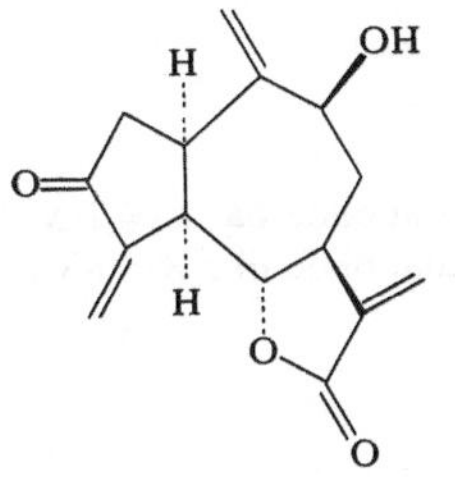

(1226)   Zaluzanin C, dehydro, 9β-hydroxy;

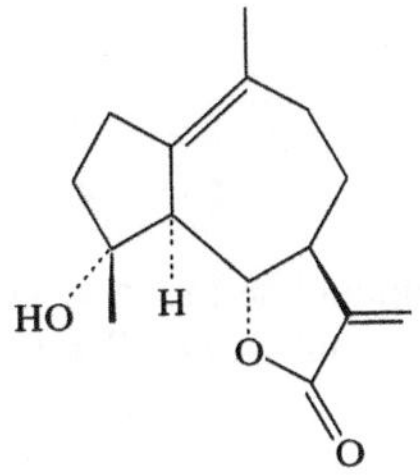

(1227)   Micheliolide;

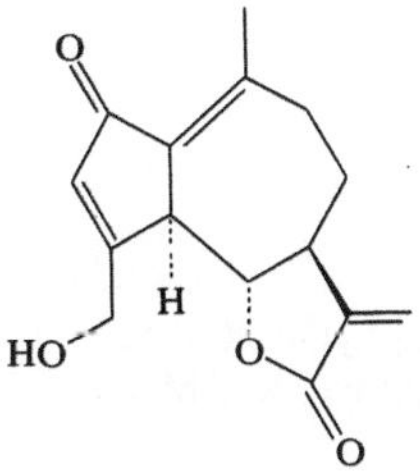

(1228)   Lactucin, 8-deoxy;

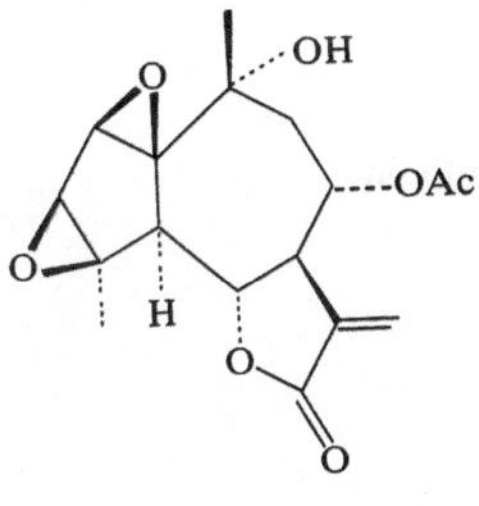

(1229)   Yomogiartemin;

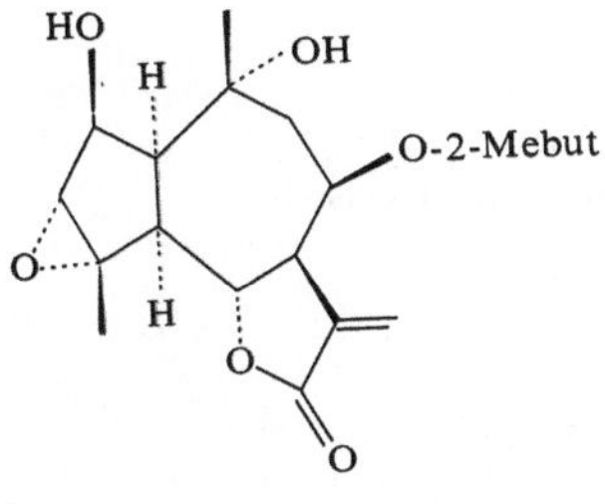

(1230)   Guaianolide 4a;

Addendum to Chart V-1 *(continued)*

(**1231**)   Guaianolide **6a**; R = 2-Mebut
(**1232**)   Guaianolide **6b**; R = *i*-Val

(**1233**)   Guainanolide **1a**; R = 2-Mebut
(**1234**)   Guainanolide **1a′**; R = *i*-Val

(**1235**)   Malaphylin; R = Ac
(**1236**)   Malaphyl; R = Mac

(**1239**)   Achillicin;

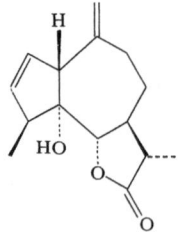

(**341a**)   Guaianolide (revised);

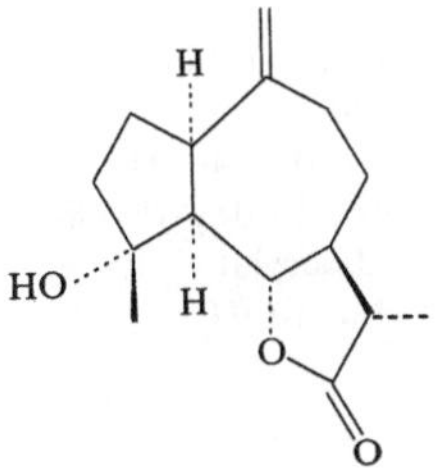

(1240) Compressanolide;

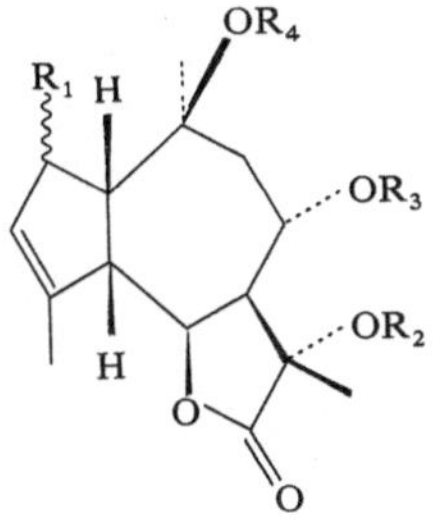

(699a) Montanolide, iso; $R_1 = R_4 = H$, $R_2 = Ac$, $R_3 = Ang$

(698a) Montanolide; $R_1 = R_4 = H$, $R_2 = Ac$, $R_3 = Sen$

(700a) Montanolide, isoacetyl; $R_1 = H$, $R_2 = R_4 = Ac$, $R_3 = Ang$

(1241) Polhovolide; $R_1 = H$, $R_2 = R_4 = Ac$, $R_3 = i\text{-But}$

(1242) Gradolide; $R_1 = R_4 = H$, $R_2 = R_3 = Ang$

(667a) Archangolide; $R_1 = OAng$, $R_2 = R_4 = Ac$, $R_3 = 2\text{-Mebut}$

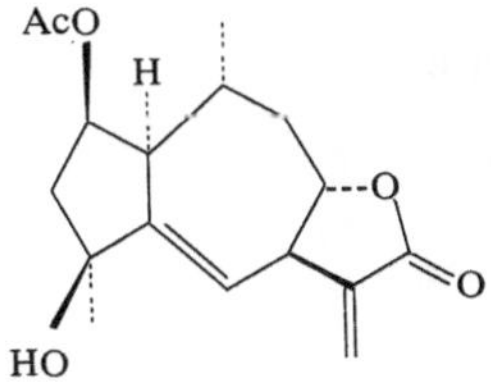

(1243) Puberolide;

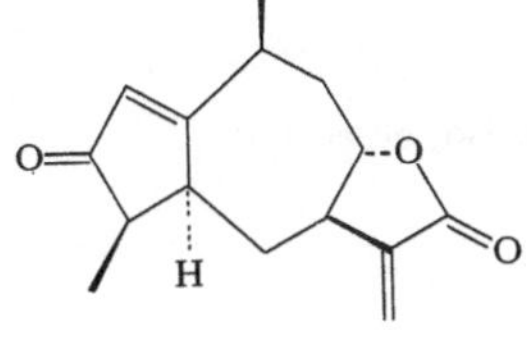

(1244) Thieleanin;

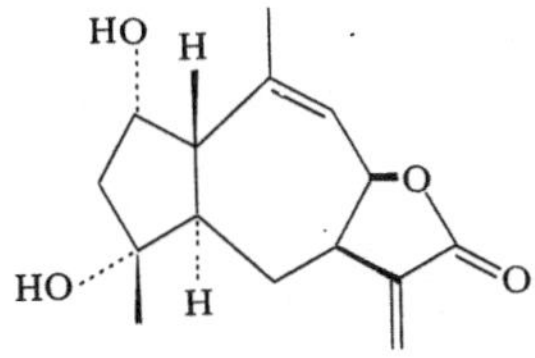

(722a) Pleniradin (revised)

Addendum to Chart VI-1

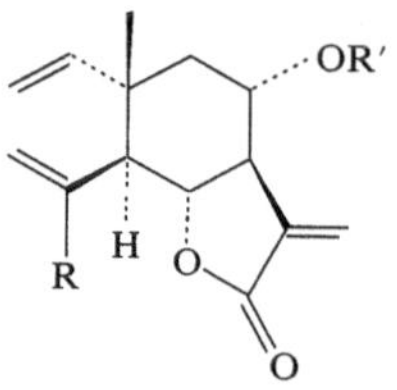

(1245) Melitensin, dehydro-8-(O)-[4'-hydroxymethacrylate];
R = CH₂-OH, R' = Mac-4-OH

(1246) Melitensin, dehydro, 15-dehydro-8-(O)-[4'-hydroxymethacrylate;
R = CHO, R' = Mac-4-OH

Addendum to Chart VII-1

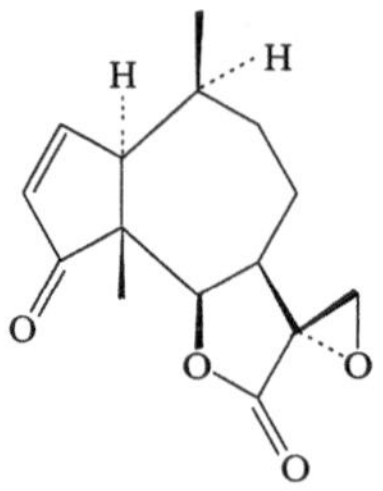

(1247) Stramonin B;

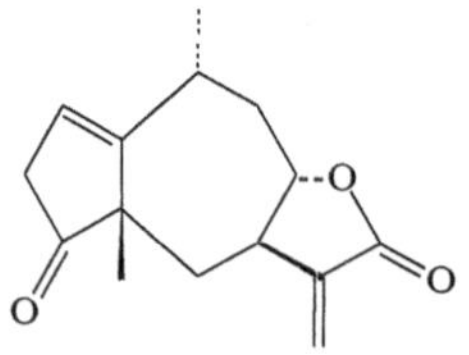

(1248) Bigelovin, iso, desacetyl;

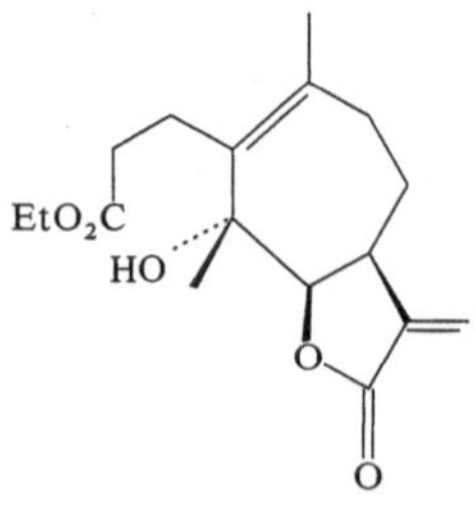

(1249) Altamisin;

Addendum to Chart VIII-4

(1250) Eremophila-1,7-dien-8,12-olide,
3-oxo-8α-H; R = H
(1251) Eremophila-1,7-dien-8,12-olide,
3-oxo-8α-hydroxy; R = OH
(1252) Eremophila-1,7-dien-8,12-olide,
3-oxo-8α-methoxy; R = OMe
(1253) Eremophila-1,7-dien-8,12-olide,
3-oxo-8α-ethoxy; R = OEt

(1254) Istanbulin C;

(1255) Eremophilenolide, 3β-hydroxy-6β-
angeloyloxy-7,8-epoxy; R = Ang
(1256) Eremophilenolide, 3β-hydroxy-6β-
tigloyloxy-7,8-epoxy; R = Tig

Addendum to Chart IX-4

(1257) Aplysistatin;

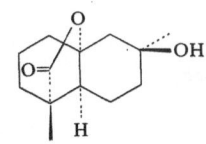

(1258) Jhanilactone;

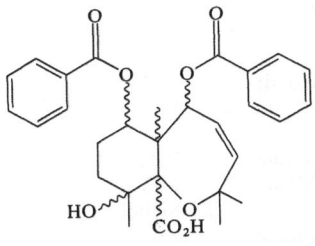

(1259) Mortonin C;

Addendum to Chart IX-4 *(continued)*

(1106)   Mortonin A; R = H (revised)
(1260)   Mortonin B; R = OAc

(1261)   Mortonin D;

(1262)   Plagiochilide;

(1263)   Quadrone;

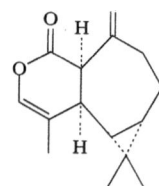

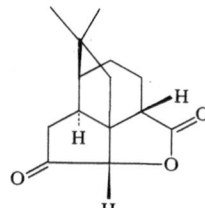

Trixikingolides
(1264) R_1 = 2-Mebut, R_2 = H
(1265) R_1 = i-Val, R_2 = H
(1266) R_1 = 2-Mebut, R_2 = OSen
(1267) R_1 = 2-Mebut, R_2 = O-i-Val
(1268) R_1 = i-Val, R_2 = O-2-Mebut
(1269) R_1 = 2-Mebut, R_2 = OAc

(1270) Lactarolide A; R = H
(1271) Lactarolide A, 3-O-ethyl; R = Et

(1272) Lactarolide B; R = H
(1273) Lactarolide B, 3-O-ethyl: R = Et

Addendum to Table III-3. *Naturally Occurring Germacranolides and Biogenetic Derivatives*

Structure Number	Name of Compound	Type[c]	Formula	m.p. °C	[α]$_D$	Plant Source[d]	References[a,b]	Comments
(1143)	Albicolide, 8α-[2--methylbutyryloxy]	GG	$C_{20}H_{28}O_6$	gum		*Cnicothamnus lorentzii*	(1209)	
(1152)	Atripliciolide, 9α-[angeloxy]-15-hydroxy-8-O-[2-methylacrylate]	GH	$C_{24}H_{26}O_9$	gum		*Calea urticifolia*	(1205)	
(1151)	Atripliciolide, 9α-hydroxy,8-O-[2-methyl-acrylate]	GH	$C_{19}H_{20}O_7$	gum		*Calea urticifolia*	(1205)	
(1153)	Atripliciolide, 9α-[isovaleryloxy]-15-hydroxy-8-O-[2-methyl-acrylate]	GH	$C_{24}H_{28}O_9$	gum		*Calea urtificolia*	(1205)	
(1154)	Atripliciolide, 9α-[senecioyloxy]-15-hydroxy-8-O-[2-methyl-acrylate]	GH	$C_{24}H_{26}O_9$	gum		*Calea urticifolia*	(1205)	
(1193)	Baileyin	GM					(1209b^a)	
(1178)	Calein A	G	$C_{22}H_{28}O_8$	180—2		*Calea zacatechichi*	(1283, 1283a^a)	mixed with Calein B
(1179)	Calein B	G	$C_{22}H_{28}O_8$	180—2		*Calea zacatechichi*	(1283, 1283a^a)	mixed with Calein A
(178)	Chromolaenide	GH				*Eupatorium glaberrimum*	(1214)	
(63)	Cnicin	GG				*Centaurea calitrapa*	(1251)	
(17)	Costunolide	GG				*Zaluzania montaguifolia*	(1306, 1260^b)	
						Michelia compressa	(1274)	Magnoliaceae
(1146)	Costunolide, diepoxide	GG	$C_{15}H_{20}O_4$	168—9	−30	*Magnolia grandiflora*	(1223)	Magnoliaceae, reported before Michelenolide
(1131)	Costunolide, 3β,9β-dihydroxy,8β-[2-methyl-butanoyloxy]	GG	$C_{20}H_{28}O_6$	gum		*Eupatorium mohrii*	(1242)	

No.	Name	Code	Formula	M.p.	[α]	Source	Ref.	Notes
(1136b)	Costunolide, 15-hydroxy-8α-isobutyryl	GG	$C_{21}H_{28}O_6$			*Onopordon leptolepis*	(1288a)	Mixture with (1136a)
(1136a)	Costunolide, 15-hydroxy-8α-[α-methylacryloyl]	GG	$C_{21}H_{26}O_6$			*Onopordon leptolepis*	(1288a)	Mixture with (1136b)
(1130a)	Costunolide, 3β-isovaleryloxy	GG	$C_{20}H_{28}O_4$	oil		*Cotula hispida*	(1209b)	
(1195)	Elephantopin, dihydro	GG				*Elephantopus tomentosus*	(1288c)	
(158)	Enhydrin	GM				*Polymnia maculata* var. *maculata*	(1259)	
(222a)	Eremantholide A	G				*Eremanthus elaeganus*	(1258)[b]	
(1154a)	Eremantholide B	G	$C_{20}H_{26}O_6$	233	+60°	*Eremanthus elaeganus*	(1258)[b]	
(1154b)	Eremantholide C	G	$C_{19}H_{22}O_6$	229–30	–12.7	*Eremanthus elaeganus*	(1258)	
(35)	Eupasserin	GG				*Helianthus mollis*	(1276)	
(34)	Eupasserin, desacetyl	GG				*Helianthus mollis*	(1276)	
(1183)	Eurecurvin	G	$C_{22}H_{30}O_8$	185–6	+42.3	*Eupatorium recurvans*	(1239)	
						Eupatorium anomalum	(1242)	
						E. mohrii	(1242)	
(1182)	Eurecurvin, 15 deshydroxy	G	$C_{22}H_{30}O_7$	113–4	+52.9°	*Eupatorium mohrii*	(1242)	
						Eupatorium recurvans	(1239)[b]	
(1186)	Fasciculide B	G	$C_{24}H_{30}O_{10}$	55–8		*Vernonia fasciculata*	(1273)	
(1185)	Gallicin	G	$C_{15}H_{22}O_3$	114–6	+121°	*Artemisia maritima gallica*	(1231)	
(47)	Hanphyllin	GG				*Pseudohandelia umbellifera*	(1303)	
(1150)	Heliangolide from *E. recurvans*	GH	$C_{22}H_{30}O_7$	129–31	–82.0	*Eupatorium recurvans*	(1239)[b]	
(1190)	Hirsutinolide-13-O-acetate, 8β-propionyl-oxy-10β-hydroxy	G	$C_{20}H_{26}O_9$	oil	+63.5	*Vernonia scorpioides*	(1206a)	
(1188)	Hirsutinolide, 13-O-acetate, 8β-acetoxy-10β-hydroxy	G	$C_{19}H_{24}O_9$	oil	+36.3	*Vernonia saltensis*	(1206a)	

Addendum to Table III-3 *(continued)*

Structure Number	Name of Compound	Type[c]	Formula	m.p. °C	$[\alpha]_D$	Plant Source[d]	References	Comments
(1192)	Hirsutinolide, 13-O-acetate, 8β-propionyl-oxy-10β-hydroxy-1-O-methyl	G	$C_{21}H_{28}O_9$	oil		*Vernonia scorpioides*	*(1206a)*	
(1189)	Hirsutinolide, 1,13-O-diacetate, 8β-acetoxy-10β-hydroxy	G	$C_{21}H_{26}O_{10}$	oil	−3.0	*Vernonia saltensis*	*(1206a)*	
(1191)	Hirsutinolide-1,13-O-diacetate, 8β-propionyl-oxy-10β-hydroxy	G	$C_{22}H_{28}O_{10}$	oil		*V. scorpioides* *Vernonia scorpioides*	*(1206a)* *(1206a)*	
(1147)	Inunolide, dihydro	GG	$C_{15}H_{22}O_2$	128−9	+120	*Inula racemosa*	*(850a)*	
(112)	Lanuginolide	GG				*Michelia compressa*	*(1274)*	
(1137)	Linearilobin A	GG	$C_{21}H_{28}O_7$	gum		*Melampodium linearilobum*	*(1291)*	
(1138)	Linearilobin B	GG	$C_{22}H_{30}O_7$	gum		*Melampodium linearilobum*	*(1291)*	
(1139)	Linearilobin C	GG	$C_{22}H_{28}O_8$	gum		*Melampodium linearilobum*	*(1291)*	
(1140)	Linearilobin D	GG	$C_{21}H_{26}O_7$	gum		*Melampodium linearilobum*	*(1291)*	
(1141)	Linearilobin E	GG	$C_{21}H_{26}O_7$	gum		*Melampodium linearilobum*	*(1291)*	$[\theta]_{225} + 2.69 \times 10^4$ $[\theta]_{204} - 2.77 \times 10^4$
(1142)	Linearilobin F	GG	$C_{20}H_{28}O_6$	gum		*Melampodium linearilobum*	*(1291)*	
(1144)	Linearilobin G	GG	$C_{20}H_{26}O_7$	gum		*Melampodium linearilobum*	*(1291)*	
(1148)	Linearilobin H	GH	$C_{21}H_{26}O_7$	gum		*Melampodium linearilobum*	*(1291)*	

(1149)	Linearilobin I	GH	$C_{22}H_{28}O_7$	gum		*Melampodium linearilobum*	(1291)	$[\theta]_{247} + 3.4 \times 10^3$ $[\theta]_{210} - 7.7 \times 10^4$
(154)	Longipilin	GM				*Tetragonotheca repanda*	(952)
(1187)	Maroniolide	G	$C_{15}H_{20}O_3$	160	−45	*Munnozia maronii*	(1204a)
(152)	Melampodinin	GM				*Melampodium diffusum*	(1282)
(1145)	Melfusin	GG	$C_{23}H_{26}O_{11}$	gum		*Melampodium diffusum*	(1282)	$[\bar{e}]_{215} - 1.57 \times 10^5$ $[\bar{e}]_{270} + 5.6 \times 10^4$
(1146)	Michelenolide					*Michelia compressa*	(1274)	Magnoliaceae, see Costunolide, diepoxide
(1129)	Mollisorin A	GG	$C_{20}H_{26}O_5$	oil		*Helianthus mollis*	(1276)
(1130)	Mollisorin B	GG	$C_{20}H_{26}O_6$	165−6		*Helianthus mollis*	(1276)
(1180)	Neurolenin A	G	$C_{20}H_{28}O_6$	127−8	−257.7	*Neurolaena lobata*	(1264)[b]
(1181)	Neurolenin B	G	$C_{22}H_{30}O_8$	165−6	−350°	*Neurolaena lobata*	(1264)[b]
(61)	Onopordopicrin					*Onopordon leptolepis*	(1288a)
(80)	Parthenolide	GG				*Michelia compressa*	(1274)
						Inula aschersoniana	(1271)
						Michelia compressa	(1274)
						Tetragonotheca ludoviciana	(1281a)	$[\bar{e}]_{276} - 852; [\bar{e}]_{247} + 2555;$ $[\bar{e}]_{214} - 27680$
(109)	Parthenolide, dihydro	GG					
(143)	Polydalin	GM					
(1155)	Polymatin A	GM	$C_{21}H_{26}O_7$	gum		*Polymnia maculata* var. *maculata*	(1259)
(1156)	Polymatin B	GM	$C_{23}H_{28}O_8$	gum		*Polymnia maculata* var. *maculata*	(1259)
(1157)	Polymatin C	GM	$C_{23}H_{28}O_{10}$	183		*Polymnia maculata* var. *maculata*	(1259)
(1184)	Ridentin, iso	G	$C_{15}H_{20}O_4$	197−9	+181	*Achillea biebersteinii*	(1311)
(1135)	Salonitenolide, 8-desoxy, 15-(2,3-dihydroxy-isobutaryl-oxy)	GG	$C_{19}H_{26}O_6$	gum		*Mikania cordifolia*	(1207)
(1134)	Salonitenolide, 8-desoxy, 15-(3-hydroxy-isobutyryloxy)	GG	$C_{19}H_{26}O_5$	gum	+55	*Mikania cordifolia*	(1207)

Addendum to Table III-3 (continued)

Structure Number	Name of Compound	Type[c]	Formula	m.p. °C	$[\alpha]_D$	Plant Source[d]	References	Comments
(1132)	Salonitenolide, 8-desoxy, 15-(3-hydroxy-2-methylacryloxy)	GG	$C_{19}H_{24}O_5$	gum		Mikania cordifolia	(1207)	
(1133)	Salonitenolide, 8-desoxy, 15-(2,3-epoxy-isobutyryloxy)	GG	$C_{19}H_{24}O_5$	144	+68.5	Mikania cordifolia	(1207)	
(1136)	Salonitenolide-8(O)-[2-methylbutyrate]	GG	$C_{20}H_{28}O_5$	gum		Cnicothamnus lorentzii	(1209)	
(1194)	Tagitinin E	G	$C_{19}H_{26}O_6$	207–10	–101	Tithonia rotundifolia	(1200[a])	
(1158)	Tetrahelin A	GM	$C_{27}H_{34}O_{12}$	gum		Tetragonotheca helianthoides	(1292)	$[\theta]_{260} - 2.6 \times 10^3$ $[\theta]_{214} - 7.7 \times 10^4$
(1176)	Tetrahelin B	GM	$C_{27}H_{34}O_{13}$	gum		Tetragonotheca helianthoides	(1292)	
(1159)	Tetrahelin C	GM	$C_{28}H_{38}O_{11}$	gum		Tetragonotheca helianthoides	(1292)	
(1160)	Tetrahelin D	GM	$C_{25}H_{32}O_{11}$	gum		Tetragonotheca helianthoides	(1292)	
(1161)	Tetrahelin E	GM	$C_{28}H_{38}O_{12}$	gum		Tetragonotheca helianthoides	(1292)	
(1177)	Tetrahelin F	GM	$C_{25}H_{32}O_{12}$	gum		Tetragonotheca helianthoides	(1292)	
(1162)	Tetraludin A	GM	$C_{21}H_{26}O_9$	oil		Tetragonotheca ludoviciana	(1281)	$[\theta]_{260} - 1165$ $[\theta]_{213} - 4.37 \times 10^4$
(1163)	Tetraludin B	GM	$C_{26}H_{36}O_{11}$	164–5		Tetragonotheca ludoviciana	(1281)	Diastereomer of Tetraludin C
(1164)	Tetraludin C	GM	$C_{26}H_{36}O_{11}$	172–3		Tetragonotheca ludoviciana	(1281)	Diastereomer of Tetraludin D
(1165)	Tetraludin D	GM	$C_{26}H_{34}O_{10}$	139–40		Tetragonotheca ludoviciana	(1281a)	Diastereomer of Tetraludin E

(1166)	Tetraludin E	GM	$C_{26}H_{34}O_{10}$	132–3	*Tetragonotheca ludoviciana*	(1281a)	Diastereomer of Tetraludin D
(1167)	Tetraludin F	GM	$C_{25}H_3O_{10}$	gum	*Tetragonotheca ludoviciana*	(1281a)	Diastereomer of Tetraludin C
(1168)	Tetraludin G	GM	$C_{25}H_{32}O_{10}$	gum	*Tetragonotheca ludoviciana*	(1281a)	Diastereomer of Tetraludin F
(1169)	Tetraludin H	GM	$C_{23}H_{28}O_{10}$	172–3	*Tetragonotheca ludoviciana*	(1281a)	C-2′-epimer polydalin
(1170)	Tetraludin I	GM	$C_{26}H_{34}O_{10}$	gum	*Tetragonotheca ludoviciana*	(1281a)	
(1171)	Tetraludin J	GM	$C_{26}H_{36}O_{10}$	171–2	*Tetragonotheca ludoviciana*	(1281a)	Diastereomer of Tetraludin K
(1172)	Tetraludin K	GM	$C_{26}H_{36}O_{10}$	gum	*Tetragonotheca ludoviciana*	(1281a)	Diastereomer of Tetraludin J
(1173)	Tetraludin L	GM	$C_{25}H_{34}O_{10}$	gum	*Tetragonotheca ludoviciana*	(1281a)	Diastereomer of Tetraludin M
(1174)	Tetraludin M	GM	$C_{25}H_{34}O_{10}$	gum	*Tetragonotheca ludoviciana*	(1281a)	Diastereomer of Tetraludin L
(1175)	Tetraludin N	GM	$C_{26}H_{34}O_{11}$	gum	*Tetragonotheca ludoviciana*	(1281a)	
(217)	Tirotundin	GH			*Tithonia diversifolia*	(1212a⁰)	
(38)	Tulipinolide	GG			*Conocephalum conicum*	(1197a)	Hepaticae
(39)	Tulipinolide, epi	GG			*Zaluzania pringlei*	(1306)	
(1150a)	Viguiepinin	GH	$C_{19}H_{22}O_{7}$	175–6 –82.4	*Viguiera pinnatilobata*	(1286a)	
(226)	Viguiestenin	GH				(1286aᵃ, 1200ᵃ)	
(225)	Viguiestenin, desacetyl	GH				(1286aᵃ, 1200ᵃ)	

Notes to this Table are the same as to Table III-3.

Addendum to Table IV-2. *Naturally Occurring Eudesmanolides and Biogenetic Derivatives*

Structure Number	Name of Compound	Formula	m.p. °C	$[\alpha]_D$	Plant Source[d]	References	Comments
(1206)	Alkhanin	$C_{15}H_{20}O_4$	201–3		*Artemisia fragrans*	(1295)	
(331)	Cyclocostunolide				*Zaluzania montagnifolia*	(1306)	
(461)	Encelin				*Baltimora recta*	(1240)	seco-eudesmanolide
(487)	Eriolanin					(1238)	total synthesis
(1203)	Gazaniolide	$C_{15}H_{18}O_2$	oil		*Gazania krebsiana*	(1209a)	
(1204)	Gazaniolide, 8α-isovaleryloxy	$C_{20}H_{26}O_4$	oil		*Gazania krebsiana*	(1209a)	
(1207)	Meridianone	$C_{15}H_{20}O_4$	184		*Artemisia feddei*	(1267)	
(1212)	Ocotealactol	$C_{15}H_{20}O_4$	199–200		*Ocotea guianensis*	(1287)	Lauraceae
(374)	Reynosin				*Michelia compressa*	(1274)	Magnoliaceae
					Magnolia grandiflora	(1223)	Magnoliaceae
(1205)	Reynosin, dihydro	$C_{15}H_{22}O_3$	129		*Michelia compressa*	(1274)	Magnoliceae
(365)	Santamarin				*Magnolia grandiflora*	(1223)	Magnoliaceae
					Michelia compressa	(1274)	Magnoliaceae
(405)	α-Santonin					(1265)	total synthesis
(406)	β-Santonin					(1265)	total synthesis
(1208)	Telekin, 3-epi, iso-1,2-dehydro,	$C_{15}H_{18}O_3$	146–7	+160	*Baltimora recta*	(1240)	
(1209)	Telekin, 3-epi, iso-1,2-dehydro, acetate	$C_{17}H_{20}O_4$	166–7	+120	*Baltimora recta*	(1240)	
(1211)	Telekin, 3-epi, iso, 11,13-dihydro	$C_{15}H_{22}O_3$	178–80	+24.5	*Baltimora recta*	(1240)	
(1210)	Telekin, iso, dehydro	$C_{15}H_{18}O_3$	144–5	+151.5	*Baltimora recta*	(1240)	
(390)	Vahlenin				*Centaurea linifolia*	(1228)	

Addendum to Table V-2. Naturally Occurring Guaianolides and Biogenetic Derivatives

Structure Number	Name of Compound	Formula	m.p. °C	$[\alpha]_D$	Plant Source[d]	References	Comments
(1239)	Achillicin	$C_{17}H_{22}O_5$			Achillea millefolium ssp. collina	(1199a)	
(1220)	Aguerin A	$C_{19}H_{24}O_5$	oil	+89	Centaurea cannariensis	(1230)	
					C. linifolia	(1228)	
(1221)	Aguerin B	$C_{19}H_{22}O_5$	oil	+96	Centaurea canariensis	(1230)	
					C. canariensis var. subspinnata	(1230)	
					C. sventenii	(1230)	
					C. linifolia	(1230)	
(640)	Arborescin					(1196)	total synthesis
(576)	Chlorohyssopifolin A				Centaurea linifolia	(1228)	
(573)	Chlorohyssopifolin B				Centaurea linifolia	(1228)	
(570)	Chlorohyssopifolin C				Centaurea linifolia	(1228)	
(575)	Chlorohyssopifolin D				Centaurea linifolia	(1228)	
(574)	Chlorohyssopifolin E				Centaurea linifolia	(1228)	
(1240)	Compressanolide	$C_{15}H_{22}O_3$	oil		Michelia compressa	(1274)	Magnoliaceae: old name: formosanolide
(580)	Costuslactone, dehydro				Munnozia gigantea	(1204a)	
(599)	Cumambrin A				Tanacetum santolina	(1310a)	
(597)	Cumambrin B				Tanacetum santolina	(1310a)	
(587)	Cynaropicrin				Centaurea canariensis	(1230)	
(1217)	Cynaropicrin, deacyl	$C_{15}H_{18}O_4$			Centaurea canariensis	(1230)	
					Amberboa muricata	(329)	
(1225)	Elegin	$C_{19}H_{23}O_6Cl$	158–9		Saussurea elegans	(982, 1297)	
(551)	Estafiatin					(1221)	synthesis
(729)	Ferulin				Ferulia tenuisecta	(1289)	
(728)	Ferulidin				Ferula tenuisecta	(1289)	
(1240)	Formosanolide					(1274)	see Compressanolide

Addendum to Table V-2 (continued)

Structure Number	Name of Compound	Formula	m.p. °C	$[\alpha]_D$	Plant Source[d]	References	Comments
(1242)	Gradolide	$C_{25}H_{34}O_7$	154	−50.9	*Laserpitium siler*	(1243)	Umbelliferae
(1233)	Guaianolide **1a**	$C_{20}H_{26}O_8$	158−9		*Eupatorium anomalum*	(1242)[b]	$[\theta]_{270} -320$ $[\theta]_{228} +3470$
(1234)	Guaianolide **1a**						
(1230)	Guaianolide **4a**	$C_{20}H_{28}O_7$	gum		*E. mohrii* *Eupatorium mohrii*	(1242) (1242)	$[\theta]_{250} -1540$ $[\theta]_{235} -1360$
(1231 + 1232)	Guaianolides **6a** and **b**	$C_{20}H_{26}O_7$	gum		*E. anomalum* *E. anomalum*	(1242) (1242)	mixture of isomers
(681)	Lactucin				*Lactuca serriola*	(1010a, 907a)	
(1228)	Lactucin, 8-deoxy	$C_{15}H_{16}O_4$			*Lactuca serriola*	(1010a)	
(592)	Grossheimin					(1272)	
(682)	Lactucopicrin				*Lactuca serriola*	(1010a)	
(535)	Leucodin, dehydro (Lidbeckia lactone)				*Cotula hispida*	(1209b)	
(1223)	Linichlorin A	$C_{19}H_{23}O_6Cl$	153−5	+89	*Centaurea linifolia*	(1228)	
(1222)	Linichlorin B	$C_{19}H_{23}O_6Cl$	144−6	+95	*Centaurea linifolia*	(1228)	
(1224)	Linichlorin C	$C_{21}H_{27}O_9Cl$	160−2	+83			*Centaurea linifolia* (1228)
(1236)	Malaphyl	$C_{28}H_{30}O_9$			*Ferula malacophylla*	(1199)	
(1235)	Malaphylin	$C_{26}H_{28}O_9$			*Ferula malacophylla*	(1199)	
(1227)	Micheliolide	$C_{15}H_{20}O_3$	141		*Michelia compressa*	(1274)	Magnoliaceae *trans*-guaianolide
(722a)	Pleniradin					(1241)[b]	
(1241)	Polhovolide	$C_{23}H_{32}O_8$	98−9	−83.5	*Laserpitium siler*	(1243)	Umbelliferae
(1243)	Puberolide	$C_{17}H_{22}O_5$	oil		*Helenium puberulum*	(1206)	$\Delta\varepsilon_{205} -12.0$
(558)	Spicatin					(1252)[b]	
(637)	Talassin				*Ferula malacophylla*	(1199)	
	Tenuferidin	$C_{22}H_{30}O_5$	164−5	+75	*Ferula tenuisecta*	(1289)	
	Tenuferin	$C_{23}H_{32}O_6$	176−8	+135	*Ferula tenuisecta*	(1289)	possibly guaianolides;

Structure Number	Name of Compound	Formula	m.p. °C	[α]D	Plant Source	References	Comments
(1244)	Thieleanin	$C_{15}H_{18}O_3$	175–7		Decachaeta thieleana	(1195)	
(1229)	Yomogiartemin	$C_{17}H_{20}O_7$	242		Artemisia feddei	(1267)[b]	
(730)	Zaluzanin A				Zaluzania augusta	(1306)	
					Z. montagnifolia	(1306)	
(731)	Zaluzanin B				Zaluzania augusta	(1306)	
					Z. montagnifolia	(1306)	
(583)	Zaluzanin C				Zaluzania triloba	(1306)	
					Cnicothamus lorentzii	(1209)	
					Conocephalum conicum	(1197a)	Hepaticeae
(1218)	Zaluzanin C, 8α-acetoxy	$C_{17}H_{20}O_5$			Conocephalum conicum	(1197a)	Hepaticae
(1226)	Zaluzanin C, dehydro, 9β-hydroxy	$C_{15}H_{16}O_4$			Vernonia scorpioides	(1206a)	
(584)	Zaluzanin D				Conocephalum conicum	(1197a)	Hepaticae
					Zaluzania triloba	(1306)	
(1219)	Zaluzanin D, 8α-acetoxy	$C_{19}H_{22}O_6$	liquid	–174	Conocephalum conicum	(1197a)	Hepaticae

Notes to this Table are the same as to Table V-2.

Addendum to Table VI-1. Naturally Occurring Elemanolides and Biogenetic Derivatives

Structure Number	Name of Compound	Formula	m.p. °C	[α]D	Plant Source	References	Comments
(1246)	Melitensin, dehydro, 15-dehydro-8-(O)-[4'-hydroxy-methacrylate]	$C_{19}H_{22}O_6$	gum		Onopordon leptolepis	(1248b)	
(1245)	Melitensin, dehydro, -8-(O)-[4'-hydroxymethacrylate]	$C_{19}H_{24}O_6$	gum	+74	Onopordon leptolepis	(1248b)	
(790)	Saussurealactone					(1266) (1263)	total synthesis oxidation
(787)	Vernolepin					(1247, 1254)	total synthesis

Addendum to Table VII-2. *Naturally Occurring Pseudoguaianolides and Seco-Derivatives*

Structure Number	Name of Compound	Type[c]	Formula	m.p. °C	$[\alpha]_D$	Plant Source	References	Comments
(1249)	Altamisin	PA	$C_{17}H_{24}O_5$	101—2	+34.9	*Ambrosia cumanensis*	(1211)	3,4-seco compound
(876)	Bigelovin	PH				*Helenium puberulum*	(1206)	
(1248)	Bigelovin, iso, desacetyl	PH	$C_{15}H_{18}O_4$	oil		*Helenium puberulum*	(1206)	
(853)	Confertin	PA					(1293)	total synthesis
(898)	Helenalin	PH				*Helenium puberulum*	(1206, 1257)	total synthesis
							(1275)	
(906)	Hymenoratin	PH				*Baileya pauciradiata*	(1242a)	
(895)	Paucin	PH				*Baileya pauciradiata*	(1242a)	
(863)	Psilostachyin	PA				*Ambrosia cumanensis*	(1211)	
(866)	Psilostachyin B	PA				*Ambrosia cumanensis*	(1211)	
(865)	Psilostachyin C	PA				*Ambrosia cumanensis*	(1211)	
(1247)	Stramonin B	PA	$C_{15}H_{18}O_4$	175—6	−125	*Parthenium tomentosum*	(1237[b])	total synthesis
(922)	Tenulin	PH				*Helenium puberulum*	(1206)	
(880)	Pulchellin	PH					(1246)	

Addendum to Table VIII-1. *Naturally Occurring Eremophilanolides and Biogenetic Derivatives*

Structure Number	Name of Compound	Formula	m.p. °C [α]$_D$	Plant Source[d]	References	Comments
(1250)	Eremophila-1,7-dien-8,12-olide, 3-oxo-8αH	$C_{15}H_{18}O_3$	186−7	*Senecio nemorensis* var. *bulgaricus*	(1249)	
(1253)	Eremophila-1,7-dien-8,12-olide, 3-oxo-8α-ethoxy	$C_{17}H_{22}O_4$	149−50	*Senecio nemorensis* var. *bulgaricus*	(1249)	
(1251)	Eremophilia-1,7-dien-8,12-olide, 3-oxo-8α-hydroxy	$C_{15}H_{18}O_4$	208−11	*Senecio nemorensis* var. *bulgaricus*	(1249)	
(1252)	Eremophila-1,7-dien-8,12-olide, 3-oxo-8α-methoxy	$C_{16}H_{20}O_4$	217−9	*Senecio nemorensis* var. *bulgaricus*	(1249)	
(1255)	Eremophilenolide, 3β-hydroxy-6β-angeloyloxy-7,8-epoxy	$C_{20}H_{28}O_6$	gum	*Senecio pyramidatus*	(1210)	
(1256)	Eremophilenolide, 3β-hydroxy, 6β-tigloyloxy, 7,8-epoxy	$C_{20}H_{28}O_6$	gum	*Senecio pyramidatus*	(1210)	
(1254)	Istanbulin C	$C_{15}H_{18}O_3$	178−80	*Smyrnium connatum*	(1302a)	Umbelliferae
	Fukinanolide F			*Petasites japonicus*	(1298)	Fukinanolide
	Fukinolide F			*Petasites japonicus*	(1298)	Fukinanolide

Addendum to Table IX-3. Special Structural Types of Sesquiterpene Lactones

Structure Number	Name of Compound	Formula	m.p. °C	$[\alpha]_D$	Plant Source[d]	References	Comments
(1257)	Aplysistatin	$C_{15}H_{21}O_3Br$	173—5	−375	Aplysia angasi	(1279)	Aplysiidae, seahare, Drimanolide Biosynthesis
(1108)	Fomannosin					(1213)	possibly a degradation product of a sesquiterpenoid
(1258)	Jhanilactone	$C_{13}H_{20}O_3$			Eupatorium jhanii	(1226a)	
(1270)	Lactarolide A	$C_{15}H_{22}O_5$	153—5	+59.8	Lactarius scrobiculatus L. blennius L. pallidus	(1217a) (1217a) (1217a)	
(1271)	Lactarolide A, 3-O-ethyl	$C_{17}H_{26}O_5$	171—3	+18.5	Lactarius blennius	(1217a)	possibly an artifact Russulaceae
(1272)	Lactarolide B	$C_{15}H_{22}O_5$	212—6	−3.5	Lactarius scrobiculatus L. blennius L. pallidus	(1217a) (1217a) (1217a)	Russulaceae Russulaceae possibly an artifact
(1273)	Lactarolide B, 3-O-ethyl	$C_{17}H_{26}O_5$	180—3	+2.1	Lactarius blennius L. pallidus	(1217a) (1217a)	
(1106a)	Mortonin A				Mortonia gregii	(899, 1220)	Celastraceae
(1260)	Mortonin B	$C_{24}H_{28}O_8$	174—6		Mortonia gregii	(1285)	Celastraceae
(1259)	Mortonin C	$C_{29}H_{32}O_8$	201—4		Mortonia gregii	(1285, 1220, 1284b, 1284c)	Celastraceae
(1261)	Mortonin D	$C_{29}H_{32}O_8$	256—7		Mortonia gregii	(1285)	Celastraceae
(1107)	Pentalenolactone					(1217, 1296)	total synthesis; biosynthesis
(1262)	Plagiochilide	$C_{15}H_{20}O_2$	110—1	−5	Plagiochila yokogurensis	(1198)	Plagiochilaceae; seco-aromadendrane
(1263)	Quadrone	$C_{15}H_{20}O_3$	185—6		Trixis species	Aspergillus terreus (1208)	(1284[a])
(1264— 1269)	Trixikingolides						
(1095)	Vallerolactone (Lactarius Lactone I)					(1224[a])	
(1096)	Vellerolactone, pyro (Lactarius Lactone II)					(1224[a])	

Acknowledgements: Parts of this work were supported by Grant Number 1-R01-CA-19800, awarded by the National Cancer Institute, DHEW and by a grant from the National Science Foundation (DEB-76-20585). We are indebted to Professor WERNER HERZ whose contributions to this review went far beyond his editorial duties.

References

1. ABE, Y., T. HARUKAWA, H. ISHIKAWA, T. MIKI, M. SUMI, and T. TOGA: Santonin. I. The Synthesis of Two Optically Inactive Stereoisomerides of Santonin. J. Amer. Chem. Soc. **75**, 2567 (1953).

2. — — — — — — The Synthesis of New Stereoisomers of Santonin. J. Amer. Chem Soc. **78**, 1416 (1956).

3. ABE, N., R. ONODA, K. SHIRAHATA, T. KATO, M. C. WOODS, Y. KITAHARA, K. RO, and T. KURIHARA: The Structures of Bakkenolides-B, -C and -D as Determined by the Use of a Nuclear Overhauser Effect. Tetrahedron Letters **1968**, 1993.

4. ABE, N., R. ONODA, K. SHIRAHATA, T. KATO, M. C. WOODS, and Y. KITAHARA: The Structure of Bakkenolide-A. Tetrahedron Letters **1968**, 369.

5. ABDEL-BASET, Z. H., L. SOUTHWICK, W. G. PADOLINA, H. YOSHIOKA, T. J. MABRY, and S. B. JONES, JR.: Sesquiterpene Lactones: A Survey of 21 United States Taxa from the Genus *Vernonia* (Compositae). Phytochem. **10**, 2201 (1971).

6. ABU-SHADY, H., and T. O. SOINE: The Chemistry of *Ambrosia maritima* L. The Isolation and Preliminary Characterization of Ambrosin and Damsin. J. Amer. Pharm. Assoc. **42**, 387 (1953).

7. ACHARI, R. G., W. E. COURT, and F. NEWCOMBE: Biosynthesis of the isovaleryl and senecioyl moieties of tropane alkaloids. Planta Med. **22**, 38 (1972); Chem. Abstr. **77**, 149747W (1972).

8. ADAMS, R., and W. HERZ: Helenalin. I. Isolation and Properties. J. Amer. Chem. Soc. **71**, 2546 (1949).

9. — — Helenalin. III. Reduction and Dehydrogenation. J. Amer. Chem. Soc. **71**, 2554 (1949).

10. AGAFONOVA, N. V., L. E. KUSHNIR, A. D. KUZOVKOV, A. I. SHRETER, and M. G. PIMENOV: Chemical Study of *Saussurea pulchella* Fisch. Aptechn. Delo. **15**, 36 (1966); Chem. Abstr. **65**, 3681 d (1966).

11. AKYEV, B.: Artemin from *Artemisia Kemrudica*. Izv. Akad. Nauk. Turkm. SSR. Ser. Biol. Nauk. **6**, 85 (1976); Chem. Abstr. **87**, 19053n (1977).

12. AKYEV, B., S. Z. KASYMOV, and G. P. SIDYAKIN: Sesquiterpene Lactones of *Artemisia santolina*. Khim. Prir. Soedin. **7**, 531 (1971) Engl. edit.: p. 514; Chem. Abstr. **75**, 148501 h (1971).

13. — — — Arsanin — A New Sesquiterpene Lactone from *Artemisia santolina*. Khim. Prir. Soedin **8**, 461 (1972) Engl. edit.: p. 458; Chem. Abstr. **78**, 13748 k (1973).

14. — — — Structure and Configuration of Arsanin. Khim. Prir. Soedin. **8**, 730 (1972) Engl. edit.: p. 713; Chem. Abstr. **78**, 84565j (1973).

15. — — — Artesin — A New Sesquiterpene Lactone from *Artemisia santolina*. Khim. Prir. Soedin. **8**, 733 (1972) Engl. edit.: p. 715; Chem. Abstr. **78**, 84563g (1973).

16. — — — Structure of Artabin. Khim. Prir. Soedin. **6**, 691 (1970) Engl. edit.: p. 703; Chem. Abstr. **74**, 112239u (1971).

17. — — — Arabsin — A New Lactone from *Artemisia absinthium*. Khim. Prir. Soedin. **8**, 245 (1972) Engl. edit.: p. 245; Chem. Abstr. **77**, 85567c (1972).

18. — — — Artabin — A New Lactone from *Artemisia absinthium*. Khim. Prir. Soedin. **6**, 622 (1970). Engl. edit.: p. 634.

18a. Aleman, E., A. Rosado, M. Rodriguez, and J. F. Bertran: Composition of *Wedelia rugosa* Greenm. Rev. Cubana Farm. **11**, 47 (1977); Chem. Abstr. **88**, 3076g (1978).

19. Ali, E., P. P. Ghosh Dastidar, S. C. Pakrashi, L. J. Durham, and A. M. Duffield: Sesquiterpene Lactones of *Enhydra fluctuans* Lour. Structures of Enhydrin, Fluctuanin and Fluctuadin. Tetrahedron **28**, 2285 (1972).

19a. Aliev, R. K., and I. A. Damirov: Zur Kenntnis einiger Arzneipflanzen der Flora Aserbeidshans. Pharmazie **21**, 457 (1966).

20. Allison, A. J., D. N. Butcher, J. D. Connolly, and K. H. Overton: Paniculides A, B and C, Bisabolenoid Lactones from Tissue Cultures of *Andrographis paniculata*. Chem. Commun. **1968**, 1493.

21. Alper, H., and E. C. H. Keung: Beckmann Rearrangement of a-Santonin Oxime. J. Heterocycl. Chem. **10**, 637 (1973); Chem. Abstr. **79**, 137300y (1973).

22. Anderson, L. A. P., W. T. de Kock, K. G. R. Pachler, and C. V. D. M. Brink: The Structure of Vermeerin. A Sesquiterpenoid Dilactone from *Geigeria africana* Gries. Tetrahedron **23**, 4153 (1967).

23. Anderson, L. A. P., W. T. de Kock, W. Nel, K. G. R. Pachler, and G. van Tonder: Gafrinin, A Sesquiterpenoid Lactone from *Geigeria africana* Gries — I. Revised Structure. Tetrahedron **24**, 1687 (1968).

24. Anderson, G. D., R. Gitany, R. S. McEwen, and W. Herz: Relative and Absolute Configuration of Pseudoivalin. Tetrahedron Letters **1973**, 2409.

25. Anderson, G. D., R. S. McEwen, and W. Herz: Relative and Absolute Configuration of Axivalin and its Congeners. Tetrahedron Letters **1972**, 4423.

26. Ando, M., and K. Takase: Chemical Transformation of α-Santonin into Reynosin, Santamarin, Epoxysantamarin and 1β-Hydroxyarbusculin A. Tetrahedron **33**, 2785 (1977).

27. Aota, K., C. N. Caughlan, M. T. Emerson, W. Herz, S. Inayama, and M. Ul-Haque: Structure and Absolute Configuration of Pulchellin. Crystal and Molecular Structure of 3-Bromoanhydrodehydrodihydropulchellin. J. Organ. Chem. (U.S.A.) **35**, 1448 (1970).

28. Appel, H. H., R. P. M. Bond, and K. H. Overton: The Constitution and Stereochemistry of Valdiviolide, Fuegin, Winterin and Futronolide. Tetrahedron **19**, 635 (1963).

29. Appel, H. H., J. D. Connolly, K. H. Overton, and R. P. M. Bond: The Constitution and Stereochemistry of Drimenin, Isodrimenin and Confertifolin. J. Chem. Soc. (London) **1960**, 4685.

30. Arigoni, D., H. Bosshard, H. Bruderer, G. Buechi, O. Jeger, and L. J. Krebaum: Über gegenseitige Beziehungen und Umwandlungen bei Bestrahlungsprodukten des Santonins. Helv. Chim. Acta **40**, 1732 (1957).

31. Arkhipova, L. I., S. Z. Kasymov, and G. P. Sidyakin: Sesquiterpenoid Lactones from *Artemisia halophila*. Khim. Prir. Soedin. **6**, 480 (1970); Chem. Abstr. **74**, 10349n (1971).

32. Asahina, Y., and T. Momose: Über das α-Oxysantonin. Ber. dtsch. Chem. Ges. **70**, 812 (1937).

33. Asahina, Y., and T. Ukita: Über das Temisin. Ber. dtsch. Chem. Ges. **74B**, 952 (1941).

34. Asaka, Y., T. Kubota, and A. B. Kulkarni: Studies on a Bitter Principle from *Vernonia anthelmintica*. Phytochem. **16**, 1838 (1977); Chem. Abstr. **88**, 47467q (1978).

35. Asakawa, Y., and T. Aratani: Occurrence of Sesquiterpene Lactones in *Frullania* and *Diplophyllum*. J. Hattori. Bot. Lab. **41**, 377 (1976); Chem. Abstr. **86**, 117570c (1977).

36. ASAKAWA, Y., J. C. MULLER, G. OURISSON, J. FOUSSEREAU, and G. DUCOMBS: New Sesquiterpene Lactones of *Frullania* (Hepaticae). Isolation, Structures, Allergenic Properties. Bull. Soc. Chim. France **1976**, 1465; Chem. Abstr. **86**, 190227a (1977).

37. ASAKAWA, Y., G. OURISSON, and T. ARATANI: New Sesquiterpene Lactone and Aldehyde of *Frullania tamarisci* Subsp. *Obscura* (Hepaticae). Tetrahedron Letters **1975**, 3957.

38. ASAKAWA, Y., M. TOYOTA, M. UEMOTO, and T. ARATANI: Sesquiterpenes of Six *Porella Species* (Hepaticae). Phytochem. **15**, 1929 (1976).

38a. ASHER, J. D. M., and G. A. SIM: Sesquiterpenoids. Part III. The Stereochemistry of Santonin: X-Ray Analysis of 2-Bromo-α-santonin. J. Chem. Soc. (London) **1965**, 6041.

39. ASPLUND, R. O., M. MCKEE, and P. BALASUBRAMANIYAN: Artevasin; a New Sesquiterpene Lactone from *Artemesia tridentata*. Phytochem. **11**, 3542 (1972).

40. AUMEER, P. S., and T. B. H. MCMURRY: Molecular Rearrangement of 4β-Acetoxyl-1,2-dihydrosantonene. J. Chem. Soc. D., Chem. Commun. **1971**, 641.

41. AYER, W. A., and ST. FUNG: Metabolites of Birds Nest Fungi — VII. Bicyclofarnesane Sesquiterpenes of *Mycocalia reticulata* Petch. Tetrahedron **33**, 2771 (1977).

42. BABAKHODZHAEV, A., S. Z. KASYMOV, and G. P. SIDYAKIN: Xanthatin from *Xanthium spinosum*. Khim. Prir. Soedin. **9**, 559 (1973); Chem. Abstr. **80**, 45621w (1974).

42a. BACHELOR, F. W., and S. ITÔ: Revision of the Stereochemistry of Lactucin. Can. J. Chem. **51**, 3626 (1973).

43. BANKAR, N. S., and G. H. KULKARNI: Reactions of Sant-4(15)-enolide. Indian J. Chem. **12**, 1012 (1974); Chem. Abstr. **82**, 140322z (1975).

44. — — Double Bond Migration in Dihydrodehydrocostus Lactone During Hydrogenation. Indian J. Chem. **10**, 952 (1972); Chem. Abstr. **78**, 111538m (1973).

45. — — Monoepoxidation of Saussurea Lactone. Indian J. Chem. **13**, 744 (1975); Chem. Abstr. **83**, 164386v (1975).

46. BARANOWSKA, E., and W. M. DANIEWSKI: Mass Spectrometric Investigations of Lactarorufin A and its Derivatives. Bull. Acad. Pol. Sci. Ser. Sci. Chim. **20**, 313 (1972); Chem. Abstr. **77**, 886912 (1972).

47. BARTON, D. H. R., and P. T. GILHAM: The Stereochemistry of Lumisantonin. J. Chem. Soc. (London) **1960**, 4596.

48. BARTON, D. H. R., O. C. BÖCKMAN, and P. DE MAYO: Sesquiterpenoids. Part XII. Further Investigations on the Chemistry of Pyrethrosin. J. Chem. Soc. (London) **1960**, 2263.

49. BARTON, D. H. R., and P. DE MAYO: The Constitution of Pyrethrosin. J. Chem. Soc. (London) **1957**, 150.

49a. — — Recent Advances in Sesquiterpenoid Chemistry. Quart. Rev. **11**, 189 (1957).

50. BARTON, D. H. R., J. E. D. LEVISALLES, and J. T. PINHEY: Some Analogues of Isophotosantonic Lactone. J. Chem. Soc. (London) **1962**, 3472.

51. BARTON, D. H. R., G. P. MOSS, and J. A. WHITTLE: Biosynthesis of Santonin. J. Chem. Soc. C **1968**, 1813.

52. BARTON, D. H. R., and C. R. NARAYANAN: The Constitution of Lactucin. J. Chem. Soc. (London) **1958**, 963.

53. BARTON, D. H. R., and J. T. PINHEY: The Stereochemical Correlation of Artemisin and Geigerin. Proc. Chem. Soc. **1960**, 279.

54. BARTON, D. H. R., J. T. PINHEY, and R. J. WELLS: Synthetic Studies on Geigerin and its Derivatives. J. Chem. Soc. (London) **1964**, 2518.

55. BASEY, K., and J. G. WOOLLEY: Biosynthesis of the Tigloyl Esters in *Datura*: Role of 2-Methylbutyric Acid. Phytochem. **12**, 2197 (1973).

56. BATES, R. B., Z. CEKAN, V. PROCHAZKA, and V. HEROUT: The Structure of Arborescin. Tetrahedron Letters **1963**, 1127.

57. BATES, R. B., C. J. CHEER, and T. C. SNEATH: Crystal Structure of Hydroxypelenolide p-Bromobenzoate. J. Organ. Chem. (U. S. A.) 35, 3960 (1970).

58. BATTERHAM, T. J., N. K. HART, and J. A. LAMBERTON: Guaianolides from the Wax of *Calocephalus brownii* F. Muell. (Compositae). Austral. J. Chem. 19, 143 (1966).

59. BAWDEKAR, A. S., G. R. KELKAR, and S. C. BHATTACHARYYA: Absolute Configuration of Parthenolide. Tetrahedron Letters 1966, 1225.

60. BEECHAM, A. F.: The CD of α,β-Unsaturated Lactones. Tetrahedron 28, 5543 (1972).

61. BEGLEY, M. J., G. PATTENDEN, T. J. MABRY, M. MIYAKADO, and H. YOSHIOKA: Constitution of Peucephyllin, a New Type of Germacranolide from *Peucephyllum schottii*. Tetrahedron Letters 1975, 1105.

62. BEHARE, E. S.: Synthesis of a-Methylene-γ-Butyrolactones and Application to the Total Synthesis of (±)-Isotelekin. Diss. Abstr. Int. B 1974, 35, 1563; Chem. Abstr. 82, 98174a (1975).

63. BENESOVA, V., and V. HEROUT: Neutral Substances from *Telekia speciosa* (Schreb) Baumg. Collect. Czech. Chem. Comm. 26, 2916 (1961).

64. BENESOVA, V., V. HEROUT, and F. SORM: Isolation and Structure of Nobilin, a Sesquiterpenic Lactone with a Ten-Membered Ring. Collect. Czech. Chem. Comm. 29, 3097 (1964).

65. — — — Structure of Telekin and Isotelekin, New Sesquiterpenic Lactones from *Telekia speciosa* (Schreb) Baumg. Collect. Czech. Chem. Comm. 26, 1350 (1961).

66. BENESOVA, V., M. V. NAZARENKO, and L. V. SLEPTSOVA: Sesquiterpenic γ-Lactones from *Artemisia jacutica*. Khim. Prir. Soedin. 5, 186 (1969); Chem. Abstr. 72, 39763n (1970).

67. BENESOVA, V., Z. SAMEK, V. HEROUT, and F. SORM: The Structure of Nobilin. Tetrahedron Letters 1970, 5017.

68. BENESOVA, V., Z. SAMEK, and S. VASICKOVA: Structure of the Sesquiterpenic Lactone Diplophyllolide and other Components from the Liverwort, *Diplophyllum albicans* (L.) Dum. Collect. Czech. Chem. Comm. 40, 1966 (1975).

69. BENTLEY, R. K., T. G. S. C. BUCHANAN, T. G. HALSALL, and V. THALLER: Urospermal A and Urospermal B, Conformers of a Germacranolide Aldehyde from *Urospermum dalechampii* F. W. Schmidt. Chem. Commun. 1970, 435.

70. BERMEJO BARRERA, J., C. BETANCOV, J. L. BRETON FUNES, and A. G. GONZALEZ: Lactonas Sesquiterpenicas De La *Amberboa lipii* D. C. Anal. Real. Soc. Esp. Fis. Quim. LXV, 285 (1969).

71. BERMEJO BARRERA, J., J. L. BRETON-FUNES, and A. G. GONZALEZ: Terpenoids of the *Sonchus*. Part III. Sesquiterpene Lactones of *S. Jacquini* D. C., *S. pinnatus* Ait, and *S. radicatus* Ait. J. Chem. Soc. (London) 1966, 1298.

72. BERMEJO BARRERA, J., J. L. BRETON-FUNES, M. FAJARDO, and A. G. GONZALEZ: Terpenoids from the *Sonchus*. VI. Tuberiferine from *Sonchus tuberifer Svent*. Tetrahedron Letters 1967, 3475.

73. — — — — Tubiferine, a New Sesquiterpene Lactone from *Sonchus tuberifer*. An. Quim. 64, 183 (1968); Chem. Abstr. 69, 27548w (1968).

74. BERMEJO BARRERA, J., J. L. BRETON-FUNES, A. G. GONZALEZ, and A. VILLAR DEL FRESNO: Lactonas Sesquiterpenicas de *Sonchus hierrensis* (Pit) Svent. stat. nov. var. *benehoavensis* Svent. var. *nova*. Anal. Real. Soc. Esp. Fis. Quim. 1968, 893.

75. BERNAL, I., and S. F. WATKINS: Direct Phase Determination from Neutron Diffraction Data of the Structure of Melampodin. Science 178, 1282 (1972); Chem. Abstr. 78, 63598g (1973).

76. BERTELLI, D. J., and J. H. CRABTREE: Naturally Occurring Fulvene Hydrocarbons. Tetrahedron 24, 2079 (1968).

77. BETKOUSKI, M., T. J. MABRY, I. F. TAYLOR, and W. H. WATSON: Glaucolide-D and -E, Two New Germacranolides from *Verononia Uniflora* Sch.-Bip. (Compositae). Rev. Latinoamer. Quim. **6**, 191 (1975).
78. BETKOUSKI, M., T. J. MABRY, T. W. ADAMS, W. H. WATSON, and S. B. JONES, JR.: Glaucolide G, a New Germacranolide Sesquiterpene Lactone from *Vernonia leiocarpa* (Compositae). Rev. Latinoamer. Quim. **1976**, 111; Chem. Abstr. **86**, 13822v (1977).
79. BHACCA, N. S., and N. H. FISCHER: The Determination of the Conformation of a Germacranolide (Dihydrotamaulipin A Acetate) with the Aid of Nuclear Overhauser Effects. Chem. Commun. **1969**, 68.
79a. BHACCA, N. S., F. W. WEHRLI, and N. H. FISCHER: Carbon-13 Magnetic Resonance Study of Terpenoids. I. An Application of Heteronuclear Selective Decoupling Experiments to the Spectral Assignments of Nonproton-Bearing Carbon-13 Resonances of a Germacranolide, Melampodin. J. Organ. Chem. (U.S.A.) **38**, 3618 (1973).
80. BHACCA, N. S., R. A. WILEY, N. H. FISCHER, and F. W. WEHRLI: Carbon-13 and Proton Magnetic Resonance Study of the Structure and Conformation of a New Germacranolide Sesquiterpene Dilactone. Chem. Commun. **1973**, 614.
81. BHADANE, N. R., and F. SHAFIZADEH: Sesquiterpene Lactones of Sagebrush: The Structure of Artecanin. Phytochem. **14**, 2651 (1975).
82. BHADANE, N. R., R. G. KELSEY, and F. SHAFIZADEH: Sesquiterpene Lactones of *Artemisia tridentata* SSP. *vaseyana*. Phytochem. **14**, 2084 (1975).
83. BHANOT, O. S., and P. C. DUTTA: Syntheses of Lactones Related to Psilostachyins. Chem. Commun. **1968**, 122.
84. BHATTACHARYYA, S. C.: Sesquiterpene Lactones. J. Indian Chem. Soc. **47**, 299 (1970); Chem. Abstr. **73**, 25669v (1970).
85. BHEDI, D. N., A. M. SHALIGRAM, and A. S. RAO: Acid-Catalyzed Rearrangement of $1\alpha,2\alpha$-Epoxy-α-tetrahydrosantonin. Indian J. Chem. Sect B **1976**, 657; Chem. Abstr. **86**, 140270p (1977).
86. BIALECKI, M., E. BLOSZYK, B. DROZDZ, B. HLADON, and S. SZWEMIN: Cytostatic Activity of Grosheimin. Pol. J. Pharmacol. Pharm. **25**, 195 (1973).
87. BIALECKI, M., B. HLADON, B. DROZDZ, E. BLOSZYK, S. SZWEMIN, and T. BOBKIEWICZ: Studies on the Cytotoxic Effect of Alatolide on the Human Lymphocytes and Neoplastic Cells in Tissue Culture. Pol. J. Pharmacol. Pharm. **26**, 511 (1974).
88. BIERNER, M. W.: Sesquiterpene Lactones and the Systematics of *Helenium quadridentatum* and *H. elegans*. Biochem. Syst. **1**, 95 (1973); Chem. Abstr. **79**, 102768c (1973).
89. BIRNBAUM, G. I., C. P. HUBER, M. L. POST, J. B. STOTHERS, J. R. ROBINSON, A. STOESSL, and E. W. B. WARD: Sesquiterpenoid Stress Compounds of *Datura stramonium*: Biosynthesis of the Three Major Metabolites from $[1,2^{-13}C]$ acetate and the X-Ray Structure of 3-Hydroxylubimin. Chem. Commun. **1976**, 330.
90. BJELDANES, L. F., and T. A. GEISSMAN: Constituents of an F_1 Hybrid *Encelia farinosa, Encelia californica*. Phytochem. **10**, 1079 (1971).
91. BLOSZYK, E., B. DROZDZ, Z. SAMEK, J. TOMAN, and M. HOLUB: Linifolin A and Helenalin from *Helenium aromaticum*. Phytochem. **14**, 1444 (1975).
92. BOGUCKA-LEDOCHOWSKA, M., A. HEMPEL, Z. DAUTER, A. KONIK, E. BOROWSKI, W. DANIEWSKI, and M. KOCOR: The Structure of Lactarorufin B-3,8-Ether 14-p-Bromobenzoate. Tetrahedron Letters **1976**, 2267.
93. BOHLMANN, F.: priv. commun.
94. BOHLMANN, F., G. BRINDÖPKE, and R. C. RASTOGI: A New Type of Germacranolide from *Vernonia* Species. Phytochem. **17**, 475 (1978).
95. BOHLMANN, F., and H. CZERSON: A New Glaucolide Derivative from *Erlangea remifolia*. Phytochem. **17**, 1190 (1978).
96. — — Neue Guajanolid-Derivate aus *Erlangea inyangana* (N. E. Br.) B. L. Burtt. Phytochem. **17**, 568 (1978).

97. Bohlmann, F., H. Czerson, and S. Schöneweiss: Neue Inhaltsstoffe aus *Inula viscosa* Ait. Chem. Ber. **110**, 1330 (1977).

98. Bohlmann, F., and D. Ehlers: Ein Neues *cis,cis*-Germacranolid aus *Chrysanthemum poteriifolium*. Phytochem. **16**, 137 (1977).

99. Bohlmann, F., D. Ehlers, C. Zdero, and M. Grenz: Über Inhaltsstoffe der Gattung *Ligularia*. Chem. Ber. **110**, 2640 (1977).

100. Bohlmann, F., and L. Fiedler: Notiz über ein neues Germacrolid aus *Chromolaena glaberrima* (D. C.) K. et R. Chem. Ber. **111**, 408 (1978).

101. Bohlmann, F., and M. Grenz: New Methylsalicylic Acid Derivative from *Othonna cylindrica*. Tetrahedron Letters **1974**, 1681.

102. — — Neue Sesquiterpenlactone aus *Athanasia*-Arten. Chem. Ber. **108**, 357 (1975).

103. — — Über ein neues Sesquiterpenlacton aus *Pluchea dioscorides DC*. Tetrahedron Letters **1969**, 5111.

104. Bohlmann, F., M. Grenz, and C. Zdero: Naturally Occurring Terpene Derivatives. Constituents of the *Liabum* Group. Phytochem. **16**, 285 (1977).

105. Bohlmann, F., J. Jakupovic, and M. Lonitz: Über Inhaltsstoffe der *Eupatorium*-Gruppe. Chem. Ber. **110**, 301 (1977).

106. Bohlmann, F., K. H. Knoll, C. Zdero, P. K. Mahanta, M. Grenz, A. Suwita, D. Ehlers, N. LeVan, W. R. Abraham, and A. A. Natu: Terpen-Derivate aus *Senecio*-Arten. Phytochem. **16**, 965 (1977).

107. Bohlmann, F., and N. LeVan: Sesquiterpenlactone und Polyine aus der Gattung *Arctotis*. Phytochem. **16**, 487 (1977).

108. — — New Germacranolides from *Dicoma anomala*. Phytochem. **17**, 570 (1978).

109. — — Neue Guajanolide aus *Podachaenium eminens*. Phytochem. **16**, 1304 (1977).

109a. — — Neue Sesqui- und Diterpene aus *Bedfordia salicina*. Phytochem. **17**, 1173 (1978).

110. Bohlmann, F., and M. Lonitz: Neue Sandaracopimardien-Derivate, Sesquiterpene und Sesquiterpenlactone aus *Zexmenia*-Arten. Chem. Ber. **111**, 843 (1978).

111. Bohlmann, F., P. K. Mahanta, J. Jakupovic, R. C. Rastogi, and A. A. Natu: New Sesquiterpene Lactones from *Inula* Species. Phytochem. **17**, 1165 (1978).

112. Bohlmann, F., P. K. Mahanta, A. Suwita, An. Suwita, A. A. Natu, C. Zdero, W. Dorner, D. Ehlers, and M. Grenz: Neue Sesquiterpenelactone und andere Inhaltsstoffe aus Vertretern der *Eupatorium*-Gruppe. Phytochem. **16**, 1973 (1977).

113. Bohlmann, F., P. K. Mahanta, A. A. Natu, R. M. King, and H. Robinson: New Germacranolides from *Isocarpha* Species. Phytochem. **17**, 471 (1978).

114. Bohlmann, F., A. Suwita, A. A. Natu, H. Czerson, and An. Suwita: Über weitere α-Longipinen-Derivate aus Compositen. Chem. Ber. **110**, 3572 (1977).

115. Bohlmann, F., and C. Zdero: Neue Sesquiterpenlactone und Thymol-Derivate aus *Inula*-Arten. Phytochem. **16**, 1243 (1977).

116. — — Zwei neue Sesquiterpenlactone aus *Lidbeckia pectinata Berg.* und *Pentzia elegans DC*. Tetrahedron Letters **1972**, 621.

117. — — New Sesquiterpenes and Acetylenes from *Athanasia* and *Pentzia* Species. Phytochem. **17**, 1595 (1978).

118. — — Ein neues Guajanolid aus *Matricaria zuurbergensis*. Phytochem. **16**, 136 (1977).

119. — — New Germacrolides from *Calea zacatechichi*. Phytochem. **16**, 1065 (1977).

120. — — Über neue Inhaltsstoffe der Gattung *Anthemis*. Chem. Ber. **108**, 1902 (1975).

121. — — Über Inhaltsstoffe der *Tribus Mutisieae*. Phytochem. **16**, 239 (1977).

122. — — Ein neues Sesquiterpenlactone aus *Matricaria suffructicosa* var. *leptoloba*. Chem. Ber. **108**, 437 (1975).

123. — — Ein neues Sesquiterpenlacton aus *Dugesia mexicana Gray*. Chem. Ber. **109**, 2651 (1976).

124. — — Neue Germacranolide aus *Platycarpha glomerata*. Phytochem. **16**, 1832 (1977).

125. — — Ein neues Germacranolid aus *Simsia dombeyana*. Phytochem. **16,** 776 (1977).

126. — — Über ein neues Diterpen aus *Melampodium perfoliatum* (Carv.) A. Gray. Chem. Ber. **109,** 1670 (1976).

127. — — Neue Sesquiterpenlactone aus *Osmitopsis asteriscoides* (L.) Cass. Chem. Ber. **107,** 1409 (1974).

128. — — Inhaltsstoffe aus *Vernonia*-Arten. Phytochem. **16,** 778 (1977).

129. — — Prutenin, 4-Angeloyloxy-pruteninon sowie 4-Acetoxy-pruteninon und -iso-pruteninon. Vier Neue Sesquiterpenlactone aus *Laserpitium prutenicum* L. Chem. Ber. **104,** 1611 (1971).

129 a. BOHLMANN, F., C. ZDERO, and M. GRENZ: Über ein neues Sesquiterpen aus *Anthemis Cotula* L. Tetrahedron Letters **1969,** 2417.

130. BOHLMANN, F., C. ZDERO, and M. LONITZ: Neue Guajen-Derivate aus *Parthenium hysterophorus* und ein weiteres Pseudoguajanolid aus *Ambrosia cumanensis*. Phytochem. **16,** 575 (1977).

131. BORISOV, V. N., M. G. PIMENOV, and A. I. BAN'KOVSKII: Distribution of Some Biologically Active Compounds in the Genus *Ferula* L. According to IR and UV Spectroscopic and Thin-Layer Chromatographic Data. Rastit. Resur. **1977,** 276; Chem. Abstr. **87,** 2380u (1977).

132. BOVILLE, M. J., P. J. COX, P. D. CRADWICK, M. H. P. GUY, G. A. SIM, and D. N. J. WHITE: X-Ray Crystal Structure Analysis of Costunolide. Acta Crystallogr. Sect. B **1976,** 3203. Corrigendum: ibid. **1978,** 349.

133. BRECKNELL, D. J., and R. M. CARMAN: Callitrin, Callitrisin, Dihydrocallitrisin, Columellarin and Dihydrocolumellarin, New Sesquiterpene Lactones from the Heartwood of *Callitris columellaris*. Tetrahedron Letters **1978,** 73.

134. BRETON-FUNES, J. L.: Sesquiterpene Lactones: Their Investigation in Compositae, the Canary Islands, and the Iberian Peninsula. Caja Gen. Ahorros: Santa Cruz de Tenerife, Spain, **1974,** 119 pp.

135. BRETON-FUNES, J. L., B. GARCIA MARRERO, and A. G. GONZALEZ: Las Lactonas Sesquiterpenicas De La *Amberboa lippii* D. C. (Sobre al formula de la grosshemina). Anal. Real. Soc. Esp. Fis. Quim. **LXIV,** 1015 (1968).

135 a. BREWSTER, J. W.: Some Applications of the Conformational Dissymmetry Rule. Tetrahedron **13,** 106 (1961).

136. BRIEGER, G.: The Stereospecific Synthesis of D,L-Winterin. Tetrahedron Letters **1965,** 4429.

137. BROOKS, C. J. W., and G. H. DRAFFAN: Sesquiterpenoids of *Warburgia* Species. Ugandensolide and Ugandensidial (Cinnamodial). Tetrahedron **25,** 2887 (1969).

138. — — Sesquiterpenoids of *Warburgia* Species. Warburgin and Warburgiadione. Tetrahedron **25,** 2865 (1969).

139. BROWN, E. D., and J. K. SUTHERLAND: The Conversion of a Germacrane into Guaiane Derivatives. Chem. Commun. **1968,** 1060.

140. BRYAN, R. F., and C. J. GILMORE: Crystal and Molecular Structures of Dehydro-eriolanin and Dehydroeriolangin in a Co-crystalline Mix. Acta Crystallogr., Sect. B **31,** 2213 (1975).

141. BUI, A. M., A. CAVE, M. M. JANOT, J. PARELLO, and J. POTIER: Isolement Et Analyse Structurale Du Collybolide, Nouveau Sesquiterpene Extrait De *Collybia maculata* ALB. Et. SCH. Ex Fries (Basidiomycetes). Tetrahedron **30,** 1327 (1974).

142. BUI, A. M., J. PARELLO, P. POTIER, and M. M. JANOT: Structure of Collybolide, a New Sesquiterpene Extracted from *Collybia maculata*. C. R. Acad. Sci. Ser. C **1970,** 270, 1022; Chem. Abstr. **72,** 133002k (1970).

143. BUKREEVA, T. V., N. P. KIR'YALOV, and V. A. GINDIN: Structure of Lactones of α- and β-Reolones. Khim. Prir. Soedin. **1977,** 30.

144. Büchi, G., and D. Rosenthal: The Structures of Helenalin and Isohelenalin. J. Amer. Chem. Soc. 78, 3860 (1956).
145. Caine, D., and G. Hasenhuettl: The Total Synthesis of d,1-Yomogin. Tetrahedron Letters 1975, 743.
146. Cane, D. E., R. B. Nachbar, J. Clardy, and J. Finer: The Absolute Configuration of Formannosin. Tetrahedron Letters 1977, 4277.
147. Canonica, L., A. Corbella, P. Gariboldi, G. Jommi, J. Krepinsky, G. Ferrari, and C. Casagrande: Sesquiterpenoids of Cinnamosma fragrans Baillon. Structure of Cinnamolide, Cinnamosmolide and Cinnamodial. Tetrahedron 25, 3895 (1969).
148. — — — — — — — Sesquiterpenoids of Cinnamosma fragrans Baillon. Structure of Bemarivolide, Bemadienolide and Fragrolide. Tetrahedron 25, 3903 (1969).
149. Canonica, L., A. Corbella, G. Jommi, J. Krepinsky, G. Ferrari, and C. Casagrande: The Structure of Cinnamolide, Cinnamosmolide and Cinnamodial, Sesquiterpenes with Drimane Skeleton from Cinnamosma fragrans Baillon. Tetrahedron Letters 1967, 2137.
150. Castille, A.: Pyrethrol. An. Real. Acad. Farm. 32, 121 (1966); Chem. Abstr. 66, 65658c (1967).
151. Caughlan, C. N., M. Ul-Haque, and M. T. Emerson: The Molecular and Crystal Structure of Bromomexicanin E ($C_{14}H_{15}O_3Br$). Chem. Commun. 1966, 151.
152. Cavallito, C. J., J. H. Bailey, and F. K. Kirchner: The Antibacterial Principle of Arctium minus. I. Isolation, Physical Properties and Antibacterial Action. J. Amer. Chem. Soc. 67, 948 (1945).
153. Cekan, Z., V. Herout, and F. Sorm: Die Struktur von Matricin, ein Guajanolid aus der Kamille (Matricaria chamomilla L.). Collect. Czech. Chem. Commun. 22, 1921 (1957).
154. — — — Isolation and Properties of the Pro-Chamazulene from Matricaria chamomilla L., a Further Compound of the Guaianolide Group. Collect. Czech. Chem. Comm. 19, 798 (1954).
155. Cekan, Z., V. Prochazka, V. Herout, and F. Sorm: Isolation and Constitution of Matricarin, another Guaianolide from Camomile (Matricaria chamomilla L.). Collect. Czech. Chem. Comm. 24, 1554 (1959).
156. — — — Isolation of Globicin, a Guaianolide from Matricaria globifera (Thunb.) Druce. Collect. Czech. Chem. Comm. 25, 2553 (1960).
157. Chao, O., J. Romo, A. Romo de Vivar, S. Leon, and R. Cetina: Momentos dipolares II. Helenalina. Rev. Latinoamer. Quim. 1, 7 (1970).
157a. Chapman, O. L.: Photochemical Rearrangements of Organic Molecules. In: Advances in Photochemistry (Noyes, W. A., G. S. Hammond, and J. N. Pitts, eds.), 1, p. 323. New York: Interscience Publishers. 1963.
158. Chien, M. K., C. H. Chen, and K. F. Tseng: The Constituents of Yejuhua; the Flower of Chrysanthemum indicum. Yao Hsueh Hsueh Pao 10, 129 (1963); Chem. Abstr. 1963, 15326b.
159. Chikamatsu, H., and W. Herz: Ivalbatin, a New Xanthanolide from Iva dealbata. J. Organ. Chem. (U. S. A.) 38, 585 (1973).
160. Chikamatsu, H., M. Maeda, and M. Nakazaki: Structure of Torilin. Tetrahedron 25, 4751 (1969).
161. Chugunov, P. V., D. A. Pakaln, and A. M. Shreter: Lactone from Inula germanica. Khim. Prir. Soedin. 6, 478 (1970); Engl. edit: p. 495; Chem. Abstr. 74, 1052h (1971).
162. Chugunov, P. V., K. S. Rybalko, and A. I. Shreter: The Structure of the Sesquiterpene Lactone Saupirin. Khim. Prir. Soedin. 7, 727 (1971); Engl. edit: p. 706; Chem. Abstr. 76, 127168k (1972).
163. Chugunov, P. V., V. I. Sheichenko, A. I. Ban'kovskii, and K. S. Rybalko: The Structure of Britannin. A Sesquiterpene Lactone from Inula britannica. Khim. Prir. Soedin. 7, 276 (1971); Chem. Abstr. 75, 110438e (1971).

164. CIMINO, G., S. DE STEFANO, A. GUERRIERO, and L. MINALE: Furanosesquiterpenoids in Sponges — I: Pallescensin-1, -2, and -3 from *Disidea pallescens*. Tetrahedron Letters **1975**, 1417.

165. CLARKE, E. P.: The Preparation of Picrotoxin. J. Amer. Chem. Soc. **57**, 1111 (1935).

166. CLEMO, G. R.: β-Santonin. J. Chem. Soc. (London) **1934**, 1343.

167. CLEMO, G. R., and W. COCKER: The Constitution of ψ-Santonin. J. Chem. Soc. (London) **1946**, 30.

167a. CLEMO, G. R., R. D. HAWORTH, and E. WALTON: The Constitution of Santonin. Part I. The Synthesis of d1-Santonous Acid. J. Chem. Soc. **1929**, 2368.

168. COATES, R. M.: Biogenetic-Type Rearrangements of Terpenes. In: Progress in the Chemistry of Organic Natural Products (HERZ, W., H. GRISEBACH, and G. W. KIRBY, eds.), Vol. 33, p. 73. Wien-New York: Springer. 1976.

168a. COCKER, W., and T. B. H. MCMURRY: Stereochemical Relationships in the Eudesmane (Selinane) Group of Sesquiterpenes. Tetrahedron **8**, 181 (1960).

169. COCKER, W., and A. NISBET: The Stereochemistry of Tetrahydroalantolactones. J. Chem. Soc. (London) **1963**, 534.

169a. COGGON, P., and G. A. SIM: Sesquiterpenoids. Part VIII. Stereochemistry of Santonin: X-ray Analysis of 2-Bromo-β-santonin. J. Chem. Soc. B **1969**, 237.

170. CONROY, H.: The Skeleton of Picrotoxinin. The Total Synthesis of d1-Picrotoxadiene. J. Amer. Chem. Soc. **74**, 3046 (1952).

171. — The Skeleton of Picrotoxinin. The Degradation to Picrotoxadiene. J. Amer. Chem. Soc. **74**, 491 (1952).

172. — Picrotoxin. V. Conformational Analysis and Problems of Structure. J. Amer. Chem. Soc. **79**, 5550 (1957).

173. CONNOLLY, J. D., and I. M. S. THORNTON: Sesquiterpenoid Lactones from the Liverwort *Frullania tamarisci*. Phytochem. **12**, 631 (1973).

174. CORBELLA, A., P. GARIBOLDI, G. JOMMI, Z. SAMEK, M. HOLUB, and B. DROZDZ: Absolute Stereochemistry of Cynaropicrin and Related Guaianolides. Chem. Commun. **1972**, 386.

175. CORBELLA, A., P. GARIBOLDI, G. JOMMI, and G. FERRARI: Structure and Absolute Stereochemistry of Vanillosmin, A Guaianolide from *Vanillosmopsis erythropappa*. Phytochem. **13**, 459 (1974).

176. CORBELLA, A., P. GARIBOLDI, and G. JOMMI: Biosynthesis of Tutin from (4R)-[4-^{3}H$_1$]-Mevalonic Acid. Chem. Commun. **1972**, 600.

177. CORBELLA, A., P. GARIBOLDI, G. JOMMI, and M. SISTI: Biosynthesis of the Terpenoid Dendrobine. Early Stages of the Path. Chem. Commun. **1975**, 288.

178. CORBELLA, A., G. JOMMI, B. RINDONE, and C. SCOLASTICO: Structure of Pretoxin and Lambicin. Tetrahedron **25**, 4833 (1969).

179. CORBELLA, A., G. JOMMI, B. RINDONE, and C. SCOLASTICO: Constituents of *Hyaenanche globosa*. Structure of Substance C and Correlation Between Picrotoxinin and Tutine. Ann. Chim. (Rome) **57**, 758 (1967); Chem. Abstr. **68**, 3008y (1968).

180. CORBELLA, A., G. JOMMI, and C. SCOLASTICO: Structural Correlation between Capenicin and Tutin. Tetrahedron Letters **1966**, 4819.

181. CORBETT, R. E., and T. L. CHEE: Extractives from *Pseudowintera colorata*. A New Sesquiterpene Lactone, Colorata-4(13), 8-dienolide, J. Chem. Soc. Perkin Trans. I **1976**, 850.

182. CORDELL, G. A.: Biosynthesis of Sesquiterpenes. Chem. Rev. **76**, 425 (1976).

183. CORDELL, G. A., and N. R. FARNSWORTH: Experimental Antitumor Agents from Plants, 1974-6. Lloydia **1977**, 1; Chem. Abstr. **87**, 11487m (1977).

184. CORREA, J., M. L. CERVERA, and R. M. MAINERO: Synthesis of Tithonine. Rev. Latinoamer. Quim. **2**, 67 (1971).

185. Corey, E. J., and A. G. Hortmann: Total Synthesis of Dihydrocostunolide. J. Amer. Chem. Soc. **85**, 4033 (1963).
186. Cortes, E., R. Miranda, and J. Romo: Mass-Spectrometry of the Sesquiterpene Lactones of the Pseudoguaianolide Series. Rev. Latinoamer. Quim. **8**, 39 (1977).
187. Cox, P. J., and G. A. Sim: X-Ray Crystallographic Determination of the Stereochemistry and Conformation of Deacetylneotenulin. J. Chem. Soc. Perkin Trans. II **9**, 990 (1976).
188. — — X-Ray Crystallographic Determination of the Stereochemistry and Conformation of the Germacranolide Glaucolide A. J. Chem. Soc. Perkin Trans. II **1975**, 455.
189. — — X-Ray Crystallographic Determination of the Molecular Conformation of the Germacranolide Alatolide Monohydrate. J. Chem. Soc. Perkin Trans. II **1977**, 255.
190. — — X-Ray Crystallographic Determination of the Stereochemistry and Conformation of Miscandenin, an Elemanediolide with a Dihydro-oxepin Ring. J. Chem. Soc. Perkin Trans. II **1974**, 1359.
191. — — X-Ray Crystallographic Analysis of Paucin Monohydrate, a Pseudoguaianolide Glucoside. J. Chem. Soc. Perkin Trans. II **1977**, 259.
192. — — X-Ray Crystallographic Determination of the Stereochemistry and Conformation of Dihydromikanolide, a Germacranolide Diepoxide. J. Chem. Soc. Perkin Trans. II **1974**, 1355.
193. Cox, P. J., G. A. Sim, and W. Herz: X-Ray Crystallographic Determination of the Molecular Structure of Berlandin, A Guaianolide Epoxide. Comments on the Circular Dichroism of Sesquiterpenoid α-Methylene-γ-Lactones with α,β-Unsaturated Ester Side Chains. J. Chem. Soc. Perkin Trans. II **1975**, 459.
194. Cox, P. J., G. A. Sim, J. S. Roberts, and W. Herz: Stereochemistry of the Biological Divinyloxiran-Dihydro-Oxepin Rearrangement: X-Ray Studies of the Sesquiterpenoids Miscandenin and Dihydromikanolide. Chem. Commun. **1973**, 428.
195. Crabbe, P., S. Burnstein, and C. Djerassi: Isolation and Structure of Three Sesquiterpenes Related to Iresin. Bull. Soc. Chim. Belges. **67**, 632 (1958).
196. Cradwick, P. D., and G. A. Sim: Crystallographic Determinations of Partial Stereochemistries of the Sesquiterpenoids Illudol and Marasmic Acid. Chem. Commun. **1971**, 431.
197. Craven, B. M.: Molecular Structure of Tutin. Nature **197**, 1193 (1963).
198. — The Molecular Structure and Absolute Configuration of Picrotoxinin. Tetrahedron Letters **1960**, No. 19, p. 21.
199. Culvenor, C. C. J.: The Structure of Jacozine, an Alkaloid of *Senecio jacobaea* L. Austral. J. Chem. **17**, 233 (1964).
200. Currie, M., and G. A. Sim: X-Ray Study of 13β-(p-Bromophenyl)-11α,13-dihydropulchellin C diacetate [2α,3β-di-acetoxy-13β-(p-bromophenyl)thio]-11α,13-dihydro isolantolactone. Stereochemistry of Addition of Thiols to Sesquiterpenoid α-Methylene γ-Lactones. J. Chem. Soc. Perkin Trans. II **1973**, 400.
200a. Czerson, H., F. Bohlmann, T. F. Stuessy, and N. H. Fischer: Sesquiterpenoid and Acetylenic Constituents of Seven *Clibadium* Species. Phytochem. **18**, 257 (1979).
201. Daloze, D., J. C. Brackman, P. Georget, and B. Tursch: Chemical Studies of Marine Invertebrates. XXII. Two Novel Sesquiterpenes from Soft Corals of the Genera *Lemnalia* and *Paralemnalia* (Coelenterata, Octocorallia, Alcyonacea). Bull. Soc. Chim. Belg. **86**, 47 (1977); Chem. Abstr. **87**, 6219w.
202. Daniewski, W. M., and M. Kocor: Isolation and Structure of Some New Sesquiterpenes from *Lactarius rufus*. Bull. Acad. Pol. Sci. Ser. Chim. **18**, 585 (1970); Chem. Abstr. **74**, 112238t (1971).
203. — — Constituents of Higher Fungi. II. Structure of Lactarorufin A. Bull. Acad. Pol. Sci. Ser. Chim. **19**, 553 (1971); Chem. Abstr. **76**, 14738c (1972).

204. DANIEWSKI, W. M., M. KOCOR, and J. KROL: Two New Sesquiterpenoic Lactones from *Lactarius necator*. Bull. Acad. Pol. Sci. Ser. Sci. Chim. **23**, 637 (1975); Chem. Abstr. **84**, 135849 p (1976).

205. — — — Constituents of Higher Fungi. Structure of 3-Deoxylactarorufin A. Rocz. Chem. **51**, 1395 (1977); Chem. Abstr. **88**, 3070 a (1978).

206. — — — Lactarorufin N and Revised Structures of Lactarorufins. Rocz. Chem. **50**, 2095 (1976); Chem. Abstr. **87**, 152402 s (1977).

207. DANIEWSKI, W. M., M. KOCOR, and S. THOREN: Constituents of Higher Fungi. Isolactarorufin, a Novel Tetracyclic Sesquiterpene Lactone from *Lactarius rufus*. Heterocycles **1976**, 77; Chem. Abstr. **86**, 90059 d.

208. DANIEWSKI, W. M., M. KOCOR, and B. ZOLTOWSKA: Structure of Lactarorufin B. Bull. Acad. Pol. Sci. Ser. Sci. Chim. **21**, 785 (1973); Chem. Abstr. **81**, 23071 k (1974).

209. DANISHEFSKY, S., P. F. SCHUDA, T. KITAHARA, and S. J. ETHEREDGE: The Total Synthesis of dl-Vernolepin and dl-Vernomenin. J. Amer. Chem. Soc. **99**, 6066 (1977).

210. DAUBEN, W. G., W. K. HAYES, J. S. P. SCHWARZ, and J. W. MCFARLAND: The Stereochemistry of ψ-Santonin. J. Amer. Chem. Soc. **82**, 2232 (1960).

211. DAUBEN, W. G., J. S. P. SCHWARZ, W. K. HAYES, and P. D. HANCE: The Structures of Two New Sesquiterpene Lactones from *Artemisia*. J. Amer. Chem. Soc. **82**, 2239 (1960).

212. DE BERNADI, M., G. FRONZA, G. VIDARI, and P. VITA-FINZI: New Sesquiterpenes from *Lactarius srobiculatus* Scop. (Russulaceae). Chim. Ind. (Milan) **58**, 177 (1976); Chem. Abstr. **85**, 17100 k (1976).

213. DE CLERCQ, P., and M. VANDEWALLE: Total Synthesis of (±)-Damsin. J. Organ. Chem. (U. S. A.) **42**, 3447 (1977).

214. DE KOCK, W. T., K. G. R. PACHLER, W. F. ROSS, P. L. WESSELS, and I. C. DU PREEZ: Griesenin and Dihydrogriesenin, Two New Sesquiterpenoid Lactones from *Geigeria africana* Gries — I. Structures. Tetrahedron **24**, 6037 (1968).

214a. DE KOCK, W. T., and K. G. R. PACHLER: Gafrinin, a Sesquiterpenoid Lactone from *Geigeria africana* Gries — II. Conformation. Tetrahedron **24**, 1701 (1968).

214b. DE KOCK, W. T., K. G. R. PACHLER, and P. L. WESSELS: Griesenin and Dihydro-griesenin, Two New Sesquiterpenoid Lactones from *Geigeria africana* Gries — II. Nuclear Magnetic Resonance Studies and Conformation. Tetrahedron **24**, 6045 (1968); Chem. Abstr. **69**, 96888 u (1968).

215. DERMANOVIC, M., S. MLADENOVIC, and M. STEFANOVIC: Chemical Investigation of Yugoslav Species of *Artemisia absinthium* L. Glas. Hem. Drus. Beograd. **41**, 287 (1976); Chem. Abstr. **87**, 98796 h (1977).

215a. DE SILVA, L. B., and T. A. GEISSMAN: Sesquiterpene Lactone, Psilotropin from *Psilostrophe cooperi*. Phytochem. **9**, 59 (1969).

215b. DE OLIVEIRA, A. B., G. G. DE OLIVEIRA, C. T. M. LIBERALLI, O. R. GOTTLIEB, and M. T. MAGALHAES: Structure and Absolute Configuration of the Sesquiterpenoid Emmotins. Phytochem. **15**, 1267 (1976).

215c. DE OLIVEIRA, A. B., M. DE LOURDES MOREIRA FERNANDES, O. R. GOTTLIEB, E. W. HAGAMAN, and E. WENKERT: Aromatic Sesquiterpenoids from *Emmotum nitens*. Phytochem. **13**, 1199 (1974).

216. DEUEL, P. G., and T. A. GEISSMAN: Xanthinin. II. The Structures of Xanthinin and Xanthatin. J. Amer. Chem. Soc. **79**, 3778 (1959).

216a. DE VILLIERS, J.: The Isolation and Structure of Gafrinin, a Sesquiterpenoid Lactone from *Geigeria africana*. J. Chem. Soc. (London) **1961**, 2049.

216b. DE VILLIERS, J. P., and K. PACHLER: The Revised Structure of Geigerinin. J. Chem. Soc. (London) **1963**, 4989.

217. DEVON, T. K., and A. I. SCOTT: Handbook of Naturally Occurring Compounds, Vol. II. Terpenes. New York and London: Academic Press. 1972.

218. DJERASSI, C., and S. BURSTEIN: Iresin (Part 5). Complete Structure and Absolute Configuration. Tetrahedron 7, 37 (1959).
219. DJERASSI, C., P. SENGUPTA, J. HERRAN, and E. WALLS: The Isolation of Iresin, a New Sesquiterpene Lactone. J. Amer. Chem. Soc. 76, 2966 (1954).
220. DOLEJS, L., and V. HEROUT: Constitution of Eupatoriopicrin, a Germacranolide from Eupatorium cannabinum L. Collect. Czech. Chem. Comm. 27, 2654 (1962).
221. DOLEJS, L., M. SOUCEK, M. HORAK, V. HEROUT, and F. SORM: The Structure of Lactucin. Collect. Czech. Chem. Comm. 23, 2195 (1958).
222. DOMINGUEZ, X. A., D. BUTRUILLE ET AQUILINO, and Y. AUBAD: La neoleonine un nouveau pseudoguaianolide. Rev. Latinoamer. Quim. 1, 136 (1970).
223. DOMINGUEZ, X. A., and E. CARDENAS G.: Achillin and Deacetylmatricarin from Two Artemisia Species. Phytochem. 14, 2511 (1975).
224. DOMINGUEZ, X. A., M. GUTIERREZ, and R. ARAGON: Isolation of Baileyolin a Tumor Inhibitory and Antibiotic Sesquiterpene Lactone from Baileya multiradiata. Planta. Med. 1976, 356; Chem. Abstr. 86, 86150j (1977).
225. DOMINGUEZ, X. A., and S. J. JIMENEZ: Aislamiento y Estructuras Del Julslimtetrol, nor-Julslimdiolona y La Isocumambrina, Metabolitos Secundarios Del Croptilon divaricatum (Compuesta). Rev. Latinoamer. Quim. 3, 177 (1972).
226. DOMINGUEZ, X. A., F. M. PEREZ, and L. LEYTER: Xanthinin and β-Sitosterol from Xanthium orientale. Phytochem. 10, 2828 (1971).
227. DOMINGUEZ, E., and J. ROMO: Mexicanin I. A New Sesquiterpene Lactone Related to Tenulin. Tetrahedron 19, 1415 (1963).
228. DOSKOTCH, R. W., and F. S. EL-FERALY: The Structure of Tulipinolide and Epitulipinolide. Cytotoxic Sesquiterpenes from Liriodendron tulipifera L. J. Organ. Chem. (U.S.A.) 35, 1928 (1970).
229. — — Isolation and Characterization of (+)-Sesamin and β-Cyclopyrethrosin from Pyrethrum Flowers. Canad. J. Chem. 47, 1139 (1969).
230. DOSKOTCH, R. W., F. S. EL-FERALY, E. H. FAIRCHILD, and C. HUANG: Peroxyferolide: a Cytotoxic Germacranolide Hydroperoxide from Liriodendron tulipifera. Chem. Commun. 1976, 402.
231. DOSKOTCH, R. W., F. S. EL-FERALY, and C. D. HUFFORD: Sesquiterpene Lactones from Pyrethrum Flowers. Canad. J. Chem. 49, 2103 (1971).
232. DOSKOTCH, R. W., and C. D. HUFFORD: The Structure of Damsinic Acid, a New Sesquiterpene from Ambrosia ambrosioides (Cav.) Payne. J. Organ. Chem. (U.S.A.) 35, 486 (1970).
233. DOSKOTCH, R. W., C. D. HUFFORD, and F. S. EL-FERALY: Further Studies on the Sesquiterpene Lactones Tulipinolide and Epitulipinolide from Liriodendron tulipifera L. J. Organ. Chem. (U.S.A.) 37, 2740 (1972).
234. DOSKOTCH, R. W., S. L. KEELY, JR., and C. D. HUFFORD: Lipiferolide, a Cytotoxic Germacranolide, and γ-Liriodenolide, Two New Sesquiterpene Lactones from Liriodendron tulipifera. Chem. Commun. 1972, 1137.
235. DOSKOTCH, R. W., S. L. KEELY, C. D. HUFFORD, and F. S. EL-FERALY: New Sesquiterpene Lactones from Liriodendron tulipifera. Phytochem. 14, 769 (1975).
235a. VON DREELE, R. B., G. R. PETIT, G. M. CRAGG, and R. H. ODE: The Crystal and Molecular Structure of the Pseudoguaianolide Autumnolide. J. Amer. Chem. Soc. 97, 5256 (1975).
236. DROZDZ, B.: Isolation of Cnicin from the Herbs of Centaurea diffusa Lam. Dissertation Pharm et Pharmacol. (Poland) 18, 281 (1966).
237. — The Occurrance of Cnicin in the Herbs of Centaurea alcitrata L., C. iberica Trev., and C. ovina Pal. Dissertation Pharm. et Pharmacol. (Poland) 19, 223 (1967).
238. — Bitter Sesquiterpene Lactones in Species of the Tribe Cynareae. Dissertation Pharm. et Pharmacol. (Poland) 20, 93 (1968).

239. DROZDZ, B., and G. BIALEK-GRYGIEL: Composition of Lactone Fraction of Leaves and Inflorescence of *Eupatorium cannabinum* L. Dissertation Pharm. et Pharmacol. (Poland) **23**, 537 (1971); Chem. Abstr. **76**, 56629y (1972).

240. DROZDZ, B., H. GRABARCZYK, Z. SAMEK, M. HOLUB, V. HEROUT, and F. SORM: Sesquiterpenic Lactones from *Eupatorium cannabinum* L. Revision of the Structure of Eupatoriopicrin. Collect Czech. Chem. Comm. **37**, 1546 (1972).

241. DROZDZ, B., M. HOLUB, Z. SAMEK, V. HEROUT, and F. SORM: The Constitution and Absolute Configuration of Onopordopicrine, a Sesquiterpenic Lactone from *Onopordon acanthium* L. Collect. Czech. Chem. Comm. **33**, 1730 (1968).

242. DROZDZ, B., Z. SAMEK, M. HOLUB, and V. HEROUT: The Structure of Alatolide, a Sesquiterpenic Lactone from *Jurinea alata* Cass. Collect. Czech. Chem. Comm. **38**, 727 (1973).

243. DROZDZ, B., and J. PIOTROWSKI: Lactones of Carduinae Subtribe. Pol. J. Pharmacol. Pharm. **25**, 91 (1973); Chem. Abstr. **78**, 156610v (1973).

244. DUGAN, J. J., P. DE MAYO, M. NISBET, J. R. ROBINSON, and M. ANCHEL: The Constitution and Biogenesis of Marasmic Acid. J. Amer. Chem. Soc. **88**, 2838 (1966).

245. DULLFORCE, T. A., G. A. SIM, D. N. J. WHITE, J. E. KELSEY, and S. M. KUPCHAN: The Stereochemistry of Gaillardin. Tetrahedron Letters **1969**, 973.

246. EAST, D. S. R., T. B. H. MCMURRY, and R. R. TALEKAR: The Chemistry of Santonene. Products of Photolysis of 4-Hydroxysantonene and Its 4-epimer, Their Structures, and Some Pyrolysis Studies. J. Chem. Soc. Perkin Trans. I **1976**, 433.

247. EASTERFIELD, T. H., and B. C. ASTON: Part I. Tutin and Coriamyrtin. J. Chem. Soc. (London) **79**, 120 (1901).

248. EDWARDS, O. E., J. L. DOUGLAS, and B. S. MOOTOO: Biosynthesis of Dendrobine. Canad. J. Chem. **48**, 2517 (1970).

248a. EINCK, J. J., C. L. HERALD, G. R. PETIT, and R. B. VON DREELE: Antineoplastic Agents. 53. The Crystal Structure of Radiatin. J. Amer. Chem. Soc. **100**, 3544 (1978).

249. EL-FERALY, F. S., Y. M. CHAN, E. H. FAIRCHILD, and R. W. DOSKOTCH: Peroxycostunolide and Peroxyparthenolide: Two Cytotoxic Germacranolide Hydroperoxides from *Magnolia grandiflora*. Structural Revision of Verlotorin and Artemorin. Tetrahedron Letters **1977**, 1973.

250. EMERSON, M. T., C. N. CAUGHLAN, and W. HERZ: The Crystal and Molecular Structure of Bromohelenalin. Tetrahedron Letters **1964**, 621.

251. EMERSON, M. T., W. HERZ, C. N. CAUGHLAN, and R. W. WITTERS: The Crystal and Molecular Structure of Bromoambrosin. Tetrahedron Letters **1966**, 6151.

252. EVANS, D. A., and C. L. SIMS: The Total Synthesis of (±)-Bakkenolide-A. Tetrahedron Letters **1973**, 4691.

253. EVANS, D. A., C. L. SIMS, and G. C. ANDREWS: Applications of [2,3]-sigmatropic Rearrangements to Natural Products Synthesis. The Total Synthesis of ±-Bakkenolide-A (Fukinanolid). J. Amer. Chem. Soc. **99**, 5453 (1977).

254. EVSTRATOVA, R. I., A. I. BAN'KOVSKII, V. I. SHEICHENKO, and K. S. RYBALKO: The Structure of Arnifolin. Khim. Prir. Soedin. **7**, 270 (1971); Engl. edit.: p. 260; Chem. Abstr. **75**, 110442b (1971).

255. EVSTRATOVA, R. I., and P. V. CHUGUNOV: Crystal Substance from *Artemisia rutaefolia*. Khim. Prir. Soedin. **5**, 445 (1969); Chem. Abstr. **72**, 97297z (1970).

256. EVSTRATOVA, R. I., E. Y. KISELEVA, V. I. SHEICHENKO, and K. S. RYBALKO: The Structure of Acroptilin — A Sesquiterpene Lactone from *Acroptilon repens*. Khim. Prir. Soedin. **7**, 272 (1971); Engl. edit.: p. 262; Chem. Abstr. **75**, 110451d (1971).

257. EVSTRATOVA, R. I., K. S. RYBALKO, and R. Y. RZAZADE: Acroptilin — A New Sesquiterpene Lactone from *Acroptilon repens*. Khim. Prir. Soedin. **3**, 284 (1967); Engl. edit.: p. 239; Chem. Abstr. **68**, 3005v (1968).

258. Evstratova, R. I., K. S. Rybalko, and V. I. Sheichenko: The Structure of the Sesquiterpene Lactone Repin. Khim. Prir. Soedin. **8**, 451 (1972); Engl. edit.: p. 450; Chem. Abstr. **77**, 152369e (1972).

259. Evstratova, R. I., V. I. Sheichenko, and K. S. Rybalko: Sesquiterpene Lactones from *Inula japonica*. Khim. Prir. Soedin. **10**, 730 (1974); Engl. edit.: p. 752.

260. Evstratova, R. I., V. I. Sheichenko, K. S. Rybalko, and A. I. Ban'kovskii: Structure of Arnifoline, A Sesquiterpene Lactone from *Arnica foliosa* and *Arnica montana*. Khim.-Farm. Zh. **3**, 39 (1969); Chem. Abstr. **72**, 32047y.

261. Evstratova, R. I., V. I. Sheichenko, A. I. Ban'kovskii, and K. S. Rybalko: Structure of Erivanin. Khim. Prir. Soedin. **5**, 239 (1969); Chem. Abstr. **72**, 32051v.

262. Evstratova, R. I., V. I. Sheichenko, and K. S. Rybalko: The Structure of Acroptilin, A Sesquiterpene Lactone from *Acroptilon repens*. Khim. Prir. Soedin. **9**, 161 (1973); Engl. edit.: p. 156; Chem. Abstr. **79**, 5466x (1973).

263. Evstratova, R. I., and A. M. Sinyukhin: Ketopelenolide B from *Artemisia anethifolia*. Khim. Prir. Soedin. **7**, 839 (1971); Engl. edit.: p. 819.

264. Fairchild, E. H.: Germacranolides of *Liriodendron tulipifera*. Compounds of *Gutierrezia dracunculoides* and *Gutierrezia sarothrae*. From Diss. Abstr. Int. B, 1976, 37 (5), 2181-2; Chem. Abstr. **86**, 121596h.

264a. Farkas, L., M. Nogradi, V. Sudarsanam, and W. Herz: Constituents of *Iva* Species. V. Isolation, Structure and Synthesis of Nevadensin, A New Flavone from *Iva nevadensis* M. E. Jones and *Iva acerosa* (Nutt.) Jackson. J. Organ. Chem. (U.S.A.) **31**, 3228 (1966).

265. Fischer, N. H.: On the Biogenesis of Pseudoguaianolides. Rev. Latinoamer. Quim. **9**, 41 (1978).

265a. — — — unpublished.

266. Fischer, N. H., and D. R. DiFeo: Three New Melampolides from *Melampodium americanum* L. (Heliantheae, Compositae). To be published.

267. Fischer, N. H., and T. J. Mabry, New Pseudoguaianolides from *Ambrosia confertiflora* DC. (Compositae). Tetrahedron **23**, 2529 (1967).

268. — — The Structure of Tamaulipin-B, A New Germacranolide and the Thermal Conversion of a Trans-1,2-Divinylcyclohexane Derivative into a Cyclodeca-1,5-Diene System. Chem. Commun. **1967**, 1235.

269. Fischer, N. H., T. J. Mabry, and H. B. Kagan: The Structure of Tamaulipin A. A New Germacranolide from *Ambrosia confertiflora* DC. (Compositae). Tetrahedron **24**, 4091 (1968).

270. Fischer, N. H., F. C. Seaman, R. A. Wiley, and K. D. Haegele: Melcanthin A, B, and C, Three New *cis*-1(10),*cis*-4,5-Germacranolides from *Melampodium leucanthum* (Heliantheae, Compositae). J. Organ. Chem. (U.S.A.) **43**, 4984 (1978).

271. Fischer, N. H., R. A. Wiley, D. L. Perry, and K. D. Haegele: Melampodinin, Leucanthinin and Melampolidin, Three New Melampolides from *Melampodium* (Compositae). J. Organ. Chem. (U.S.A.) **41**, 3956 (1976).

272. Fischer, N. H., R. A. Wiley, H. N. Lin, K. Karimian, and S. M. Politz: New Melampolide Sesquiterpene Lactones from *Melampodium leucanthum*. Phytochem. **14**, 2241 (1975).

273. Fischer, N. H., R. A. Wiley, and D. L. Perry: Sesquiterpene Lactones from *Melampodium* (Compositae, Heliantheae); Structural and Biosynthetic Considerations. Rev. Latinoamer. Quim. **7**, 87 (1976).

274. Fischer, N. H., R. A. Wiley, and J. D. Wander: Melampodin, A New Germacranolide from *Melampodium leucanthum* Torr. and Gray var. *leucanthum* (Compositae). Chem. Commun. **1972**, 137.

275. Frazer, R. R.: Long Range Coupling Constants in the NMR Spectra of Olefins. Canad. J. Chem. **38**, 549 (1960).

276. FRÖHLICH, A., K. ISHIKAWA, and T. B. H. MCMURRY: Chemistry of Santonene. Products of the Action of Phosphorus Pentachloride on Santonin. J. Chem. Soc. Perkin Trans. I 1975, 726.

277. — — — The Action of Phosphorus Pentachloride and Thionyl Chloride on Santonin. Tetrahedron Letters 1973, 995.

277a. FU, FENG-YUNG, YEN-YANG CHEN, and TIEN-MING SHANG: Isolation of Santonin from Artemisia incana. Yao Hsue Hsue Pao 10, 140 (1963); Chem. Abstr. 59, 14028h (1965).

278. FUJIMOTO, Y., T. SHIMIZU, and T. TATSUNO: 10-Membered Dienones. Japan. Kokai 76, 125, 355; Chem. Abstr. 87, 53463f (1977).

279. — — — Modification of a-Santonin. Synthesis of Dihydrocostunolide. Tetrahedron Letters 1976, 2041.

280. — — — Azulene Derivatives. Japan. Kokai 115, 454 (1976); Chem. Abstr. 86, 106828s.

280a. FUKUI, T.: Structure of Mibulactone. Yakugaku Zasshi 78, 712 (1958).

281. FURUKAWA, H., K. H. LEE, T. SHINGU, R. MECK, and C. PIANTADOSI: Carolenin and Carolenalin, Two New Guaianolides in Helenium autumnale L. from North Carolina. J. Organ. Chem. (U.S.A.) 38, 1722 (1973).

282. GABE, E. J., S. NEIDLE, D. ROGERS, and C. E. NORDMAN: X-Ray Determination of the Molecular Structure of a 3-O-Chlorophenylisoxazoline Derivative of Pyrothrosin. Chem. Commun. 1971, 559.

283. GARCIA, M., A. J. R. DASILVA, P. M. BAKER, B. GILBERT, and J. A. RABI: Absolute Stereochemistry of Eremanthine, A Schistosomicidal Sesquiterpene Lactone from Eremanthus elaegnus. Phytochem. 15, 331 (1976).

284. GAWELL, L., and K. LEANDER: The Constitution of Aduncin, A Sesquiterpene Related to Picrotoxinin, Found in Dendrobium aduncum. Phytochem. 15, 1991 (1976).

285. GEISSMAN, T. A.: Sesquiterpene Lactones of Artemisia — A. verlotorum and A. vulgaris. Phytochem. 9, 2377 (1970).

286. — Sesquiterpenoid Lactones of Artemisia Species. I. Artemisia princeps Pamp. J. Organ. Chem. (U.S.A.) 31, 2523 (1966).

287. — The Structure of Xanthinin. J. Organ. Chem. (U.S.A.) 27, 2692 (1962).

288. — The Biosynthesis of Sesquiterpene Lactones of the Compositae. In: Recent Advances in Phytochemistry (RUNECKLES, V. C., and T. J. MABRY, eds.), Vol. 6, p. 65. New York and London: Academic Press. 1973.

289. — Private communication to T. J. MABRY.

290. GEISSMAN, T. A., and D. H. G. CROUT: Organic Chemistry of Secondary Plant Metabolism. San Francisco: Freeman, Cooper and Co. 1969.

291. GEISSMAN, T. A., P. DEUEL, E. K. BONDE, and F. A. ADDICOTT: Xanthinin: A Plant Growth-regulating Compound from Xanthium pennsylvanicum. J. Amer. Chem. Soc. 76, 685 (1954).

292. GEISSMAN, T. A., and G. A. ELLESTAD: Vulgarin, a Sesquiterpene Lactone from Artemisia vulgaris L. J. Organ. Chem. (U.S.A.) 27, 1855 (1962).

293. GEISSMAN, T. A., and T. S. GRIFFIN: unpublished.

294. — — Sesquiterpene Lactones of Artemisia carruthii. Phytochem. 11, 833 (1972).

295. — — Sesquiterpene Lactones. Tomentosin from Montanoa tomentosa Cerv. Rev. Latinoamer. Quim. 2, 81 (1971).

296. — — Sesquiterpene Lactones: Acid-Catalysed Color Reactions as an Aid in Structure Determination. Phytochem. 10, 2475 (1971).

297. GEISSMAN, T. A., T. S. GRIFFIN, and M. A. IRWIN: Sesquiterpene Lactones of Artemisia. Artecalin from A. californica and A. tripartita SSP. rupicola. Phytochem. 8, 1297 (1969).

298. Geissman, T. A., S. Griffin, T. G. Waddell, and H. H. Chen: Some New Constituents of *Ambrosia* Species: *A. psilostachya* and *A. acanthicarpa.* Phytochem. **8,** 145 (1969).
299. Geissman, T. A., and M. A. Irwin: Chemical Contributions to Taxonomy and Phylogeny in the Genus *Artemisia.* Pure Appl. Chem. **21,** 167 (1970); Chem. Abstr. **73,** 117117s (1970).
300. Geissman, T. A., and K. H. Lee: Sesquiterpene Lactones of *Artemisia:* Artemorin and Dehydroartemorin (Anhydroverlotorin). Phytochem. **10,** 419 (1971).
301. — — Sesquiterpene Lactones of *Artemisia verlotorum.* Phytochem. **10,** 663 (1971).
302. Geissman, T. A., and J. Levy: Chamissonin from *Ambrosia acanthicarpa* Hook. Phytochem. **6,** 899 (1967).
303. Geissman, T. A., A. J. Lucas, T. Saitoh, and W. W. Payne: Sesquiterpene Lactones of *Ambrosia chamissonis.* Biochem. Syst. **1,** 13 (1973); Chem. Abstr. **78,** 145204q (1973).
304. Geissman, T. A., and S. Matsueda: Constituents of Diploid and Polyploid *Ambrosia dumosa* Gray. Phytochem. **7,** 1613 (1968).
305. Geissman, T. A., and R. Murkherjee: Sesquiterpene Lactones of *Encelia farinosa* Gray. J. Organ. Chem. (U. S. A.) **33,** 656 (1968).
306. Geissman, T. A., and T. Saitoh: Ludalbin, A New Lactone from *Artemisia ludoviciana.* Phytochem. **11,** 1157 (1972).
307. Geissman, T. A., T. Stewart, and M. A. Irwin: Sesquiterpene Lactones of Artemisia Species II. *Artemisia tridentata* Nutt. ssp. *tridentata.* Phytochem. **6,** 901 (1967).
308. Geissman, T. A., and F. P. Toribio: Constituents of *Hymenoclea salsola* T. and G. Phytochem. **6,** 1563 (1967).
309. Geissman, T. A., and R. Turley: Sesquiterpene Lactones. Coronopilic Acid. J. Organ. Chem. (U. S. A.) **29,** 2553 (1964).
310. Geissman, T. A., R. J. Turley, and S. Murayama: Chamissonin, A Germacranolide from an *Ambrosia* Species. J. Organ. Chem. (U. S. A.) **31,** 2269 (1966).
311. Geissman, T. A., and T. E. Winters: The Structure of Artabsin. Tetrahedron Letters **1968,** 3145.
312. Gitany, R., G. D. Anderson, and R. S. McEwen: Crystal and Molecular Structure of Pseudoivalin Bromoacetate. Acta. Crystallogr. Sect. B **1974,** B 30, 1900; Chem. Abstr. **81,** 112347x (1974).
313. Gnecco, S., J. P. Poyser, M. Silva, P. G. Sammes, and T. W. Tyler: Sesquiterpene Lactones from *Podanthus ovatifolius.* Phytochem. **12,** 2469 (1973).
314. Goodlet, V. W.: Use of *in situ* Reactions for Characterization of Alcohols and Glycols by Nuclear Magnetic Resonance. Analyt. Chemistry 37, 431 (1965).
315. Gonzalez Gonzalez, A., J. M. Arteaga, J. Bermejo Barrera, and J. L. Breton Funes: Melitensin, New Sesquiterpene Lactone from *Centaurea melitensis.* An. Quim. **67,** 1243 (1971); Chem. Abstr. **77,** 34722s (1972).
316. Gonzalez Gonzalez, A., J. M. Arteaga, and J. L. Breton Funes: Germacranolides from *Centaurea seridis.* Phytochem. **12,** 2997 (1973).
317. — — — Structure and Partial Synthesis of Melitensin. An. Quim. **70,** 158 (1974); Chem. Abstr. **81,** 78104d (1974).
318. — — — Constituents of the Compositae. 27. Elemanolides from *Centaurea melitensis.* Phytochem. **14,** 2039 (1975).
319. Gonzalez Gonzalez, A., J. Bermejo Barrera, J. L. Breton Funes, and M. Fajardo: Compounds from Plants of the Canary Islands. Sesquiterpene Lactones of *Artemisia canariensis.* An. Quim. **69,** 667 (1973); Chem. Abstr. **79,** 92415w (1973).
320. Gonzalez Gonzalez, A., J. Bermejo Barrera, J. L. Breton Funes, G. M. Massanet, B. Dominguez, and J. M. Amaro: Absolute Configuration of the Sesqui-

terpene Lactones Centaurepensin (Chlorohyssopifolin A), Acroptilin (Chlorohyssopifolin C), and Repin. J. Chem. Soc. Perkin Trans. I **1976**, 1663.

321. GONZALEZ GONZALEZ, A., J. BERMEJO BARRERA, J. L. BRETON FUNES, G. M. MASSANET, and J. TRIANA: Chlorohyssopifolin C, D, E and Vahlenin, Four New Sesquiterpene Lactones from *Centaurea hyssopifolia*. Phytochem. **13**, 1193 (1974).

322. GONZALEZ GONZALEZ, A., J. BERMEJO BARRERA, J. L. BRETON FUNES, and M. RODRIGUEZ RINCONES: Configuration of C-1 in Grosshemin. An. Quim. **69**, 563 (1973); Chem. Abstr. **79**, 92406u (1973).

323. GONZALEZ GONZALEZ, A., J. BERMEJO BARRERA, J. L. BRETON FUNES, and J. TRIANA: Chlorohyssopifolin A and B, Two New Sesquiterpene Lactones Isolated from *Centaurea hyssopifolia* Vahl. Tetrahedron Letters **1972**, 2017.

323a. GONZALEZ GONZALEZ, A., J. BERMEJO, I. CABRERA, A. GALINDO, and G. M. MASSANET: Principos Activos de la *Centaurea janeri* Graells. An. Quim. **73**, 86 (1977).

324. GONZALEZ GONZALEZ, A., J. BERMEJO BARRERA, I. CABRERA, and G. M. MASSANET: 11,13-Dehydromelitensin and Chlorohyssopifolin A, Sesquiterpene Lactones Isolated from *Centaureas pullata* and *nigra*. An. Quim. **70**. 74 (1974); Chem. Abstr. **80**, 130489d (1974).

325. GONZALEZ GONZALEZ, A., J. BERMEJO BARRERA, A. D. DE LA ROSA, and G. M. MASSANET: Sesquiterpene Lactones from *Artemisia lanata* Willd. An. Quim. **1976**, 695; Chem. Abstr. **86**, 72910v (1977).

326. GONZALEZ GONZALEZ, A., J. BERMEJO BARRERA, G. M. MASSANET, J. M. AMARO, B. DOMINGUEZ, and J. FAYOS: Hypochaerin, $C_{15}H_{20}O_3H_2O$. Cryst. Struct. Commun. **6**, 373 (1977); Chem. Abstr. **87**, 175925d (1977).

327. GONZALEZ GONZALEZ, A., J. BERMEJO BARRERA, G. M. MASSANET, M. M. AMARO, B. DOMINGUEZ, and A. MORALES: Hypochaerin: A New Sesquiterpene Lactone from *Hypochaeris setosus*. Phytochem. **15**, 991 (1976).

328. GONZALEZ GONZALEZ, A., J. BERMEJO BARRERA, H. MANSILLA, G. M. MASSANET, I. CABRERA, J. M. AMARO, and A. GALINDO: The Structure and Stereochemistry of Artemin. Phytochem. **16**, 1836 (1977).

329. GONZALEZ GONZALEZ, A., J. BERMEJO BARRERA, G. M. MASSANET, and J. PEREZ: Muricatin, A New Sesquiterpene Lactone Isolated from *Amberboa muricata*. An. Quim. **69**, 1333 (1973); Chem. Abstr. **81**, 4086c.

330. GONZALEZ GONZALEZ, A., J. BERMEJO BARRERA, G. M. MASSANET, and J. TRIANA: Picridin and Dihydropicridin, Revised Structures. Phytochem. **13**, 611 (1974).

331. GONZALEZ GONZALEZ, A., J. BERMEJO BARRERA, and R. M. RODRIGUEZ: Dihydroestafietone Isolated from *Centaurea webbiana*. An. Quim. **68**, 333 (1972); Chem. Abstr. **77**, 123813q (1972).

332. GONZALEZ GONZALEZ, A., J. L. BRETON FUNES, A. GALINDO, and I. CABRERA: Componentes De Umbelliferas. XIII. Estudio De las Decipieninas G y H. Rev. Latinoamer. Quim. **7**, 37 (1976).

332a. GONZALEZ GONZALEZ, A., J. L. BRETON FUNES, A. GALINDO, and L. F. RODRIGUEZ: Components of Umbelliferae. I. Two New Sesquiterpene Lactones of *Melanoselinum decipiens*. An. Quim. **69**, 1339 (1974). Chem. Abstr. **80**, 70985a (1974).

333. — — — — Sesquiterpenoid Lactones in Umbelliferae. Rev. R. Acad. Ciens. Exactas. Fis. Nat. Madrid **69**, 547 (1976); Chem. Abstr. **85**, 33196b (1976).

334. GONZALEZ GONZALEZ, A., J. B. BRETON FUNES, and B. GARCIA MARRERO: Lipidiol. Page 341 in ref. *1172*.

335. GONZALEZ GONZALEZ, A., J. L. BRETON FUNES, and J. STOCKEL: Sesquiterpenic Lactones from *Artemisia granatensis*. An. Quim. **70**, 231 (1974); Chem. Abstr. **81**, 117046h (1974).

336. GONZALEZ GONZALEZ, A., R. ESTEVEZ REYES, and J. HERRERA VELAZQUEZ: Sesquiterpene Lactones and Coumarins of *Artemisia ramosa*. An. Quim. **71**, 437 (1975); Chem. Abstr. **83**, 128692h (1975).

337. Gonzalez Gonzalez, A., B. Garcia Marrero, and J. L. Breton Funes: Structure of Grosshemin, Lipidiol and Isolipidiol. Lactones of *Amberboa lipii* and Their Stereochemistry. An. Quim. **66**, 799 (1970); Chem. Abstr. **74**, 64315s (1971).

338. Gopalakrishna, E. M., T. W. Adams, W. H. Watson, M. Betkouski, and T. J. Mabry: Glaucolide-E and -D, (0.7) $C_{23}H_{28}O_9$ (0.3) $C_{23}H_{28}O_{10}$. Cryst. Struct. Commun. **6**, 201 (1977); Chem. Abstr. **86**, 198336f (1977).

339. Gopalakrishna, E. M., W. H. Watson, M. Hoeneisen, and M. Silva: Ovatifolin, A Sesquiterpene Lactone. J. Cryst. Mol. Struct. **7**, 49 (1977); Chem. Abstr. **87**, 109805p (1977).

339a. Goryaev, M. I., R. N. Sazonova, and P. P. Polyakov: The Results of Work of 1952 Expedition Investigating Essential Oil Bearers of Kazakstan and Central Asia. Trody Inst. Khim. Nauk, Akad. Nauk SSR **4**, 24 (1959); Chem. Abstr. **54**, 11387 (1960).

340. Govindachari, T. R., B. S. Joshi, and V. N. Kamat: Structure of Parthenolide. Tetrahedron **21**, 1509 (1965).

341. Govindachari, T. R., A. R. Sidhaye, and N. Viswanathan: Deoxyelephantopin, A New Sesquiterpene from *Elephantopus scaber*. Indian J. Chem. **8**, 762 (1970).

342. Govindachari, T. R., N. Viswanathan, and H. Fuehrer: Isodeoxyelephantopin, A New Germacranediolide from *Elephantopus scaber*. Indian J. Chem. **10**, 272 (1972).

343. Grabarczyk, H.: The Search for Sesquiterpene Lactones in the Species of Genera *Ursinia*, *Venidium* and *Arctotis*. Pol. J. Pharmacol. Pharm. **25**, 469 (1973); Chem. Abstr. **80**, 45661j (1974).

344. — The New Sesquiterpene Lactones from the Herb of *Venidium hirsutum* Berol. Pol. J. Pharmacol. Pharm. **27**, 107 (1975); Chem. Abstr. **83**, 4995q (1975).

345. Grabarczyk, H., B. Drozdz, and A. Mozdzanowska: Lactones in Aerial Parts of *Tanacetum vulgare* L. Pol. J. Pharmacol. Pharm. **25**, 95 (1973); Chem. Abstr. **78**, 156612x (1973).

346. Grabarczyk, H., and B. Makowska: New Sesquiterpene Lactones from the Herb of *Venidium hirsutum*. Pol. J. Pharmacol. Pharm. **27**, 107 (1975); Chem. Abstr. **83**, 4995q (1975).

347. — — Isolation of Grosheimin from the Herb of *Venidium decurens* Less. Pol. J. Pharmacol. Pharm. **25**, 477 (1973); Chem. Abstr. **80**, 45662k (1974).

348. Granelli, I., and K. Leander: Studies on Orchidaceae Alkaloids XIX. Synthesis and Absolute Configuration of Dendrine. Acta Chem. Scand. **24**, 1108 (1970).

349. Granelli, I., K. Leander, and B. Luning: New Alkaloid, 2-Hydroxydendrobine, from *Dendrobium findleyanum*. Acta Chem. Scand. **24**, 1209 (1970); Chem. Abstr. **73**, 117121p (1970).

350. Greene, A. E., J. C. Muller, and G. Ourisson: Conversions of α-Methyl to α-Methylene γ-Lactones. Synthesis of Two Allergenic Sesquiterpene Lactones. (−)-Frullanolide and (+)-Arbusculin B. J. Organ. Chem. (U.S.A.) **39**, 186 (1974).

351. — — — A New Approach to α-Methylene-γ-Butyrolactones. Synthesis of (−)-Frullanolide. Tetrahedron Letters **1972**, 2489.

352. — — — A New Approach to α-Methylene-γ-Butyrolactones. Synthesis of (+)-Arbusculin-B. Tetrahedron Letters **1972**, 3375.

353. Grieco, P. A.: Methods for the Synthesis of α-Methylene Lactones. Synthesis **1975**, 67.

354. Grieco, P. A., and M. Nishizawa: Application of Organoselenium Chemistry to the Total Synthesis of (±)-Tuberiferins. Chem. Commun. **1976**, 582.

355. — — Total Synthesis of (+)-Costunolide. J. Organ. Chem. (U.S.A.) **42**, 1717 (1977).

356. Grieco, P. A., M. Nishizawa, S. D. Burke, and N. Marinovic: Total Synthesis of (±)-Vernolepin and (±)-Vernomenin. J. Amer. Chem. Soc. **98**, 1612 (1976).

357. GRIECO, P. A., M. NISHIZAWA, T. OGURI, S. D. BURKE, and N. MARINOVIC: Sesquiterpene Lactones: Total Synthesis of (±)-Vernolepin and (±)-Vernomenin. J. Amer. Chem. Soc. **99,** 5773 (1977).
358. GRIECO, P. A., T. OGURI, C. L. J. WANG, and E. WILLIAMS: Stereochemistry and Total Synthesis of (±)-Ivangulin. J. Organ. Chem. (U. S. A.) **42,** 4113 (1977).
359. GRIECO, P. A., Y. OHFUNE, and G. MAJETICH: Pseudoguaianolides. Stereospecific Total Synthesis of (±)-Ambrosin, (±)-Damsin, and (±)-Psilostachyin C. J. Amer. Chem. Soc. **99,** 7393 (1977).
360. GRIECO, P. A., Y. YOKOYAMA, S. GILMAN, and Y. OHFUNE: Conversion of Ketones into Lactones with Benzeneseleninic Acid and Hydrogen Peroxide (Benzeneperoxyseleninic Acid): A New Reagent for the Baeyer-Villiger Reaction. Chem. Commun. **1977,** 870.
361. GRIFFIN, T. S.: Acid Color Reactions and Structures of Sesquiterpene Lactones. From Diss. Abstr. Int. B **31,** 5875 (1971); Chem. Abstr. **75,** 77057e (1971).
362. GRIFFIN, T. S., T. A. GEISSMAN, and T. W. WINTERS: The Chemistry of a Structurally Diagnostic Color Reaction of Xanthinin and Related Sesquiterpene Lactones. Phytochem. **10,** 2487 (1971).
363. GUERRERO, C., E. DIAZ, M. MARTINEZ, J. TABOADA, S. MIRANDA PLATA, M. GONZALEZ DIDDI, and J. TELLEZ: Determination of the Structure of Euparhombin and Its Cytotoxic Activity in Two Cellular Lines. Rev. Latinoamer. Quim. **8,** 123 (1977).
364. GUERRERO, J., R. ENRIQUEZ, I. SALAZAR, and E. DIAZ: Desplazamientos Inducidos Por Lantanidos En RMP De Guayanolidos y Pseudoguayanolidos. Rev. Latinoamer. Quim. **5,** 169 (1974).
365. GUERRERO, C., A. IRIARTE, E. DIAZ, J. TABOADA, M. GONZALEZ DIDDI, and J. TELLEZ: Determinacion De La Estructura y Estereoquimica De La Elemenolida Verafinina C y Aislamiento De La Verafinina B, Dos Substancias Citotoxicas Aisladas De *Verbesina* aff. *coahuilensis.* Rev. Latinoamer. Quim. **6,** 119 (1975).
366. GUERRERO, C., M. MARTINEZ, E. DIAZ, and A. ROMO DE VIVAR: Estructura Y Estereoquimica De La Verafinina, Una Nueva Elemenolida Aislada De *Verbesina* aff. *coahuilensis* Gray. Rev. Latinoamer. Quim. **6,** 53 (1975).
367. GUERRERO, C., A. ORTEGA, E. DIAZ, and A. ROMO DE VIVAR: Estructura De La Viguiestenina y De La Desacetil Viguiestenina. Rev. Lationoamer. Quim. **4,** 118 (1973).
368. GUERRERO, C., M. SANTANA, and J. ROMO: Estudio Quimico De La *Viguiera augustifolia* HBK Blake. Rev. Latinoamer. Quim. **7,** 41 (1976).
369. GUY, M. H. P., G. A. SIM, and D. N. J. WHITE: Molecular Mechanics Calculations on Lactones; An Interpretation of the Preferential C-8 Relactonization of Germacranolides Containing C-6 and C-8 Lactonizable α-Oxygen Functions. J. Chem. Soc. Perkin Trans. II **15,** 1917 (1976).
370. GÜREN, K. C., A. ULUBELEN, and S. ÖKSÜZ: Compositae: Santonin from *Artemisia fragrans.* Phytochem. **11,** 3542 (1972).
371. HABIB, A. A. M., and A. M. METWALLY: New Sesquiterpene Keto Lactone from *Senecio.* Planta Med. **23,** 88 (1973); Chem. Abstr. **78,** 156620y (1973).
371a. HALL, I. H., K. H. LEE, E. C. MAR, C. O. STARNES, and T. G. WADDELL: A Proposed Mechanism for Inhibition of Cancer Growth by Tenulin and Helenalin and Related Cyclopentenones. J. Medicin. Chem. **20,** 333 (1977).
372. HAMILTON, J. A., A. T. MCPHAIL, and G. A. SIM: The Structure of Acetylbromogeigerin. Proc. Chem. Soc. **1960,** 278.
373. HARLEY-MASON, J., A. T. HEWSON, O. KENNARD, and R. C. PETTERSEN: Isolation of Centaurepensin, A Guaianolide Sesquiterpene Lactone Ester Containing Two Chlorine Atoms; Determination of Structure and Absolute Configuration by X-Ray Crystallography. Chem. Commun. **1972,** 460.

374. Harmatha, J., Z. Samek, M. Synackova, L. Novotny, V. Herout, and F. Sorm: Neutral Components of the Extract from *Homogine alpina* (L.) Cass. Collect. Czech. Chem. Comm. **41,** 2047 (1976).

375. Hausen, B. M., K. H. Schulz, O. Jarchow, K. H. Klaska, and H. Schmalle: First Allergenic Sesquiterpene Lactone from *Chrysanthemum indicum.* Arteglasin-A. Naturwissenschaften **62,** 585 (1975); Chem. Abstr. **84,** 57278k (1976).

376. Hayashi, S., N. Hayashi, and T. Matsuura: Sericealactone and Deoxysericealactone, Methyl Esters of Sesquiterpene Acids from *Neolitsea sericea* Koidz. Tetrahedron Letters **22,** 2647 (1968).

377. Hayashi, K., H. Nakamura, and H. Mitsuhashi: Sesquiterpenes from *Cacalia hastata.* Phytochem. **12,** 2931 (1973).

377a. — — — Synthesis of Bakkenolide A. Chem. Pharm. Bull. Soc. Japan **21,** 2806 (1973).

378. Heathcock, C. H.: Total Synthesis of Sesquiterpenes. In: The Total Synthesis of Natural Products (Ap Simon, J. W., ed.), Vol. 2. New York-London: Wiley Interscience Publ. 1973.

379. Heathcock, C. H., and Y. Amano: Sesquiterpenoids: The Synthesis and Formolysis of 4αβ,8β-Dimethyl-8α,α-Hydroxydecahydronaphth-2β-oic Acid and 4αβ-8β-Dimethyl-8α-α-Hydroxydecahydronaphth-2α-oic Acid Lactone. Tetrahedron **24,** 4917 (1968).

380. Heathcock, C. H., and T. Ross Kelly: Sesquiterpenoids: Acid-Catalyzed Methyl Migration in 9-Methyl Decalins. Tetrahedron **24,** 3753 (1968).

381. Heckel, E., and F. Sohlagdenhauffen: De l'*Artemisia gallica* Willd., comme plante a santonin et de sa composition chimique. Compt. rend. **100,** 804 (1885).

382. Hendrickson, J. B.: Stereochemical Implications in Sesquiterpene Biogenesis. Tetrahedron **7,** 82 (1959).

383. Hendrickson, J. B., C. Ganter, D. Dorman, and H. Link: The Stereospecific Synthesis of Pseudoguaianolide Sesquiterpenes. Tetrahedron Letters **1968,** 2235.

383a. Hernandez, R., A. Sandoval, A. Jetzer, and J. Romo: Chemical Study of *Helenium quadridentatum.* Bol. Inst. Quim. Univ. Nacl. Auton. Mex. **20,** 81 (1968); Chem. Abstr. **71,** 64041n (1969).

384. Herout, V.: Chemotaxonomy of the Family Compositae (Asteraceae). In: Pharmacognosy and Phytochemistry (Wagner, H., and L. Hörhammer, eds.). Wien: Springer. 1971. Pharmacogn. Phytochem. Int. Congr. 1st, **1970** (Pub. 1971) 93; Chem. Abstr. **76,** 124092v (1972).

385. Herout, V., L. Novotny, and F. Sorm: Die Isolierung von weiteren kristallinen Substanzen aus Wermut *(Artemisia absinthium* L.). Collect. Czech. Chem. Comm. **21,** 1485 (1956).

386. Herout, V., and F. Sorm: Chemotaxonomy of the Sesquiterpenoids of the Compositae. In: Perspectives in Phytochemistry (Harborne, J. B., and T. Swain, eds.). London and New York: Academic Press. 1969.

387. — — On the Components of Wormwood *(Artemisia absinthium* L.) and the Isolation of a Crystalline Pro-Chamazulenogen. Collect. Czech. Chem. Comm. **18,** 854 (1953).

388. — — Contribution to the Constitution of Pro-Chamazulenogen, The Natural Precursor of Chamazulene in *Artemisia absinthium* L. Collect. Czech. Chem. Comm. **19,** 792 (1954).

389. — — Isolation and Structure of Costunolide from *Artemisia balchanorum.* Chem. & Ind. **1959,** 1067.

390. — — Monocyclische Lactone aus Wermut *(Artemisia absinthium* L.). Collect. Czech. Chem. Comm. **21,** 1494 (1956).

391. Herout, V., M. Suchy, and F. Sorm: Isolation and Structure of Costunolide, Balchanolide, Isobalchanolide and Hydroxybalchanolide, Sesquiterpenic Lactones of

Germacrane Type from *Artemisia balchanorum*, H. Krasch. Collect. Czech. Chem. Comm. **26**, 2612 (1961).

392. HERZ, W.: Sesquiterpene Lactones in Compositae. In: Pharmacognosy and Phytochemistry (WAGNER, H., and L. HÖRHAMMER, eds.). Wien: Springer. 1971. Chem. Abstr. **76**, 59761 w (1972).

393. — Biogenetic Aspects of Sesquiterpene Lactone Chemistry. Israel J. Chem. **16**, 32 (1977).

394. — Pseudoguaianolides in Compositae. In: Recent Advances in Phytochemistry (MABRY, T. J., R. E. ALSTON, and V. C. RUNECKLES, eds.), Vol. 1, p. 229. New York: Appleton-Century-Crofts. 1968.

395. — Pseudoguaianolides in Compositae. In: Chemistry in Botanical Classification (Nobel Foundation, Stockholm), Nobel Symposium **25**. New York and London: Academic Press. 1973.

396. — Constituents of *Helenium* Species. XII. Sesquiterpene Lactones of Some Southwestern Species. J. Organ. Chem. (U. S. A.) **27**, 4043 (1962).

397. — In: Biology and Chemistry of the Compositae (HEYWOOD, V., B. L. TURNER, and J. B. HARBORNE, eds.). New York and London: Academic Press. 1978.

398. HERZ, W., G. ANDERSON, S. GIBAJA, and D. RAULAIS: Sesquiterpene Lactones of Some *Ambrosia* Species. Phytochem. **8**, 877 (1969).

399. HERZ, W., K. AOTA, and A. HALL: New Pseudoguaianolides from *Hymenoxys* Species. A New Type of Lactone Closure. J. Organ. Chem. (U. S. A.) **35**, 4117 (1970).

400. HERZ, W., K. AOTA, A. L. HALL, and A. SRINIVASAN: Antileukemic Pseudoguaianolides from *Hymenoxys grandiflora* (T. & G.) Parker. Application of Lanthanide-Induced Shifts to Structure Determination. J. Organ. Chem. (U. S. A.) **39**, 2013 (1974).

401. HERZ, W., K. AOTA, M. HOLUB, and Z. SAMEK: Sesquiterpene Lactones and Lactone Glycosides from *Hymenoxys* Species. J. Organ. Chem. (U. S. A.) **35**, 2611 (1970).

402. HERZ, W., and S. V. BHAT: Woodhousin, a New Germacranolide from *Bahia woodhousei* (Gray) Gray. J. Organ. Chem. (U. S. A.) **37**, 906 (1972).

403. — — Isolation and Structure of Two New Germacranolides from *Polymnia uvedalia* (L.) L. J. Organ. Chem. (U. S. A.) **35**, 2605 (1970).

404. — — Maculatin: An Isomer of Uvedalin Epoxide from *Polymnia maculata*. Phytochem. **12**, 1737 (1973).

405. HERZ, W., S. V. BHAT, H. CRAWFORD, H. WAGNER, G. MAURER, and L. FARKAS: Bahifolin, A New Sesquiterpene Lactone, and 5,7-Dihydroxy-3,3′,4′,6-Tetramethoxyflavone, a New Flavone, from *Bahia oppositifolia*. Phytochem. **11**, 371 (1972).

406. HERZ, W., S. V. BHAT, and A. L. HALL: Parthemollin, A New Xanthanolide from *Parthenice mollis* Gray. J. Organ. Chem. (U. S. A.) **35**, 1110 (1970).

407. HERZ, W., S. V. BHAT, and A. SRINIVASAN: Berlandin and Subacaulin, Two New Guaianolides from *Berlandiera subacaulis*. J. Organ. Chem. (U. S. A.) **37**, 2532 (1972).

408. HERZ, W., S. V. BHAT, and V. SUDARSANAM: Sesquiterpene Lactones and Flavones of *Iva frutescens*. Phytochem. **11**, 1829 (1972).

408a. HERZ, W., and J. F. BLOUNT: Structure of Tirotundin. J. Organ. Chem. (U. S. A.) **43**, 1268 (1978).

408b. — — Stereochemistry of Woodhousin. J. Organ. Chem. (U. S. A.) **43**, 4887 (1978).

409. HERZ, W., H. CHIKAMATSU, N. VISWANTHAN, and V. SUDARSANAM: Structure of Ivalbin, a Modified Guaianolide from *Iva dealbata* Gray. J. Organ. Chem. (U. S. A.) **32**, 682 (1967).

409a. HERZ, W., and R. DE GROOTE: Desacetyleupaserrin and Nevadensin From *Helianthus pumilus*. Phytochem. **16**, 1307 (1977).

410. HERZ, W., C. M. GAST, and P. S. SUBRAMANIAM: Sesquiterpene Lactones of *Helenium alternifolium* (Spreng.) Cabrera. Structures of Brevilin A, Linifolin A, and Alternilin. J. Organ. Chem. (U. S. A.) **33**, 2780 (1968).

411. Herz, W., and G. Högenauer: Ivalin, a New Sesquiterpene Lactone. J. Organ. Chem. (U.S.A.) **27**, 905 (1962).

412. — — Isolation and Structure of Coronopilin, A New Sesquiterpene Lactone. J. Organ. Chem. (U.S.A.) **26**, 5011 (1961).

413. Herz, W., G. Högenauer, and A. Romo de Vivar: Constituents of *Iva* Species. III. Structure of Microcephalin, A New Sesquiterpene Lactone. J. Organ. Chem. (U.S.A.) **29**, 1700 (1964).

414. Herz, W., and S. Inayama: Constituents of *Gaillardia* Species. II. Structures of Pulchellin B and Pulchellin C. Tetrahedron **20**, 341 (1964).

414a. Herz, W., and H. B. Kagan: Determination of the Absolute Configuration of Hydroxylated Sesquiterpene Lactones by Horeau's Method of Asymmetric Esterification. J. Organ. Chem. (U.S.A.) **32**, 216 (1967).

414b. Herz, W., P. Jayaraman, and H. Watanabe: Constituents of *Helenium* Species. IX. The Sesquiterpene Lactones of *H. flexuosum* Raf. and *H. campestre* Small. J. Amer. Chem. Soc. **82**, 2276 (1960).

415. Herz, W., and P. S. Kalyanaraman: Mikanokryptin in *Melampodium divaricatum*. Phytochem. **14**, 1664 (1975).

416. — — Acanthospermal A and Acanthospermal B, Two New Melampolides from *Acanthospermum* Species. J. Organ. Chem. (U.S.A.) **40**, 3486 (1975).

417. Herz, W., P. S. Kalyanaraman, G. Ramakrishnan, and J. F. Blount: Sesquiterpene Lactones of *Eupatorium perfoliatum*. J. Organ. Chem. (U.S.A.) **42**, 2264 (1977).

418. Herz, W., Y. Kishida, and M. V. Lakshmikantham: Constituents of *Helenium* Species. XVI. Structures of Flexuosin A and Flexuosin B. Tetrahedron **20**, 979 (1964).

419. Herz, W., and M. V. Lakshmikantham: Constituents of *Helenium* Species. XVII. Sesquiterpene Lactones of *Helenium thurberi* Gray and the Stereochemistry of Bigelovin. Tetrahedron **21**, 1711 (1965).

420. Herz, W., M. V. Lakshmikantham, and R. N. Mirrington: Constituents of *Helenium* Species. XVIII. 1-Epiisotenulin and Its Transformations. Tetrahedron **22**, 1709 (1966).

420a. Herz, W., and R. B. Mitra: Correlation of Helenalin and Alloisotenulin. J. Amer. Chem. Soc. **80**, 4876 (1958).

421. Herz, W., R. B. Mitra, K. Rabindran, and W. A. Rohde: Constituents of *Helenium* Species. VII. Bitter Principles of *H. pinnatifidum* (Nutt.) Rydb., *H. vernale* Walt., *H. brevifolium* (Nutt.) A. Wood, and *H. flexuosum* Raf. J. Amer. Chem. Soc. **81**, 1481 (1959).

422. Herz, W., R. B. Mitra, K. Rabindran, and N. Viswanathan: Constituents of *Helenium* Species. XI. The Structure of Pinnatifidin. J. Organ. Chem. (U.S.A.) **27**, 4041 (1962).

423. Herz, W., M. Miyazaki, and Y. Kishida: Structures of Parthenin and Ambrosin. Tetrahedron Letters **1961**, 82.

424. Herz, W., J. Poplawski, and R. P. Sharma: New Guaianolides from *Liatris* Species. J. Organ. Chem. (U.S.A.) **40**, 199 (1975).

425. Herz, W., S. Rajappa, M. V. Lakshmikantham, D. Raulais, and J. J. Schmid: Constituents of *Gaillardia* Species. V. Isolation and Structure of Spathulin. J. Organ. Chem. (U.S.A.) **32**, 1042 (1967).

426. Herz, W., S. Rajappa, M. V. Lakshmikantham, and J. J. Schmid: Constituents of *Gaillardia* Species. III. The Structure of Gaillardilin, a New Pseudoguaianolide. Tetrahedron **22**, 693 (1966).

427. Herz, W., S. Rajappa, S. K. Roy, J. J. Schmid, and R. N. Mirrington: Constituents of *Gaillardia* Species. IV. The Sesquiterpene Lactones of *Gaillardia fastigiata* Greene. Tetrahedron **22**, 1907 (1966).

428. HERZ, W., D. RAULAIS, and G. D. ANDERSON: Sesquiterpene Lactones of *Ambrosia cordifolia.* Phytochem. **12,** 1415 (1973).
429. HERZ, W., W. A. ROHDE, K. RABINDRAN, P. JAYARAMAN, and N. VISWANATHAN: Revised Structure of Tenulin. J. Amer. Chem. Soc. **84,** 3857 (1962).
430. HERZ, W., A. ROMO DE VIVAR, and M. V. LAKSHMIKANTHAM: Structure of Pseudoivalin, a New Guaianolide. J. Organ. Chem. (U. S. A.) **30,** 118 (1965).
431. HERZ, W., A. ROMO DE VIVAR, J. ROMO, and N. VISWANATHAN: Constituents of *Helenium* Species. XV. The Structure of Mexicanin C, Relative Stereochemistry of its Congeners. Tetrahedron **19,** 1359 (1963).
432. — — — — Constituents of *Helenium* Species. XIII. The Structure of Helenalin and Mexicanin A. J. Amer. Chem. Soc. **85,** 19 (1963).
433. HERZ, W., and S. K. ROY: New Pseudoguaianolides from *Gaillardia pulchella.* Phytochem. **8,** 661 (1969).
434. HERZ, W., and P. S. SANTHANAM: Virginolide, a New Guaianolide from *Helenium virginicum* Blake. J. Organ. Chem. (U. S. A.) **32,** 507 (1967).
435. — — Arbiglovin. A New Guaianolide from *Artemisia bigelovii* Gray. J. Organ. Chem. (U. S. A.) **30,** 4340 (1965).
436. HERZ, W., and R. P. SHARMA: A trans-1,2-cis-4,5-Germacradienolide and Other New Germacranolides from *Tithonia* Species. J. Organ. Chem. (U. S. A.) **40,** 3118 (1975).
437. — — Complete Stereochemistry of Tenulin. Carbon-13 Nuclear Magnetic Resonance Spectra of Tenulin Derivatives. J. Organ. Chem. (U. S. A.) **40,** 2557 (1975).
438. — — Pycnolide, a seco-Germacradienolide from *Liatris pycnostachya,* and other Antitumor Constituents of *Liatris* Species. J. Organ. Chem. (U. S. A.) **41,** 1248 (1976).
439. — — New Germacranolides from *Liatris* Species. Phytochem. **14,** 1561 (1975).
440. — — Seco-germacradienolide from *Liatris pycnostachya.* J. Organ. Chem. (U. S. A.) **40,** 392 (1975).
441. — — Sesquiterpene Lactones of *Eupatorium hyssopifolium.* A Germacranolide with an Unusual Lipid Ester Side Chain. J. Organ. Chem. (U. S. A.) **41,** 1015 (1976).
442. HERZ, W., and A. SRINIVASAN: Reexamination of *Gaillardia amblyodon.* Isolation of New Pseudoguaianolides. Phytochem. **13,** 1187 (1974).
443. — — Pseudoguaianolides of *Gaillardia amblyodon.* Phytochem. **11,** 2093 (1972).
444. — — Stereochemistry of Spathulin. Phytochem. **13,** 1171 (1974).
445. HERZ, W., A. SRINIVASAN, and P. S. KALYANARAMAN: Mikanokryptin, a New Guaianolide from *Mikania.* Phytochem. **14,** 233 (1975).
446. HERZ, W., and P. S. SUBRAMANIAM: Pseudoguaianolides in *Helenium autumnale* from Pennsylvania. Phytochem. **11,** 1101 (1972).
447. HERZ, W., P. S. SUBRAMANIAM, and N. DENNIS: Solvent Shift Studies on Pseudoguaianolides of the Helenalin Series. J. Organ. Chem. (U. S. A.) **34,** 3691 (1969).
448. — — — Stereochemistry of Flexuosin A and Related Compounds. J. Organ. Chem. (U. S. A.) **34,** 2915 (1969).
449. HERZ, W., P. S. SUBRAMANIAM, and T. A. GEISSMAN: 3-Epiisotelekin from *Gaillardia aristata* Pursh. and the Structure of Farinosin. J. Organ. Chem. (U. S. A.) **33,** 3743 (1968).
450. HERZ, W., P. S. SUBRAMANIAM, R. MURARI, N. DENNIS, and J. F. BLOUNT: Microdilin, a Complex Elemenolide from *Mikania cordifolia.* J. Organ. Chem. (U. S. A.) **42,** 1720 (1977).
451. HERZ, W., P. S. SUBRAMANIAM, P. S. SANTHANAM, K. AOTA, and A. L. HALL: Structure Elucidation of Sesquiterpene Dilactones from *Mikania scandens* (L.) Willd. J. Organ. Chem. (U. S. A.) **35,** 1453 (1970); Errata: J. Organ. Chem. (U. S. A.) **38,** 4217 (1973).
452. HERZ, W., V. SUDARSANAM, and J. J. SCHMID: New Guaianolides from *Iva axillaris* Pursh. ssp. *robustior.* J. Organ. Chem. (U. S. A.) **31,** 3232 (1966).

453. Herz, W., and Y. Sumi: Constituents of *Ambrosia hispida* Pursh. J. Organ. Chem. (U.S.A.) **29,** 3438 (1964).

454. Herz, W., Y. Sumi, V. Sudarsanam, and D. Raulais: Constituents of *Iva* Species. X. Ivangulin, a Novel seco-Eudesmanolide from *Iva angustifolia* Nutt. J. Organ. Chem. (U.S.A.) **32,** 3658 (1967).

455. Herz, W., and K. Ueda: The Sesquiterpene Lactones of *Artemisia tilesii* Ledeb. J. Amer. Chem. Soc. **83,** 1139 (1961).

456. Herz, W., K. Ueda, and S. Inayama: Constituents of *Gaillardia* Species. I. The Structure of Pulchellin. Tetrahedron **19,** 483 (1963).

457. Herz, W., and N. Viswanathan: Constituents of *Iva* Species. II. The Structures of Asperilin and Ivasperin, Two New Sesquiterpene Lactones. J. Organ. Chem. (U.S.A.) **29,** 1022 (1964).

458. Herz, W., and I. Wahlberg: Provincialin, a Cytotoxic Germacradienolide from *Liatris provincialis* Godfrey with an Unusual Ester Side Chain. J. Organ. Chem. (U.S.A.) **38,** 2485 (1973).

459. — — Punctatin: A New Germacradienolide from *Liatris punctata*. Phytochem. **12,** 1421 (1973).

460. Herz, W., I. Wahlberg, C. S. Stevens, and P. S. Kalyanaraman: Three New 5,10-Epoxygermacranolides from *Liatris chapmanii* and *Liatris gracilis*. Phytochem. **14,** 1803 (1975).

461. Herz, W., H. Watanabe, M. Miyazaki, and Y. Kishida: The Structures of Parthenin and Ambrosin. J. Amer. Chem. Soc. **84,** 2601 (1962).

462. Heywood, V. H.: Role of Chemistry in Plant Systematics. Pure Appl. Chem. **34,** 355 (1973); Chem. Abstr. **79,** 102687c (1973).

463. Higo, A., Z. Hammam, B. N. Timmermann, H. Yoshioka, J. Lee, T. J. Mabry, and W. W. Payne: Sesquiterpene Lactones from the Genus *Ambrosia*. Phytochem. **10,** 2241 (1971).

463a. Hikino, H., Y. Hikino, and T. Yosioka: Structure and Autoxidation of Atractylon. Chem. Pharm. Bull. Japan **10,** 641 (1962).

464. Hikino, H., D. Kuwano, and T. Takemoto: Structure of Helenium Lactone. Chem. Pharm. Bull. Japan **16** (8), 1601 (1968); Chem. Abstr. **70,** 20233b (1969).

465. Hikino, H., K. Meguro, G. Kusano, and T. Takemoto: Structure of Mokko Lactone and Dehydrocostus Lactone. Chem. Pharm. Bull. Japan **12,** 632 (1964); Chem. Abstr. **61,** 5698d (1964).

466. Hill, D. W., H. L. Kim, C. L. Martin, and B. J. Camp: Identification of Hymenoxone in *Baileya multiradiata* and *Helenium hoopsii*. J. Agric. Food Chem. **25,** 1304 (1977); Chem. Abstr. **87,** 180648v (1977).

467. Hladon, B., T. Bobkiewicz, and B. Drozdz: Sesquiterpene Lactones, Preliminary Studies on the Mode of Action. Inhibition of Synthesis of Tumor Cell Protein and RNA. Archiv. Immunol. et Therap. Experiment. **25,** 243 (1977).

468. Hladon, B., and A. Chodera: Sesquiterpene Lactones, Cytostatic and Pharmacological Activity. Archiv. Immunol. et Therap. Experiment. **23,** 857 (1975).

469. Hladon, B., B. Drozdz, H. Grabarczyk, T. Bobkiewicz, and J. Olszewski: Cytotoxic Activity of Eupatolide and Eupatoriopicrin on Human and Animal Malignant Cells in Tissue Culture *in vitro*. Pol. J. Pharmacol. Pharm. **27,** 429 (1975).

470. Hladon, B., B. Drozdz, M. Holub, and T. Bobkiewicz: In Vitro Studies on Cytotoxic Properties of Sesquiterpene Lactones in Tissue Cultures of Human and Animal Malignant Cells. Archiv. Immunol. Therap. Experiment. **23,** 345 (1975).

471. Ho, C. M., and R. Toubiana: Lactones Sesquiterpeniques. Determination de la Position du Cycle Lactonique dans le Vernolide et L'Hydroxyvernolide. Tetrahedron **26,** 941 (1970).

472. Hochmannova, J., V. Herout, and F. Sorm: Isolation and Structure of Sesqui-

terpenic Lactones from Common Yarrow *(Achillea millefolium* L.). Collect Czech. Chem. Comm. **26,** 1826 (1961).

473. HOCHMANNOVA, J., L. NOVOTNY, and V. HEROUT: Sesquiterpenic Hydrocarbons from Coltsfoot Rhizomes *(Petasites officinalis* Moench.). Collect. Czech. Chem. Comm. **27,** 1870 (1962).

474. HODGES, R., E. P. WHITE, and J. S. SHANNON: The Structure of Mellitoxin. Tetrahedron Letters **1964,** 371.

475. HOLUB, M., R. DE GROOTE, V. HEROUT, and F. SORM: Oxygen-Containing Components of Light Petroleum Extract of *Laser trilobum* (L.) Borkh. Root. Structure of Laserine. Collect. Czech. Chem. Comm. **33,** 2911 (1968).

476. HOLUB, M., and V. HEROUT: unpublished.

477. — — Isolation of Desacetoxymatricarin from *Artemisia leukodes* Schrenk. Collect. Czech. Chem. Comm. **27,** 2980 (1962).

478. HOLUB, M., O. MOTL, Z. SAMEK, and V. HEROUT: The Structure of Two Sesquiterpenic Lactones, Isomontanolide and Acetylisomontanolide from *Laserpitium siler* L. Collect. Czech. Chem. Comm. **37,** 1186 (1972).

479. HOLUB, M., D. P. POPA, V. HEROUT, and F. SORM: Isolation of a Guaianolide from *Laser trilobum* L. Borkh. Root. Collect. Czech. Chem. Comm. **29,** 938 (1964).

480. HOLUB, M., D. P. POPA, Z. SAMEK, V. HEROUT, and F. SORM: Neutral Components of the Light Petroleum Extract of the Root of *Laserpitium siler* L. The Structure of the Sesquiterpenic Lactone Montanolide. Collect. Czech. Chem. Comm. **35,** 3296 (1970).

481. HOLUB, M., J. POPLAWSKI, P. SEDMERA, and V. HEROUT: N-Ethoxycarbonyl-L-Prolinamide, a New Alkaloid from the Leaves of *Arnica montana* L. Collect. Czech. Chem. Comm. **42,** 151 (1977).

482. HOLUB, M., and Z. SAMEK: Isolation and Structure of 3-Epinobilin, 1,10-Epoxynobilin and 3-Dehydronobilin — Other Sesquiterpenic Lactones from the Flowers of *Anthemis nobilis* L. Revision of the Structure of Nobilin and Eucannabinolide. Collect. Czech. Chem. Comm. **42,** 1053 (1977).

483. — — The Structure of Lasolide. A Sesquiterpenic Lactone from *Laser trilobum* (L.) Borkh. Collect. Czech. Chem. Comm. **38,** 1428 (1973).

484. — — The Structure of Archangelolide, A Sesquiterpenic Lactone from *Laserpitium archanglica* Wulf. Collect. Czech. Chem. Comm. **38,** 731 (1973).

485. HOLUB, M., Z. SAMEK, R. DE GROOTE, V. HEROUT, and F. SORM: The Structure of the Sesquiterpenic Triester Lactone Trilobolide. Collect. Czech. Chem. Comm. **38,** 1551 (1973).

486. HOLUB, M., Z. SAMEK, and V. HEROUT: Structure of Isolaserolide from *Laser trilobum.* Phytochem. **11,** 3053 (1972).

487. HOLUB, M., Z. SAMEK, D. P. POPA, and V. HEROUT: The Structure of the Sesquiterpenic Lactone *Silerolide.* Collect. Czech. Chem. Comm. **38,** 1804 (1973).

488. HOLUB, M., Z. SAMEK, D. P. POPA, V. HEROUT, and F. SORM: The Structure of Laserolide, A Sesquiterpenic Lactone from the Roots of *Laser trilobum* L. Borkh. Collect. Czech. Chem. Comm. **35,** 284 (1970).

489. HOLUB, M., Z. SAMEK, and J. TOMAN: Carabrone from *Arnica foliosa.* Phytochem. **11,** 2627 (1972).

490. HOLZER, K., and A. ZINKE: Über die Bitterstoffe der Zichorie *(Cichorium intybus* L.). Monatsschr. Chem. **84,** 901 (1953).

491. HONWAD, V. K., E. SISKOVIC, and A. S. RAO: Synthesis of Tauremisin (Vulgarin) and Saussurea Lactone. Tetrahedron **23,** 1273 (1967).

492. HOREAU, A.: Principe et applications d'une nouvelle methode de determination des configurations dites »par dedoublement partiel«. Tetrahedron Letters **1961,** 506.

493. HOREAU, A., and H. B. KAGAN: Determination des configurations par "dédoublement partiel" — III. Alcools Steroides. Tetrahedron **20,** 2431 (1964).

494. Huber, C. P., and K. J. Watson: The Crystal and Molecular Structure of 2-Bromo-lumisantonin. J. Chem. Soc. (London) **1968**, 2441.
495. Hudson, C. S.: A Relation Between the Chemical Constitution and the Optical Rotatory Power of the Sugar Lactones. J. Amer. Chem. Soc. **32**, 338 (1910).
496. Inayama, S., T. Kawamata, and M. Yanagita: Sesquiterpene Lactones of *Gaillardia pulchella*. Phytochem. **12**, 1741 (1973).
497. Inayama, S., T. Ohkura, and Y. Iitaka: The Complete Structure of *Spathulin*, a Crystallographic Study of Diacetylspathulin. Chem. Pharm. Bull (Japan) **25**, 1928 (1977); Chem. Abstr. **88**, 23171n (1978).
498. Inayama, S., T. Ohkura, T. Kawamata, and M. Yanagita: Ambrosic acid. New Irritant Principle from *Ambrosia arthemisiifolia*. Chem. Pharm. Bull. (Japan) **22**, 1435 (1974); Chem. Abstr. **81**, 132853n (1974).
499. Ingham, C. F., and R. A. Massy-Westropp: Stereochemistry of Freelingyne and the Synthesis of Dihydrofreelingyne. Austral. J. Chem. **27**, 1491 (1974); Chem. Abstr. **81**, 105715n (1974).
500. Ingham, C. F., R. A. Massy-Westropp, and G. D. Reynolds: A Synthesis of Freelingyne. Austral. J. Chem. **27**, 1477 (1974); Chem. Abstr. **81**, 105718r (1974).
501. Inubushi, Y., Y. Tsuda, and E. Katarao: The Structure of Dendramine. Chem. Pharm. Bull. (Japan) **14**, 668 (1966); Chem. Abstr. **65**, 10632f (1966).
502a. Iriuchijima, S., S. Kuyama, N. Takahashi, and S. Tamura: Chemistry of Heliangine. Part I. Structure of Heliangine. Agr. Biol. Chem. **30**, 1152 (1966).
503. Iriuchijima, S., and S. Tamura: Stereochemistry of Pyrethrosin, Cyclopyrethrosin Acetate, and Isocyclopyrethrosin Acetate. Agr. Biol. Chem. **34**, 204 (1970); Chem. Abstr. **73**, 15028w (1970).
504. — — Stereochemical Structures of Pyrethrosin, Cyclopyrethrosin Acetate and Iso-cyclopyrethrosin Acetate. Tetrahedron Letters **1967**, 1965.
505. Irwin, M. A.: Sesquiterpene Lactones of *Artemisia*. Ph. D. Dissertation, University of California, Los Angeles: 1971; Chem. Abstr. **75**, 137459z (1971).
506. Irwin, M. A., and T. A. Geissman: Rupicolin-A and -B, Rupin-A and -B, Cumambrin-B Oxide from *Artemisia tripartita* ssp. *rupicola*. Phytochem. **12**, 863 (1971).
507. — — Sesquiterpene Lactones from *Artemisia:* Arbusculin-C, Rothin-A and Rothin-B. Phytochem. **10**, 637 (1971).
508. — — Sesquiterpene Lactones of *Artemisia* Species. New Lactones from *A. arbuscula* ssp. *arbuscula* and *A. tripartita* ssp. *rupicola*. Phytochem. **8**, 2411 (1969).
509. — — Novanin: A Germacranolide from *Artemisia nova*. Phytochem. **12**, 875 (1973).
510. — — Sesquiterpene Lactones. Constituents of *Artemisia nova* Nels. and *A. tripartita* A Gray ssp. *rupicola*. Phytochem. **8**, 305 (1969).
511. — — Arbusculin-D from *Artemisia arbuscula* ssp. *arbuscula*. Phytochem. **12**, 853 (1973).
512. Irwin, M. A., K. H. Lee, R. F. Simpson, and T. A. Geissman: Sesquiterpene Lactones of *Artemisia*. Ridentin. Phytochem. **8**, 2009 (1969).
513. Irwin, M. A., and T. A. Geissman: Ridentin-B: An Eudesmanolide from *Artemisia tripartita* ssp. *rupicola*. Phytochem. **12**, 871 (1973).
514. Ishii, H., T. Tozyo, and H. Minato: Components of the Root of *Ligularia fischeri* Turcz. J. Chem. Soc. C (London) **17**, 1545 (1966).
515. Ishikawa, K., and T. B. H. McMurry: Chemistry of Santonene. Photoreaction of 4αH-pyrosantonin. J. Chem. Soc. Perkin Trans. I **1973**, 914.
516. Ishizaki, Y., Y. Tanahashi, Y. Moriyama, T. Takahashi, and H. Koyama: Sesqui-terpenes from the Roots of *Ligularia hodgsonii*. Phytochem. **13**, 674 (1974).
517. Ishizaki, Y., Y. Tanahashi, and T. Takahashi: Furanoeremophilan-14β,6α-olide, a New Furanosesquiterpene Lactone from *Ligularia hodgsonii* Hook. f. The Structure and Nuclear Overhauser Effects. Chem. Commun. **1969**, 551.

518. ISHIZAKI, Y., Y. TANAHASHI, T. TAKAHASHI, and K. TORI: The Structure of Ligularenolide. A New Sesquiterpene Lactone of Eremophilane Type. Tetrahedron **26**, 5387 (1970).

518 a. IVIE, G. W., D. A. WITZEL, W. HERZ, R. KANNAN, J. O. NORMAN, D. D. RUSHING, J. H. JOHNSON, L. D. ROWE, and J. A. VEECH: Hymenovin. Major Toxic Constituents of Western Bitterweed *(Hymenoxys odorata* DC.). J. Agric. Food Chem. **23**, 841 (1975).

518 b. IVIE, G. W., D. A. WITZEL, W. HERZ, R. P. SHARMA, and A. E. JOHNSON: Isolation of Hymenovin from *Hymenoxys richardsonii* (Pingue) and *Dugaldia hoopesii* (Orange Sneezeweed). J. Agric. Food Chem. **24**, 681 (1976).

519. JAIN, T. C., C. M. BANKS, and J. E. McCLOSKEY: Novel Cyclization of trans-1,2-Divinylcyclohexane-3,4-trans-γ-Lactone Unit. Tetrahedron Letters **1970**, 2387.

520. — — — Reversible Dimethylamine Addition as a Protecting Reaction for α,β-Unsaturated Methylene Groups of γ-Lactones and its Regeneration by Basic Elimination of Quaternary Ammonium Salts. Tetrahedron **32**, 765 (1976).

521. — — — Dehydrosaussurea Lactone from Costunolide and Reversibility in the Germacranolide-Cope Reaction. Tetrahedron Letters **1970**, 841.

522. JAIN, T. C., and J. E. McCLOSKEY: Thermolysis of Dihydrocostunolide. Tetrahedron Letters **1969**, 4525.

523. — — Bromotrifluoride Catalyzed Cyclization of Costunolide. Synthesis of 4α-Hydroxycyclocostunolide. Tetrahedron Letters **1971**, 1415.

524. — — A Facile and Stereospecific Cyclization of Costunolide. Tetrahedron Letters **1969**, 2917.

525. — — Carbocyclization in Natural Products. Amberlite IR-120 Cation Exchange Resin-Catalyzed Cyclization of Costunolide. Structure of β-Cyclocostunolide. Tetrahedron **31**, 2211 (1975).

526. JAMIESON, G. R., E. H. REID, B. P. TURNER, and A. T. JAMIESON: Bakkenolide-A. Its Distribution in Petasites Species and Cytotoxic Properties. Phytochem. **15**, 1713 (1976).

527. JEREMIC, D., A. JOKIC, A. BEHBUD, and M. STEFANOVIC: A New Type of Sesquiterpene Lactone Isolated from *Artemisia annua* L. Tetrahedron Letters **1973**, 3039.

528. JIZBA, J., Z. SAMEK, and L. NOVOTNY: A Sesquiterpene Alkaloid, Eremophilene Lactam, from the Rhizomes of *Petasites hybridus*. Collect. Czech. Chem. Comm. **42**, 2438 (1977).

528 a. JOLAD, S. D., R. M. WIEDHOPF, and J. R. COLE: Tumor-Inhibitory Agent from *Zaluzania robinsonii* (Compositae). J. Pharm. Sci. **1974**, 1321.

529. JOHNS HOPKINS UNIV., patent. Chemical Excitation of Santonin. Chem. Abstr. **82**, 73249 r (1975).

530. JOMMI, G., Sesquiterpenoids [from *Toxicodendrum capense, motroantinana* and *Valeriana officinalis*]. Corsi Semin. Chim. **11**, 44 (1968); Chem. Abstr. **72**, 90652 c (1970).

531. JOMMI, G., P. MANITTO, F. PELIZZONI, and C. SCOLASTICO: Bitter Constituents of *Hyenanche globosa*. Chimica Industria Ital. **46**, 549 (1964); Chem. Abstr. **61**, 5697 d (1964).

532. — — — — Constituents of *Hyaenanche globosa*. Structure of Capenicin. Chim. Ind. **47**, 1328 (1965); Chem. Abstr. **64**, 6699 f (1966).

533. JOMMI, G., P. MANITTO, and C. SCOLASTICO: Correlation Between Futin and Hyenanchin. Chim. Ind. **47**, 407 (1965); Chem. Abstr. **63**, 14917 b (1965).

534. JOMMI, G., F. PELIZZONI, and C. SCOLASTICO: Constituents of *Hyenanche globosa* — Structure of Substance D. Chim. Ind. **47**, 865 (1965); Chem. Abstr. **63**, 11467 c (1965).

535. JOSEPH-NATHAN, P.: Carbon-13 Nuclear Magnetic Resonance Studies of Helenalin. Rev. Soc. Quim. Mex. **20**, 255 (1976); Chem. Abstr. **86**, 121546 s (1977).

536. Joseph-Nathan, P., and E. Diaz: La Conformacion De La Helenalina Deducida De Su Espectro De Resonancia Magnetica Nuclear. Rev. Latinoamer. Quim. 2, 34 (1971).

537. Joseph-Nathan, P., and J. Romo: Isolation and Structure of Peruvin. Tetrahedron 22, 1723 (1966).

538. Joshi, B. S.: Some Recent Work on Germacranolides. J. Sci. Ind. Res. 1976, 239; Chem. Abstr. 86, 72888u (1977).

539. Joshi, B. S., A. S. Bawdekar, G. H. Kulkarni, A. S. Rao, G. R. Kelkar, and S. C. Bhattacharyya: Examination of Costus Root Oil. Perfumery Essent. Oil Record. 52, 773 (1961); Chem. Abstr. 56, 11727d (1962).

540. Joshi, B. S., and V. N. Kamat: Structure of Enhydrin, a Germacranolide from Enhydra fluctuans Lour. Ind. J. Chem. 10, 771 (1972).

541. Joshi, B. S., V. N. Kamat, and T. R. Govindachari: Sesquiterpenes of Neolitsea zeylanica Merr. III. Structure of Zeylanine, Zeylanicine and Zeylanidine. Tetrahedron 23, 273 (1967).

542. — — — Sesquiterpenes of Neolitsea zeylancia Merr. I. Isolation of Some Constituents. Tetrahedron 23, 261 (1967).

543. — — — Structure of Neolinderane. Tetrahedron 23, 267 (1967).

544. Joshi, B. S., V. N. Kamat, and H. Fuhrer: Revised Structure of Enhydrin. Tetrahedron Letters 1971, 2373.

545. Joshi, B. S., G. H. Kulkarni, G. R. Kelkar, and S. C. Bhattacharyya: Transformation Products of Costunolide. Stereochemistry of (−)-Santonin and Solid Dihydrocostunolide at C-11. Tetrahedron 22, 2331 (1966).

546. Kabuto, C., N. Takada, S. Maeda, and Y. Kitahara: X-Ray Structure Determination of Eremophilenolide. Chemistry Letters 1973, 371.

547. Kadival, M. V., and G. H. Kulkarni: Conversion of Costunolide into Tetrahydrosantonins. Chem. and Ind. (London) 1967, 2084; Chem. Abstr. 68, 49804h (1968).

548. Kagan, J., and H. B. Kagan: Dehydration of Coronopilin. J. Organ. Chem. 33, 2807 (1968).

549. Kagan, H. B., H. E. Miller, W. Renold, M. V. Lakshmikantham, L. R. Tether, W. Herz, and T. J. Mabry: The Structure of Psilostachyin C, a New Sesquiterpene Dilactone from Ambrosia psilostachyia DC. J. Organ. Chem. (U. S. A.) 31, 1629 (1966).

550. Kagan, J., S. P. Singh, K. Warden, and D. A. Harrison: The Photochemistry of Parthenin and Coronopilin. Tetrahedron Letters 1971, 1849.

551. Kallen, J.: Über Helenin und Alantkampher. Ber. dtsch. chem. Ges. 6, 1506 (1873).

552. Kaneko, H., S. Naruto, and S. Takahashi: Sesquiterpenes of Achillea sibirica. Phytochem. 10, 3305 (1971).

553. Kariyone, T., T. Fukui, M. Ishikawa, and T. Imawaki: Constituents of Artemisia monogyna. Structure of Mibulactone. J. Pharmac. Soc. Japan 69, 310 (1949).

553a. Kariyone, T., T. Fukui, and T. Omoto: Constituents of Artemisia monogyna. Structure of Monogynin. Yakugaku Zasshi 78, 710 (1958).

554. Kariyone, T., and S. Naito: Components of Carpesium abrotanoides. Chemical Structure of Carpesia Lactone. J. Pharmac. Soc. Japan 75, 39 (1955); Chem. Abstr. 50, 890g (1956).

555. Kariyone, T., S. Naito, and J. Chatani: Studies on the Components of Carpesium abrotanoides. Chemical Constitution of Carpesia Lactone. Pharm. Bull. (Japan) 2, 339 (1954); Chem. Abstr. 50, 7089g (1956).

556. Karlsson, B., A. M. Pilotti, A. C. Wiehager, I. Wahlberg, and W. Herz: Crystal Structure of Spicatin Hydrobromide. Revision of the Structure of Spicatin. Tetrahedron Letters 1975, 2245.

557. Kartha, G., and K. T. Go: Structure and Absolute Configuration of Enhydrin Bromohydrin. J. Cryst. Mol. Struct. 1976, 31; Chem. Abstr. 86, 63990u (1977).

558. KARTHA, G., K. T. Go, and B. S. JOSHI: X-Ray Crystal Structure and Configuration of Enhydrin Bromohydrin. Chem. Commun. 1972, 1327.

559. KARUBE, A.: Hydrogenation Products of Santonin. Nippon Kagaku Kaishi 7, 1026 (1977); Chem. Abstr. 87, 201795h (1977).

560. KASHMAN, Y., D. LAVIE, and E. GLOTTER: Sesquiterpene Lactones from Inula helenium. Israel J. Chem. 5, 23 (1967).

561. KASYMOV, SH. Z., and G. P. SIDYAKIN: Lactones of Achillea millefolium. Khim. Prir. Soedin. 8, 246 (1972); Engl. edit.: p. 246.

561a. — — Lactones from Artemisia tenuisecta. Khim. Prir. Soedin. 5, 445 (1969); Chem. Abstr. 72, 75666s (1970).

562. KATAYAMA, C., A. FURUSAKI, I. NITTA, M. HAYASHI, and K. NAYA: Crystal Structure of Fukinolidol Di-Bromoacetate. Bull. Chem. Soc. Japan 43, 1976 (1970).

563. KATO, T., T. IIDA, T. SUZUKI, Y. KITAHARA, and K. H. OVERTON: Structure and Synthesis of Futronolide. Tetrahedron Letters 1972, 4257.

564. KATO, T., K. SHIRAHATA, and Y. KITAHARA: Mass Spectra of Bakkenolides and their Derivatives. Recent Develop. Mass Spectrosc. Proc. Int. Conf. Mass Spectrosc. 1969, 1259; Chem. Abstr. 75, 110446f (1971).

564a. KAWATANI, T.: Discovery of Artemisias Containing α-β-Santonin. J. Pharm. Soc. Japan 73, 873 (1953).

565. KAWATANI, T., and T. TAKEUCHI: Two Crystalline Compounds from Artemisia finita; Further Discovery of an Artemisia containing 1-β-Santonin. J. Pharm. Soc. Japan 74, 793 (1954); Chem. Abstr. 49, 11607f (1955).

566. KECHATOVA, N. A., K. S. RYBALKO, V. I. SHEICHENKO, and L. P. TOLSTYKH: Sesquiterpene Lactones from Artemisia taurica. Khim. Prir. Soedin. 4, 205 (1968); Engl. edit.: p. 177.

567. KECHATOVA, N. A., K. S. RYBALKO, P. N. SMIRNOV, and M. I. KAMNEV: Tauremisin Content of Artemisia taurica. Tauremisin Distribution in the Organs. Rast. Resur. 5, 444 (1969); Chem. Abstr. 72, 39786x (1970).

567a. KECHATOVA, N. A., and M. I. VLASOV: Sesquiterpene Lactone from Artemisia juncea. Khim. Prir. Soedin. 2, 216 (1966); Chem. Abstr. 65, 17360e (1966).

568. KELSEY, R. G., M. S. MORRIS, N. R. BHADANE, and F. SHAFIZADEH: Sesquiterpene Lactones of Artemisia. TLC [Thin-Layer Chromatography] Analysis and Taxonomic Significance. Phytochem. 12, 1345 (1973).

569. KELSEY, R. G., J. W. THOMAS, T. J. WATSON, and F. SHAFIZADEH: Population Studies in Artemisia tridentata ssp. vaseyana: Chromosome Numbers and Sesquiterpene Lactone Races. Biochem. Syst. Ecol. 3, 209 (1975); Chem. Abstr. 84, 118461p (1976).

570. KEPLER, J. A., M. E. WALL, J. E. MASON, C. BASSET, A. T. McPHAIL, and G. A. SIM: The Structure of Fomannosin, a Novel Sesquiterpene Metabolite of the Fungus Fomes annosus. J. Amer. Chem. Soc. 89, 1260 (1967).

571. KERIMOV, S. SH., and O. S. CHIZHOV: Sesquiterpene Lactones of Inula helenium. Khim. Prir. Soedin. 10, 254 (1974); Chem. Abstr. 81, 60894b (1974).

572. KHAFAGY, S. M., and A. METWALLY: Isolation of a Crystalline Sesquiterpenic Keto-Lactone from Xanthium occidentale. Planta Med. 18, 318 (1970); Chem. Abstr. 73, 95439u (1970).

572a. KHAFAGY, S. M., S. A. GHARBO, and T. M. SARG: Phytochemical Investigation of Artemisia herba-alba. Planta Med. 20, 90 (1971); Chem. Abstr. 75, 85176h (1971).

573. KHAFAGY, S. M., A. M. METWALLY, and A. A. OMAR: Isolation of Three Lactones from Onopordon alexandrinum. Pharmazie 1977, 123; Chem. Abstr. 86, 117665n (1978).

574. KHVOROST, P. P., and N. F. KOMISSARENKO: Lactones of Inula helenium. Khim. Prir. Soedin. 12, 820 (1976); Chem. Abstr. 86, 136307p (1978).

575. Kim, H. L., L. D. Rowe, and B. J. Camp: Hymenoxon, a Poisonous Sesquiterpene Lactone from *Hymenoxys odorata* (Bitterweed). Res. Comm. Chem. Pathol. Pharmacol. **11**, 647 (1975); Chem. Abstr. **83**, 160803f (1975).

576. King, T. J., J. C. Roberts, and D. J. Thompson: Isolation and Structure of Purpuride, a Metabolite of *Penicillium purpurogenum*. J. Chem. Soc. Perkin Trans I **1973**, 78.

577. Kinoshita, K.: Constituents of *Coriaria japonica* Gray. II. Coriarine, a Poisonous Constituent. J. Chem. Soc. (Japan) **51**, 99 (1930).

578. Kir'yalov, N. P., T. V. Bukreeva, and V. A. Gindin: The Structure of Grilactone. Khim. Prir. Soedin. **8**, 446 (1972); Engl. edit.: p. 445; Chem. Abstr. **77**, 152371z (1972).

579. Kir'yalov, N. P., and S. V. Serkerov: Guaianolides from Plants of the Umbelliferae Family. Mezhdunar. Kongr. Efirnym. Maslam [Mater]. 4th, 147 (1968); Chem. Abstr. **78**, 159896d (1973).

580. — — The Structure of Badkhysin. Khim. Prir. Soedin. **4**, 341 (1968); Engl. edit.: p. 287.

581. Kiseleva, E. Y., O. A. Konovalova, and K. S. Rybalko: Talassin — A New Sesquiterpene Lactone. Khim. Prir. Soedin. **7**, 668 (1971); Engl. edit.: p. 650; Chem. Abstr. **76**, 96990r (1972).

582. Kiseleva, E. Y., A. I. Saidkhodzhaev, and G. K. Nikonov: Sesquiterpene Lactones from the Roots of *Ferula diversivittata*. Izv. Akad. Nauk. Turkm. SSR, Ser. Fiz.-Tekh. Khim. Geol. Nauk. **1975**, 126; Chem. Abstr. **84**, 5166c (1976).

583. Kiseleva, E. Y., V. I. Sheichenko, K. S. Rybalko, G. A. Kalabin, and A. I. Ban'kovskii: The Structure of Inulicin — A New Sesquiterpene Lactone from *Inula japonica*. Khim. Prir. Soedin. **7**, 263 (1971); Engl. edit.: p. 254; Chem. Abstr. **75**, 110445e (1971).

583a. Kiseleva, E. Y., V. I. Sheichenko, K. S. Rybalko, A. I. Shretev, and D. A. Pakaln: Separation of Gaillardin from *Inula oculus-Christi*. Khim. Prir. Soedin. **5**, 444 (1969); Chem. Abstr. **72**, 82893s (1970).

584. Kisiel, W.: Phytochemical Investigation of *Vernonia flexuosa*. Vernoflexuoside and Vernoflexin, New Sesquiterpene Lactones. Pol. J. Pharmacol. Pharm. **27**, 461 (1975); Chem. Abstr. **84**, 14634f (1976).

585. — Sesquiterpene Lactones of *Vernonia* Species. New Group of Biologically Active Compounds. Postepy. Biochem. **20**, 67 (1974); Chem. Abstr. **81**, 101782j (1974).

586. — New Germacranolide from *Zinnia haageana*. Phytochem. **17**, 1059 (1978).

587. Kitagawa, I., H. Shibuya, H. Takeno, T. Nishino, and I. Yosioka: Biogenetically Patterned Transformation of Eudesmanolide to Eremophilanolide. Structures of Minor Products Obtained by Acid Treatment of 5α,6α-epoxyeudesman-8β,12-olide. Chem. Pharm. Bull. **24**, 56 (1976); Chem. Abstr. **84**, 135857q (1976).

588. Kitagawa, I., H. Shibuya, Y. Yamazoe, H. Takeno, and I. Yosioka: Conversion of Dihydroalantolactone to Tetrahydroligularenolide. A Biogenetic-type Transformation of Eudesmanolide to Eremophilanolide. Tetrahedron Letters **1974**, 111.

589. Kitagawa, I., Y. Yamazoe, H. Shibuya, R. Takeda, H. Takeno, and I. Yosioka: Biogenetically Patterned Transformation of Eudesmanolide to Eremophilanolide. I. Angular Methyl Migration of 5α,6α-epoxydihydroalantolactone. Chem. Pharm. Bull. **22**, 2662 (1974); Chem. Abstr. **82**, 125488q (1975).

590. Kitagawa, I., Y. Yamazoe, R. Takeda, and I. Yosioka: Conversion of Dihydroalantolactone to Eremophilane-Type Derivatives: A Biogenetic-Type Transformation. Tetrahedron Letters **48**, 4843 (1972).

591. Kitahara, Y., T. Kato, T. Suzuki, S. Kanno, and M. Tanemura: Biogenetic-type Synthesis of (±)-Drimenin. Chem. Commun. **1969**, 342.

592. Kitahara, Y., S. Maeda, U. Masako, F. Makoto, K. Tadahiro, L. Novotny, V. Herout, and F. Sorm: Revision of Structure of Petasitolides by the Synthesis of

3β- and 3α-Hydroxy- and 2β-Senecioyloxyeremophilenolides. Chemistry Letters **1977**, 1031.

593. KLYNE, W.: The Application of Hudson's Lactone Rule to the Molecular Rotations of Polycyclic Compounds. Chem. and Ind. **1954**, 1198.

594. KNIGHT, D. W., and G. PATTENDEN: Total Synthesis of the Acetylenic Sesquiterpene Freelingyne. Chem. Commun. **1974**, 188.

595. — — Synthesis of Freelingyne and Acetylenic Sesquiterpene from *Eremophila freelingii*. J. Chem. Soc. Perkin Trans. I **1975**, 641.

596. — — Freelingnite, a New Furanosesquiterpene from *Eremophila freelingii*. Tetrahedron Letters **1975**, 1115.

597. KNOCHE, H., G. OURISSON, G. W. PEROLD, J. FOUSSEREAU, and J. MALEVILLE: Allergenic Component of a Liverwort: Sesquiterpene Lactone. Science **166**, 239 (1969); Chem. Abstr. **72**, 719c (1970).

598. KONDO, Y., F. HAMADA, and F. YOSHIZAKI: 2-Methoxydihydrohelenalin from the Rhizoma of Sendai *Helenium autumnale* L. Heterocycles **1976**, 373; Chem. Abstr. **86**, 90060x (1977).

599. KONDO, Y., T. TOMIMORI, N. HIRAGA, and T. TAKEMOTO: Picrohelenin, a New Cytotoxic and Bitter Pseudoguaianolide from Sendai *Helenium autumnale* L. Heterocycles **1977**, 19; Chem. Abstr. **86**, 68371c (1977).

600. KONDO, Y., F. YOSHIZAKI, F. HAMADA, J. IMAI, and G. KUSANO: Sulferalin, A Novel Sulfonyl Pseudoguaianolide Sesquiterpene Lactone from Sendai *Helenium atumnale* L. Tetrahedron Letters **1977**, 2155.

601. KONITZ, A., M. BOGUCKA-LEDO'CHOWSKA, Z. DAUTER, A. HEMPEL, and E. BOROWSKI: The Structure of Isolactarorufin. Tetrahedron Letters **1977**, 3401.

602. KONOVALOVA, O. A., A. I. BAN'KOVSKII, K. S. RYBALKO, V. I. SHEICHENKO, P. I. ZAKHAROV, and M. G. PIMENOV: Sesquiterpene Lactones from *Ferula olgae*. Khim. Prir. Soedin. **8**, 651 (1972); Engl. edit.: p. 623; Chem. Abstr. **78**, 121364r (1973).

603. KONOVALOVA, O. A., K. S. RYBALKO, and V. S. KABANOV: Sesquiterpene Lactones from *Carpesium eximium*. Khim. Prir. Soedin. **8**, 721 (1972); Engl. edit.: p. 705; Chem. Abstr. **78**, 94805b (1973).

604. KONOVALOVA, O. A., K. S. RYBALKO, and M. G. PIMENOV: Sesquiterpene Lactones from *Talassia transiliensis*. Khim. Prir. Soedin. **9**, 122 (1973); Engl. edit.: p. 120; Chem. Abstr. **78**, 148082j (1973).

605. KONOVALOVA, O. A., K. S. RYBALKO, and V. I. SHEICHENKO: Structure of New Sesquiterpene Lactones from *Inula germanica*. Khim. Prir. Soedin. **10**, 578 (1974); Engl. edit.: p. 591; Chem. Abstr. **82**, 73219f (1975).

606. — — — Sesquiterpene Lactones of *Ferula olgae*. Khim. Prir. Soedin. **11**, 590 (1975); Engl. edit.: p. 620; Chem. Abstr. **84**, 105802a (1976).

607. KONOVALOVA, O. A., K. S. RYBALKO, V. I. SHEICHENKO, and D. A. PAKALN: Sesquiterpene Lactones from *Artemisia caucasica*. Khim. Prir. Soedin. **7**, 741 (1971); Engl. edit.: p. 719; Chem. Abstr. **76**, 141061b (1972).

607a. KORTE, F., H. BARKEMEYER, and I. KORTE: Pflanzliche Bitterstoffe. In: Fortschr. d. Chemie Organ. Naturstoffe (ZECHMEISTER, L., ed.), Vol. 17, p. 124. Wien: Springer. 1959.

608. KOSUGI, H., and H. UDA: Studies Directed Towards the Synthesis of Fomannosin. A Synthesis of (±)-Dihydrofomannosin Acetate. Chem. Letters **1977**, 1491.

609. KOVACS, Ö., V. HEROUT, M. HORAK, and F. SORM: On Terpenes. LXVII. Hydrogenation Products of Santonin and Alantolactone. Collect. Czech. Chem. Comm. **21**, 225 (1956).

610. KOYAMA, H., and Y. MIZUNO-TSUKUDA: Crystal and Molecular Structure of the Germacrane Furanosesquiterpenoid Linderalactone. J. Chem. Soc. Perkin Trans. II **1977**, 646.

611. Kozuka, M., K. H. Lee, A. T. McPhail, and K. D. Onan: Structure and Absolute Stereochemistry of Dihydrofloriolenalin, a New Sesquiterpene Lactone from Florida *Helenium autumnale*. Chem. Pharm. Bull. **23**, 1895 (1975); Chem. Abstr. **83**, 144542 q (1975).

612. Kretchmer, R. A., and W. J. Thompson: The Total Synthesis of (±)-Damsin. J. Amer. Chem. Soc. **98**, 3379 (1976).

613. Krishnaswamy, N. R., T. R. Seshadri, and T. N. C. Vedantham: Chemistry of Enhydrin, a New Germacranolide from *Enhydra fluctuans*. Indian J. Chem. **10**, 249 (1972); Chem. Abstr. **77**, 88700 b (1972).

614. Kulkarni, G. H.: Allylic Oxidations of Santenolides. Chem. Ind. (London) **1970**, 1498; Chem. Abstr. **74**, 31860 n (1971).

615. Kulkarni, G. H., G. R. Kelkar, and S. C. Bhattacharyya: Cyclocostunolides. Tetrahedron **20**, 2639 (1964).

616. Kupchan, S. M.: Novel Natural Products of Biological Interest. Rev. Latinoamer. Quim. **5**, 133 (1974).

617. — Recent Advances in the Chemistry of Terpenoid Tumor Inhibitors. Pure Appl. Chem. **21**, 227 (1970).

618. Kupchan, S. M., J. W. Ashmore, and A. T. Sneden: Eriofertopin and 2-O-Acetyl-eriofertopin, New Tumor Inhibitory Germacradienolides from *Eriophyllum confertiflorum*. Phytochem. **16**, 1834 (1977).

619. Kupchan, S. M., Y. Aynehchi, J. M. Cassady, A. T. McPhail, G. A. Sim, H. K. Schnoes, and A. L. Burlingame: The Isolation and Structural Elucidation of Two Novel Sesquiterpenoid Tumor Inhibitors from *Elephantopus elatus*. J. Amer. Chem. Soc. **88**, 3674 (1966).

620. Kupchan, S. M., Y. Aynehchi, J. M. Cassady, H. K. Schnoes, and A. L. Burlingame: Tumor Inhibitors XL. The Isolation and Structural Elucidation of Elephantin and Elephantopin, Two Novel Sesquiterpenoid Tumor Inhibitors from *Elephantopus elatus*. J. Organ. Chem. (U. S. A.) **34**, 3867 (1969).

621. Kupchan, S. M., R. L. Baxter, C.-K. Chiang, C. J. Gilmore, and R. F. Bryan: Eriolangin and Eriolanin, Novel Antileukaemic seco-Eudesmanolides from *Eriophyllum lanatum*. Chem. Commun. **1973**, 842.

622. Kupchan, S. M., J. M. Cassady, J. E. Kelsey, H. K. Schnoes, D. H. Smith, and A. L. Burlingame: Structural Elucidation and High-Resolution Mass Spectrometry of Gaillardin, a New Cytotoxic Sesquiterpene Lactone. J. Amer. Chem. Soc. **88**, 5292 (1966).

623. Kupchan, S. M., V. H. Davies, T. Fujita, M. R. Cox, and R. F. Bryan: Liatrin, a Novel Antileukemic Sesquiterpene Lactone from *Liatris chapmanii*. J. Amer. Chem. Soc. **93**, 4916 (1971).

624. Kupchan, S. M., V. H. Davies, T. Fujita, M. R. Cox, R. J. Restivo, and R. F. Bryon: The Isolation and Structural Elucidation of Liatrin, a Novel Antileukemic Sesquiterpene Lactone from *Liatris chapmanii*. J. Organ. Chem. (U. S. A.) **38**, 1853 (1973).

625. Kupchan, S. M., T. Fujita, M. Maruyama, and R. W. Britton: The Isolation and Structure Elucidation of Eupaserrin and Desacetyleupaserrin, New Antileukemic Sesquiterpene Lactones from *Eupatorium semiserratum*. J. Organ. Chem. (U. S. A.) **38**, 1260 (1973).

626. Kupchan, S. M., T. J. Giacobbe, and I. S. Krull: Reversible Thiol Addition as a Protecting Reaction for Conjugated α-Methylene Groups of Lactones. Tetrahedron Letters **1970**, 2859.

627. Kupchan, S. M., R. J. Hemingway, A. Karim, and D. Werner: Vernodalin and Vernomygdin, Two New Cytotoxic Sesquiterpene Lactones from *Vernonia amygdalina* Del. J. Organ. Chem. (U. S. A.) **34**, 3908 (1969).

628. KUPCHAN, S. M., R. J. HEMINGWAY, D. WERNER, and A. KARIM: Vernolepin, a Novel Sesquiterpene Dilactone Tumor Inhibitor from *Vernonia hymenolepis* A. Rich. J. Organ. Chem. (U.S.A.) **34**, 3903 (1969).

629. KUPCHAN, S. M., R. J. HEMINGWAY, D. WERNER, A. KARIM, A. T. MCPHAIL, and G. A. SIM: Vernolepin, a Novel Elemanolide Dilactone Tumor Inhibitor from *Vernonia hymenolepis*. J. Amer. Chem. Soc. **90**, 3596 (1968).

630. KUPCHAN, S. M., J. E. KELSEY, M. MARUYAMA, J. M. CASSADY, J. C. HEMINGWAY, and J. R. KNOX: Tumor Inhibitors XLI. Structural Elucidation of Tumor-Inhibitory Sesquiterpene Lactones from *Eupatorium rotundifolium*. J. Organ. Chem. (U.S.A.) **34**, 3876 (1969).

630 a. KUPCHAN, S. M., J. E. KELSEY, and G. A. SIM: The Stereochemistry of Germacranolide Sesquiterpenes. Tetrahedron Letters **1967**, 2863.

631. KUPCHAN, S. M., and M. MAYURAMA: Reductive Elimination of Epoxides to Olefins with Zinc-Copper Couple. J. Organ. Chem. (U.S.A.) **36**, 1187 (1971).

632. KUPCHAN, S. M., M. MARUYAMA, R. J. HEMINGWAY, J. C. HEMINGWAY, S. SHIBUYA, and T. FUJITA: Structural Elucidation of Novel Tumor-Inhibitory Sesquiterpene Lactones from *Eupatorium cuneifolium*. J. Organ. Chem. (U.S.A.) **38**, 2189 (1973).

633. KUPCHAN, S. M., M. MARUYAMA, R. J. HEMINGWAY, J. C. HEMINGWAY, S. SHIBUYA, T. FUJITA, P. D. CRADWICK, A. D. HARDY, and G. A. SIM: Eupacunin, a Novel Antileukemic Sesquiterpene Lactone from *Eupatorium cuneifolium*. J. Amer. Chem. Soc. **93**, 4914 (1971).

634. KUROKAWA, T., K. NAKANISHI, W. WU, H. Y. HSU, M. MARUYAMA, and S. M. KUPCHAN: Deoxyelephantopin and its Interrelation with Elephantopin. Tetrahedron Letters **1970**, 2863.

635. LANE, J. F., W. T. KOCH, N. S. LEEDS, and G. GORIN: On the Toxin of *Illicium anisatum*. I. The Isolation and Characterization of a Convulsant Principle: Anisatin. J. Amer. Chem. Soc. **74**, 3211 (1952).

636. LEE, K. H., D. C. ANUFORO, E. S. HUANG, and C. PIANTADOSI: Angustibalin, a New Cytotoxic Sesquiterpene Lactone from *Balduina angustifolia*. J. Pharm. Sci. **61**, 626 (1972); Chem. Abstr. **77**, 28809j (1972).

637. LEE, K. H., C. M. COWHERD, and M. T. WOLO: Deoxyelephantopin, an Antitumor Principle from *Elephantopus carolinianus*. J. Pharm. Sci. **64**, 1572 (1975); Chem. Abstr. **83**, 160810f (1975).

638. LEE, K. H., H. FURUKAWA, and E. S. HUANG: Synthesis and Cytotoxic Activity of Helenalin amine. Adducts and Related Derivatives. J. Med. Chem. **15**, 609 (1972); Chem. Abstr. **77**, 56404s (1972).

639. LEE, K. H., H. FURUKAWA, S. H. KIM, and C. PIANTADOSI: Reaction of Helenalin with Hydrogen Chloride. J. Pharm. Sci. **62**, 987 (1973); Chem. Abstr. **79**, 53610m (1973).

640. LEE, K. H., H. FURUKAWA, M. KOZUKA, H. C. HUANG, P. A. LUHAN, and A. T. MCPHAIL: Molephantin, a Novel Cytotoxic Germacranolide from *Elephantopus mollis*. X-Ray Crystal Structure. Chem. Commun. **1973**, 476.

641. LEE, K. H., and T. A. GEISSMAN: Sesquiterpene Lactones of *Artemisia* Species. Artefransin from *A. franserioides*. Phytochem. **10**, 205 (1971).

642. — — Sesquiterpene Lactones of *Artemisia*. Constituents of *A. ludoviciana* ssp. *mexicana*. Phytochem. **9**, 403 (1970).

643. LEE, K. H., M. HARUNA, H. C. HUANG, B. S. WU, and I. H. HALL: Helenalin, an Antitumor Principle from *Anaphalis morrisonicola* Hay. J. Pharm. Sci. **66**, 1194 (1977); Chem. Abstr. **87**, 145893s (1977).

644. LEE, K. H., H. C. HUANG, E. S. HUANG, and H. FURUKAWA: Eupatolide, a New Cytotoxic Principle from *Eupatorium formosanum*. J. Pharm. Sci. **61**, 629 (1972); Chem. Abstr. **77**, 28810c (1972).

354   N. H. Fischer, E. J. Olivier, and H. D. Fischer:

645. Lee, K. H., T. Ibuka, H. C. Huang, and D. L. Harris: Molephantinin, a New Potent Antitumor Sesquiterpene Lactone from *Elephantopus mollis*. J. Pharm. Sci. **64,** 1077 (1975); Chem. Abstr. **83,** 172545 (1975).

646. Lee, K. H., T. Ibuka, M. Kozuka, A. T. McPhail, and K. D. Onan: The Structure and Absolute Configuration of Florilenalin, A New Cytotoxic Guaianolide from Florida *Helenium autumnale* L. Tetrahedron Letters **1974,** 2287.

647. Lee, K. H., T. Ibuka, A. T. McPhail, K. D. Onan, T. A. Geissman, and T. G. Waddell: The Structure and Absolute Configuration of Plenolin, A Cytotoxic Sesquiterpene Lactone. Tetrahedron Letters **1974,** 1149.

648. Lee, K. H., Y. Imakura, and D. Sims: Structure and Stereochemistry of Microhelenin-A, a New Antitumor Sesquiterpene Lactone from *Helenium microcephalum*. J. Pharm. Sci. **65,** 1410 (1976).

649. Lee, K. H., Y. Imakura, D. Sims, A. T. McPhail, and K. D. Onan: Structure and Stereochemistry of Microlenin, a Novel Antitumor Dimeric Sesquiterpene Lactone from *Helenium microcephalum;* X-Ray Crystal Structure. Chem. Commun. **1976,** 341.

650. — — — — — Antitumor Sesquiterpene Lactones from *Helenium microcephalum:* Isolation of Mexicanin-E and Structural Characterization of Microhelenin-B and -C. Phytochem. **16,** 393 (1977).

651. Lee, K. H., T. Kimura, M. Haruna, A. T. McPhail, K. D. Onan, and H. C. Huang: Structure and Stereochemistry of Eupaformosanin, a New Antileukemic and Antisarcoma Germacranolide from *Eupatorium formosanum*. Phytochem. **16,** 1068 (1977).

652. Lee, K. H., T. Kimura, M. Okamoto, C. M. Cowherd, A. T. McPhail, and K. D. Onan: The Structure and Stereochemistry of Eupahyssopin, a New Antitumor Germacranolide from *Eupatorium hyssopifolium*. Tetrahedron Letters **1976,** 1051.

653. Lee, K. H., S. Matsueda, and T. A. Geissman: Sesquiterpene Lactones of Artemisia: New Guaianolides from Fall Growth of *A. douglasiana*. Phytochem. **10,** 405 (1971).

654. Lee, K. H., R. Meck, C. Piantadosi, and E. S. Huang: Cytotoxicity and In Vivo Activity of Helenalin Esters and Related Derivatives. J. Med. Chem. **16,** 299 (1973); Chem. Abstr. **78,** 154756y (1973).

655. Lee, K. H., R. F. Simpson, and T. A. Geissman: Sesquiterpenoid Lactones of *Artemisia*. Constituents of *Artemisia cana* ssp. *cana*. The Structure of Canin. Phytochem. **8,** 1515 (1969).

656. Leppard, D. C., M. Rey, A. S. Dreiding, and R. Grieb: Structure of Arteannuin B and its Acid Hydrolysis Product. Helv. Chim. Acta **57,** 602 (1974); Chem. Abstr. **81,** 49864w (1974).

657. L'Homme, M. F., T. A. Geissman, H. Yoshioka, T. H. Porter, W. Renold, and T. J. Mabry: The Structure of Chamissonin. Tetrahedron Letters **1969,** 3161.

657a. Linde, H. H. A., and M. S. Ragab: Components of *Achillea santolina* I. Stereochemistry of Deacetoxymatricarin. Helv. Chim. Acta **50,** 1961 (1967); Chem. Abstr. **68,** 13189x (1968).

658. Lucas, R. A., S. Rovinski, R. J. Kiesel, L. Dorfman, and H. B. MacPhillamy: A New Sesquiterpene Lactone with Analgesic Activity from *Helenium amarum*. J. Organ. Chem. (U.S.A.) **29,** 1549 (1964).

659. Lucas, R. A., R. G. Smith, and L. Dorfman: The Isolation of Dihydromexicanin E from *Helenium autumnale* L. J. Organ. Chem. (U.S.A.) **29,** 2101 (1964).

660. Lugovskaya, S. A., N. V. Plekhanova, and K. Y. Orovbaev: Alantolactone from *Inula grandis*. Khim. Prir. Soedin. **1976,** 110.

661. — — — Alantolactone and Isoalantolactone from *Inula grandis* Chrenk. Izv. Akad. Nauk. Kirg. SSR **1976,** 54; Chem. Abstr. **86,** 167866z (1977).

662. Mabry, T. J.: Intraspecific Variation of Sesquiterpene Lactones in *Ambrosia* (Compositae). Applications to Evolutionary Problems at the Populational Level. Phytochem. Phylogeny. Proc. Phytochem. Soc. Symp. **1969,** 269; Chem. Abstr. **76,** 96924x (1972).

663. — Chemistry of Geographical Races. Pure Appl. Chem. **34,** 377 (1973).

663a. MABRY, T. J., Z. ABDEL-BASET, W. G. PADOLINA, and S. B. JONES: Systematic Implications of Flavonoids and Sesquiterpene Lactones in Species of *Vernonia.* Biochem. System. Ecol. **2,** 185 (1975).

664. MABRY, T. J., H. B. KAGAN, and H. E. MILLER: Psilostachyin B, a New Sesquiterpene Dilactone from *Ambrosia psilostachya* DC. Tetrahedron **22,** 1943 (1966).

665. MABRY, T. J., H. E. MILLER, H. B. KAGAN, and W. RENOLD: The Structure of Psilostachyin, a New Sesquiterpene Dilactone from *Ambrosia psilostachyia.* Tetrahedron **22,** 1139 (1966).

666. MABRY, T. J., W. RENOLD, H. E. MILLER, and H. B. KAGAN: The Structure of Ambrosiol. A New Sesquiterpene Lactone from *Ambrosia psilostachyia.* J. Organ. Chem. (U.S.A.) **31,** 681 (1966).

667. MACAIRA, L. A., M. GARCIA, and J. A. RABI: Chemical Transformations of Abundant Natural Products. Modifications of Eremanthin Leading to Other Naturally Occurring Guaianolides. J. Organ. Chem. (U.S.A.) **42,** 4207 (1977).

668. MAGNUSSON, G., and S. THOREN: The Stereostructure of Two Sesquiterpene Lactones from *Lactarius.* Acta Chem. Scand. **27,** 2396 (1973); Chem. Abstr. **80,** 37313g (1974).

669. — — Two Sesquiterpene Lactones from *Lactarius.* Acta Chem. Scand. **27,** 1573 (1973); Chem. Abstr. **79,** 126650m (1973).

670. — — Structure of Lactaral, a New Sesquiterpene Furan-3-Aldehyde from *Lactarius,* by Spectroscopic Methods. Tetrahedron **30,** 1431 (1974).

671. MARSHALL, J. A., N. COHEN, and K. R. ARENSON: Synthesis of 4-Demethyltetrahydroalantolactone. J. Organ. Chem. (U.S.A.) **30,** 762 (1965).

672. MARSHALL, J. A., and R. H. ELLISON: The Stereoselective Total Synthesis of Pseudoguaianolides: Confertin. J. Amer. Chem. Soc. **98,** 4312 (1976).

673. MARTIN, S. S., J. H. LANGENHEIM, and E. ZAVARIN: Biosynthesis of Sesquiterpenes in *Hymenaea* Inferred from their Quantitative Co-Occurrence. Phytochem. **15,** 113 (1976).

674. MARTIN, D. G., G. SLOMP, S. MIZSAK, D. J. DUCHAMP, and C. G. CHIDESTER: The Structure and Absolute Configuration of Pentalenolactone (PA 132). Tetrahedron Letters **1970,** 4901.

675. MARTIN-SMITH, M., P. DE MAYO, S. J. SMITH, J. B. STENLAKE, and W. D. WILLIAMS: Revised Structure of Aristolactone. Tetrahedron Letters **1964,** 2391.

676. MARUYAMA, M., and S. OMURA: Carpesiolin from *Carpesium abrotanoides.* Phytochem. **16,** 782 (1977).

676a. MARUYAMA, M., and A. KARUBE: Phytochemical Reports. Phytochem. **15,** 2026 (1976).

677. MARUYAMA, M., and F. SHIBATA: Stereochemistry of Granilin Isolated from *Carpesium abrotanoides.* Phytochem. **14,** 2247 (1975).

678. MARX, J. N., and S. M. McGAUGHEY: Synthesis of Dihydroarbiglovin and Stereochemistry of Arbiglovin. Tetrahedron **28,** 3583 (1972).

679. MARX, J. N., and E. H. WHITE: The Stereochemistry and Synthesis of Achillin. Tetrahedron **25,** 2117 (1969).

680. MASSY-WESTROPP, R. A., G. D. REYNOLDS, and T. M. SPOTSWOOD: Freelingyne, an Acetylenic Sesquiterpenoid. Tetrahedron Letters **1966,** 1939.

681. MATHUR, S. B.: Composition of Punjab Costus Root Oil. Phytochem. **11,** 449 (1972).

682. MATHUR, S. B., and C. M. FERMIN: Terpenes of *Mikania mongenansis.* Phytochem. **12,** 226 (1973).

683. MATHUR, S. B., P. GARCIA TELLO, C. MARCANO FERMIN, and V. MORA-ARELLANO: Terpenoids of *Mikania monagasensis* and their Biological Activities. Rev. Latinoamer. Quim. **6,** 201 (1975).

684. MATHUR, S. B., S. V. HIREMATH, G. H. KULKARNI, G. R. KELKAR, and S. C. BHATTACHARYYA: Terpenoids — LXX Structure of Dehydrocostus Lactone. Tetrahedron **21,** 3575 (1965).

356   N. H. Fischer, E. J. Olivier, and H. D. Fischer:

685. Matsueda, S.: Chemical Constitution of Diploid and Polyploid *Ambrosia*. Sci. Rep. Hirooaki Univ. **16**, 51 (1969); Chem. Abstr. **73**, 63223c (1970).
686. — Chemical Constitution of the Young Growth of *Artemisia* Species. Yakugaku Zasshi. **90**, 1024 (1970); Chem. Abstr. **73**, 127783 v (1970).
687. — Guaiane Type Sesquiterpene Lactone Formation from Germacranolide by Chemical Procedure. Sci. Rep. Hirosaki Univ. **18**, 8 (1971); Chem. Abstr. **76**, 14739 d (1972).
688. — Mass Spectra of Pseudoguaiane-Sesquiterpene Lactones. Yakugaku Zasshi **92**, 905 (1972); Chem. Abstr. **77**, 101921 w (1972).
689. Matsueda, S., and T. A. Geissman: Sesquiterpene Lactones of *Artemisia* Species. III. Arglanine from *Artemisia douglasiana* Bess. Tetrahedron Letters **1967**, 2013.
690. — — Sesquiterpene Lactones of *Artemisia* Species. IV. Douglanine from *Artemisia douglasiana* Bess. Tetrahedron Letters **1967**, 2159.
691. — — Stereochemistry of Sesquiterpene Lactones of the Eudesmane Type. Yakugaku Zasshi **91**, 59 (1971); Chem. Abstr. **74**, 88153 p (1971).
692. Matsueda, S., A. Karube, and S. Sasaki: Mass Spectra of Eudesmane Sesquiterpene Lactones. Yakugaku Zasshi **97**, 1046 (1977); Chem. Abstr. **88**, 23173 q (1978).
693. Matsueda, S., and Y. Otaki: Chemical Composition of Mature Plants of the *Artemisia* Species. Yakugaku Zasshi **90**, 1140 (1970); Chem. Abstr. **74**, 1108 f (1971).
694. Matsuura, T., Y. Sata, K. Ogura, and M. Mori: A Novel Photorearrangement of Santonin in the Solid State. Tetrahedron Letters **1968**, 4627.
695. Mazur, Y., and A. Meisels: The Structure of Arborescine, a New Sesquiterpene from *Artemisia arborescens* L. Chem. and Ind. **1956**, 492.
696. McClure, R. J., G. A. Sim, P. Coggon, and A. T. McPhail: Conformations of Ten-Membered Ring Sesquiterpenes by X-Ray Crystallography. Chem. Commun. **1970**, 128.
697. McMillan, C., P. I. Chavez, and T. J. Mabry: Sesquiterpene Lactones of *Xanthium strumarium* in a Texas Population and in Experimental Hybrids. Biochem. System. Ecol. **3**, 137 (1975); Chem. Abstr. **84**, 56625 j (1976).
697a. McMillan, C., P. I. Chavez, S. G. Plettman, and T. J. Mabry: Systematic Implications of the Sesquiterpene Lactones in the "Strumarium" Morphological Complex *(Xanthium strumarium,* Asteraceae) of Europe, Asia and Africa. Biochem. System. Ecol. **1975**, 181.
698. McMurry, T. B. H., and D. F. Rane: Chemistry of Pyrosantonin. J. Chem. Soc. C **1971**, 1389.
699. — — Action of Phosphorus Trihalides on the Santonins, Santonene and Related Compounds. J. Chem. Soc. C **1971**, 3851.
700. McMurry, T. B. H., D. F. Rane, and S. G. Traynor: 1-Nordesmotroposantonins. Chem. and Ind. (London) **1971**, 658.
701. McMurry, T. B. H., and R. R. Talekar: The Chemistry of Santonene. Photolysis of 4-acetoxysantonene and its 4-epimer: an Example of Photodecarboxylation of Acetates. J. Chem. Soc. Perkin Trans. I **1976**, 442.
702. McPhail, A. T., P. A. Luhan, K. H. Lee, H. Furukawa, R. Meck, C. Piantadosi, and T. Shingu: Crystal Structure of Carolenalin Monoacetate: Revision of the Stereochemistry of Carolenalin and Carolenin. Tetrahedron Letters **1973**, 4087.
703. McPhail, A. T., R. W. Miller, B. Mompon, and R. Toubiana: Structure and Stereochemistry of Vernodesmine, a Novel Phenyl Bearing Sesquiterpene Lactone from *Vernonia pectoralis* Baker. Tetrahedron Letters **1975**, 3675.
704. McPhail, A. T., and K. D. Onan: Crystal and Molecular Structure of Eupaformonin, a Cytotoxic Germacranolide from *Eupatorium formosanum* Hay. J. Chem. Soc. Perkin Trans II **1976**, 578.
705. — — Structure and Absolute Configuration of Florilenalin. X-Ray Analysis of 4-O-acetyl-2-O-(p-iodobenzoyl) Florilenalin. J. Chem. Soc. Perkin Trans. II **1975**, 492.

706. — — X-Ray Determination of the Structure and Conformation of an Oxide from Helenalin. J. Chem. Soc. Perkin Trans. II **1975**, 496.

707. — — Crystal and Molecular Structure of Carolenalone, a Cycloheptenone Sesquiterpene Lactone from North Carolina *Helenium autumnale* L. J. Chem. Soc. Perkin Trans. II, **1976**, 332.

708. — — Crystal and Molecular Structure of Eupatolide, the Major Cytotoxic Principle from *Eupatorium formosanum*. J. Chem. Soc. Perkin Trans. II **1975**, 1798.

709. — — Structure and Absolute Configuration of Plenolin. X-Ray Analysis of Plenolin p-iodobenzoate. J. Chem. Soc. Perkin Trans. II **1975**, 487.

710. McPHAIL, A. T., K. D. ONAN, H. FURUKAWA, and K. H. LEE: Structure and Stereochemistry of Carolenalone, a New Sesquiterpene Lactone from North Carolina *Helenium autumnale* L. Tetrahedron Letters **1975**, 1229.

711. McPHAIL, A. T., K. D. ONAN, K. H. LEE, H. FURUKAWA, S. H. KIM, and C. PIANTADOSI: Transformations of Helenalin: Crystal Structure Analysis of the Oxide from the Reaction with Hydrogen Chloride. Tetrahedron Letters **1973**, 4641.

712. McPHAIL, A. T., K. D. ONAN, K. H. LEE, T. IBUKA, and H.-CH. HUANG: Structure and Stereochemistry of Eupaformonin, a Novel Cytotoxic Sesquiterpene Lactone from *Eupatorium formosanum* Hay. Tetrahedron Letters **1974**, 3203.

713. McPHAIL, A. T., K. D. ONAN, K. H. LEE, T. IBUKA, M. KOZUKA, T. SHINGU, and H. C. HUANG: Structure and Stereochemistry of the Epoxide of Phantomolin, a Novel Cytotoxic Sesquiterpene Lactone from *Elephantopus mollis*. Tetrahedron Letters **1974**, 2739.

714. McPHAIL, A. T., and G. A. SIM: Conformations of Cycloheptane and a-Methylene-γ-Lactone Rings in Sesquiterpenoids: X-Ray Determination of the Constitution, Absolute Stereochemistry, and Conformation of Euparotin Bromoacetate. Tetrahedron **29**, 1751 (1973).

715. — — Constitution and Absolute Stereochemistry of Elephantol: X-Ray Analysis of Elephantol p-Bromobenzoate. J. Chem. Soc. Perkin Trans. II **1972**, 1313.

716. — — X-Ray Determination of the Structure of Vernolepin p-Bromobenzene sulphonate. J. Chem. Soc. (London) **1971**, 198.

716a. — — Fungal Metabolites. Part VIII. Sesquiterpenoids. Part VII. The Structure of Fomannosin: X-ray Analysis of Dihydrofomannosia *p*-Bromobenzoylurethane. J. Chem. Soc. **B 1968**, 1104.

717. MILLER, R. B., and E. S. BEHARE: Stereoselective Synthesis of Sesquiterpene Lactones. Total Synthesis of (±)-Isotelekin. J. Amer. Chem. Soc. **96**, 8102 (1974).

718. MILLER, H. E., and T. J. MABRY: 3-Hydroxydamsin, a New Pseudoguaianolide from *Ambrosia psilostachya* (Compositae). J. Organ. Chem. **32**, 2929 (1967).

718a. MILLER, H. E., T. J. MABRY, B. L. TURNER, and W. W. PAYNE: Infraspecific Variation of Sesquiterpene Lactones in *Ambrosia psilostachya*. Amer. J. Bot. **55**, 316 (1968).

719. MILLER, R. B., and R. D. NASH: Highly Stereoselective Total Synthesis of (±)-Isoalantolactone. Tetrahedron **30**, 2961 (1974).

720. MILLS, R. W., and T. MONEY: Sesquiterpenoids. In: Terpenoids and Steroids, Specialist Periodical Reports (OVERTON, K. H., ed.), Vol. 4, p. 77. London: The Chemical Society. 1974.

721. MINATO, H., and I. HORIBE: Total Synthesis of (±)-Carabrone. J. Chem. Soc. (London) **1968**, 2131.

722. — — Structure and Stereochemistry of Xanthumin, a Stereoisomer of Xanthinin. J. Chem. Soc. (London) **1965**, 7009.

723. — — A New Synthetic Method for α-Methylene-γ-Lactones; Total Synthesis of (±)-Isoalantolactone. J. Chem. Soc. C (London) **1967**, 1575.

724. MINATO, H., S. NOSAKA, and I. HORIBE: The Structure of Carabrone, a New Component of *Carpesium abrotanoides*, Linn. J. Chem. Soc. (London) **1964**, 5503.

725. Mincione, E., and C. Iavarone: Terpenes of the *Commiphora myrra*. Chim. Ind. **54**, 525 (1972); Chem. Abstr. **77**, 111467s (1972).

725a. Mitchell, D. R., and R. O. Asplund: Pulchellins C and E in *Gaillardia aristata*. Phytochem. **12**, 2541 (1973).

726. Miyano, K., Y. Ohfune, S. Azuma, and T. Matsumoto: Synthetic Approach to Fomannosin. Tetrahedron Letters **1974**, 1545.

727. Mnatsakanyan, V. A., and L. V. Revazova: Tamirine from *Tanacetum chiliophyllum*. Khim. Prir. Soedin. **1974**, 396; Chem. Abstr. **82**, 13988w (1975).

728. — — Phytochemical Analysis of *Tanacetum myriocophyllum*. II. Structure of Tanamyrin. Arm. Khim. Zh. **27**, 208 (1974); Chem. Abstr. **81**, 91747k (1974).

729. — — Phytochemical Analysis of the Plant *Tanacetum myriophyllum*. I. Sesquiterpenic Lactones. Arm. Khim. Zh. **26**, 914 (1973); Chem. Abstr. **80**, 121134g (1974).

730. Moiseeva, G. P., B. Akyev, S. Z. Kasymov, and G. P. Sidyakin: The Structure of the Lactones Arsanin and Arsantin. Khim. Prir. Soedin. **9**, 167 (1973); Engl. edit.: p. 162; Chem. Abstr. **79**, 5470u (1973).

731. Mompon, B., C. M. Ho, and R. Toubiana: Sur la Structure du Pectorolide, Nouvelle Lactone Sesquiterpenique Isolee du *Vernonia pectoralis* Baker. C. R. Acad. Sci. Paris, Series C **276**, 1799 (1973); Chem. Abstr. **79**, 105430z (1973).

732. Mompon, B., G. Massiot, and R. Toubiana: Structure of Subluteolide, a New Guaianolide Isolated from *Vernonia sublutea* (Compositae). C. R. Hebd. Seances Acad. Sci. Ser. C **279**, 907 (1974); Chem. Abstr. **82**, 171223x (1975).

733. Mompon, B., and R. Toubiana: Lactones Sesquiterpeniques Du *Vernonia Pectoralis* Baker (Comp.). Stereochimie Du Pectorolide, et Structure Des Vernopectolides-A Et B. Tetrahedron **32**, 2545 (1976).

734. — — Configuration Du Subluteolide; Nouveau Guaianolide Isole Du *Vernonia sublutea* Scott Elliot (Composees). Tetrahedron **33**, 2199 (1977).

735. Money, T.: Sesquiterpenoids. In Terpenoids and Steroids (Overton, K. H., ed.), Vol. 5, p. 46. London: The Chemical Society. 1975.

736. Morikawa, K., and Y. Hirose: Thermal Isomerization Between (−)-δ-Elemol and (+)-Epi-δ-Elemol. Tetrahedron Letters **1969**, 869.

737. Morikawa, K., Y. Hirose, and S. Nozoe: Biosynthesis of Germacrene-C. Tetrahedron Letters **1971**, 1131.

738. Morimoto, H., Y. Sanno, and H. Oshio: Chemical Studies on Heliangine. A New Sesquiterpene Lactone Isolated from the Leaves of *Helianthus tuberosus* L. Tetrahedron **22**, 3173 (1966).

739. Moriyama, Y., and T. Takahashi: New Sesquiterpene Lactones of the Eremophilane-Type from *Ligularia fauriei* (Fr.) Koidz. Bull. Chem. Soc. Japan **1976**, 3196; Chem. Abstr. **86**, 140267t (1977).

740. — — New Eremophilane-Type Lactones from *Ligularia fauriei* (Fr.) Koidz. Chem. Pharm. Bull. **24**, 360 (1976); Chem. Abstr. **84**, 147721d (1976).

741. Moss, G. P., P. S. Pregosin, and E. W. Randall: Assignments in the Carbon-13 Fourier Spectra of Eudesmanolides. J. Chem. Soc. Perkin Trans. I **1974**, 1525.

742. Motl, O., and B. Drozdz: Components of *Libanotis intermedia* Rupr. Fruits. Collect. Czech. Chem. Comm. **42**, 2815 (1977).

743. Mukhametzhanov, M. N., A. I. Shreter, and D. Pakalns: Stizolycin from *Centaurea solstilialis*. Khim. Prir. Soedin. **5**, 590 (1969); Chem. Abstr. **73**, 73825q (1970).

744. Mukhametzhanov, M. N., V. I. Sheichenko, A. I. Ban'kovskii, and K. S. Rybalko: Structure of Stizolin — A Sesquiterpene Lactone from *Stizolophus balsamita*. Khim. Prir. Soedin. **7**, 405 (1971); Engl. edit.: p. 386; Chem. Abstr. **75**, 151925e (1971).

745. — — — — Structure of the Sesquiterpene Lactone Stizolicin. Khim. Prir. Soedin. **6**, 505 (1970); Engl. edit.: p. 525.

746. MUKHAMETHANOV, M. N., V. I. SHEICHENKO, K. S. RYBALKO, and K. I. BORYAEV: Isolation of Grosshemin from *Chartolepis intermedia*. Khim. Prir. Soedin. 5, 184 (1969); Chem. Abstr. 72, 12898y (1970).

747. MUKHAMETHANOV, M. N., V. I. SHEICHENKO, K. S. RYBALKO, and D. A. PAKALN: Stizolicin — A New Sesquiterpene Lactone from *Stizolophus coronopifolius*. Khim. Prir. Soedin. 5, 125 (1969); Engl. edit.: p. 108.

748. NAGAKURA, I., S. MAEDA, M. UENO, F. MAKOTO, and Y. KITAHARA: Eremophilanes: Total Syntheses of (±)-Eremophilenolide and (±)-Furanoeremophilane. Chemistry Letters 1975, 1143.

749. NAIR, M. S. R., H. TAKESHITA, T. C. McMORRIS, and M. ANCHEL: Metabolites of *Clitocybe illudens*. Illudalic Acid, a Sesquiterpenoid, and Illudinine, a Sesquiterpenoid Alkaloid. J. Organ. Chem. (U.S.A.) 34, 240 (1969).

749a. NAKAMURA, H., T. OHTA, and G. FUKUCHI: Constituent of Santonin-free Wormseed. J. Pharm. Soc. Japan 53, 1265 (1933).

750. NAKANISHI, K., K. HABAGUCHI, Y. NAKADAIRA, M. C. WOODS, M. MARUYAMA, R. T. MAJOR, M. ALAUDDIN, A. R. PATEL, K. WEINGES, and W. BAEHR: Structure of Bilobalide, a Rare Tert-butyl Containing Sesquiterpenoid Related to the C_{20}-ginkgolides. J. Amer. Chem. Soc. 93, 3544 (1971).

751. NAKAZAKI, M., and K. NAEMURA: Total Synthesis of (−)-Artemisin. Bull. Chem. Soc. Japan 42, 3366 (1969); Chem. Abstr. 72, 55677t (1970).

752. NANAO, H.: Sesquiterpenoids in the Essential Oil of *Neolitsea sericea*. J. Sci. Hiroshima Univ., Ser. A-2 33, 107 (1969); Chem. Abstr. 72, 70563w (1970).

752a. NANO, G. M., A. MARTELLI, and P. SANCIN: Investigations of *Artemisia* Grown in Piedmont. Riv. Ital. Essenze-Profumi, Piante Offic. Aromi-Saponi Cosmet.-Aerosol. 98, 409 (1966); Chem. Abstr. 66, 31930v (1967).

753. NARAYANAN, C. R., and N. K. VENKATASUBRAMANIAN: Simple Methods to Find the Stereochemistry of the Side Chain of γ-Lactones. J. Organ. Chem. (U.S.A.) 33, 3156 (1968).

754. — — Stereochemical Studies by PMR Spectroscopy V — a Simple Method to Find the Stereochemistry of the Side Chain of γ-Lactones. Tetrahedron Letters 1966, 5865.

755. NAYA, K., M. HAYASHI, I. TAKAGI, S. NAKAMURA, and M. KOBAYASHI: The Structural Elucidation of Sesquiterpene Lactones from *Petasites japonicus* Maxim. Bull. Chem. Soc. Japan 45, 3673 (1972); Chem. Abstr. 78, 124744a (1973).

756. NAYA, K., R. KANAZAWA, and M. SAWADA: The Photosensitized Oxygenation of Furanoeremophilanes. I. The Isomeric Hydroperoxides from Petasalbin and Their Transformations to Lactones. Bull. Chem. Soc. Japan 48, 3220 (1975); Chem. Abstr. 84, 90329a (1976).

757. NAYA, K., M. KAWAI, M. NAITO, and T. KASAI: Structures of Eremofukinone, 9-Acetoxyfukinanolide, and S-japonin from *Petasites japonicus*. Chem. Letters 1972, 241.

758. NAYA, K., Y. MIYOSHI, H. MORI, K. TAKAI, and M. NAKANISHI: The Sesquiterpenes of *Cacalia* Species. Chem. Letters 1976, 73.

759. NAYA, K., N. NOGI, Y. MAKIYAMA, H. TAKASHINA, and T. IMAGAWA: The Preparation and Stereochemistry of the Isomeric Hydroperoxides and the Corresponding Lactones from Furanofukinin and Furanoeremophilane. Bull. Chem. Soc. (Japan) 50, 3002 (1977).

760. NAYA, K., I. TAKAGI, M. HAYASHI, S. NAKAMURA, M. KOBAYASHI, and S. KATSUMURA: Structures of New Skeletal Sesquiterpenoids from *Petasites japonicus* Maxim. Chem. and Ind. 1968, 318.

761. NAYA, K., I. TAKAGI, Y. KAWAGUCHI, Y. ASADA, Y. HIROSE, and N. SHINODA: The Structure of Fukinone, a Constituent of *Petasites japonicus* Maxim. Tetrahedron 24, 5871 (1968).

761a. NAZARENKO, M. V.: Proazulenes from Pollen of *Artemisia macrophylla.* Zh. Prikl. Khim. **34,** 1633 (1961); Chem. Abstr. **55,** 27244 (1961).

762. — Guaienolides of Sievers Wormwood, *Artemisia sieversiana.* Zh. Prikl. Khim. **38,** 2372 (1965); Chem. Abstr. **64,** 3608 g (1966).

762a. NAZARENKO, M. V., and L. I. LEONTEVA: Sieversinin, A New Sesquiterpenic γ-Lactone. Khim. Prir. Soedin. **2,** 399 (1966); Chem. Abstr. **67,** 32803 p (1967).

763. NEGRON, G., L. RODRIGUEZ-HAHN, and J. ROMO: Una Reaccion Anormal De Mexicanina E Con N-Bromosuccinimida. (Abnormal reaction of mexicanin E with N-bromosuccinimide.) Rev. Latinoamer. Quim. **5,** 116 (1974).

764. NEIDLE, S., and D. ROGERS: X-Ray Determination of the Structure and Absolute Configuration of a Novel Sesquiterpenoid, Melampodin. Chem. Commun. **1972,** 140.

765. NES, W. R., and M. L. MCKEAN: Biochemistry of Steroids and Other Isopentenoids. Baltimore, London, Tokyo: University Park Press. 1977.

766. NESHTA, I. D., and N. A. KALOSHINA: Leucomisin from *Achillea cartilaginae.* Khim. Prir. Soedin. **8,** 652 (1972); Engl. edit.: p. 625.

767. NESHTA, I. D., K. S. RYBALKO, O. A. KONOVALOVA, and O. F. IVANENKO: Lactones of *Achillea micrantha* and *A. vermicularis.* Khim. Prir. Soedin. **1976,** 395; Chem. Abstr. **85,** 106653 k (1976).

768. NIKONOVA, L. P.: Chemosystematic Study of *Inula* Species of the Corvisartia (Merat) Dumort Section. Rastit. Resur. **12,** 252 (1976); Chem. Abstr. **85,** 30636j (1976).

769. — Lactones of *Inula magnifica.* Khim. Prir. Soedin. **9,** 558 (1973); Engl. edit.: p. 528; Chem. Abstr. **80,** 45620 v (1974).

770. NIKONOVA, L. P., and G. K. NIKONOV: Granilin — A New Lactone from *Inula grandis.* Khim. Prir. Soedin. **8,** 289 (1972); Engl. edit.: p. 286; Chem. Abstr. **78,** 2009 u (1973).

771. — — Grandulin and Grandicin. New Sesquiterpene Lactones from *Inula grandis.* Khim. Prir. Soedin. **6,** 133 (1970); Chem. Abstr. **73,** 117175j (1970).

772. — — The Structure of Grandicin and Grandulin. Khim. Prir. Soedin. **8,** 679 (1972).

773. — — Igalan, a New Sesquiterpene Lactone from *Inula grandis.* Khim. Prir. Soedin. **6,** 508 (1970).

774. — — Sesquiterpene Lactones from *Ligularia macrophylla* and *L. thomsonii.* Khim. Prir. Soedin. **6,** 742 (1976); Chem. Abstr. **86,** 1030556 (1977).

775. — — Sesquiterpenes of the Roots of *Inula grandis.* Khim. Prir. Soedin. **7,** 17 (1971); Engl. edit.: p. 14.

776. NISHIZAWA, M., K. KAMIYA, A. TAKABATAKE, H. OSHIO, Y. TOMIIE, and I. NITTA: The X-Ray Analysis of Dihydroheliangine Monochloroacetate. Tetrahedron **22,** 3601 (1966).

776a. NISHIZAWA, M., P. A. GRIECO, S. D. BURKE, and W. METZ: Structure and Total Synthesis of Temisin. J. Chem. Soc. Chem. Comm. **1978,** 76.

777. NOVIKOV, V. I., F. N. FOROSTYAN, and D. P. POPA: A Sesquiterpene Lactone of *Cyclachaena xanthifolia* Anhydrocoronopilin. Khim. Prir. Soedin. **6,** 29 (1970); Chem. Abstr. **73,** 45615 c (1970).

778. — — — Structure of the Sesquiterpene Lactones from *Cyclachaena xanthifolia.* Khim. Prir. Soedin. **5,** 487 (1969); Chem. Abstr. **73,** 15019 u (1970).

779. NOVOTNÝ, L., and V. HEROUT: Constituents of Rhizoms of *Petasites japonicus* (Sieb. et Zucc.) Maxim Ssp. *giganteus* Kitamura. Collect. Czech. Chem. Comm. **30,** 3579 (1965).

780. — — unpublished.

781. — — The Composition of *Artemisia sieversiana* Willd. Collect. Czech. Chem. Comm. **27,** 1508 (1962).

782. NOVOTNÝ, L., V. HEROUT, and F. SORM: Constitution of Petasalbine, Albopetasine and Hydroxyeremophilenolide, the Components of *Petasites albus* L. Rhizomes. Collect. Czech. Chem. Comm. **29,** 2189 (1964).

783. — — — Substances from *Petasites officinalis* Moench. and *Petasites albus* (L.) Gaertn. Tetrahedron Letters **1961**, 697.

784. — — — A Contribution to the Structure of Absinthin and Anabsinthin. Collect. Czech. Chem. Comm. **25**, 1492 (1960).

785. — — — A Contribution to Stereochemistry of Absinthin and Artabsin. Collect. Czech. Chem. Comm. **25**, 1500 (1960).

786. — — — Constitution of Petasitolides and S-Petasitolides. Collect. Czech. Chem. Comm. **29**, 2182 (1964).

787. NOVOTNÝ, L., J. JIZBA, V. HEROUT, and F. SORM: The Constituents of Coltsfoot Rhizomes *(Petasites officinalis* Moench). Collect. Czech. Chem. Comm. **27**, 1393 (1962).

788. — — — Constitution and Absolute Configuration of Eremophilenolide. Tetrahedron **19**, 1101 (1963).

789. NOVOTNÝ, L., K. KOTVA, J. TOMAN, and V. HEROUT: Sesquiterpenes from *Petasites*. Phytochem. **11**, 2795 (1972).

790. NOVOTNÝ, L., Z. SAMEK, and F. SORM: Isolation and Structure of Dimethoxydihydrofuroeremophilane. Collect. Czech. Chem. Comm. **31**, 371 (1966).

791. OGURA, H., and H. TAKAYANAGI: Mass Spectra of 2- and 14-Chlorosantonin. Shitsuryo Bunseki **17**, 665 (1969); Chem. Abstr. **73**, 66739 (1970).

792. OGURA, H., H. TAKAYANAGI, A. YOSHINO, and T. OKAMOTO: Chlorinated α-Santonins. Chem. Pharm. Bull. **22**, 1433 (1974); Chem. Abstr. **81**, 169642u (1974).

793. OHNO, N., H. HIRAI, H. YOSHIOKA, X. A. DOMINGUEZ, and T. J. MABRY: Cynaropicrin, a Sesquiterpene Lactone from *Centaurea americana*. Phytochem. **12**, 221 (1973).

794. OHTA, Y., N. H. ANDERSEN, and C. B. LIU: Sesquiterpene Constituents of Two Liverworts of Genus *Diplophyllum*. Novel Eudesmanolides and Cytotoxicity Studies for Enantiomeric Methylene Lactones. Tetrahedron **33**, 617 (1977).

795. OKAMOTO, T., M. NATSUME, T. ONAKA, F. UCHIMARU, and M. SHIMIZU: The Structure of Dendroxine, the Third Alkaloid from *Dendrobium nobile*. Chem. Pharm. Bull. **14**, 672 (1966); Chem. Abstr. **65**, 10633a (1966).

796. — — — — — The Structure of Dendramine (6-Hydroxydendrobine) and 6-Hydroxydendroxine, the Fourth and Fifth Alkaloid from *Dendrobium nobile*. Chem. Pharm. Bull. **14**, 676 (1966); Chem. Abstr. **65**, 10633c (1966).

797. OKIGAWA, M., and N. KAWANO: The Structure of Pseudoanisatin. Tetrahedron Letters **1971**, 75.

797a. OKUDA, T.: Components of *Coriaria japonica* XIII. Pharm. Bull. Japan **2**, 185 (1954).

798. OKUDA, T., and T. YOSHIDA: Structure of Corianin. Tetrahedron Letters **1971**, 4499.

799. — — Structure of Coriamyrtin. Tetrahedron Letters **1964**, 439; errata **1964**, 694.

800. OLIVIER, E. J., and N. H. FISCHER: unpublished.

801. ONAN, K. D.: Structural and Conformational Studies by X-Ray Diffraction Methods. I. Phosphorus and Sulfur Heterocycles. II. Sesquiterpene Lactones. Diss. Abstr. Int. B 1976, **36**, 3413; Chem. Abstr. **84**, 105821f (1976).

802. OSAWA, T., A. SUZUKI, S. TAMURA, Y. OHASHI, and Y. SASADA: Structure of Chlorochrymorin, a Novel Sesquiterpene Lactone from *Chrysanthemum morifolium*. Tetrahedron Letters **1973**, 5135.

803. OSAWA, T., D. TAYLOR, A. SUZUKI, and S. TAMURA: Revised Structure and Stereochemistry of Chrysartemin B. Tetrahedron Letters **1977**, 1169.

804. ORTEGA, A., C. GUERRERO, and J. ROMO: Estructura de la Zexbrevina D. Rev. Latinoamer. Quim. **4**, 91 (1973).

805. ORTEGA, A., C. GUERRERO, A. ROMO DE VIVAR, J. ROMO, and A. PALAFOX: La Orizabina y la Zexbrevina B, Nuevos Germacranolidos Furanicos. Rev. Latinoamer. Quim. **2**, 38 (1971).

805a. Ortega, A., R. Martinez, and A. Romo de Vivar: Las Elemanolidas de *Verbesina* aff. *stricta*. Estructura de las Zempoalinas A y B. Rev. Latinoamer. Quim. **8**, 166 (1977).

806. Ortega, A., A. Romo de Vivar, E. Diaz, and J. Romo: Determinacion de las estructuras de la calaxina y de la ciliarina, nuevos germacranolidos furanonicos. Rev. Latinoamer. Quim. **1**, 81 (1970).

807. Ortega, A., A. Romo de Vivar, and J. Romo: Odoratin: A New Pseudoguaianolide Isolated from *Hymenoxis odorata* DC. Canad. J. Chem. **46**, 1539 (1968).

808. Ortega, A., C. Vargas, C. Guerrero, and A. Romo de Vivar: Los Componentes De *Zexmenia brevifolia* III. Estructura De Zexbrevina C. Rev. Latinoamer. Quim. **4**, 1, (1973).

809. Padolina, W. G., Chemistry and Distribution of New Germacranolide-Type Sesquiterpene Lactones in the North American, Mexican, and South American Taxa of the Genus *Vernonia* (Compositae). From Diss. Abstr. Int. B 1973, **34**, 1907; Chem. Abstr. **80**, 57421u (1974).

810. Padolina, W. G., N. Nakatani, H. Yoshioka, T. J. Mabry, and S. A. Monti: Marginatin, a New Germacranolide from *Vernonia* Species. Phytochem. **13**, 2225 (1974).

811. Padolina, W. G., H. Yoshioka, N. Nakatani, T. J. Mabry, S. A. Monti, R. E. Davis, P. J. Cox, G. A. Sim, W. H. Watson, and I. Beth Wu: Glaucolide-A and -B, New Germacranolide-Type Sesquiterpene Lactones from *Vernonia* (Compositae). Tetrahedron **30**, 1161 (1974).

812. Pakaln, D. A., R. I. Evstratova, L. P. Tolstykh, and K. S. Rybalko: Content of Tauremisin in *Artemisia taurica*. Regional Procurement of Raw Materials from *Artemisia taurica* in the Caucasus. Rastit. Resur. **11**, 376 (1975); Chem. Abstr. **83**, 128703n (1975).

813. Pal, R., D. K. Kulshreshtha, and R. P. Rastogi: Chemical Constituents of *Tithonia tagitiflora* Desf.: Constitution of Tagitinin A. Indian J. Chem. Sect. B **14B**, 259 (1976).

814. — — — Chemical Constituents of *Tithonia tagitiflora* Desf. Structure of Tagitinin-B by Application of Homonuclear Indor Spectroscopy. Indian J. Chem. Sect B **1976**, 77.

815. — — — Antileukemic and Other Constituents of *Tithonia tagitiflora* Desf. J. Pharm. Sci. **65**, 918 (1976); Chem. Abstr. **85**, 72205y (1976).

816. — — — Chemical Constituents of *Tithonia tagitiflora* Desf. Part IV — Tagitinins C, D, and F. Indian J. Chem. Sect. B **1977**, 208.

817. Parker, B. A., and T. A. Geissman: The Sesquiterpenoid Lactones of *Helenium bigelovii* Gray. J. Organ. Chem. (U.S.A.) **27**, 4127 (1962).

818. Parker, W., J. S. Roberts, and R. Ramage: Sesquiterpene Biogenesis. Quart. Rev. (Chem. Soc. London) **21**, 311 (1967).

819. Pascard, C.: X-Ray Crystallographic Determination of the Structure of the Sesquiterpenoid Isocollybolide. Chem. Comm. **1970**, 1722.

820. — Determination par les Rayons X de la Structure d'un Ester Sesquiterpenique: le Vernolide. Tetrahedron Letters **1970**, 4131.

820a. Pashchenko, M. M., G. P. Pivnenko, O. V. Chuiko, and J. O. Kholyupak: Sesquiterpene Lactones of *Xanthium riparium* and their microbiological activity. Farm. Zh. (Kiev) **1969**, 24; Chem. Abstr. **71**, 67924p (1969).

821. Pathak, S. P., B. V. Bapat, and G. H. Kulkarni: Conversion of Costunolide into Santamarin and Reynosin. Indian J. Chem. **8**, 471 (1970); Chem. Abstr. **73**, 35532j (1970).

822. — — — Structure of Balchanin. Chem. Ind. (London) **1970**, 1147; Chem. Abstr. **73**, 99058s (1970).

823. — — — Reactions of Sant-3-enolide. Indian J. Chem. **9**, 85 (1971); Chem. Abstr. **74**, 100233z (1971).

824. PATHAK, S. P., and G. H. KULKARNI: An Example of Selenium Dioxide Oxidation at a Tertiary Center. Chem. and Ind. (London) **1968**, 913; Chem. Abstr. **69**, 59408s (1968).

825. PELLETIER, S. W., and S. PRABHAKAR: The Total Synthesis of Isoiresin, Dihydroiresin, and Isodihydroiresin. J. Amer. Chem. Soc. **90**, 5318 (1968).

826. PEROLD, G. W., J. C. MULLER, and G. OURISSON: Structure D'Une Lactone Allergisante: Le Frullanolide — I. Tetrahedron **28**, 5797 (1972).

827. PEROLD, G. W., and G. OURISSON: Photolysis of 1,2-dihydro-(6β, 11α and 11β)-Santonins. Stereospecific Intramolecular Photosensitization. Tetrahedron Letters **1969**, 3871.

828. PERRY, D. L., D. M. DESIDERIO, and N. H. FISCHER: Mass Spectra of Germacranolide Type Sesquiterpene Dilactones and Derivatives. Org. Mass Spectr. **13**, 325 (1978).

829. PERRY, D. L., and N. H. FISCHER: New Germacranolide Sesquiterpene Dilactones from the Genus *Melampodium* (Compositae). J. Organ. Chem. (U.S.A.) **40**, 3480 (1975).

830. PETTERSEN, R. C., and H. L. KIM: X-Ray Structures of Hymenoxon and Hymenolane: Pseudoguaianolides Isolated from *Hymenoxys odorata* DC. (Bitterweed). J. Chem. Soc. Perkin Trans. II **1976**, 1399.

831. PETTIT, G. R., J. C. BUDZINSKI, G. M. CRAGG, P. BROWN, and L. D. JOHNSTON: Antineoplastic Agents. *Helenium autumnale*. J. Med. Chem. **17**, 1013 (1974); Chem. Abstr. **82**, 25674d (1975).

831a. PETTIT, G. R., C. L. HERALD, G. F. JUDD, G. BOLLIGER, and P. S. THAYER: Antineoplastic and Cytotoxic Components of desert *Baileya*. J. Pharm. Sci. **64**, 2023 (1975).

831b. PETTIT, G. R., C. L. HERALD, D. GUST, D. L. HERALD, and L. D. VANELL: Antineoplastic Agents. Isolation and Structure of Multigilin and Multistatin. J. Organ. Chem. (U.S.A.) **42**, 1092 (1978).

832. PIERS, E., and M. B. GERAGHTY: Total Synthesis of Eremophilane-Type Sesquiterpenoids. (±)-Eremophilenolide, (±)-Tetrahydroligularenolide and (±)-Aristolochone. Canad. J. Chem. **51**, 2166 (1973); Chem. Abstr. **79**, 92401p (1973).

833. PIERS, E., M. B. GERAGHTY, and R. D. SMILLIE: Total Synthesis of (±)-Eremophilenolide. Chem. Commun. **1971**, 614.

834. PINDER, A. R.: The Chemistry of the Eremophilane and Related Sesquiterpenes. In: Progress in the Chemistry of Natural Products (HERZ, W., H. GRISEBACH, and G. W. KIRBY, eds.), Vol. 34. Wien: Springer. 1977.

835. PINHEY, J. T., and S. STERNHELL: Structure of a-Hydroxysantonin and Some Aspects of the Stereochemistry of Related Eudesmanolides and Guaianolides. Austral. J. Chem. **18**, 543 (1965).

836. POPLAWSKI, J., M. HOLUB, Z. SAMEK, and V. HEROUT: Arnicolides — Sesquiterpenic Lactones from the Leaves of *Arnica montana* L. Collect. Czech. Chem. Comm. **36**, 2189 (1971).

837. POPOVA, A. I., K. S. RYBALKO, and R. I. EVSTRATOVA: Sesquiterpene Lactones from *Artemisia lagocephala, Artemisia schrenkiana,* and *Grossheimia ossica*. Khim. Prir. Soedin. **1974**, 528; Chem. Abstr. **82**, 28594p (1975).

838. PORTER, T. H., and T. J. MABRY: Sesquiterpene Lactones. Constituents of *Ambrosia artemisiifolia* L. (Compositae). Phytochem. **8**, 793 (1969).

839. PORTER, T. H., T. J. MABRY, H. YOSHIOKA, and N. H. FISCHER: The Isolation and Structure Determination of Artemisiifolin, A New Germacranolide from *Ambrosia artemisiifolia* L. (Compositae). Phytochem. **9**, 199 (1970).

840. POTTER, J. L., and T. J. MABRY: Origin of the Texas Gulf Coast Island Populations of *Ambrosia psilostachya:* A Numerical Study using Terpenoid Data. Phytochem. **11**, 715 (1972).

841. Power, F. B., and H. Browning: The Constituents of the Flowers of *Anthemis nobilis.* J. Chem. Soc. (London) **1914,** 1829.

842. Pregosin, P. S., E. W. Randall, and T. B. H. McMurry: Carbon-13 Fourier Studies. Configurational Dependence of the Carbon-13 Chemical Shifts in Santonin Derivatives. J. Chem. Soc. Perkin Trans. I **1972,** 299.

843. Prochazka, V., Z. Cekan, and R. B. Bates: Structure of Globicin, a Guaianolide from *Matricaria globifera* (Thunb.) Druce. Collect. Czech. Chem. Comm. **28,** 1202 (1963).

844. Quick, A., and D. Rogers: Crystal and Molecular Structure of Parthenolide [4,5-Epoxy-germacra-1(10),11(13)-dien-12,6-olactone]. J. Chem. Soc. Perkin Trans. II **1976,** 465.

845. Quijano, L., A. Ortega, T. Rios, and A. Romo de Vivar: Estructura De Las Zinaflorinas, Nuevas Lactonas Sesquiterpenicas, Aisladas De *Zinnia pauciflora* L. Rev. Latinoamer. Quim. **6,** 94 (1975).

846. Quing Hau Sau: A New Type of Sesquiterpene Lactone — Quing Hau Sau. Co-ordinating Group for Research on the Structure of Quing Hau Sau (Peop. R. China). K'o Hsueh T'ung Pao **22,** 142 (1977); Chem. Abstr. **87,** 98788 g (1977).

847. Raghavan, R., K. R. Ravindranath, G. K. Trivedi, S. K. Pakmikar, and S. C. Bhattacharyya: Inunolide — A New Sesquiterpene Lactone from *Inula racemosa* Root. Indian J. Chem. **7,** 310 (1969); Chem. Abstr. **70,** 115345.

848. Raffauf, R. F., P. K. C. Huang, P. W. LeQuesne, S. B. Levery, and T. F. Brennan: Eremantholide A, a Novel Tumor-Inhibiting Compound from *Eremanthus elaeagus* Schultz-Bip (Compositae). J. Amer. Chem. Soc. **97,** 6884 (1975).

849. Rao, A. S., G. R. Kelkar, and S. C. Bhattacharyya: The Structure of Costunolide, A New Sesquiterpene Lactone from Costus Root Oil. Tetrahedron **9,** 275 (1960).

850. Rao, A. S., A. P. Sadgopal, and S. C. Bhattacharyya: Structure of Saussurea Lactone. Tetrahedron **13,** 319 (1961).

850a. Ravindranath, K. R., S. K. Paknikar, G. K. Trivedi, and S. C. Bhattacharyya: Structure and Stereochemistry of Inunolide, Dihydroinunolide and Neoalantolactone. Ind. J. Chem. **16 B,** 27 (1978).

850b. Raszeja, W., and St. Gill: Isolation and Identification of Psilostachyin B from *Ambrosia artemisiifolia* L. Planta Med. **32,** 319 (1977).

851. Renold, W., H. Yoshioka, and T. J. Mabry: Chihuahuin, a New Germacranolide from *Ambrosia confertiflora* DC. (Compositae). J. Organ. Chem. (U.S.A.) **35,** 4264 (1970).

852. Revazova, L. V., P. V. Chugunov, and D. Pakaln: Deacetylmatricarin from *Artemisia incana.* Khim. Prir. Soedin. **6,** 372 (1970); Chem. Abstr. **73,** 117180 g (1970).

853. Rios, T., A. Romo de Vivar, and J. Romo: Stevin, a New Pseudoguaianolide Isolated from *Stevia rhombifolia* H. B. K. Tetrahedron **23,** 4265 (1967).

854. Roberts, J. S.: Sesquiterpenoids. In: Terpenoids and Steroids (Overton, K. H., ed.), Vol. 2, p. 65. London: The Chemical Society. 1974.

855. — Sesquiterpenoids. In: Terpenoids Steroids (Overton, K. H., ed.), Vol. 3, p. 92. London: The Chemical Society. 1973.

856. — Sesquiterpenoids. In: Terpenoids Steroids (Overton, K. H., ed.), Vol. 1, p. 51. London: The Chemical Society. 1971.

857. Robinson, D. L., and D. W. Theobald: Sesquiterpenoids: The Synthesis of Some 2-Ketoeudesmanes. Tetrahedron **24,** 5227 (1968).

858. Rodrigues, A. A. S., M. Garcia, and J. A. Rabi: Facile Biomimetic Synthesis of Costunolide-1,10-epoxide, Santamarin and Reynosin. Phytochem. **17,** 953 (1978).

859. Rodriguez, E.: Sesquiterpene Lactones: Chemotaxonomy, Biological Activity and Isolation. Rev. Latinoamer. Quim. **8,** 56 (1977).

860. — The Chemistry and Distribution of Sesquiterpene Lactones and Flavonoids in

Parthenium (Compositae). Systematic and Ecological Implications. Diss. Abstr. Int. B **1976**, 37, 49; Chem. Abstr. **85**, 90194y (1976).

861. — Ecogeographic Distribution of Secondary Constituents in *Parthenium* (Compositae). Biochem. System. Ecol. **5**, 207 (1977).

862. RODRIGUEZ, E., M. O. DILLON, T. J. MABRY, J. C. MITCHELL, and G. H. N. TOWERS: Dermatologically Active Sesquiterpene Lactones in Trichomes of *Parthenium hysterophorus* L. (Compositae). Experientia **32**, 236 (1976); Chem. Abstr. **84**, 131199d (1976).

863. RODRIGUEZ, E., W. EPSTEIN, and J. C. MITCHELL: The Role of Sesquiterpene Lactones in Contact Sensitivity of Some North and South American Species of Feverfew (*Parthenium* — Compositae). J. Contact Dermatitis 3, 155 (1977).

864. RODRIGUEZ, E., G. H. N. TOWERS, and J. C. MITCHELL: Biological Activities of Sesquiterpene Lactones. Phytochem. **15**, 1573 (1976).

865. RODRIGUEZ, E., H. YOSHIOKA, and T. J. MABRY: Sesquiterpene Lactone Chemistry of *Parthenium fruticosum* and *P. schottii*. Rev. Latinoamer. Quim. **2**, 184 (1972); Chem. Abstr. **76**, 153956k (1972).

866. — — — The Sesquiterpene Lactone Chemistry of the Genus *Parthenium* (Compositae). Phytochem. **10**, 1145 (1971).

867. RODRIGUEZ-HAHN, L., M. JIMENEZ, E. DIAZ, C. GUERRERO, A. ORTEGA, and A. ROMO DE VIVAR: The Revised Structure of Mortonin. Tetrahedron **33**, 657 (1977).

868. ROGERS, D., G. P. MOSS, and S. NEIDLE: Proposed Conventions for Describing Germacranolide Sesquiterpenes. Chem. Commun. **1972**, 142.

869. ROGERS, D., and M. UL-HAGUE: The Structure of Bromoisotenulin. Proc. Chem. Soc. (London) **1963**, 92.

870. ROJAS GARCIDUENAS, M., X. A. DOMINGUEZ, J. FERNANDEZ, and G. ALANIS: New Growth Inhibitors from *Parthenium hysterophorus*. Rev. Latinoamer. Quim. **3**, 52 (1972).

871. ROMANUK, M., V. HEROUT, and F. SORM: The Constitution of Dehydrocostuslactone. Collect. Czech. Chem. Comm. **21**, 894 (1956).

872. ROMO, J.: Recent Studies on Sesquiterpenes. Pure Appl. Chem. **21**, 123 (1970).

873. ROMO, J., P. JOSEPH-NATHAN, and F. DIAZ: The Constituents of *Helenium aromaticum* (Hook) Bailey. The Structures of Aromatin and Aromaticin. Tetrahedron **20**, 79 (1964).

874. ROMO, J., P. JOSEPH-NATHAN, A. ROMO DE VIVAR, and C. ALVAREZ: The Structure of Peruvinin, a Pseudoguaianolide Isolated from *Ambrosia peruviana* Willd. Tetrahedron **23**, 529 (1967).

875. ROMO, J., P. JOSEPH-NATHAN, and G. SIADE: The Structure of Cumanin, a Constituent of *Ambrosia cumanensis*. Tetrahedron **22**, 1499 (1966).

876. ROMO, J., and C. LOPEZ VANEGAS: Stereochemistry Determinations of the Substituents in Position 1 of Some Guaianolides. Bol. Inst. Quim. Univ. Nac. Auton. Mex. **21**, 82 (1969); Chem. Abstr. **73**, 35540k (1970).

877. ROMO, J., T. RIOS, and L. QUIJANO: Ligustrin, A Guaianolide Isolated from *Eupatorium ligustrinum* DC. Tetrahedron **24**, 6087 (1968).

878. ROMO, J., and L. RODRIGUEZ-HAHN: Canambrin, A New Sesquiterpene Dilactone from *Ambrosia canescens*. Phytochem. **9**, 1611 (1970).

879. ROMO, J., and A. ROMO DE VIVAR: The Pseudoguaianolides. In: Fortschr. Chem. Organ. Naturstoffe (ZECHMEISTER, L., ed.), Vol. 25, p. 90. Wien: Springer. 1967.

880. ROMO, J., A. ROMO DE VIVAR, and M. AGUILAR: Natural Products from Compositae Species. Neohelenalin. Bol. Inst. Quim. Univ. Nac. Auton. Mex. **21**, 66 (1969); Chem. Abstr. **73**, 25685x (1970).

881. ROMO, J., A. ROMO DE VIVAR, and E. DIAZ: The Guaianolides of *Ambrosia cumanensis* HBK. The Structures of Cumambrins A and B. Tetrahedron **24**, 5625 (1968).

882. Romo, J., A. Romo de Vivar, E. Diaz, A. Velez, E. Leon, and E. Urbina: Distribution of Sesquiterpene Lactones in Several *Ambrosia* Species. In: Recent Advan. Phytochem. (Steelink, C., and Y. C. Runeckles, eds.), Vol. 3, p. 249. New York: Appleton Century Croft. 1970.

883. Romo, J., A. Romo de Vivar, and P. Joseph-Nathan: The Structure of Mexicanin H. Tetrahedron Letters **1966**, 1029.

884. — — — The Constituents of *Zaluzania augusta*. The Structures of Zaluzanins A and B. Tetrahedron **23**, 29 (1967).

885. Romo, J., A. Romo de Vivar, and W. Herz: Constituents of *Helenium* Species. XIV. The Structure of Mexicanin E. Tetrahedron **19**, 2317 (1963).

886. Romo, J., A. Romo de Vivar, A. Ortega, E. Diaz, and M. A. Carino: Las Estructuras De La Zinarosina y De La Dihidrozinarosina Componentes De La *Zinnia acerosa* DC. (Gray). Rev. Latinoamer. Quim. **2**, 24 (1971).

887. Romo, J., and H. Tello: Estudia De La *Artemisia mexicana* Armexina, Un Nuevo Santanolido Cuya Lactona Posee Fusion Cis. Rev. Latinoamer. Quim. **3**, 122 (1972); Chem. Abstr. **78**, 121339m (1973).

888. Romo, J., A. Romo de Vivar, R. Trevino, P. Joseph-Nathan, and E. Diaz: Constituents of *Artemisia* and *Chrysanthemum* Species. The Structure of Chrysartemins A and B. Phytochem. **9**, 1615 (1970).

889. Romo, J., A. Romo de Vivar, A. Velez, and E. Urbina: Franserin and Confertin; New Pseudoguaianolides Isolated from *Franseria* and *Ambrosia* Species. Canad. J. Chem. **146**, 1535 (1968).

890. Romo de Vivar, A.: Organic Structure Determination by the Double and Triple Irradiation Technique. Rev. Soc. Quim. Mex. **14**, 54 (1970); Chem. Abstr. **74**, 42501x (1971); Chem. Abstr. **73**, 88037h (1970).

891. — Sesquiterpene Lactones in Compositae. Biogenesis and Taxonomic Implications. Rev. Latinoamer. Quim. **8**, 63 (1977); Chem. Abstr. **87**, 18934p (1977).

892. Romo de Vivar, A., M. Aguilar,, H. Yoshioka, E. Rodriguez, A. Higo, J. A. Mears, and T. J. Mabry: New Pseudoguaianolides from *Parthenium confertum* Gray (Compositae). Tetrahedron **26**, 2775 (1970).

893. Romo de Vivar, A., E. A. Bratoeff, and T. Rios: Structure of Hysterin, a New Sesquiterpene Lactone. J. Organ. Chem. (U. S. A.) **31**, 673 (1966).

894. Romo de Vivar, A., A. Cabrera, A. Ortega, and J. Romo: Constituents of *Zaluzania* Species — II. Structures of Zaluzanin C and Zaluzanin D. Tetrahedron **23**, 3903 (1967).

895. Romo de Vivar, A., and G. Gonzalez: Aislamiento De La Lactona Dihidrocostus. Preparacion De Espiroepoxidos De Lactones Sesquiterpenicas. Rev. Latinoamer. Quim. **2**, 142 (1971); Chem. Abstr. **76**, 46326h (1972).

896. Romo de Vivar, A., C. Guerrero, and G. Wittgreen: Los terpenoides de *Parthenium incanum* H. B. K. Rev. Latinoamer. Quim. **1**, 39 (1970); Chem. Abstr. **74**, 88165u (1971).

897. Romo de Vivar, A., C. Guerrero, E. Diaz, and A. Ortega: Structure and Stereochemistry of Zexbrevin A₃ (2 H) Furanone Germacranolide. Tetrahedron **26**, 1657 (1970).

898. Romo de Vivar, A., C. Guerrero, E. Diaz, E. A. Bratoeff, and L. Jimenez: The Germacranolides of *Viguiera buddleiaeformis*. Structures of Budlein-A and -B. Phytochem. **15**, 525 (1976).

899. Romo de Vivar, A., J. Guevara, C. Guerrero, and A. Ortega: Los Terpenoids De *Mortonia gregii* Gray. Estructura De La Mortonina A. Rev. Latinoamer. Quim. **3**, 1 (1972); Chem. Abstr. **77**, 140326a (1972).

900. Romo de Vivar, A., and H. Jimenez: Structure of Santamarine, a New Sesquiterpene Lactone. Tetrahedron **21**, 1741 (1965).

900a. Romo de Vivar, A., and F. Olmos: Chemical Study of *Achillea millefolium*. Rev. Soc. Quim. Mex. **12**, 212A (1968); Chem. Abstr. **71**, 3493q (1969).

901. Romo de Vivar, A., and A. Ortega: Structures of Bahia-I and Bahia-II: Two New Guaianolides. Canad. J. Chem. **47**, 2849 (1969).

902. Romo de Vivar, A., A. L. Perez, H. Flores, L. Rodriguez-Hahn, and M. Jimenez: Confertdiolide, A Novel Sesquiterpenoid Lactone from *Parthenium*. Phytochemistry **17**, 279 (1978).

903. Romo de Vivar, A., L. Rodriguez-Hahn, J. Romo, M. V. Lakshmikantham, R. N. Mirrington, J. Kagan, and W. Herz: Constituents of *Helenium* Species — XIX. Further Transformations of Helenalin and its Congeners. The 1-Epihelenalin and 1-Epiambrosin Series. Tetrahedron **22**, 3279 (1966).

904. Romo de Vivar, A., and J. Romo: Constituents of *Helenium mexicanum* H. B. K. Chem. and Ind. **1959**, 882.

905. Romo de Vivar, A., F. Vazquez, and C. Zetina: Sesquiterpene Lactones of *Artemisia mexicana* var. *angustifolia*. Rev. Latinoamer. Quim. **8**, 127 (1977).

906. Ross, J. M., D. S. Tarbell, W. E. Lovett, and A. D. Cross: The Chemistry of Fumagillin. IV. The Presence of an Epoxide Grouping and Other Observations on the Nature of the Oxygen Functions. J. Amer. Chem. Soc. **78**, 4675 (1956).

907. Rossman, M. G., and W. N. Lipscomb: Molecular Structure and Stereochemistry of an Iresin Diester. J. Amer. Chem. Soc. **80**, 2592 (1958).

907a. Ruban, G., V. Zabel, K. H. Gensch, and H. Smalla: The Crystal Structure and Absolute Configuration of Lactucin. Acta Crystall **B 34**, 1163 (1978).

908. Rustaiyan, A., and L. Nazarians: Isolation of the Sesquiterpene Lactone Deacetoxymatricarin from *Achillea eriophora* DC. Fitoterapia **48**, 175 (1977).

909. Rücker, G.: Sesquiterpene. Angew. Chem. **85**, 895 (1973).

910. Rüesch, H., and T. J. Mabry: The Isolation and Structure of Tetraneurin-A, a New Pseudoguaianolide from *Parthenium alpinum* var. *tetraneuris* (Compositae). Tetrahedron **25**, 805 (1969).

911. Ruzicka, L.: a. The Isoprene Rule and the Biogenesis of Terpenic Compounds. Experientia **9**, 357 (1953).

912. — Faraday Lecture — History of the Isoprene Rule. Proc. Chem. Soc. (London) **1959**, 341.

913. Ruzicka, L., P. Pieth, T. Reichstein, and L. Ehinann: Zur Kenntnis der Alantolactone. Synthese des 1,4-Dimethyl-6-isopropyl und des 1,5-Dimethyl-7-isopropyl-naphthalins. Helv. Chim. Acta **16**, 268 (1933).

914. Rybalko, K. S.: Isolation of Mibulactone from *Artemisia taurica*. Khim. Prir. Soedin. **1965**, 142; Chem. Abstr. **63**, 6786f (1965).

914a. Rybalko, K. S., A. N. Bankovskaya, and R. I. Evstratova: Sesquiterpene Lactone from *Artemisia austriaca*. Med. Prom. SSSR. **16**, 13 (1962); Chem. Abstr. **58**, 2474 (1963).

915. Rybalko, K. S., A. I. Ban'kovskii, and P. N. Kibalcic: Grosshemine, a New Sesquiterpene Lactone from *Grossheimia macrocephala*. Zhurn. Obshchei Khimii **34**, 1358 (1964); Chem. Abstr. **61**, 1896h (1964).

916. Rybalko, K. S., A. I. Ban'kovskii, and V. I. Sheichenko: Natural Sesquiterpenoid Lactones. Medicinal Plants (Russia) **15**, 168 (1969); Chem. Abstr. **75**, 20632j (1971).

917. Rybalko, K. S., and L. Dolejs: Structure of Tauremisin, A Sesquiterpenic Lactone of Santonine Type from *Artemisia taurica* Willd. Collect. Czech. Chem. Comm. **26**, 2909 (1961).

918. Rybalko, K. S., O. A. Konovalova, N. D. Orishchenko, and A. I. Shreter: Lactones of Some Species of the Genus *Saussurea* DC. Rastit. Resur. **12**, 387 (1976); Chem. Abstr. **85**, 174252d (1976).

919. Rybalko, K. S., O. A. Konovalova, and E. F. Petrova: Orientin, a New Sesquiterpene Lactone from *Siegesbeckia orientalis*. Khim. Prir. Soedin. 1976, 394; Chem.
Abstr. 85, 124182k (1976).

920. Rybalko, K. S., M. N. Mukhametzhanov, V. I. Sheichenko, and O. A. Konovalova: Sesquiterpene Lactones of *Stizolophus balsamita*. Khim. Prir. Soedin. 12,
467 (1976); Chem. Abstr. 86, 5643x (1977).

921. Rybalko, K. S., V. I. Sheichenko, G. A. Maslova, E. Y. Kiseleva, and I. A.
Gubanov: Britanin, a Lactone from *Inula britannica*. Khim. Prir. Soedin. 4, 251
(1968); Chem. Abstr. 70, 47618t (1969).

922. Rybalko, K. S., E. A. Trutneva, and P. N. Kibal'chich: A Crystalline Substance
Isolated from *Arnica foliosa*. Aptechn. Delo 14, 32 (1965); Chem. Abstr. 63,
18541a (1965).

922a. Saber, A. H., H. Abu Shady, and M. S. El-Antably: Judaicin. Bull. Fac. Pharm.
Cairo Univ. 3, 119 (1964); Chem. Abstr. 64, 12729d (1966).

923. Saitbaeva, I. M., and G. P. Sidyakin: Artemisin from *Artemisia cina*. Khim. Prir.
Soedin. 7, 120 (1971); Engl. edit.: p. 113.

924. Saitoh, T., T. A. Geissman, T. G. Waddell, W. Herz, and S. V. Bhat: Sesquiterpene Lactones of *Eriophyllum confertiflorum* (DC.) Gray. Rev. Latinoamer. Quim.
2, 69 (1971); Chem. Abstr. 75, 129952q (1971).

925. Sakabe, N., Y. Hirata, A. Furusaki, Y. Tomiie, and I. Nitta: X-Ray Structure
Determination of Bromonoranisatinone, A Derivative of Anisatin. Tetrahedron
Letters 1965, 4795.

926. Saleh, A. A., G. A. Cordell, and N. R. Farnsworth: Acanthamolide, a Melampolide Amide from *Acanthospermum glabratum* (Compositae). Chem. Commun. 1977, 376.

927. Salmon, M., E. Diaz, and A. Ortega: Christinine, a New Epoxyguaianolide from
Stevia serrata Cav. J. Organ. Chem. (U. S. A.) 38, 1759 (1973).

927a. — — — Epoxylactonas de *Stevia serrata* Cav. Rev. Latinoamer. Quim. 8, 172
(1977).

928. Salmon, M., A. Ortega, and E. Diaz: Structure and Stereochemistry of a New
Germacrane Sesquiterpene Lactone. Rev. Latinoamer. Quim. 6, 45 (1975); Chem.
Abstr. 83, 193528y (1975).

929. Samek, Z.: The Determination of the Stereochemistry of Five-membered α,β-Un-
saturated Lactones with an Exomethylene Double Bond, Based on the Allylic Long-
Range Couplings of Exomethylene Protons. Tetrahedron Letters 1970, 671.

929a. — On the Validity of the "cis/trans" Lactone Rule for Allylic Coupling Constants
of the α-Exomethylene Protons in Natural Sesquiterpenic α-Exomethylene γ-Lactones. Collect. Czech. Chem. Comm. 43, 3210 (1978).

930. Samek, Z., M. Holub, E. Bloszyk, B. Drozdz, and V. Herout: Relative and
Absolute Configuration of the Sesquiterpenic Lactone Erivanin. Collect. Czech.
Chem. Commun. 40, 2676 (1975).

931. Samek, Z., M. Holub, B. Drozdz, and H. Grabarczyk: Arctolide — A New
Sesquiterpenic Lactone from *Arctotis grandis* Thunb. Collect. Czech. Chem. Comm.
42, 2217 (1977).

932. Samek. Z., M. Holub, B. Drozdz, H. Grabarczyk, and B. Hladon: Xerantholide — A New Cytotoxically Active Sesquiterpene Lactone from *Xeranthenum
cylindraceum*. Collect. Czech. Chem. Comm. 42, 2441 (1977).

933. Samek, Z., M. Holub, B. Drozdz, G. Jommi, A. Corbella, and P. Gariboldi:
Sesquiterpenic Lactones of the *Cynara scolymus* L. Species. Tetrahedron Letters
1971, 4775.

934. Samek, Z., M. Holub, H. Grabarczyk, B. Drozdz, and V. Herout: Structure of
Sesquiterpenic Lactones from *Tanacetum vulgare* L. Collect. Czech. Chem. Comm.
38, 1971 (1973).

935. — — — — — The Structure of Hydroxyisonobilin — A Cytostatically Active Sesquiterpene Lactone from the Leaves of *Anthemis mobilis* L. Collect. Czech. Chem. Comm. **42**, 1065 (1977).

936. SAMEK, Z., M. HOLUB, V. Herout, and F. SORM: Revision of the Structure of Cnicin. Tetrahedron Letters **1969**, 2931.

937. SAMEK, Z., M. HOLUB, V. J. NOVIKOV, J. N. FOROSTJAN, and D. P. POPA: The Structure of Ivoxanthin, a Sesquiterpenic Lactone from *Cyclachaena xanthifolia* Fresen. Collect. Czech. Chem. Comm. **35**, 3818 (1970).

938. SAMEK, Z., M. HOLUB, K. VOKAC, B. DROZDZ, G. JOMMI, P. GARIBOLDI, and A. COR-BELLA: The Structure of Grosheimin. Collect. Czech. Chem. Comm. **37**, 2611 (1972).

939. SANCHES PARAREDA, I., J. SANCHES PARAREDA, and J. M. VIGUERA: Stenophyllolide, a New Sesquiterpene Lactone. An. Chim. **64**, 633 (1968); Chem. Abstr. **70**, 37933c (1969).

940. SANCHEZ-VIESCA, F., and J. ROMO: Estafiatin, a New Sesquiterpene Lactone Isolated from *Artemisia mexicana* (Willd). Tetrahedron **19**, 1285 (1963).

941. SASAKI, T., and S. EGUCHI: Reactions of Isoprenoids. Revised Structure of Pyro-lumisantonin. Bull. Chem. Soc. Japan **42**, 2736 (1969); Chem. Abstr. **72**, 32044v (1970).

942. SATHE, R. N., N. R. DESHPANDE, G. H. KULKARNI, G. R. KELKAR, and K. G. DAS: Correlation of Structure and Fragmentation Modes of Costunolide and Its Derivatives. Org. Mass Spectrom. **5**, 197 (1971); Chem. Abstr. **74**, 142092h (1971).

943. SATHE, R. N., G. H. KULKARNI, and G. R. KELKAR: Catalytic Hydrogenation Products of Dihydrocostunolide. Indian J. Chem. **9**, 101 (1971); Chem. Abstr. **74**, 112234p (1971).

944. SATHE, R. N., G. H. KULKARNI, G. R. KELKAR, and K. G. DAS: Fragmentation of Costunolide and Its Derivatives Under Electron-Impact. Org. Mass Spectrom. **2**, 935 (1969); Chem. Abstr. **72**, 32045w (1970).

945. SATODA, I., N. YOSHIDA, and E. YOSHII: Isolation of Lumisantonin from *Artemisia kurramensis.* Yakugaku Zasshi **79**, 267 (1959); Chem. Abstr. **53**, 11532h (1959).

945a. SCHAFFNER, K.: Photochemische Umwandlungen ausgewählter Naturstoffe. In: Fortschritte der Chemie organischer Naturstoffe (ZECHMEISTER, L., ed.), Vol. 22, p. 1. Wien: Springer. 1967.

946. SCHENCK, G., H. GRAF, and W. SCHREBER: Bitter Substances from the Latex of *Lactuca virosa.* Isolation of Lactucin and Lactucopicrin. Arch. Pharm. Deutsch. **277**, 137 (1939); Chem. Abstr. **33**, 7041-8.

947. SCHMALLE, H. W., K. H. KLASKA, and O. JARCHOW: Die Kristall- und Molekül-struktur von Arteglasin A, 8α-Acetoxy-3,4-epoxy-4ξ-guaia-1(10),11(13)-dieno-6α,12-lacton. Acta Crystallogr. **B 33**, 2213 (1977).

948. SCHNEIDER, G., and K. THIELE: Properties and Determination of the Bitter Principle Cynaropicrin in *Cynara.* Planta Med. **25**, 149 (1974); Chem. Abstr. **81**, 35534b (1974).

949. — — Distribution of the Bitter Principle Cynaropicrin in *Cynara.* Planta Med. **26**, 174 (1974); Chem. Abstr. **82**, 54198g (1975).

950. SCHULZ, K. H., B. M. HAUSEN, L. WALLHOEFER, and P. SCHMIDT-LOEFFLER: Chrysanthemum Allergy. Experimental Studies on the Causative Agents. Arch. Dermatol. Forsch. **251**, 235 (1975); Chem. Abstr. **82**, 110222z (1975).

951. SEAMAN, F. C., and N. H. FISCHER: Longipin, a New Melampolide from *Melampodium longipes* (Heliantheae, Compositae). Phytochem. **18**, 1065 (1979).

952. — — Longipilin, A New Melampolide from *Melampodium longipilum* (Heliantheae, Compositae). Phytochem. **17**, 2131 (1978).

953. SEAMAN, F. C., G. P. JUNEAU, D. R. DiFEO, S. JUNGK, D. F. BLOOMENSTIEL, and N. H. FISCHER: Repandin A, B, C, and D, Four New Germacranolides from *Tetragonotheca repanda* (Compositae). J. Organ. Chem. (U.S.A.) **44**, 3400 (1979).

954. Seaman, F. C., T. F. Stuessy, and N. H. Fischer: The Chemistry and Biochemical Systematics of the Subtribe Melampodiinae (Heliantheae, Compositae). To be published.

955. Segal, R., S. Sokoloff, B. Haran, D. V. Zaitschek, and D. Lichtenberg: New Sesquiterpene Lactones from *Artemisia herba alba*. Phytochem. **16,** 1237 (1977).

956. Sekita, T., and S. Inayama: The Complete Structures of Pulchellidine and Pulchellin; A Crystallographic Study of 11,13-Dibromopulchellin. Tetrahedron Letters **1970,** 135.

956 a. — — The Crystal Structure and Absolute Configuration of 11,13-Dibromopulchellin. Acta Cryst. **B 27,** 877 (1971).

957. Semmler, F. W., and J. Feldstein: Zur Kenntnis der Bestandteile der ätherischen Öle. (Über das Vorkommen einer Säure $C_{15}H_{22}O_2$ und zweier Lactone $C_{15}H_{22}O_2$ und $C_{15}H_{20}O_2$ im Costuswurzelöl.) Ber. dtsch. chem. Ges. **47,** 2433 (1914).

958. — — Zur Kenntnis der Bestandteile ätherischer Öle. (Über Bestandteile des Costuswurzel-Öles.) Ber. dtsch. chem. Ges. **47,** 2687 (1914).

959. Serkerov, S. V.: 11,13-Dehydroopodine from *Ferula oopoda* roots. Khim. Prir. Soedin. **5,** 490 (1969); Chem. Abstr. **73,** 25677w (1970).

960. — A New Sesquiterpene Hydroxylactone from *Ferula oopoda*. Khim. Prir. Soedin. **7,** 838 (1971); Engl. edit.: p. 817; Chem. Abstr. **76,** 151049t (1972).

961. — Structure of Feropodin. Khim. Prir. Soedin. **1971,** 667; Chem. Abstr. **77,** 5623e (1972).

962. — Feropodin, Sesquiterpene Lactone from *Ferula oopoda* roots. Khim. Prir. Soedin. **5,** 245 (1969); Chem. Abstr. **72,** 63618q (1970).

963. — Unpublished.

964. — Sesquiterpene Lactone, Semopodin, from *Ferula oopoda* seeds. Khim. Prir. Soedin. **5,** 241 (1969); Chem. Abstr. **72,** 63626r (1970).

965. — The Structure of Badkhysinin. Khim. Prir. Soedin. **7,** 590 (1971); Engl. edit.: p. 570; Chem. Abstr. **77,** 5622d (1972).

966. — Badkhysidin — A New Sesquiterpene Lactone from the Roots of *Ferula oopoda*. Khim. Prir. Soedin. **8,** 176 (1972); Engl. edit.: p. 181; Chem. Abstr. **77,** 58787t (1972).

967. — Lactones of *Ferula badghysi*. Khim. Prir. Soedin. **1976,** 393; Chem. Abstr. **85,** 139718p (1976).

968. — The Structure of Oopodin and Dehydro-oopodin. Khim. Prir. Soedin. **8,** 63 (1972); Engl. edit.: p. 57; Chem. Abstr. **77,** 48652f (1972).

969. — The Structure of Ferulin. Khim. Prir. Soedin. **6,** 371 (1970).

970. — The Sesquiterpene Lactone Ferulin from the Roots of *Ferula oopoda*. Khim. Prir. Soedin. **6,** 134 (1970); Chem. Abstr. **73,** 73111d (1970).

971. — Structure of Semopodine and Hydroxylactone. Khim. Prir. Soedin. **1976,** 392; Chem. Abstr. **85,** 124181j (1976).

972. — Structure of Ferulidin: A New Sesquiterpene Lactone. Khim. Prir. Soedin. **6,** 428 (1970).

973. Serkerov, S. V., and R. M. Abbasov: Sesquiterpene Lactones of *Artemisia chasarica*. Khim. Prir. Soedin. **11,** 657 (1975); Engl. edit.: p. 691; Chem. Abstr. **84,** 71431m (1976).

974. Serkerov, S. V., R. M. Abbasov, and A. N. Aleskerova: Sesquiterpene Lactones from *Artemisia hanseniana*. Khim. Prir. Soedin. **12,** 661 (1976); Chem. Abstr. **86,** 103081g (1977).

975. Serkerov, S. V., and V. I. Sheichenko: Structure of Isobadkhysin. The Stereochemistry of Badkhysin and Isobadkhysin. Khim. Prir. Soedin. **6,** 425 (1970); Chem. Abstr. **74,** 17843c (1971).

976. Seshadri, V., A. K. Batta, and S. Rangaswami: A New Crystalline Lactone from *Bombax malabaricum*. Indian J. Chem. Sect. B **14 B,** 616 (1976); Chem. Abstr. **86,** 27674k (1977).

977. SHAFIZADEH, F., and N. R. BHADANE: Sesquiterpene Lactones of Sagebrush. New Guaianolides from *Artemisia cana* ssp. *viscidula.* J. Organ. Chem. (U.S.A.) **37,** 3168 (1972).

978. — — Sesquiterpene Lactones of *Artemisia arbuscula* and *A. tridentata.* Phytochem. **12,** 857 (1973).

979. — — Badgerin, a New Germacranolide from *Artemisia arbuscula* ssp. *arbuscula.* J. Organ. Chem. (U.S.A.) **37,** 274 (1972).

980. SHAFIZADEH, F., N. R. BHADANE, and R. G. KELSEY: Sesquiterpene Lactones of Sagebrush. Constituents of *Artemisia tripartita.* Phytochem. **13,** 669 (1974).

981. SHAFIZADEH, F., N. R. BHADANE, M. S. MORRIES, R. G. KELSEY, and S. N. KHANNA: Sesquiterpene Lactones of Big Sagebrush. Phytochem. **10,** 2745 (1971).

982. SHAM'YANOV, I. D., A. MALLABAEV, V. RAKHMANKULOV, and G. P. SIDYAKIN: Sesquiterpene Lactones of *Saussurea elegans.* Khim. Prir. Soedin. **12,** 819 (1976); Chem. Abstr. **86,** 136305 m (1977).

983. SHEICHENKO, V. I., and K. S. RYBALKO: NMR-Spectra, Structure and Stereochemistry of Grossheimin. Khim. Prir. Soedin. **8,** 724 (1972); Engl. edit.: p. 708; Chem. Abstr. **78,** 84562 f (1973).

984. — — The Structure of Grossheimin. Khim. Prir. Soedin. **6,** 687 (1970); Engl. edit.: p. 699; Chem. Abstr. **74,** 112237 s (1971).

985. SHEICHENKO, V. I., V. F. ZAKHAROV, V. P. ZVOLINSKII, R. I. EVSTRATOVA, and K. S. RYBALKO: Use of the Paramagnetic Shift Agent Eu(DPM)₃ to Study the PMR Spectra and Stereochemistry of Arnifolin. Khim. Prir. Soedin. **9,** 630 (1973); Engl. edit.: p. 595.

986. SHIRAHATA, K., T. KATO, Y. KITAHARA, and N. ABE: Constituents of Genus *Petasites.* Bakkenolide-A, A Sesquiterpene of Novel Carbon Skeleton. Tetrahedron **25,** 3179 (1969).

987. — — — — Mass Spectra of Bakkenolides and Their Derivatives. Tetrahedron **25,** 4671 (1969).

987a. SILVA, M.: Constituents of *Helenium plantagineum.* J. Pharm. Sci. **56,** 922 (1967).

988. SIM, G. A.: X-Ray Diffraction. Natural Products and Related Compounds. Mol. Struct. Diffr. Methods **4,** 134 (1976); Chem. Abstr. **86,** 138854 p (1977).

989. — X-Ray Diffraction. Natural Products and Small Biological Molecules. Mol. Struct. Diffr. Methods **2,** 131 (1974); Chem. Abstr. **82,** 169551 j (1975).

990. SIMS, J. J., and K. A. BERRYMAN: Virginin. A Sesquiterpene Lactone from *Encelia virginensis.* Phytochem. **11,** 444 (1972).

991. SIMONOVIC, D. M., A. S. RAO, and S. C. BHATTACHARYYA: The Synthesis of Tetrahydrosaussurea Lactone. Tetrahedron **19,** 1061 (1963).

991a. SIMONSEN, J., and D. H. R. BARTON: The Terpenes, Vol. III, The Sesquiterpenes, Diterpenes and their Derivatives. Cambridge: At the University Press. 1952.

991b. SMOLENSKI, C. L. BELL, and L. BAUER: Isolation of Achillin from *Achillea millefolium.* Lloydia **30,** 144 (1967).

992. SNATZKE, G.: Stereochemistry of Sesquiterpenes and Circular Dichroism. Mezhdunar. Kongr. Efirnym. Maslam. [Mater.], 4th, **1968,** 316; Chem. Abstr. **78,** 124746 c (1973).

992a. — Application of Circular Dichroism to the Stereochemistry of Essential Oils. Riechstoffe, Aromen, Körperpflegemittel **19,** 98 (1969); Chem. Abstr. **71,** 42157 c (1969).

993. SOKOLOFF, S., and R. SEGAL: Eudesmanolides Derived from Herbolide B. Tetrahedron **33,** 2837 (1977).

994. SORIA, E. L.: Analytical Study of the Distribution of Sesquiterpene Lactones in Different Species of *Ambrosia.* Bol. Soc. Quim. Peru **41,** 52 (1975); Chem. Abstr. **84,** 28052 b (1976).

995. SORM, F.: Medium Ring Terpenes. In: Fortschritte der Chemie organischer Naturstoffe (ZECHMEISTER, L., ed.), Vol. 19, p. 1. Wien: Springer. 1961.

996. Sorm, F.: Sesquiterpenes with Ten-Membered Carbon Rings. J. Agric. Food. Chem. **19,** 6 (1971).

997. — Advances in Terpene Chemistry. Pure Appl. Chem. **21,** 263 (1970).

998. Sorm, F., and L. Dolejs: Guaianolides and Germacranolides. In: Chimie des substances naturelles (Lederer, E., ed.). Paris: Herman. 1966.

999. Sorm, F., M. Suchy, M. Holub, A. Linek, I. Hadinec, and C. Novak: Crystalline Structure of the Addition Compound of Costunolide with Silver Nitrate. Tetrahedron Letters **1970,** 1893.

1000. Soucek, M., V. Herout, and F. Sorm: Constitution of Parthenolide. Collect. Czech. Chem. Comm. **26,** 803 (1961).

1001. Srivastava, S. C., S. K. Paknikar, and S. C. Bhattacharyya: Simple Procedure for the Regeneration of α-Methylene-γ-Lactones from the Amine-Adducts. Indian J. Chem. **8,** 201 (1970); Chem. Abstr. **72,** 121727n (1970).

1002. — — — Biogenetic-Type Synthesis of Santamarin and Its Identity with Balchanin. Indian J. Chem. **8,** 850 (1970); Chem. Abstr. **73,** 131145f (1970).

1003. Stagno d'Alcontres, G., M. Gattuso, M. C. Aversa, and C. Caristi: The Structure of Graveolide, A New Sesquiterpene Lactone. Gazz. Chim. Ital. **1973,** 239; Chem. Abstr. **79,** 53614r (1973).

1004. Stahl, E., and S. N. Datta: New Sesquiterpenoids of the Ground Ivy *(Glechoma hederacea)*. Liebigs Ann. Chem. **1972,** 757, 23; Chem. Abstr. **77,** 98721y (1972).

1005. Steele, J. W., J. B. Stenlake, and W. D. Williams: The Structure of Aristolactone. J. Chem. Soc. (London) **1959,** 3289.

1006. Steelink, C., and J. C. Spitzer: Sesquiterpene Lactones in Chemotaxonomy. Phytochem. **5,** 357 (1966).

1007. Stefanovic, M., A. Jokic, A. Behbud, and D. Jeremic: New Sesquiterpenic Lactones from *Ambrosia artemisiifolia* L. 8-acetoxy-3-oxopseudoguaian-6,12-olide and 4-hydroxy-3-oxopseudoguaian-6,12-olide. Bull. Acad. Serbe. Sci. Arts Cl. Sci. Math. Nat. Sci. Nat. **1976,** 54; Chem. Abstr. **87,** 35891q (1977).

1008. Stefanovic, M., S. Solujic, D. Jeremic, A. Jokic, D. Miljkovic, and S. Velimirovic: Chemical Transformations of Arteannuin B — Cadinane Sesquiterpenic Lactone — Isolated from *Artemisia annua* L. Glas. Hem. Drus. Beograd. **42,** 227 (1977); Chem. Abstr. **87,** 168213s (1977).

1008a. Sternhell, S.: Correlation of Interproton Spin-Spin-Coupling Constants with Structure. Quart. Rev. (Chem. Soc. London) **23,** 236 (1969).

1009. Stevenson, D. S., and C. J. W. Brooks: Separation and Characterization of Δ^3- and $\Delta^{4(15)}$-Cyclopyrethrosin Acetates *Via* Chromatography on a Lipophilic Dextran Gel. J. Chromatogr. **75,** 308 (1973); Chem. Abstr. **78,** 84558j (1973).

1010. St. Pyrek, J.: Isolation of Gaillardin from Flowers of *Inula britannica* L. Rocz. Chem. **51,** 1277 (1977); Chem. Abstr. **87,** 197230a (1977).

1010a. — Terpenes of Compositae Plants. Part V. Sesquiterpene Lactones of *Lactuca serriola* L. The Structure of 8-Deoxylactucin and the Site of Esterification of Lactupicrin. Rocz. Chem. **51,** 2165 (1977).

1011. Stöcklin, W., T. G. Waddell, and T. A. Geissman: Circular Dichroism and Optical Rotatory Dispersion of Sesquiterpene Lactones. Tetrahedron **26,** 2397 (1970).

1012. Suchy, M.: The Structure of Balchanin, a Sesquiterpenic Lactone of Santonin Type from *Artemisia balchanorum* H. Krash. Collect. Czech. Chem. Comm. **27,** 2925 (1962).

1013. Suchy, M., L. Dolejs, V. Herout, F. Sorm, G. Snatzke, and J. Himmelreich: The Constitution of Jurineolide, a New Germacranolide from *Jurinea cyanoides* (L.) Rchb. Collect. Czech. Chem. Comm. **34,** 229 (1969).

1014. Suchy, M., and V. Herout: Calendin — The Bitter Principle from *Calendula officinalis* L. Collect. Czech. Chem. Comm. **26,** 890 (1961).

1015. — — Identity of the Bitter Principle from *Centaurea stoebe* (L.) Sch. et Thell. Collect. Czech. Chem. Comm. **27,** 1510 (1962).

1016. SUCHY, M., V. HEROUT, and F. SORM: Proof of Structure of Arctiopicrin with a Note on its Stereochemistry. Collect. Czech. Chem. Comm. **24,** 1542 (1959).

1017. — — — On the Nature of Arctiopicrin — The Unsaturated Lactone from *Arctium minus* Bernh. Collect. Czech. Chem. Comm. **22,** 1902 (1957).

1018. — — — On Hydrogenation Products of Cynaropicrin, The Bitter Principle of Artichoke *(Cynara scolymus* L.). Collect. Czech. Chem. Comm. **25,** 507 (1960).

1019. — — — Structure of Cynaropicrin. Collect. Czech. Chem. Comm. **25,** 2777 (1960).

1020. — — — Lactones of the Germacranolide Group and Their Stereochemical Relationship. Collect. Czech. Chem. Comm. **28,** 1715 (1963).

1021. — — — The Proof of Existence and Structure of Hydroxycostunolide, a Sesquiterpenic Lactone of Germacrane Type in *Artemisia balchanorum* H. Krash. Collect. Czech. Chem. Comm. **28,** 1618 (1963).

1022. — — — On Terpenes CLV. Structure of Damsine, a Sesquiterpenic Lactone from *Ambrosia maritima* L. Collect. Czech. Chem. Comm. **28,** 2257 (1963).

1023. — — — Isolation and Structure of Scabiolide, Further Sesquiterpene Lactone with a Ten-Membered Ring in Molecule. Collect. Czech. Chem. Comm. **27,** 1905 (1962).

1024. — — — Absolute Configuration of Cnicin and Scabiolide. Collect. Czech. Chem. Comm. **27,** 2398 (1962).

1025. — — — Geometry of Double Bonds in the Ten-Membered Ring of Costunolide. Collect. Czech. Chem. Comm. **31,** 2899 (1966).

1026. — — — Proof of Structure of Guaianolides *Artabsin* and *Arborescin*. Collect. Czech. Chem. Comm. **29,** 1829 (1964).

1027. — — — The Structure of Salonitolide, A Sesquiterpenic Lactone of Germacrane Type from *Centaurea salonitana* Vis. Collect. Czech. Chem. Comm. **30,** 2863 (1965).

1028. SUCHY, M., V. HEROUT, F. SORM, P. DE MAYO, A. N. STARRATT, and J. B. STOTHERS: The Constitution of Arctiopicrin. Tetrahedron Letters **1964,** 3907.

1029. SUCHY, M., Z. SAMEK, V. HEROUT, and F. SORM: The Structure of Salonitenolide, a Sesquiterpenic Lactone of Germacrane Type from *Centaurea salonitana* Vis. Collect. Czech. Chem. Comm. **32,** 2016 (1967).

1030. — — — — The Structure and Absolute Configuration of Scabiolide. Collect. Czech. Chem. Comm. **33,** 2238 (1968).

1031. — — — — Revision of Structure of Artiopicrin, Cnicin and Scabiolide. Collect. Czech. Chem. Comm. **30,** 3473 (1965).

1032. — — — — Constitution and Configuration of Albicolide, a New Germacranolide from *Jurinea albicaulis*. Collect. Czech. Chem. Comm. **32,** 3934 (1967).

1033. SUCHY, M., Z. SAMEK, V. HEROUT, R. B. BATES, G. SNATZKE, and F. SORM: Constitution and Configuration of Pelenolides, a New Group of Sesquiterpene Lactone Germacranolides. Collect. Czech. Chem. Comm. **32,** 3917 (1967).

1034. SUMI, M.: The Constitution and Stereochemistry of Artemisin. J. Amer. Chem. Soc. **80,** 4869 (1958).

1035. SUNDARARAMAN, P., and R. S. McEWEN: Crystal and Molecular Structure of Parthemollin [3,3a,4,5,6,8α-Hexahydro-7-(1-hydroxy-3-oxobutyl)-6-methyl-3-methylenecyclohepta[b]furan-2-one. J. Chem. Soc. Perkin Trans. II **1975,** 440.

1036. SUNDARARAMAN, P., R. S. McEWEN, and W. HERZ: Constituents of *Iva* Species. Stereochemistry of Parthemollin and Related Xanthanolides. Tetrahedron Letters **1973,** 3809.

1037. SUTHERLAND, J. K.: Regio- and Stereo-Specificity in the Cyclization of Medium Ring 1,5-Dienes. Tetrahedron **30,** 1651 (1974).

1038. SUZUKI, T., M. TANEMURA, T. KATO, and Y. KITAHARA: Synthesis of Cinnamolide. Bull. Chem. Soc. Japan **43,** 1268 (1970); Chem. Abstr. **73,** 25681 t (1970).

1039. Sykora, V., and M. Romanuk: Die Hudson-Klynesche Lactonregel und ihre Anwendung in der Terpenchemie. Collect. Czech. Chem. Comm. **22**, 1909 (1957).

1040. Tada, H., and K. Takeda: Germacranolides from *Laurus nobilis* L. Chem. Pharm. Bull. **24**, 667 (1976); Chem. Abstr. **85**, 17129b (1976).

1041. — — Structure of the Sesquiterpene Lactone Laurenobiolide. Chem. Commun. **1971**, 1391.

1042. Takeda, K.: Stereospecific Cope Rearrangement of the Germacrene-Type Sesquiterpenes. Tetrahedron **30**, 1525 (1974).

1043. — Sesquiterpenes Having a Five-Membered Ether-Ring in the Molecule. Pure Appl. Chem. **21**, 181 (1970).

1044. Takeda, K., and I. Horibe: Cope Rearrangement of Some Germacrane-Type Furan Sesquiterpenes. Part V. Preparation and Thermal Rearrangement of Some cis,trans-Germacranolides. J. Chem. Soc. Perkin Trans. I **1975**, 870.

1045. Takeda, K., I. Horibe, and H. Minato: Cope Rearrangement of Some Germacrane-Type Furan Sesquiterpenes Rearrangement of cis,trans-Cyclodeca-1,5-diene Derivatives. J. Chem. Soc. C **1970**, 2704.

1046. — — — Sesquiterpene Lactones from the Root of *Lindera strychnifolia* Vill. J. Chem. Soc. (London) **1968**, 569.

1046a. — — — Cope Rearrangement of Some Germacrane-Type Furan Sesquiterpenes. J. Chem. Soc. C **1970**, 1142.

1047. — — — Absolute Configuration of Neolinderane, Pseudoneolinderane and Linderadine. Chem. Commun. **1968**, 1168.

1048. Takeda, K., I. Horibe, M. Teraoka, and H. Minato: Structures and Absolute Configuration of Litsealactone, Litseaculane, Zeylanine, and Zeylanane. Chem. Commun. **1968**, 940.

1049. — — — — Absolute Configuration of Linderalactone and Linderane. Chem. Commun. **1968**, 637.

1050. — — — — Absolute Configuration of Linderalactone, Linderane, and Isolinderalactone and its Derivatives. J. Chem. Soc. C (London) **1969**, 1491.

1051. — — — — Components of *Neolitsea aciculata* Koidz. J. Chem. Soc. C (London) **1970**, 973.

1052. — — — — Components of the Root of *Lindera strychnifolia*. Structures of Neolindera Lactone and Lindenenone. J. Chem. Soc. C **1969**, 2786.

1053. Takeda, K., H. Minato, and I. Horibe: Components of the Root of *Lindera strychnifolia* Vill. — VII. Structure of Linderane. Tetrahedron **19**, 2307 (1953).

1054. Takeda, K., H. Minato, and M. Ishikawa: Structures of Linderalactone and Isolinderalactone. J. Chem. Soc. (London) **1964**, 4578.

1055. Takeda, K., H. Tada, and H. Minato: Components of the Root of *Lindera strychnifolia*. Neosericenyl Acetate and Dehydrolindestrenolide. J. Chem. Soc. C **1971**, 1070.

1056. Takeda, K., K. Tori, I. Horibe, M. Ohtsuru, and H. Minato: Cope Rearrangement of Some Germacrane-Type Furan Sesquiterpenes Relation between Conformations of Carbocyclic Ten-Membered Rings in Germacrane Type Furan Sesquiterpenes and the Stereochemistry of their Cope Rearrangement Products. J. Chem. Soc. C **1970**, 2697.

1057. Takeuchi, S., Y. Ogawa, and H. Yonehara: The Structure of Pentalenolactone (Pa-132). Tetrahedron Letters **1969**, 2737.

1058. Talapatra, S. K., A. Patra, and B. Talapatra: Lanuginolide and Dihydroparthenolide, Two New Sesquiterpenoid Lactones from *Michelia lanuginosa*. The Structure, Absolute Configuration, and a Novel Rearrangement of Lanuginolide. Chem. Commun. **1970**, 1534.

1059. — — Parthenolide and a New Germacranolide, 11,13-Dehydrolanuginolide, from *Michelia lanuginosa*. Phytochem. **12**, 1827 (1973).

1060. TANAHASHI, Y., Y. ISHIZAKI, T. TAKAHASHI, and K. TORI: Ligularenolide. A New Sesquiterpene Lactone of Eremophilane Type. Tetrahedron Letters 1968, 3739.

1061. TANAKA, N., T. YAZAWA, K. AOYAMA, and T. MURAKAMI: Chemical Studies on the Constituents of Xanthium canadense Mill. Chem. Pharm. Bull. 24, 1419 (1976); Chem. Abstr. 85, 106647 m (1976).

1062. TARASOV, V. A., N. D. ABDULLAEV, S. Z. KASYMOV, and G. P. SIDYAKIN: Chrysartemin B, a Sesquiterpene Lactone from Handelia trichophylla. Khim. Prir. Soedin. 12, 667 (1976); Chem. Abstr. 86, 117598 t (1977).

1063. — — — — Hanphyllin, a New Germacranolide from Handelia trichophylla. Khim. Prir. Soedin. 1976, 263; Chem. Abstr. 85, 143313 v (1976).

1064. — — — — Cumambrin A — A Sesquiterpene Lactone from Handelia trichophylla. Khim. Prir. Soedin. 10, 799 (1974); Engl. edit.: p. 826; Chem. Abstr. 82, 152142 d (1975).

1065. TARASOV, V. A., N. D. ABDULLAEV, S. Z. KASYMOV, G. P. SIDYAKIN, and M. R. YAGUDAEV: Structure of Handelin, a New Diguaianolide from Handelia trichophylla. Khim. Prir. Soedin. 6, 745 (1976); Chem. Abstr. 88, 7079 h (1978).

1066. TARASOV, V. A., S. Z. KASYMOV, and G. P. SIDYAKIN: Sesquiterpene Lactones from Handelia trychophylla. Khim. Prir. Soedin. 1976, 113; Chem. Abstr. 85, 59573 x (1976).

1067. — — — The Structure of the Sesquiterpene Lactone Arsubin. Khim. Prir. Soedin. 7, 745 (1971); Engl. edit.: p. 722; Chem. Abstr. 76, 127172 g (1972).

1068. — — — Structure and Configuration of Arsubin. Khim. Prir. Soedin. 9, 676 (1973); Engl. edit.: p. 649.

1069. TATEE, T., and T. TAKAHASHI: Total Synthesis of ($\pm$)-Tetrahydroligularenolide. Bull. Chem. Soc. Japan 48, 281 (1975); Chem. Abstr. 82, 125487 p (1975).

1070. — — Synthesis of ($\pm$)-Tetrahydroligularenolide. Chem. Letters 1973, 929.

1071. TAYLOR, I. F., JR., W. H. WATSON, M. BETKOUSKI, W. G. PADOLINA, and T. J. MABRY: The Structure of Glaucolide-D, $C_{23}H_{28}O_{10}$. A Sesquiterpene Lactone. Acta Crystallogr. Sect. B B32, 107 (1976); Chem. Abstr. 84, 98067 j (1976).

1072. TETTWEILER, K., O. ENGEL, and E. WEDEKIND: Über die Konstitution des Artemisins. Ann. Chem. Deutsch. 492, 105 (1932).

1073. THIESSEN, W. E., and H. HOPE: Structure and Absolute Configuration of Solstitialin, $C_{15}H_{20}O_5$. Acta Crystallogr. Sect. B 26, 554 (1970); Chem. Abstr. 73, 49628 g (1970).

1074. THIESSEN, W. E., H. HOPE, N. ZARGHAMI, D. E. HEINZ, P. DEUEL, and E. A. HAHN: A New Sesquiterpene Lactone from Centaurea solstitialis L. (Yellow Star Thistle). Chem. and Ind. 1969, 460.

1075. THOMS, H.: Über einige Chemische Bestandteile der Blüten von Chrysanthemum cinerariaefolium. Ber. dtsch. pharm. Ges. 1891, (471); Chem. Zentralbl. 2, 670 (1891).

1076. TILLYAEV, K. S., K. K. KHALMATOV, I. PRIMUKHAMEDOV, and M. A. TALIPOVA: Chemical Characterization of Achillea millefolium growing in Uzbekistan. Rast. Resur. 9, 58 (1973); Chem. Abstr. 78, 121284 q (1973).

1077. TOLSTYKH, L. P., O. A. KONOVALOVA, K. S. RYBALKO, and A. I. SHRETER: Sesquiterpene Lactones of Species of the Genus Artemisia. Rast. Resur. 10, 275 (1974); Chem. Abstr. 81, 60770 h (1974).

1078. TOLSTYKH, L. P., V. I. SHEICHENKO, A. I. BAN'KOVSKII, and K. S. RYBALKO: Artemin — A New Sesquiterpene Lactone from Artemisia taurica. Khim. Prir. Soedin. 4, 384 (1968); Engl. edit.: p. 326.

1079. TOMASSINI, T. C. B., and B. GILBERT: a-Cyclocostunolide and Dihydro-β-Cyclo-costunolide from Moquinea velutina. Phytochem. 11, 1177 (1972).

1080. TOMCZYK, H., and W. KISIEL: Sesquiterpene Lactones of Helenium tenuifolium. Pol. J. Pharmacol. Pharm. 27, 101 (1975); Chem. Abstr. 83, 4994 p (1975).

1081. TOMITA, Y., A. UOMORI, and H. MINATO: Sesquiterpenes and Phytosterols in the Tissue Cultures of Lindera strychnifolia. Phytochem. 8, 2249 (1969).

1082. Torrance, S. J., T. A. Geissman, and M. R. Chedekel: Sesquiterpene Lactones. The Constituents of *Eriophyllum confertiflorum*. Phytochem. **8,** 2381 (1969).

1083. Tori, K., and I. Horibe: Intramolecular Nuclear Overhauser Effect [NOE] and Conformation of Isolinderalactone, a Furan Sesquiterpene. An Example of Large NOE Values Between Olefinic Geminal Protons. Tetrahedron Letters **1970,** 2881.

1084. Tori, K., I. Horibe, K. Kuriyama, H. Tada, and K. Takeda: Conformational Isomers of Laurenobiolide, a New Ten-Membered Ring Sesquiterpene Lactone. Chem. Commun. **1971,** 1393.

1085. Tori, K., I. Horibe, K. Kuriyama, and K. Takeda: Conformational Isomers and Ring Inversion of Neolinderalactone, a Ten-Membered Ring Furanosesquiterpene. Chem. Commun. **1970,** 957.

1086. Tori, K., I. Horibe, Y. Tamura, K. Kuriyama, H. Tada, and K. Takeda: Re-Investigation of the Conformation of Laurenobiolide, a Ten-Membered Ring Sesquiterpene Lactone by Variable-Temperature Carbon-13 NMR Spectroscopy. Evidence for the Presence of Four Conformational Isomers in Solution. Tetrahedron Letters **1976,** 387.

1087. Tori, K., I. Horibe, Y. Tamura, and H. Tada: Simultaneous Application of the Nuclear Overhauser Effect and an NMR Shift Reagent. Conformations of Costunolide and Dihydrocostunolide in Solution. Chem. Commun. **1973,** 620.

1088. Tori, K., I. Horibe, H. Yoshioka, and T. J. Mabry: Configuration of the Endocyclic Double-Bond and Conformation of the Germacranolide Dilactones, Isabelin, Iso-isabelin and Related Germacranolide Monolactones, as Studied by Nuclear Overhauser Effects. J. Chem. Soc. B (London) **1971,** 1084.

1089. Tori, K., M. Ohtsuru, I. Horibe, and K. Takeda: Conformations of Ten-Membered Carbocyclic Rings in Zeylanine and Zeylanane as Determined by Application of Nuclear Overhauser Effects. Chem. Commun. **1968,** 943.

1090. Tori, K., M. Ueyama, I. Horibe, Y. Tamura, and K. Takeda: Carbon-13 NMR Spectra of Some Furanosesquiterpenes, Major Components of *Lindera strychnifolia*. Tetrahedron Letters **1975,** 4583.

1091. Toribio, F. P., and T. A. Geissman: Sesquiterpene Lactones of *Hymenoclea monogyra*. Phytochem. **8,** 313 (1969).

1092. — — Sesquiterpene Lactones. New Lactone from *Hymenoclea salsola* T. and G. Phytochem. **7,** 1623 (1968).

1092a. Toth, J., S. Holly, L. Ferenczy, and O. Kovacš: Antibacterial Substance from *Xanthium italicum*. Rev. Chim. Acad. Pop. Rumaine **7,** 1339 (1962); Chem. Abstr. **61,** 8129 (1964).

1093. Toubiana, R.: Structure de l'hydroxyvernolide, Nouvel Ester Sesquiterpenique Isolé du *Vernonia colorata* Drake. C. R. Acad. Sci. Paris, Serie C **268,** 82 (1969).

1094. Toubiana, R., and A. Gaudemer: Structure du Vernolide, Nouvel Ester Sesqui-terpenique Isolé de *Vernonia colorata*. Tetrahedron Letters **1967,** 1333.

1095. Toubiana, R., B. Mompon, Chi Man Ho, and M. J. Toubiana: Isolement du Verno-dalin et du Vernolepin à Partir de *Vernonia guineensis:* Authenticite du Squelette Elemane. Phytochem. **14,** 775 (1975).

1096. Toubiana, R., M. J. Toubiana, and B. C. Das: Structure of Confertolide, a New Sesquiterpene Lactone Isolated from *Vernonia conferta*. C. R. Acad. Sci. Ser. C **1970,** 1033; Chem. Abstr. **72,** 133001j (1970).

1097. — — — Structure du Confertolide, Nouveau Germacranolide Isolé de *Vernonia conferta* (Composee). Tetrahedron Letters **1972,** 207.

1098. Toubiana, R., M. J. Toubiana, K. Tori, and K. Kuriyama: Absolute Configuration and Conformation of Confertolide, A Germacranolide Isolated from *Vernonia conferta*. Tetrahedron Letters **1974,** 1753.

1099. TOWERS, G. H. N.: Contact Hypersensitivity and Photodermatitis Evoked by Compositae. J. Econom. Bot., submitted.

1100. TREHAN, I. R., C. MONDER, and A. K. BOSE: Classification of Steroid Alcohols by NMR Spectrsocopy. Tetrahedron Letters **1968,** 67.

1101. TSAI, L., R. J. HIGHET, and W. HERZ: The Mass Spectra of Pseudoguaianolides Related to Helenalin. J. Organ. Chem. (U. S. A.) **34,** 945 (1969).

1102. TSUDA, K., K. TANABE, I. IWAI, and K. FUNAKOSHI: The Structure of Alantolactone. J. Amer. Chem. Soc. **79,** 5721 (1957).

1103. TURNBULL, K. W., W. ACKLIN, D. ARIGONI, A. CORBELLA, P. GARIBOLDI, and G. JOMMI: Biological Conversion of Copaborneol into Tutin. Chem. Commun. **1972,** 598.

1104. UL-HAQUE, M., and C. N. CAUGHLAN: The Molecular and Crystal Structure of Bromomexicanin E ($C_{14}H_{15}O_3Br$). J. Chem. Soc. (London) **B 1967,** 355.

1104 a. — — Crystal and Molecular Structure of Bromohelenalin ($C_{15}H_{17}O_4Br$). J. Chem. Soc. (London) **B 1969,** 956.

1104 b. UL-HAQUE, M., C. N. CAUGHLAN, M. T. EMERSON, T. A. GEISSMAN, and S. MATSUEDA: Crystal and Molecular Structure of O-(Bromoacetyl)-tetrahydrodouglanine, $C_{17}H_{25}O_4Br$. J. Chem. Soc. (London) **B 1970,** 598.

1105. UL-HAQUE, M., D. ROGERS, and C. N. CAUGHLAN: Crystal Structure and Absolute Configuration of Bromoisotenulin. J. Chem. Soc. Perkin Trans. II **1974,** 223.

1106. ULUBELEN, A., S. ÖKSÜZ, Z. SAMEK, and M. HOLUB: Sesquiterpenic Lactones from *Smyrnium olusatrum* L. Roots. Tetrahedron Letters **1971,** 4455.

1106 a. USKOKOVIC, M. R., T. H. WILLIAMS, and J. F. BLOUNT: The Structure and Absolute Configuration of Arteannuin B. Helv. Chim. Acta **57,** 600 (1974).

1107. USYNINA, R. V., L. I. OLISHEVETS, V. V. DUDKA, and E. B. MARTIN: Lactones from *Artemisia compacta*. Khim. Prir. Soedin **6,** 809 (1976); Chem. Abstr. **86,** 103058 e (1977).

1108. VANHAELEN-FASTRE, R., and M. VANHAELEN: Presence of Salonitenolide in *Cnicus benedictus*. Planta Med. **26,** 375 (1974); Chem. Abstr. **82,** 108835 h (1975).

1109. VICHNEWSKI, W., and B. GILBERT: Schistosomicidal Sesquiterpene Lactone from *Eremanthus elaeagnus*. Phytochem. **11,** 2563 (1972).

1110. VICHNEWSKI, W., J. N. C. LOPES, D. D. S. FILHO, and W. HERZ: 15-Deoxygoyazensolide, a New Heliangolide from *Vanillosmopsis erythropappa*. Phytochem. **15,** 1775 (1976).

1111. VICHNEWSKI, W., S. J. SARTI, B. GILBERT, and W. HERZ: Goyazensolide, a Schistosomicidal Heliangolide from *Eremanthus goyazensis*. Phytochem. **15,** 191 (1976).

1112. VICHNEWSKI, W., I. K. SHUHAMA, R. C. ROSANSKE, and W. HERZ: Granilin and Ivasperin from *Ambrosia polystachya*. ^{13}C-NMR Spectra of Hydroxylated Isoalantones. Phytochem. **15,** 1531 (1976).

1113. VICHNEWSKI, W., F. WELBANEIDA, L. MACHADO, J. A. RABI, R. MURARI, and W. HERZ: Eregoyazin and Eregoyazidin, Two New Guaianolides from *Eremanthus goyazensis*. J. Organ. Chem. (U. S. A.) **42,** 3910 (1977).

1114. VIDARI, G., M. DE BERNARDI, P. VITA-FINZI, and G. FRONZA: Fungal Metabolites. Sesquiterpenes from *Lactarius blennius*. Phytochem. **15,** 1953 (1976).

1115. VIDARI, G., L. GARLASCHELLI, M. DE BERNARDI, G. FRONZA, and P. VITA-FINZI: The Structure of a New Sesquiterpene Epoxylactone from *Lactarius scrobiculatus* Scop. (Russulaceae) by Spectroscopic Methods. Tetrahedron Letters **1975,** 1773.

1116. VIEHOEVER, A., and R. G. CAPEN: New Sources of Santonin. J. Amer. Chem. Soc. **45,** 1941 (1963).

1117. VOKAC, K., and Z. SAMEK: Standard Sesquiterpenic Lactones for Structural Correlations; Stereoisomeric (6R,7S,10S,11S)-10-Hydroxyguaian-6,12-olides and Related (6R,7S)-Guaienolides. Collect. Czech. Chem. Comm. **39,** 480 (1974).

1118. VOKAC, K., Z. SAMEK, V. HEROUT, and F. SORM: The Structure of Artabsin and the Origin and Structure of the Coloured Hydrocarbon Chamazulenogen from Wormwood Oil. Collect. Czech. Chem. Comm. **34**, 2288 (1969).

1119. — — — — Absolute Configuration of Artabsin. Collect. Czech. Chem. Comm. **37**, 1346 (1972).

1120. WADA, K., Y. ENOMOTO, and K. MUNAKATA: Thermal Rearrangement of Shiromodiol-Monoacetate. Tetrahedron Letters **1969**, 3357.

1121. WADDELL, T. G.: Structure, Biogenesis, and Circular Dichroism of Sesquiterpene Lactones. From Diss. Abstr. Int. B **1970**, 30, 4986; Chem. Abstr. **73**, 131146g (1970).

1122. WADDELL, T. G., and T. A. GEISSMAN: Paucin, a Sesquiterpene Lactone Glucoside. Tetrahedron Letters **1969**, 515.

1123. — — Sesquiterpene Lactones. Constituents of *Baileya* Species. Phytochem. **8**, 2371 (1969).

1124. WADDELL, T. G., W. STÖCKLIN, and T. A. GEISSMAN: Circular Dichroism of Sesquiterpene Lactones. Tetrahedron Letters **1969**, 1313.

1125. WATANABE, M., and A. YOSHIKOSHI: Transformation of 1-α-Santonin into the Ten-Membered Sesquiterpene, Dihydronovanin. Tohoku Daigaku Hisuiyoeki Kagaku Kenkyusho Hokoku **23**, 53 (1973); Chem. Abstr. **81**, 169643v (1974).

1126. — — Transformation of α-Santonin into the Germacranolide Dihydronovanin. Chem. Commun. **1972**, 698.

1127. WATKINS, S. F.: Unpublished.

1128. WATKINS, S. F., N. H. FISCHER, and I. BERNAL: Neutron Diffraction Structure of Melampodin: Its Role in the Reclassification of the Germacranolides. Proc. Nat. Acad. Sci. (U.S.A.) **70**, 2434 (1973).

1129. WATKINS, S. F., J. D. KORP, I. BERNAL, D. L. PERRY, N. S. BHACCA, and N. H. FISCHER: Molecular Structure of Two Derivatives of the Germacranolide Sesquiterpene Lactone Melnerin. J. Chem. Soc. Perkin Trans. II **1978**, 599.

1130. WATSON, W. H.: The Application of X-Ray Diffraction Techniques to the Determination of Structures of Natural Products. Rev. Latinoamer. Quim. **7**, 1 (1976).

1131. WATSON, W. H., M. G. REINECKE, and J. C. HITT: The Structure and Biological Activities of Germacranolide Lactones and Cactus Alkaloids. Rev. Latinoamer. Quim. **6**, 1 (1975).

1132. WATSON, W. H., I. B. WU, S. A. MONTI, R. E. DAVID, T. J. MABRY, and W. G. PADOLINA: Dihydrodesacetoxyglaucolide-A, $C_{21}H_{28}O_8$. Cryst. Struct. Commun. **3**, 697 (1974); Chem. Abstr. **82**, 50142m (1975).

1133. WEINGES, K., and W. BAEHR: Comparison of the NMR and Mass Spectra of Bilobalide $C_{15}H_{18}O_8$ and of the Ginkgolides $C_{20}H_{24}O_{9-11}$. Liebigs Ann. Chem. **1972**, 759, 158; Chem. Abstr. **77**, 152375d (1972).

1134. WENKERT, E., and D. P. STRIKE: Synthesis of Some Drimanic Sesquiterpenes. J. Amer. Chem. Soc. **86**, 2044 (1964).

1135. WHITE, E. H., S. EGUCHI, and J. N. MARX: The Synthesis and Stereochemistry of Desacetoxymatricarin and the Stereochemistry of Matricarin. Tetrahedron **25**, 2099 (1969).

1136. WHITE, D. N. J., and G. A. SIM: Santonins. Quantitative Conformational Analysis. Tetrahedron **29**, 3933 (1973).

1137. — — Conformations of the Episantonins. Crystal Structures of 2-bromo-6-epi-α-Santonin and 2-bromo-6-epi-β-Santonin. J. Chem. Soc. Perkin Trans. II **1975**, 1826.

1138. WHITE, E. H., and R. E. K. WINTER: Natural Products from *Achillea lanulosa*. Tetrahedron Letters **1963**, 137.

1138a. WHITE, E. H., and J. N. MARX: The Synthesis and Stereochemistry of Desacetoxymatricarin and Achillin. J. Amer. Chem. Soc. **89**, 5511 (1967).

1138b. WICHMANN, G.: Beiträge zur Biologie der Santoninpflanzen. I. Die Verbreitung des Santonins und verwandter Verbindungen im Pflanzenreich. Pharmazie **13**, 481 (1958).

1139. WIEDHOPF, R. M., M. YOUNG, E. BIANCHI, and J. R. COLE: Tumor Inhibitory Agent from *Magnolia grandiflora* (Magnoliaceae). I. Parthenolide. J. Pharm. Sci. **62,** 345 (1973); Chem. Abstr. **78,** 115152d (1973).

1140. WILLUHN, G., and H. D. HERRMANN: Studies on the Constituents of *Arnica* Species. Two Sesquiterpene Lactones from the Flowers of *Arnica congifolia.* Arch. Pharm. **309,** 333 (1976); Chem. Abstr. **85,** 17073d (1976).

1141. WINTER, R. E. K., and R. F. LINDAUER: The Photoisomerization of Dihydrocostunolide. Tetrahedron **32,** 955 (1976).

1142. WINTERS, T. E., T. A. GEISSMAN, and D. SAFIR: Sesquiterpene Lactones of Xanthium Species. Xanthanol and Isoxanthanol, and Correlation of Xanthinin with Ivalbin. J. Organ. Chem. (U. S. A.) **34,** 153 (1969).

1143. WITT, M. E., and S. F. WATKINS: Crystal and Molecular Structure of Tamaulipin A, A $\Delta^{1(10)}$-*Trans*, Δ^4-*Trans*-Germacranolide Sesquiterpene Lactone. J. Chem. Soc. Perkin Trans. II **1978,** 204.

1144. WITZEL, D. A., G. W. IVIE, and J. W. DOLLAHITE: Mammalian Toxicity of Helenalin, the Toxic Principle of *Helenium microcephalum* DC. (Smallhead Sneezeweed). Am. J. Vet. Res. **37,** 859 (1976); Chem. Abstr. **85,** 117697d (1976).

1144a. WOODWARD, R. B., F. J. BRUTSCHY, and H. BAER: The Structure of Santonic Acid. J. Amer. Chem. Soc. **70,** 4216 (1948).

1145. YAMADA, K., S. TAKADA, and Y. HIRATA: Anisatinic Acid and Isoanisatinic Acid, Isomerication Products of Anisatin. Tetrahedron **24,** 1255 (1968).

1146. YAMADA, K., S. TAKADA, S. NAKAMURA, and Y. HIRATA: Facile Acetylation of a Tertiary Hydroxyl Group and an Unusual Deshielding Phenomenon by an Acetoxyl Group in NMR Spectra. Tetrahedron **24,** 1267 (1968).

1147. — — — — The Structures of Anisatin and Neoanisatin. Toxic Sesquiterpenes from *Illicium anisatum* L. Tetrahedron **24,** 199 (1968).

1148. — — — — Unpublished.

1149. — — — — The Structure of Anisatin. Tetrahedron Letters **1965,** 4797.

1150. YAMAKAWA, K., S. KIDOKORO, N. UMINO, R. SAKAGUCHI, T. TAKAKUWA, and M. SUZUKI: Studies on the Terpenoids and Related Alicyclic Compounds. Synthesis of 5α- and 5β-2-Oxosantan-6:13-olide from Santonin. Chem. Pharm. Bull. (Japan) **21,** 296 (1973); Chem. Abstr. **78,** 136426m (1973).

1151. YAMAKAWA, K., and K. NISHITANI: Studies on the Terpenoids and Related Alicyclic Compounds. Dimerization of 14-Bromo-6-dehydroxysantoninic Acid. J. Organ. Chem. (U. S. A.) **41,** 1256 (1976).

1152. — — Studies on Terpenoids and Related Alicyclic Compounds. Brominationdehydrobromination of 2-Oxo-5β-santanolide. Chem. Pharm. Bull. **25,** 371 (1977); Chem. Abstr. **87,** 39681u (1977).

1153. — — Chemical Transformation of α-Santonin into Arsantin and Arsanin. Chem. Pharm. Bull (Japan) **1976,** 2810; Chem. Abstr. **86,** 140269v (1977).

1154. YAMAKAWA, K., K. NISHITANI, and K. AZUSAWA: Chemical Transformation of α-Santonin into Balchanin, Colartin, and Arbusculin A, B, C, and E. Heterocycles **8,** 103 (1977).

1155. YAMAKAWA, K., K. NISHITANI, and A. YAMAMOTO: Transposition of Lactone in Sesquiterpene Lactone: Chemical Transformation of α-Santonin into Sesquiterpene α-Methylene-γ-Lactone, Yomogin. Chemistry Letters **1976,** 177.

1156. YAMAKAWA, K., K. NISHITANI, E. NAGAKURA, S. KIDOKORO, and R. SAKAGUCHI: Unpublished.

1157. — — — — — Studies on Terpenoids and Related Alicyclic Compounds. Bromination-Dehydrobromination of 2-Oxo-5α-santanolide. Chem. Pharm. Bull. (Japan) **1977,** 385; Chem. Abstr. **87,** 23531j (1977).

1158. YAMAKAWA, K., K. NISHITANI, and T. TOMINAGA: Chemical Transformation of α-

Santonin into Sesquiterpene α-Methylene-γ-Lactones, Tuberiferine and Artecalin. Tetrahedron Letters **1975**, 2829.

1159. Yamakawa, K., T. Tominaga, and K. Nishitani: Chemical Transformation of α-Santonine into Sesquiterpene α-Methylene-γ-Lactones, Arglanine and Santamarine. Tetrahedron Letters **1975**, 4137.

1160. Yamamura, S., and Y. Hirata: Structures of Nobiline and Dendrobine. Tetrahedron Letters **1964**, 79.

1161. Yanagawa, H., T. Kato, and Y. Kitahara: One-Step Synthesis of Drimenin and Cinnamolide. Synthesis **1970**, 257; Chem. Abstr. **73**, 45618f (1970).

1162. Yanagita, M., S. Inayama, and T. Kawamata: The Stereostructures of Pulchellidine and Pulchellin. Tetrahedron Letters **1970**, 131.

1163. — — — Neopulchellidine and Neopulchellin. Tetrahedron Letters **1970**, 3007.

1164. Yanagita, M., S. Inayama, T. Kawamata, and T. Okuna: Pulchellidine, a Novel Sesquiterpene Alkaloid Isolated from *Gaillardia pulchella* Foug. Tetrahedron Letters **1969**, 2073.

1165. Yoshitake, A., and T. A. Geissman: Sesquiterpene Lactones of *Baileya* Species. Pleniradin and Radiatin. Phytochem. **8**, 1753 (1969).

1166. Yoshioka, H., A. Higo, T. J. Mabry, W. Herz, and G. D. Anderson: Apachin, A New Sesquiterpene Lactone, and Other Xanthanolides from *Iva ambrosiaefolia*. Phytochem. **10**, 401 (1971).

1167. Yoshioka, H., and T. J. Mabry: The Structure and Chemistry of Isabelin. A New Germacranolide Dilactone from *Ambrosia psilostachya* DC. (Compositae). Tetrahedron **25**, 4767 (1969).

1168. Yoshioka, H., T. J. Mabry, N. Dennis, and W. Herz: Structure and Stereochemistry of Pulchellin B, C, E and F. J. Organ. Chem. (U.S.A.) **35**, 627 (1970).

1169. Yoshioka, H., T. J. Mabry, and A. Higo: Photochemistry of Isabelin. J. Amer. Chem. Soc. **92**, 923 (1970).

1170. Yoshioka, H., T. J. Mabry, M. A. Irwin, T. A. Geissman, and Z. Samek: The Geminal Coupling and Paramagnetic Shift of Exomethylene Protons in the α,β-Unsaturated γ-Lactone Group of Sesquiterpene Lactones Containing C_8-α-Hydroxyl Groups. Tetrahedron **27**, 3317 (1971).

1171. Yoshioka, H., T. J. Mabry, and H. E. Miller: Isabelin, a Germacranolide Dilactone from *Ambrosia psilostachya*. Chem. Commun. **1968**, 1679.

1172. Yoshioka, H., T. J. Mabry, and B. N. Timmermann: Sesquiterpene Lactones, Chemistry, N.M.R. and Plant. Distribution. Tokyo: University of Tokyo Press. 1973.

1173. Yoshioka, H., T. H. Porter, A. Higo, and T. J. Mabry: Photocoronopilin-A, a Cleaved Pseudoguaianolide from the Photolysis of Coronopilin. J. Organ. Chem. (U.S.A.) **36**, 229 (1971).

1174. Yoshioka, H., W. Renold, N. H. Fischer, A. Higo, and T. J. Mabry: Sesquiterpene Lactones from *Ambrosia confertiflora* (Compositae). Phytochem. **9**, 823 (1970).

1175. Yoshioka, H., W. Renold, and T. J. Mabry: The Structure of Salonitenolide and the Preferential C-8 Relactonization of Germacranolides Containing C-6 and C-8 Lactonizable α-Oxygen Groups. Chem. Commun. **1970**, 148.

1176. Yoshioka, H., E. Rodriguez, and T. J. Mabry: Tetraneurin-E and -F. New C-15 Oxygenated Pseudoguaianolides from *Parthenium* (Compositae). J. Organ. Chem. (U.S.A.) **35**, 2888 (1970).

1177. Yoshioka, H., H. Rüesch, E. Rodriguez, A. Higo, T. J. Mabry, J. G. Calzada Alan, and X. A. Dominguez: Tetraneurin-B, -C and -D, New C_{14}-Oxygenated Pseudoguaianolides from *Parthenium* (Compositae). Tetrahedron **26**, 2167 (1970).

1178. Yunusov, A. I., N. D. Abdullaev, S. Z. Kasymov, and G. P. Sidyakin: Tanachin, a New Sesquiterpene Lactone from *Tanacetum pseudoachillea*. Khim. Prir. Soedin. **1976**, 263; Chem. Abstr. **85**, 177643t (1976).

1179. YUNUSOV, A. I., N. D. ABDULLAEV, S. Z. KASYMOV, G. P. SIDYAKIN, and M. R. YAGUDAEV: Structure of the Sesquiterpene Lactone Tanacin. Khim. Prir. Soedin. **1976,** 170; Chem. Abstr. **85,** 177641 (1976).

1180. — — — — — Structure of Tanachin. Khim. Prir. Soedin. **1976,** 462; Chem. Abstr. **85,** 177660w (1976).

1181. YUNUSOV, A. I., S. Z. KASYMOV, and G. P. SIDYAKIN: Tanapsin from *Tanacetum pseudoachillea.* Khim. Prir. Soedin. **1976,** 261; Chem. Abstr. **85,** 117642s (1976).

1182. — — — Lactones of *Tanacetum pseudoachillea.* Khim. Prir. Soedin. **9,** 276 (1973); Engl. edit.: p. 267; Chem. Abstr. **78,** 159898f (1973).

1183. — — — Structure of Tanapsin. Khim. Prir. Soedin. **1976,** 309; Chem. Abstr. **85,** 124180h (1976).

1184. — — — Tanacin, A New Germacranolide from *Tanacetum pseudoachillea.* Khim. Prir. Soedin. **11,** 262 (1975); Chem. Abstr. **83,** 59097d (1975).

1185. YUSUPOV, M. I., A. MALLABAEV, and G. P. SIDYAKIN: Lactones of *Achillea biebersteinii.* Khim. Prir. Soedin. **1976,** 396; Chem. Abstr. **85,** 106654m (1976).

1186. ZAKHAROV, P. I., R. I. EVSTRATOVA, and K. S. RYBALKO: A Mass-Spectrometric Study of Arnifolin, a New Sesquiterpene Lactone from *Arnica montana* and *A. foliosa.* Khim. Prir. Soedin. **7,** 587 (1971); Engl. edit.: p. 567; Chem. Abstr. **77,** 19815f (1972).

1187. ZAKHAROV, P. I., P. B. TERENT'EV, O. A. KONOVALOVA, and K. S. RYBALKO: Mass-Spectrometric Study of the Sesquiterpene Lactone Grossmizin and Its Derivatives. Khim. Prir. Soedin. **3,** 344 (1977); Chem. Abstr. **87,** 184707m (1977).

1188. ZAKIROV, S. K., S. Z. KASYMOV, N. D. ABDULLAEV, and G. P. SIDYAKIN: The Structure of Maximolide. Khim. Prir. Soedin. **11,** 261 (1975); Engl. edit.: p. 273.

1189. ZAKIROV, S. K., S. Z. KASYMOV, V. RAKHMANKULOV, and G. P. SIDYAKIN: Lactones of *Artemisia ashurbajevii.* Khim. Prir. Soedin. **1976,** 397; Chem. Abstr. **85,** 106655n (1976).

1190. ZAKIROV, S. K., S. Z. KASYMOV, and G. P. SIDYAKIN: Structure and Configuration of Jurmolide. Khim. Prir. Soedin. **1976,** 398; Chem. Abstr. **85,** 177650t (1976).

1191. — — — Structure of Ashurbin and Arabsin. Khim. Prir. Soedin. **1976,** 548; Chem. Abstr. **86,** 16805c (1977).

1192. — — — Sesquiterpene Lactones from *Jurinea maxima.* Khim. Prir. Soedin. **11,** 656 (1975); Engl. edit.: p. 690; Chem. Abstr. **84,** 71430k (1976).

1193. — — — Lactones of *Jurinea maxima.* Khim. Prir. Soedin. **10,** 255 (1974); Chem. Abstr. **81,** 74890j (1974).

1194. ZARGHAMI, N., and D. E. HEINZ: Solstitialin acetate: A Sesquiterpene Lactone from *Centaurea solstitialis* L. (Yellow Star Thistle). Chem. and Ind. **1969,** 1556.

References (Addendum)

1195. ALVARADO, S., J. F. CICCIO, J. CALZADA, V. ZABEL, and W. H. WATSON: Thieleanine, a New Guaianolide from *Decachaeta thieleana.* Phytochem. **18,** 330 (1979).

1195a. ANANTHASUBRAMANIAN, L., S. V. GOVINDAN, K. D. DEODHAR, and S. C. BHATTACHARYYA: The Use of Morpholine Adducts in the Chemistry of Sesquiterpene α-Methylene-γ-Lactones. Indian. J. Chem. Sect. B **16,** 191 (1978).

1196. ANDO, M., A. AKAHANE, and K. TAKASE: Total Synthesis of Arborescin. Chem. Letters **1978,** 727.

1197. ANDO, M., K. TAJIMA, and K. TAKASE: Studies on the Synthesis of Sesquiterpene Lactones IV. Total Synthesis of Saussurea Lactone. Chem. Letters **1978,** 617.

1197a. ASAKAWA, Y., and T. TAKEMOTO: Sesquiterpene Lactones of *Conocephalum conicum.* Phytochem. **18,** 285 (1979).

1198. Asakawa, Y., M. Toyota, and T. Takemoto: Plagiochilide et Plagiochiline A, Seco-aromadendrane — Type Sesquiterpenes de la Mousse, *Plagiochila yokogurensis* (Plagiochilaceae). Tetrahedron Letters **18**, 1553 (1978).

1199. Bagirov, V. Y., V. I. Sheichenko, R. Y. Gasanova, and M. G. Pimenov: Study of *Ferula malacophyla* Lactones. Khim. Prir. Soedin, **1978**, 445; Chem. Abstr. **89**, 211929k (1978).

1199a. Banh-Nhu, C., E. Gacs-Baitz, L. Radics, J. Tamas, K. Ujszaszy, and G. Verzar-Petri: Achillicin, the first Proazulene from *Achillea millefolium.* Phytochem. **18**, 331 (1979).

1200. Baruah, N. C., R. P. Sharma, K. P. Madhusudanan, G. Thyagarayan, W. Herz, and R. Murani: The Sesquiterpene Lactones of *Tithonia diversifolia* Stereochemistry of the Tagitinins and Related Compounds. J. Organ. Chem. (U. S. A.) **44**, 1831 (1979).

1201. Bhat, K. L., P. L. Kamat, A. M. Shaligram, and G. K. Trivedi: Hydrogenation of Sodium Santoninate. Indian J. Chem. Sect. B **16**, 358 (1978).

1202. Bhat, K. L., A. L. Shaligram, and G. K. Trivedi: Acid-Catalyzed Rearrangement of 1-beta, 2-beta-Epoxy-Gamma-Tetrahydrosantonin. Indian J. Chem. Sect. B **16**, 647 (1978).

1203. Bischt, N. P. S., and R. Singh: Chemical Investigation of Leaves of *Xanthium strumarium* Linn. J. Indian Chem. Soc. **55**, 707 (1978).

1204. Blum, S., R. Segal, S. Solokoff, and D. Lichtenberg: Photoherbolide A, A9-Acetoxyguaianolide. Lloydia **41**, 117 (1978); Chem. Abstr. **89**, 43822z (1978).

1204a. Bohlmann, F., and M. Grenz: Ein neues Germacranolid aus *Munnozia maronii.* Phytochem. **18**, 334 (1979).

1205. Bohlmann, F., and J. Jakupovic: Neue Germacranolide aus *Calea urticifolia.* Phytochem. **18**, 119 (1979).

1206. — — Zwei neue Sesquiterpenlactone und eine neue Sesquiterpensäure aus *Helenium puberulum.* Phytochem. **18**, 131 (1979).

1206a. Bohlmann, F., P. K. Mahanta, and L. N. Dutta: Weitere Hirsutinolide aus *Vernonia*-Arten. Phytochem. **18**, 289 (1979).

1207. Bohlmann, F., A. A. Natu, and P. K. Mahanta: Naturally Occurring Terpene Derivatives. Part 122. New Diterpenes and Germacranolides from *Mikania* Species. Phytochem. **17**, 483 (1978).

1208. Bohlmann, F., and C. Zdero: Über eine neue Gruppe von Sesquiterpenlactonen aus der Gattung *Trixis.* Chem. Ber. **112**, 435 (1979).

1209. — — Neue Germacranolide und andere Inhaltsstoffe aus Vertretern der *Subtribus Gochnatiinae.* Phytochem. **18**, 95 (1979).

1209a. — — Neue Eudesmanolide aus *Gazania krebsiana.* Phytochem. **18**, 332 (1979).

1209b. — — 3β-Isovaleryloxycostunolid, ein neues Germacranolid aus *Cotula hispida.* Phytochem. **18**, 336 (1979).

1210. Bohlmann, F., C. Zdero, D. Berger, A. Suwita, P. Mahanta, and C. Jeffrey: Neue Furanoeremophilane und weitere Inhaltsstoffe aus südafrikanischen *Senecio*-Arten. Phytochem. **18**, 79 (1979).

1211. Borges, J., M. T. Manresa, J. L. Martin, C. Pascual, and P. Vazquez: Altamisin, A New Sesquiterpene Lactone from *Ambrosia cumanensis* H. B. K. Tetrahedron Letters **17**, 1513 (1978).

1212. Braga de Oliveira, A., G. G. de Oliveira, F. Carazza, B. Braz Filho, C. T. Moreira Bacha, L. Bauer, G. A. de Silva, and N. C. S. Siqueira: Laevigatin, A Sesquiterpenoid Furan from *Eupatorium laevigatum* Lam. Tetrahedron Letters **1978**, 2653.

1212a. Calzada, J. G., and J. F. Ciccio': Aislamiento de Tirotundina a Partir de *Tithonia diversifolia* (Hemsl.) Gray. Rev. Latinoamer. Quim. **9**, 202 (1978).

1213. CANE, D. E., and R. B. NACHBAR: Stereochemical Studies of Isoprenoid Biosynthesis. Biosynthesis of Fomannosin from [1,2-^{13}C$_2$] Acetate. J. Amer. Chem. Soc. **100**, 3208 (1978).

1214. CICCIÓ, J. F., and J. G. CALZADA: Glabberin, a New Heliangolide from *Eupatorium glaberrimum.* Abstract Book, V. International Symposium for the Chemistry of Natural Products, Monterrey, Mexico, April 26—29, 1978.

1215. CORBET, J. P., and C. BENREZA: Allergenic α-Methylene-γ-butyrolactones. A One-Carbon Degradation of Isoalantolactone *via* Pummerer Rearrangement of Sulfoxides. Canad. J. Chem. **57**, 213 (1979).

1216. CORTES, E., M. C. ROMERO, and J. ROMO: Mass Spectrometry of Sesquiterpenic Lactones of the Pseudoguaianolide Series II. Rev. Latinoamer. Quim. **8**, 168 (1977).

1217. DANISHEFSKY, S., M. HIRAMA, K. GOMBATZ, T. HARAYAMA, E. BERMAN, and P. SCHUDA: Stereospecific Total Synthesis of DL-Pentalenolactone. J. Amer. Chem. Soc. **100**, 6536 (1978).

1217a. DE BERNARDI, M., G. FRONZA, G. MELLERIO, G. VIDARI, and P. VITA-FINZI: New Sesquiterpene Hydroxylactones from *Lactarius* Species. Phytochem. **18**, 293 (1979).

1218. DEL AMO, S., and A. L. ANAYA: Effect of Some Sesquiterpenic Lactones on the Growth of Certain Secondary Tropical Species. J. Chem. Ecol. **4**, 305 (1978). Chem. Abstr. **89**, 103933y (1978).

1219. DEL AMO, R. S., and A. GOMEZ-POMPA: Variability in *Ambrosia cumanensis* (Compositae). Syst. Bot. **1**, 363 (1976); Chem. Abstr. **89**, 39436j (1978).

1220. DOMINGUEZ, X. A., O. R. FRANCO, G. CANO, S. GARCIA, and V. PENA: Mexican Medicinal Plants. XXXII. Terpenoids of the Ether Extracts of Two Celastraceae, *Mortonia greggi* Gray and *M. palmeri* Hemsl. Rev. Latinoamer. Quim. **9**, 33 (1978).

1221. EDGAR, M. T., A. E. GREENE, and P. CRABBÉ: Stereoselective Synthesis of (−)-Estafiatin. J. Organ. Chem. (USA) **44**, 159 (1979).

1222. EDWARD, J. T., and M. J. DAVIS: Reaction of Santonin with Hydroxylamine. J. Organ. Chem. (USA) **43**, 536 (1978).

1223. EL-FERALY, F. S., and Y. M. CHAN: Isolation and Characterization of the Sesquiterpene Lactones Costunolide, Parthenolide, Costunolide Diepoxide, Santamarine, and Reynosin from *Magnolia grandiflora* L. J. Pharm. Sci. **67**, 347 (1978).

1224. FROBORG, J., and G. MAGNUSSON: Construction of the Vellerane Skeleton with Total Syntheses of Racemic Velleral, Vellerolactone and Pyrovellerolactone, Revised Structures. J. Amer. Chem. Soc. **100**, 6728 (1978).

1225. FURUKAWA, H., M. ITOIGAWA, N. KUMAGAI, K. ITO, A. T. MCPHAIL, and K. D. ONAN: Isolation and Structure Determination of 4-O-Tigloyl-11,13-Dihydroautumnolide, a New Sesquiterpene Lactone from North Carolina *Helenium autumnale* L. Chem. Pharm. Bull. **26**, 1335 (1978).

1226. GONZALEZ, A. G., J. M. ARTEAGA, B. M. FRAGA, M. G. HERNANDEZ, and J. FAYOS: The Structure of Jhanilactone. Experientia **34**, 554 (1978); Chem. Abstr. **89**, 75406h (1978).

1227. GONZALEZ, A. G., J. T. BARROSO, H. LOPEZDORTA, J. R. LUIS, and F. RODRIGUEZLUIS: Compounds from Umbelliferae 20. Components of *Heracleum pyrenaicum* Lam. An. Quim. **74**, 832 (1978).

1228. GONZALEZ, A. G., J. BERMEJO, J. M. AMARO, G. M. MASSANET, A. GALINDO, and I. CABRERA: Sesquiterpene Lactones from *Centaurea linifolia* Vahl. Canad. J. Chem. **56**, 491 (1978).

1229. GONZALEZ, A. G., J. BERMEJO, J. L. BRETON, A. GALINDO, and G. M. MASSANET: C$_{11}$-Hydroxylation of Eudesmanolides. Rev. Latinoamer. Quim. **9**, 78 (1978).

1230. GONZALES, A. G., J. BERMEJO, I. CABRERA, G. M. MASSANET, H. MANSILLA, and A. GALINDO: Two Sesquiterpene Lactones from *Centaurea canariensis*. Phytochem. **17**, 955 (1978).

1231. Gonzalez, A. G., J. Bermejo, H. Mansilla, A. Galindo, J. M. Amaro, and G. M. Massanet: Structure and Absolute Configuration of Gallicin, a New Germacranolide from *Artemisia.* J. Chem. Soc. Perkin Trans I **1978,** 1243.
1232. Gonzalez, A. G., J. Bermejo, and G. M. Massanet: Contribution to the Chemotaxonomical Study of the Genus *Centaurea.* Structural Determination of Sesquiterpenic Lactones Present in *Centaurea* of the Canary Islands and the Iberian Peninsula. Rev. Latinoamer. Quim. **8,** 176 (1977).
1233. Gonzalez, A. G., and A. Galindo: Sesquiterpene Lactones in Umbelliferae. Contrib. Pluridiscip. Syst., Actes Symp. Int., 2nd **1977,** 178, p. 365.
1234. Gonzalez, A. G., and B. Garciamarrero: Stereochemistry of Sesquiterpenic Lactones of *Amberboa lippii* DC. An. Quim. **74,** 1121 (1978).
1235. Govindan, S. V., and S. C. Bhattacharyya: Oxymercuration of Some Sesquiterpene Lactones, Costunolide, Dihydrocostunolide, Dehydrocostuslactone and Dihydrodehydrocostuslactone. Indian J. Chem. Sect. B **16,** 1 (1978).
1236. — — Transformations of Isoalantolactone and Oxidation of 8,11α-H-Eudesm-4-en-8,13-olide. Indian J. Chem. Sect. B **16,** 271 (1978).
1237. Grieco, P. A., T. Oguri, S. Burke, E. Rodriguez, G. T. De Titta, and S. Fortier: Structure, Absolute Configuration, and Synthesis of Stramonin-B, a New Cytotoxic Pseudoguaianolide. J. Organ. Chem. (USA) **43,** 4552 (1978).
1238. Grieco, P. A., T. Oguri, S. Gilman, and G. T. De Titta: Total Synthesis of (±)-Eriolanin. J. Amer. Chem. Soc. **100,** 1616 (1978).
1239. Herz, W., R. De Groote, R. Murari, and J. F. Blount: Sesquiterpene Lactones of *Eupatorium recurvans.* J. Organ. Chem. (USA) **43,** 3559 (1978).
1240. Herz, W., and N. Kumar: Sesquiterpene Lactones of *Baltimora recta.* Phytochem., submitted.
1241. Herz, W., R. Murari, and J. F. Blount: Revised Structures of Pleniradin and Baileyin and their Bearing on the Biogenesis of Helenanolides. J. Organ. Chem. (USA), **44,** 1873 (1979).
1242. Herz, W., R. Murari, and S. V. Govindan: Sesquiterpene Lactones of *Eupatorium anomalum* and *Eupatorium mohrii.* Phytochem. **18,** 1337 (1979).
1242a. Hoffmann, J. J., S. D. Jolad, S. J. Torrance, D. J. Luzbetak, R. M. Wiedhopf, and J. R. Cole: Odoratin and Paucin: Cytotoxic Sesquiterpene Lactones from *Baileya pauciradiata* (Compositae). J. Pharm. Sci. **67,** (11) 1633 (1978).
1243. Holub, M., O. Motl, and Z. Samek: The Structure and Relative and Absolute Configurations of the Sesquiterpenic Lactones Gradolide and Polhovolide from *Laserpitium siler* L. Collect. Czech. Chem. Commun. **43,** 2471 (1978).
1244. Holub, M., Z. Samek, S. Vasickova, and M. Masojidkova: 11-Hydroxy-1βH, 5βH, 6αH, 7αH-Guaian-6,12-olides: Relative and Absolute Configurations of the Sesquiterpenic Lactones Montanolide, Isomontanolide, Acetylisomontanolide and Related Substances. Collect. Czech. Chem. Commun. **43,** 2444 (1978).
1245. Imakura, Y., K. H. Lee, D. Sims, and I. H. Hall: Structural Elucidation of the Novel Antitumor Sesquiterpene Lactone, Microlenin, from *Helenium microcephalum.* J. Pharm. Sci. **67,** 1228 (1978).
1246. Inayama, S., T. Kawamata, and T. Ohkura: The Chemical Transformation to Pulchellon from Pulchellin. Tetrahedron Letters **18,** 1557 (1978).
1247. Isobe, M., H. Iio, T. Kawai, and T. Goto: Synthesis of Sesquiterpene Antitumor Lactones. A New Stereocontrolled Total Synthesis of (±)-Vernolepin. J. Amer. Chem. Soc. **100,** 1940 (1978).
1248. Ito, T., T. Shimizu, Y. Fujimoto, and T. Tatsuno: 5α-Hydroxy-4αH,1,6,11βH-guai-2,10(15)-dien-6,12-olide. Acta Crystallogr. B, **34,** 1009 (1978).
1248a. Jankowski, K.: Utilisation de la Simulation dans l'Analyse des Spectres RMN Enregistrés en Présence de Réactifs Déplaçants. Rev. Latinoamer. Quim. **9,** 175 (1978).

1249. Jizba, J., Z. Samek, L. Novotny, E. Najdenova, and A. Boeva: Components of *Senecio nemorensis* Var. *bulgaricus* (Vel. Pro. Sp.). Collect Czech. Chem. Commun. **43**, 1113 (1978).

1250. Joseph-Nathan, P.: Carbon-13 NMR of Pyrethrosin from *Chrysanthemum*. Rev. Latinoamer. Quim. **9**, 36 (1978).

1251. Karawya, M. S., S. H. Hilal, M. S. Hifnawy, and S. S. El-Hawary: Isolation and Preliminary Pharmacological and Microbiological Screening of Cnicin from *Centaurea calitrapa* L. growing in Egypt. Egypt. J. Pharm. Sci. **16**, 445 (1977); Chem. Abstr. **89**, 160096b (1978).

1252. Karlsson, B., A. M. Pilotti, and A. C. Söderholm: Structure of the Hydrogen Bromide Adduct of Spicatine, a Sesquiterpenoid Lactone. Acta Crystallogr. B, **35**, 244 (1979).

1253. Karwe, M. V., N. R. Deshpande, S. V. Hiremath, G. H. Kulkarni, and G. R. Kelkar: Preparation of some Sesquiterpenic Conjugated Endo- and Exo-α-Methylene-γ-Lactones. Indian J. Chem. Sect. B **16**, 539 (1978).

1254. Kieczykowski, G. R., and R. H. Schlessinger: Total Synthesis of (±)-Vernolepin. J. Amer. Chem. Soc. **100**, 1938 (1978).

1255. Kupchan, S. M., J. W. Ashmore, and A. T. Sneden: Structure-Activity Relationships Among *in vivo* Active Germacranolides. J. Pharm. Sci. **6**, 865 (1978).

1256. Lansbury, P. T., and A. K. Serelis: A Facile Entry to Pseudoguaianes. Total Synthesis of Damsinic Acid. Tetrahedron Letters **22**, 1909 (1978).

1257. Lee, K. H., T. Ibuka, E. C. Mar, and I. H. Hall: Antitumor Agents. 31. Helenalin Sym-Dimethylethylenediamine Reaction Products and Related Derivatives. J. Med. Chem. **21**, 698 (1978).

1258. LeQuesne, P. W., S. B. Levery, M. D. Menachery, T. F. Brennan, and R. F. Raffauf: Novel Modified Germacranolides and Other Constituents of *Eremanthus elaeagnus* Schultz — Bip (Compositae). J. Chem. Soc. Perkin Trans I **1978**, 1572.

1259. LeVan, N., and N. H. Fischer: Three New Melampolide Sesquiterpenes, Polymatin A, B and C, from *Polymnia maculata* Cav. var. *maculata*. Phytochem. **18**, 851 (1979).

1260. Linek, A., and C. Novak: Refinement of the Structure of the Costunolide-Silver Nitrate Complex. Acta Crystallogr. B **34**, 3369 (1978).

1261. Lopez de Lerma, J., J. Fayos, S. Garcia-Blanco, and M. Martinez-Ripoll: Centaurepensin. A Redetermination of its Absolute Configuration by X-Ray Crystallography. Acta Crystallogr. B **34**, 2669 (1978).

1262. Mallabaev, A., U. Rakhmankulov, and G. P. Sidyakin: Lactones from *Achillea santolina*. Khim. Prir. Soedin **1978**, (4), 530; Chem. Abstr. **89**, 193848b (1978).

1263. Mane, B. M., S. V. Hiremath, and G. H. Kulkarni: Selenium Dioxide Oxidation of Saussurealactone. Curr. Sci. **47**, 677 (1978); Chem. Abstra. **90**, 39056f (1979).

1264. Manchand, P. S., and J. F. Blount: Stereostructures of Neurolensins A and B, Novel Germacranolide Sesquiterpenes from *Neurolaena lobata* (L.). R. Br. J. Organ. Chem. (USA) **43**, 4352 (1978).

1265. Marshall, J. A., and P. G. M. Wuts: Stereocontrolled Total Synthesis of α- and β-Santonin. J. Organ. Chem. (USA) **43**, 1086 (1978).

1266. Masayoshi, A., K. Tajima, and K. Takase: Total Synthesis of Saussurea Lactone. Chem. Letters **1978**, 617.

1267. Matsueda, S., M. Nagaki, J. Ichita, Y. Masuchi, and M. Koreeda: Structure of Meridianone. Abstract Book, ACS-CSJ Chemical Congress, Honolulu, Hawaii, April 2—6, 1979. Organic Chemistry, paper 571.

1268. Mukhametzhanov, M. N., Abdrakhmanov, O. A., and S. M. Adekenov: Sesquiterpene Lactones from *Artemisia austisca*. Teor. Osnovy Pererab. Mineral'n. i Organ. Syr'ya. **4**, 73 (1977); Chem. Abstr. **89**, 39281n (1978).

1269. Murai, A., M. Ono, A. Abiko, and T. Masamune: Synthesis of Phytuberin. J. Amer. Chem. Soc. **100**, 7751 (1978).

1270. Nadgouda, S. A., G. K. Trivedi, and S. C. Bhattacharyya: Sensitized Photo-oxygenation of α-Cyclocostunolide and Dihydro α-Cyclocostunolide: A Biogenetic Type Transformation of Costunolide to Santonin. Ind. J. of Chem. Sect. B **16**, 16 (1978).

1271. Naidenova, E., and E. Bloshik: Parthenolides in *Inula aschersoniana* Ika, Variety *Aschersoniana* Staj., Stef., and Kitan. Farmatsiya (Sofia) **28**, 26 (1978); Chem. Abstr. **89**, 193899 u (1978).

1272. Naidenova, E., N. L. Dryanovska, and B. Drozdz: Nitrogen-Containing Derivatives of the Sesquiterpene Lactone Grossheimin. Probl. Farm **5**, 85 (1977); Chem. Abstr. **89**, 110014 c (1978).

1273. Narain, N. K.: Spectroscopic Studies of Fasciculide-B, a New Sesquiterpene Lactone from the Leaves of *Vernonia fasciculata* Michx. Spectrosc. Letters **11**, 267 (1978).

1274. Ogura, M., G. A. Cordell, and N. R. Farnsworth: Anticancer Sesquiterpene Lactones of *Michelia compressa* (Magnoliaceae). Phytochem. **17**, 957 (1978).

1275. Ohfune, Y., P. A. Grieco, C. L. J. Wang, and G. Majetich: Stereospecific Total Synthesis of DL-Helenalin. A General Route to Helenanolides and Ambrosanolides. J. Amer. Chem. Soc. **100**, 5946 (1978).

1276. Ohno, N., and T. J. Mabry: Germacranolides from *Helianthus mollis* Lam. Phytochem. **18**, 1003 (1979).

1276a. Ohsawa, T.: Root Formation and Physiologically Active Substances. Kagaku to Seibutsu **16** (10), 644 (1978); Chem. Abstr. **90**, 51345 g (1979).

1277. Ortega, A., R. Martinez, and A. Romo de Vivar: Elemanolides of *Verbesina* Aff *Stricta*. Structure of Zempoalines A and B. Rev. Latinoamer. Quim. **8**, 166 (1977).

1278. Paknikar, S. K., J. Veeravalli, and J. K. Kirtany: Revised Structures for Iselin and Iliensin and the Identity of the Former with Archangelin. Experientia **34**, 553 (1978).

1279. Pettit, G. R., C. L. Herald, M. S. Allen, R. B. Von Dreele, L. D. Vanell, J. P. Y. Kao, and W. Blake: The Isolation and Structure of Aplysistatin. J. Amer. Chem. Soc. **99**, 262 (1977).

1280. Picman, A. K., R. H. Elliott, and G. H. N. Towers: Insect Feeding Deterrents Property of Alantolactone. Biochem. Systematics and Ecology **6**, 333 (1978).

1281. Quijano, L., D. Bloomenstiel, and N. H. Fischer: Tetraludin A, B and C, Three New Melampolides from *Tetragonotheca ludoviciana* (Compositae, Heliantheae). Phytochem. **18**, 1529 (1979).

1281a. Quijano, L., and N. H. Fischer: New Melampolides from *Tetragonotheca ludoviciana*. To be published.

1282. — — Melfusin, a New Germacrolide from *Melampodium diffusum* (Compositae, Heliantheae). To be published.

1283. Quijano, L., A. Romo de Vivar, and T. Rios: *Calea zacatechichi* Components. Structure of Calein A and B. Rev. Latinoamer. Quim. **9**, 86 (1978).

1283a. — — — Revision of the Structures of Calein A and B. Phytochem., in press.

1284. Ranieri, R. L., and G. J. Calton: Quadrone, A New Antitumor Agent from *Aspergillus terreus*. Tetrahedron Letters **6**, 499 (1978).

1284a. Rodriguez, E.: Allergenic and Irritant Plant Constituents. Rev. Latinoamer. Quim. **9**, 125 (1978).

1284b. Rodriguez-Hahn, L., M. Jimenez, E. Diaz, C. Guerrero, A. Ortega, and A. Romo de Vivar: The Revised Structure of Mortonin. Tetrahedron **33**, 657 (1977).

1284c. — — — — — Photolysis of Mortonin. Tetrahedron **33**, 661 (1977).

1285. RODRIGUEZ-HAHN, L., M. JIMENEZ, M. OLIVEROS, and E. DIAZ: Isolation and Structure of Mortonins B, C and D. Rev. Latinoamer. Quim. **8**, 161 (1977).

1286. ROMO, J., L. RODRIGUEZ-HAHN, and C. VICHIDO: Oxidation Experiments in Sesquiterpenic Lactones. Rev. Latinoamer. Quim. **8**, 149 (1977).

1286a. ROMO DE VIVAR, A., G. DELGADO, C. GUERRERO, J. RESENDIZ, and A. ORTEGA: Estudio de Viguieras. Estructura de la Viguiepinina y Correcion de la Viguiestenina. Rev. Latinoamer. Quim. **9**, 171 (1978).

1287. ROQUE, N. F., Z. S. FERREIRA, O. R. GOTTLIEB, R. L. STEPHENS, and E. WENKERT: The Structure of Ocotealactol, a New Eudesmanolide. Rev. Latinoamer. Quim. **9**, 25 (1978).

1288. RUECKER, G., and M. SCHIKARSKI: Conversion of Isofuranodien to Germacranolide. Arch. Pharm. **311**, 754 (1978); Chem. Abstr. **90**, 23293v (1979).

1288a. RUSTAIYAN, A., L. NAZARIANS, and F. BOHLMANN: Two New Germacranolides from *Onopordon leptolepis* DC. Phytochem. **18**, 883 (1979).

1288b. — — — Two New Elemanolides from *Onopordon leptolepis*. Phytochem. **18**, 879 (1979).

1288c. RUSTAIYAN, A., A. NIKNEJAD, W. H. WATSON, V. ZABEL, T. J. MABRY, G. YABUTA, and S. B. JONES, Jr.: Dihydroelephantopin, a New Tumor Inhibitor from *Elephantopus tomentosus* L. (Compositae). Rev. Latinoamer. Quim. **9**, 200 (1978).

1289. SAIDKHODZHAEV, A. I.: Structure of Tenuferin, Tenuferinin and Tenuferidin. Khim. Prir. Soedin. **14**, 70 (1978); engl. edit.; p. 55.

1290. SAMEK, Z., and J. HARMATHA: Use of Structural Changes for Stereochemical Assignments of Natural α-Exomethylene γ-Lactones of the Germacra-1(10),4-Dienolide Type on the Basis of Allylic and Vicinal Couplings of Bridgehead Protons. Hydrogenation of Endocyclic Double Bonds. Collect. Czech. Chem. Commun. **43**, 2779 (1978).

1291. SEAMAN, F. C., and N. H. FISCHER: New Sesquiterpene Lactones from *Melampodium linearilobum* (Compositae, Heliantheae). To be published.

1292. — — New Melampolides from *Tetragonotheca helianthoides* (Compositae, Heliantheae). To be published.

1293. SEMMELHACK, M. F., A. YAMASHITA, J. C. TOMESCH, and K. HIROTSU: Total Synthesis of Confertin *via* Metal-Promoted Cyclization-Lactonization. J. Amer. Chem. Soc. **100**, 5565 (1978).

1294. SERKEROV, S. V.: A Study of Badkhyzinin. Khim. Prir. Soedin. **13**, 787 (1977); engl. edit.: p. 663.

1295. SERKEROV, S. V., and A. N. ALESKEROVA: The Structure of A Sesquiterpene Lactone from *Artemisia fragrans*. Khim. Prir. Soedin. **14**, 75 (1978); engl. edit.: p. 59.

1296. SETO, H., T. SASAKI, H. YONEHARA, and J. UZAWA: Studies on the Biosynthesis of Pentalenolactone. Part I. Application of Long Range Selective Proton Decoupling (LSPD) and Selective ^{13}C-$\{^1H\}$ Noe in the Structural Elucidation of Pentalenolactone G. Tetrahedron Letters **10**, 923 (1978).

1297. SHAM'YANOV, I. D., A. MALLABAEV, and G. P. SIDYAKIN: Structure of the Sesquiterpene Lactone Elegin. Khim. Prir. Soedin. **14**, 442 (1978); Chem. Abstr. **89**, 197740h (1978).

1298. SHIBATA, H., and S. SHIMIZU: Three Chemovars of *Petasites japonicus* Maxim. Agric. Biol. Chem. **42**, 1427 (1978); Chem. Abstr. **89**, 193827u (1978).

1299. SINGH, A. N., V. V. MHASKAR, and S. DEV: Synthesis of Jalaric Ester-I, Possible Key Compound in the Elaboration of Lac Resin by *Laccifer lacca* Kerr. Tetrahedron **34**, 595 (1978).

1300. TARASOV, V. A., N. D. ABDULLAEV, S. Z. KASYMOV, G. P. SIDYAKIN, and M. R. YAGUDAEV: Structure of Hanphyllin. Khim. Prir. Soedin. **14**, 78 (1978); engl. edit.: p. 62.

1301. TORII, S., K. UNEYAMA, and H. ICHIMURA: Synthesis of (±)-Norisoambreinolide and (+)-Isoambrox. J. Organ. Chem. (USA) 43, 4680 (1978).
1302. TOWERS, G. H. N., J. C. MITCHELL, E. RODRIGUEZ, F. D. BENNETT, and S. P. V. RAO: Biology and Chemistry of Parthenium hysterophorus L, a Problem Weed in India. J. Sci. Ind. Res. 36, 672 (1977); Chem. Abstr. 89, 72943v (1978).
1302a. ULUBELEN, A., N. ATES, and T. NISHIDA: Istanbulin C, a Novel Sesquiterpene Lactone from Smyrnium connatum. Phytochem. 18, 338 (1979).
1303. URMANOVA, F. F., S. Z. KASYMOV, and G. P. SIDYAKIN: Sesquiterpene Lactones from Pseudohandelia (Umbellifera). Khim. Prir. Soedin. 14, 530 (1978); Chem. Abstr. 89, 176474j (1978).
1304. WENDER, P. A., and J. C. LECHLEITER: A Photochemically Mediated [4C+2C] Annelation. Synthesis of (±)-10-Epijunenol. J. Amer. Chem. Soc. 100, 4321 (1978).
1305. WILSON, S. R., and D. T. MAO: An Intramolecular Diels-Alder Route to Eudesmane Sesquiterpenes. J. Amer. Chem. Soc. 100, 6289 (1978).
1306. YABUTA, G., J. S. OLSEN, and T. J. MABRY: Sesquiterpene Lactones from the Genus Zaluzania (Compositae: Heliantheae). Rev. Latinoamer. Quim. 9, 83 (1978).
1307. YAMAKAWA, K., K. NISHITANI, and K. AZUSAWA: Chemical Transformation of α-Santonin into Sesquiterpene α-Methylene-γ-Lactones, Douglanin, and Ludovicin A and B. Heterocycles 9, 499 (1978).
1308. YAMAKAWA, K., K. NISHITANI, and A. MURAKAMI: Chemical Transformation of α-Santonin into Yomogin, Pinnatifidin, and 8-Epiartimisin. Abstract Book, ACS-CSJ Chemical Congress, Honolulu, Hawaii, April 2—6, 1979. Organic Chemistry, paper 535.
1309. YAMAKAWA, K., and T. TOMINAGA: 11-Exo-Methylene-α-Tetrahydrosantonin. Japan. Kokai 78, 759 (1978); Chem. Abstr. 89, 24563z (1978).
1310. — — 11-Exo-Methylenesantonin. Japan Kokai 78, 758 (1978); Chem. Abstr. 89, 24564a (1978).
1310a. YUNUSOV, A. I., G. P. SIDYAKIN, and D. KURBANOV: Cumambrins A and B from Tanacetum santolina. Khim. Prir. Soedin 5, 656 (1978); Chem. Abstr. 90, 51410z (1979).
1311. YUSUPOV, M. I., S. Z. KASYMOV, N. D. ABDULLAEV, G. P. SIDYAKIN, and M. R. YAGUDAEV: A New Lactone, Isoridentin, from Achillea biebersteinii. Khim. Prir. Soedin. 13, 800 (1977); engl. edit.: p. 674.
1312. ZAKHAROV, P. I., P. B. TERENTLEV, O. A. KONOVALOVA, and K. S. RYBALKO: Mass-Spectrometric Study of the Sesquiterpene Lactone Grossmisin and Its Derivatives. Khim. Prir. Soedin. 13, 344 (1977); engl. edit.: p. 293.
1313. — — — — Mass-Spectrometric Study of the Sesquiterpene Lactone Germanin A and its Derivatives. Khim. Prir. Soedin 14, 337 (1978); Chem. Abstr. 89, 147083a (1978).

Index of Charts Tabulating the Different Structural Types of Sesquiterpene Lactones

Index of Tables Containing the Various Structural Types of Sesquiterpene Lactones

(Received August 4, 1978)

Author Index

By

A. SIEGEL, Wien

Page numbers printed in *italics* refer to References

Subject Index

By

A. Siegel, Wien

Satz: Austro-Filmsatz Richard Gerin, A-1020 Wien

Fortschritte der Chemie organischer Naturstoffe

Progress in the Chemistry of Organic Natural Products

All Volumes and Cumulative Index 1—20 available / Alle Bände und Generalregister 1—20 lieferbar.

Price reduction for subscribers / Preisermäßigung für Subskribenten: 10%.

Special price reduction (20% of the list price) for the Vols. 1—20 plus Cumulative Index. / Vorzugspreis (20% Nachlaß) bei Bezug der Bände 1—20 inklusive Generalregister.

Volume 34: 63 figures. X, 620 pages. 1977. ISBN 3-211-81415-9.

Contents: C. R. ENZELL, I. WAHLBERG, and A. J. AASEN, Isoprenoids and Alkaloids of Tobacco — A. R. PINDER, The Chemistry of the Eremophilane and Related Sesquiterpenes — D. GROSS, Phytoalexine und verwandte Pflanzenstoffe — K. H. OVERTON and D. J. PICKEN, Studies in Secondary Metabolism with Plant Tissue Cultures — D. P. CHAKRABORTY, Carbazole Alkaloids — J. JACOB, Bürzeldrüsenlipide — W. VOELTER, Hypothalamus-Regulationshormone — Author Index — Subject Index.

Volume 35: VIII, 589 pages. 1978. ISBN 3-211-81460-4.

Contents: O. R. GOTTLIEB, Neolignans — K. HERRMANN, Hydroxyzimtsäuren und Hydroxybenzoesäuren enthaltende Naturstoffe in Pflanzen — G. PATTENDEN, Natural 4-Ylidenebutenolides and 4-Ylidenetetronic Acids — R. D. H. MURRAY, Naturally Occurring Plant Coumarins — G. OHLOFF, Recent Developments in the Field of Naturally-Occurring Aroma Components — Author Index — Subject Index.

Volume 36: 11 figures. VII, 425 pages. 1979. ISBN 3-211-81472-8.

Contents: F. W. WEHRLI and T. NISHIDA, The Use of Carbon-13 Nuclear Magnetic Resonance Spectroscopy in Natural Products Chemistry — G. OHLOFF and I. FLAMENT, The Role of Heteroatomic Substances in the Aroma Compounds of Foodstuffs — A. J. WEINHEIMER, C. W. J. CHANG, and J. A. MATSON, Naturally Occurring Cembranes — Author Index — Subject Index.

Volume 37: 8 figures. IX, 367 pages. 1979. ISBN 3-211-81528-7.

Contents: J. M. BRAND, J. CHR. YOUNG, and R. M. SILVERSTEIN, Insect Pheromones: A Critical Review of Recent Advances in Their Chemistry, Biology, and Application — M. McNEIL, A. G. DARVILL, and P. ALBERSHEIM, The Structural Polymers of the Primary Cell Walls of Dicots — U. SCHMIDT, J. HÄUSLER, ELISABETH ÖHLER, and H. POISEL, Dehydroamino Acids, α-Hydroxy-α-amino Acids and α-Mercapto-α-amino Acids — Author Index — Subject Index.

Springer-Verlag Wien · New York